B.M. FRAEIJS DE VEUBEKE
MEMORIAL VOLUME OF SELECTED PAPERS

B.M. FRAEIJS DE VEUBEKE MEMORIAL VOLUME OF SELECTED PAPERS

edited by

Dr. Michel Geradin

Aerospace Laboratory
The University of Liège, Belgium

SIJTHOFF & NOORDHOFF 1980
Alphen aan den Rijn, The Netherlands
Rockville, Maryland, U.S.A.

SOLID MECHANICS DIVISION
University of Waterloo
Waterloo, Ontario, Canada

Softcover reprint of the hardcover 1st edition 1980

ISBN-13:978-94-009-9149-1 e-ISBN-13:978-94-009-9147-7
DOI: 10.1007/978-94-009-9147-7

Contents

SPECIAL MEMORIAL VOLUME

DEDICATED TO

PROFESSOR B.M. FRAEIJS DE VEUBEKE

"Professor Fraeijs de Veubeke's collaborators remember a man who in a natural way allied professional competence, a deep purpose in life and a sense of humour".

Michel Geradin

Professor B.M. Fraeijs de Veubeke

THE MAN

Professor B.M. Fraeijs de Veubeke graduated successively in Mechanical Engineering (1940) and in Aeronautical Engineering (1944) from the universities of Louvain and Liege, Belgium. The next six years were spent practicing engineering on aircraft design and testing at Fairey Aircraft Industries and the Belgian Aeronautical Administration. Subsequently he joined the University of Liege where he later was appointed to the chair of Aerospace Engineering and Continuum Mechanics. He acted at the same time as a Professor in the Department of Applied Mechanics of the University of Louvain. Professor Fraeijs de Veubeke was also a Visiting Professor at M.I.T., Massachusetts in 1952 and at Stanford University, California in 1955.

Many honours were bestowed upon Professor Fraeijs de Veubeke in recognition of his considerable scientific activity, including election as a full member of the Belgian Academy of Sciences in 1976, two months before his untimely death. Such an election is rare in engineering and represents an acknowledgement of the scientific rigor of his work. Indeed, his contributions to various disciplines such as unsteady aerodynamics, elasticity, structural mechanics, theory of vibrations and optimal control are significant and original. They bear the stamp of an incisive and penetrating mind, and all are of lasting value.

Although his main interest lay in research he was active in several of the international agencies devoted to co-operation in aeronautics and astronautics. In particular, he was a founder member of the Structures and Materials Panel of the Advisory Group for Aeronautical Research and Development (AGARD). He also took a very active part in the constitution of ELDO, from which the present European Space Agency (ESA) has originated.

Professor Fraeijs de Veubeke's activity took place essentially at the University of Liege, where he was the major driving force in the Department of Aerospace Engineering for more than 20 years. Around him a group of researchers rapidly grew who shared his faith in research and helped in developing his pioneering work, mainly on numerical methods for structural analysis. This work was based on the application of variational principles and their finite element discretization. Although Professor Fraeijs de Veubeke's work appears quite theoretical, he, nevertheless, possessed a remarkable physical sense and intuition. Despite an increasing administration load he always maintained a significant output of strictly individual research.

Professor Fraeijs de Veubeke's collaborators not only appreciated his outstanding scientific leadership, but were also inspired by his personality and his ability to communicate his own faith in science. Together with his relatives and friends they fondly and respectfully remember a man who in a natural way allied professional competence, a deep purpose in life and a sense of humour. His muscial gifts were precious to him and through them he communicated the joy of living a full and complete life.

The present volume has been prepared by several of Professor Fraeijs de Veubeke's colleagues. It draws together what is considered the most important part of his work as well as some relatively unknown contributions. The choice has been centered around the development of the variational methods in continuum mechanics and their finite element implementation. Interesting and significant contributions in other fields, especially in optimal control, have not been included solely for reasons of paucity.

HIS WORK

Professor Fraeijs de Veubeke's work possesses the remarkable proerty of having very few redundancies. It consists of more than 80 publications related to the following fields:

- *applied mechanics;*
- *unsteady aerodynamics;*
- *theory of vibrations;*
- *applied optimal control;*
- *fluid mechanics;*
- *structural mechanics;*
- *elasticity;*
- *variational methods in linear and nonlinear elasticity, fluid mechanics and heat conduction;*
- *foundations of the finite element method*

His deep understanding of physical phenomena led him to adopt the 'principle of least action', taken in its broadest sense, as a natural approach to many problems of applied mechanics. His contribution to the AGARD Manual of Aeroelasticity (1959), *Influence of Internal Damping on Aircraft Resonance*, is an interesting sample of this original approach: the proposed method for vibration mode excitation, based on his "Characteristic Phase Lag Theory", suggests the use of the stationarity property of the reactive energy as a criterion for pure mode excitation. This work has provided the theoretical background for the modern methods of vibration based on mode appropriation.

The most important landmark in Professor Fraeijs de Veubeke's work is probably his now famous paper, *Upper and Lower Bounds in Matrix Structural Analysis*, presented in 1962 at the 14th AGARD Meeting of the Structures and Materials Panel. This paper developed the principle of dual analysis in elastostatics, using successively purely compatible and purely equilibrium approximations to obtain upper and lower bounds to the potential energy. Conforming and equilibrium models were proposed at the same time to solve the plate stretching problem.

With his contribution *Displacement and Equilibrium Models in the Finite Element Method* in the book "Stress Analysis" edited in 1965 by Zienkiewicz and Holister, Professor Fraeijs de Veubeke established the filiation of the variational principles in Elastostatics using the method of Friedrichs' transformations. He also made a clear distinction between the kinematical or equilibrium nature of the assumptions defining the models, and the matrix methods of solution using displacements or forces as unknowns. This second paper opened the way to the development of mixed variational principles, and at the same time systematized the concept of dual analysis in a finite element context.

Five subsequent papers related to the same topics have been selected which retrace the evolution of Professor Fraeijs de Veubeke's thought in the development of the dual method in elastostatics: *Strain Energy Bounds in Finite Element Analysis by Slab Analogy* (1967, co-author O.C. Zienkiewicz); *A Conforming Finite Element for Plate Bending* (1968); *An Equilibrium Model for Plate Bending* (1968, co-author G. Sander); *Matrix Structural Analysis* (1973); *Diffusive Equilibrium Models* (1973). These papers are concerned mainly with the application of dual analysis to plate bending, the use of stress functions to derive equilibrium models, and the methods of solution at the structural level.

His 1974 paper, *Variational Principles and the Patch Test*, provided a deeper insight into the physical meaning of Irons' patch test. It is complementary to contributions of applied mathematicians in the same area. The concept of weak and strong connectivity allows for a systematic construction of nonconforming elements which pass the patch test: the elements based on the discretization of rotational equilibrium that are proposed in *Stress Function Approach* (1965) pertain to this category.

The joint paper (in collaboration with M. Hogge), *Dual Analysis for Head Conduction by Finite Elements* (1972) is a natural extension of the dual method to heat conduction problems. Functionals operating on temperature and heat flux distributions form the basis for finite element discretizations which yield to a bracketing of the dissipated energy.

In his lecture, *The Dual Variational Principles of Elastodynamics Finite Element Applications*, at the NATO lecture series on Finite Element Methods in Continuum Mechanics (Lisbon, 1971) Professor Fraeijs de Veubeke generalized the concept of dual analysis to dynamic problems. Using the method of Friedrichs' transformations as he did earlier in elastostatics, he showed that Toupin's principle can be employed to calculate the kinetic energy of an equilibrium model based on the discretization of impulses.

His paper, *The Theoretical Design Laws of Warping-Free Multicellular Box Beams* (1969), is related to its earlier, pre-finite element work on box-beam analysis. It illustrates Professor Fraeijs de Veubeke's great physical intuition. His understanding of the mechanism of deformation of the multicellular beam led him to bring significant simplifications which reduced the analysis to that of a one-dimensional beam.

Amongst his later papers, two of them have opened new paths for research and will probably have a significant impact in the future: *A New Variational Principle for Finite Elastic Displacements* (1972), based on the polar decomposition of the Jacobian of the deformation, represents a breakthrough in the development of the complementary energy approach to nonlinear elasticity. *The Dynamics of Flexible Bodies* (1976) is a lucid and quite general approach to the problem of elastic deformation in a noninertial frame of reference. The paper provides a firm background to the study of a large variety of problems such as the structural dynamics of flying vehicles, the dynamics of flexible rotors and the vibration analysis of rotor blades.

Finally, two of Professor Fraeijs de Veubeke's papers that are also very worthy for inclusion in the memorial volume are, firstly, the overview that he made of *The Variational Principles of Fluid Mechanics* (1965) in which he presented a logical derivation of purely Eulerian principles starting from Hamilton's principle and, secondly, his very concise presentation of *Nonlinear Shell Theory* (1972), a contribution which was originally published as a part of a general report on "Dual Analysis by Finite Elements - Linear and Nonlinear Applications".

M. GERADIN and G. SANDER

BIBLIOGRAPHY

I. Books

A Course in Elasticity, Series in Applied Mathematical Sciences, Vol. 29, Springer-Verlag, 1979.

II. Technical Publications

1. L'aérodynamique des planeurs, *Revue des Elèves des Ecoles Spéciales de l'Université de Vouvain,* II et IVème trimestre (1939) et IIème trimestre (1940).
2. Effect gyroscopique des rotors sur les vitesses critiques de flexion, *Bulletin technique des Ingénieurs de Louvain,* (1942).
3. Déphasages caractéristiques et vibrations forcées d'un système amorti, *Bulletin de la Classe des Sciences, XXXIV-5,* Académie Royale de Belgique, (1948).
4. Application des équations de Lagrange au calcul des forces de liaison dans les moteurs en étoile, *Bulletin 22* du Service Technique de l'Aéronautique, Bruxelles, (1949).
5. Quelques applications de la méthode des déphasages caractéristiques, Communication présentée au deuxième Congrès National des Sciences, Bruxelles, (juin 1950).
6. Crankshaft-propeller vibration modes as influenced by torsional flexibility of the engine suspension, *Journal of the Aeronautical Sciences,* (May 1950).
7. Calcul des cadres bordant une ouverture circulaire dans un champ de tension, *Mécanique,* 340-348, (Sept. 1950).
8. Diffusion des inconnues hyperstatiques dan les voilures à longerons couplés, *Bulletin 24* du Service Technique de l'Aéronautique, Bruxelles, (1951).
9. Dispositifs stabilisateurs contre les couples moteurs variables, *Annales de l'Union Belge des Ingénieurs Navals,* (1951).
10. Aérodynamique instationnaire des profils minces déformables, *Bulletin 25* du Service Technique de l'Aéronautique, Bruxelles, (1953).
11. Iteration in semi-definite eignevalue problems, *Journal of the Aeronautical Sciences,* (Oct. 1955).
12. Quelques problèmes des grandes vitesses en aéronautique, Extrait de la *Revue des Questions Scientifiques,* (Oct. 20, 1955).

13. Aspects cinématique et énergétique de la flexion sans torsion, *mémoire 8, n°1657*, Académie Royale de Belgique, (1955).

14. Matrices de projection et techniques d'itération, *Annales de la Société Scientifique de Bruxelles*, 70-1, (1956).

15. A variational approach to pure mode excitation based on characteristic phase lag theory, *Rapport 39* du Groupe de Travail Matériaux et Structures de l'AGARD, Washington (1956).

16. Méthodes variationnelles et performances optimales en aéronatique, *Bulletin de la Société Mathématique de Belgique*, VIII-2 (1956).

17. Le problème du maximum de rayon d'action dans un champ de gravitation uniforme, *Astronautica Acta, IV-I*, (1958).

18. Performances balistiques des fusées, *Revue Universelle des Mines*, (juin 1958).

19. Creep buckling, Chapitre 13 du livre *High Temperature Effects in Aircraft Structures, AGARDograph n°28*, éditeur N.J. Hoff, Pergamon Press (1958).

20. Quelques propriétés des structures ayant deux ou plusieurs fréquences propres confondues, *Compte-Rendus du 3ème Congrès Aéronautique Européen*, Bruxelles (1958).

21. Flexion et extension des plaques d'épaisseur modérée, *mémoire 8, n°1698*, Académie Royale de Belgique (1959).

22. Influence de l'amortissement interne sur la résonance d'un avion (Influence of internal damping on aircraft resonance) *Manuel d'Aéroélasticité de l'AGARD (OTAN)*, volume 1, lère partie, chapitre III (1959).

23. L'énergie potentielle complémentaire dans les problèmes dynamiques. Un principe de variation des accélérations, *Annales de la Société Scientifique de Bruxelles*, 73-III (1960).

24. La méthode des fonctions propres dans les problèmes de transmission de chaleur en régime transitoire, *Bulletin de la Société Royale des Sciences de Liège*, 7-8 (1960).

25. Principes variationnels en mécanique des fluides compressibles, *Annales de la Société Scientifique de Bruxelles*, 74-III (1960).

26. Sur certaines inégalités fondamentales et leur généralisation dans la théorie des bornes supérieures et inférieures en élasticité, *Revue Universelle des Mines*, (mai 1961).

27. Théorie des coques prismatiques minces renforcées par lisses ou diaphragmes, *mémoire 8, n°1729*, Académie Royale de Belgique (1961).

28. Optimisation des groupements de fusées en parallèle, *Astronautica Acta*, 5-6 (1961).

29. *La propulsion par fusées*, Edition Dunod, (1961), co-authors: M. Barrere, A. Jaumotte, and J. Vandenkerckove.

30. Les grandes méthodes de calcul des modes et fréquences propres dans les structures complexes, *Revue Universelle des Mines*, (février 1962).

31. The inertia tensor of a liquid bounded by walls in rigid body motion, *International Journal of Engineering Sciences*, n°1, (May 1962).

32. Optimisation des systèmes, *Quelques formes modernes des mathématiques utiles à l'ingénieur*, ed. OCDE, SRB, 615-633, Colloque (1962).

33. Upper and lower bounds in matrix structural analysis, Présenté à la 14ème réunion du groupe de travail Matériaux et Structures de l'AGARD, juillet 1962, *AGARDograph n°72*, Pergamon Press (1964).

34. Optimization of multiple impulse orbital transfers by the maximum principle, *Proc. of the XVth Congress*, Warsaw (1964), co-author: J. Geerts.

35. Displacement and equilibrium models in the finite element method, Présenté au Symposium Numerical Methods in Elasticity, University College of Swansea, January 1964, Chapter 9 in *Stress Analysis*, John Wiley & Sons (1965).

36. Considérations sur la recherche spatiale technique, extrait de la *Revue des Questions Scientifiques*, Tome XXVI (janvier 1965).

37. Les déphasages caractéristiques en présence de modes rigides et de modes non amortis, *Bulletin de la Classe des Sciences*, 5ème série, Tome 51 (1965).

38. Variational principles in fluid mechanics, 7ème conférence scientifique sur la "Mécanique des Fluides", Académie Polonaise des Sciences, Jurata (Sept. 1-8, 1965).

39. Analyse de la réponse forcée des systèmes amortis par la méthode des déphasages caractéristiques, *Revue Française de Mécanique*, n°13, 49-58 (1965).

40. Canonical transformations and the thrust-coast-thrust optimal transfer problem, *Astronautica Acta*, Vol. 11, n°4 (1965).

41. Optimal steering and cutoff-relight programs for orbital transfers, *Astronautica Acta*, Vol. 12, n°4 (July-August 1966).

42. Le problème de Newton du solide de révolution présentant une traînée minimum, Académie Royale de Belgique, *Bull. Cl. Sc.*, 171-182 (1966-2).

43. Une condition nécessaire et suffisante de minimum relatif dans le probleme des extrémés liés, Académie Royale de Belgique, *Bull. Cl. Sc.*, 512-517 (1966-4).

44. Bending and stretching of plates. Special models for upper and lower bounds, *Technical Report*, AFFDL-TR-66-80, 863-886 (1966).

45. Upper and lower bounds to structural deformation by dual analysis in finite elements, *Technical Report*, AFFDL-TR-66-199 (January 1967), co-author: G. Sander.

46. The second variation test with algebraic and differential constraints, Colloquium on Optimization in Space Flight Mechanics, Liege, (June 19-23, 1967) Pergamon Press.

47. Basis of a well conditioned force program for equilibrium models via the southwell slab analogies, *Technical Report*, AFFDL-TR-67-10 (1967).

48. Strain energy bounds in finite element analysis by slab analogy, *Journal of Strain Analysis*, Vol. 2, n°4 (1967), co-author: O.C. Zienkiewicz

49. A conforming finite element for plate bending, *Int. Jnl. of Solids and Structures*, Vol. 4, 95-108, Pergamon Press (1968).

50. An equilibrium model for plate bending, *Int. Jnl. of Solids and Structures*, Vol. 4, 447-463, Pergamon Press, (1968), co-author: G. Sander.

51. A maximum-minimum principle for bang-bang systems, *Lecture notes in Mathematics*, Vol. 112, Colloquium on Methods of Optimization at Akademgorodok, Springer-Verlag, Berlin (1969).

52. The theoretical design laws of warping-free multi-cellular box beams, *Problems of Hydrodynamics and Continuum Mechanics*, SEDOV Anniversary Volume, SIAM 287-305 (1969).

53. Natural stresses and strains for trapezoidal structures analysis, Air Force Flight Dynamics Laboratory, *Technical Report*, AFFDL-TR-69-18, (1969), in collaboration with: H.D. Nguyen.

54. Une généralisation du principe du maximum pour les systèmes bang-bang avec limitation du nombre de commutations, *Colloque sur la Théorie Mathématique du Contrôle Optimal*, Centre Belge de Recherches Mathématiques, 56-67 (1970).

55. Stability analysis by finite elements, *Technical Report*, AFFDL-TR-70-35, Wright-Patterson A.F.B., (1970), co-author: A.P. Kabaila.

56. Bifurcation of space frames, *Technical Report*, AFFDL-TR-70-36, Wright-Patterson, A.F.B., (1970), co-author: A.P. Kabaila.

57. Large displacement formulations for elastic bodies, *Technical Report*, AFFDL-TR-70-34, Wright-Patterson A.F.B., (1970).

58. The dual principles of elastodynamics. Finite element applications, *NATO Advanced Studies Institute Lecture Series on Finite Element Methods in Continuum Mechanics*, Lisbon, 1971, eds. Oden and Oliveira, UAH Press (1973).

59. Duality in structural analysis by finite elements, *NATO Advances Studies Institute Lecture Series on Finite Element Methods in Continuum Mechanics*, Lisbon, 1971, eds. Oden and Oliveira, UAH Press (1973).

60. Static-geometric analogies and finite element models, *NATO Advanced Studies Institute Lecture Series on Finite Element Methods in Continuum Mechanics*, Lisbon, 1971, eds. Oden and Oliveira, UAH Press (1973).

61. On frequency shifting by elementary modifications of inertia and stiffness, *Professor van der Neut's Anniversary Volume*, Brookhuis, Rotterdam, 413-421 (1972).

62. A new variational principle for finite elastic displacements, *International Journal of Engineering Sciences*, 10, 745-763 (1972).

63. Dual analysis for heat conduction problems by finite elements, *Int. Jnl. for Num. Methods in Engng.*, Vol. 5, 65-82, (1972), co-author: M.A. Hogge.

64. Dual analysis by finite elements linear and non linear applications, *Technical Report*, AFFDL-TR-72-93, Wright-Patterson A.F.B., (1972) co-authors: G. Sander and P. Beckers.

65. Structural dynamics - Heat conduction, Courses held at the Department of Mechanics of Solids, CISM, *Courses and Lectures*, n°126, Udine (July 1972), co-authors: M. Geradin, M. Hogge, A. Huck.

66. The numerical analysis of structures, Proc. of the Thirteenth Int. Congress of Theoretical and Applied Mechanics, Moscow, Springer-Verlag, Berlin, 20-28, (August 1972).

67. Matrix structural analysis, Lecture notes for the International Research Seminar on 'The theory and application of finite element methods', held at the University of Calgary, Alberta, Canada (July 26-August 10, 1973).

68. Diffusive equilibrium models, Lecture notes for the International Research Seminar on 'The theory and application of finite element methods' held at the University of Calgary, Alberta, Canada (July 26-August 10, 1973).

69. Variational principles in fluid mechanics and finite element applications, *Progress in Numerical Fluid Dynamics*, von Karman Institute (1974); *Lecture Notes in Physics*, Springer-Verlag, Berlin, 226-259 (1975).

70. Variational principles and the patch test, *Int. Jnl. for Num. Methods in Engng.*, Vol. 8, n°4, 783-801 (1974).

71. Structural optimization, *Proc. of the IFIP Technical Conference on Optimization Techniques*, Novosibirsk, (July 1-7, 1974), co-author: C. Fleury.

72. Some applications of computer methods in aerospace engineering, *Proc. of the XXVth Congress of the International Astronautical Federation*, Amsterdam, (Sept. 29-Oct. 5, 1974), co-authors: G. Sander, M. Hogge, C. Delcourtbon, S. Idelsohn, A. Huck, J.F. Debongnie and C. Fleury.

73. The numerical integration of laminar boundary layer equations, *Comp. and Math. with Appl.*, Vol. 1, 167-173, Pergamon Press (1975).

74. Un nouveau principe variationnel pour les déplacements élastiques finis, (en russe), *Volume anniversaire de la naissance de Galerkin*, Institut des Problèmes de Mécanique, Académie des Sciences d'U.R.S.S., Moscou, 1974-210 (1975).

75. Stress function approach, *Proc. of the World Congress of Finite Element Methods in Structural Mechanics,* Vol. 1, J.1 - J.51, Bournemouth, Dorset, England (Oct. 12-17, 1975).

76. Discretization of rotational equilibrium in the finite element method, *Proc. of the Meeting on Mathematical Aspects of Finite Element Methods,* Consiglio Nazionale delle Ricerche, Roma (Dec. 1975).

77. Non linear dynamics of flexible bodies, *Proc. of the Symposium on 'Dynamics and Control of non-rigid Spacecraft',* held in Frascati, Italy, 24-26 May 1976, ESA SP 117, 11-19 (July 1976).

78. Convexity properties in structural optimization, *Lecture Notes in Computer Science,* n°40, ed. Jean Cea, Springer-Verlag, 1-12 (1976).

79. The dynamics of flexible bodies, *Int. J. Engng. Sci.,* Vol. 14, 895-913 (1976).

80. Discretization of stress fields in the finite element method, *Journal of the Franklin Institute,* Vol. 302, n°5 & 6, 389-412 (Nov./Dec. 1976), co-author: A. Millard.

III. Internal Reports

1. Régularisation des réticences et réduction du principe du maximum de Pontryagin au calcul des variations.

2. The analysis of equilibrium bifurcation problems by finite element methods, co-author: A.P. Kabaila.

3. Non linear theory of shells.

4. Les connexions cinématiques co-déformables des éléments de coque plans.

5. Cours de théorie des structures - élasticité.

6. Approximations par le calcul des variations.

7. Coordonnées curvilignes générales.

8. Cours de Mécanique des fluides.

INFLUENCE OF INTERNAL DAMPING ON AIRCRAFT RESONANCE

B.M. Fraeijs de Veubeke

University of Liège, Belgium

1. INTRODUCTION

Assumptions concerning the effects of damping are generally based on a purely viscous model in which the energy dissipated per cycle is proportional to frequency or on a 'structural' hypothesis which makes this energy independent of frequency. The latter approach is the most widely used in flutter calculations. However, by the application of the Fourier integral technique, it is shown that the transient behaviour associated with 'structural damping' is not physically acceptable. Nevertheless, structural damping can be retained as an approximation provided that it be restricted to a given frequency band.

A Maxwell model, involving a single 'hidden' coordinate, furnishes a more satisfactory type of viscoelastic damping. Here, the transient behaviour is fully described by a heredity function that shows how the damping force depends on the past history of the motion. The concept of the heredity function extented to an eventually infinite, number of internal freedoms can provide a

---

Original publication, *AGARD Manual of Aeroelasticity, Part I*, by the Advisory Group for Aerospace Research and Development, North Atlantic Treaty Organization (AGARD/OTAN), November, 1959.

correct formulation. Experimental evidence is, however, needed to establish the exact form of the heredity function. These points are discussed in some detail in the first part of this paper.

The second part is devoted to the analysis of resonance in one degree of freedom under these various damping laws. In the case of hereditary damping a difficulty arises. The effective spring stiffness of the system becomes a function of the forcing frequency and the concept of a natural undamped frequency is not as clear-cut as it is in the simple viscous or structural damping cases. For a simple Maxwell model two slightly different undamped natural frequencies can be defined and phase resonance is shown to occur at some intermediate value of the forcing frequency.

In the third part resonance in many degrees of freedom is discussed with the help of matrix formulation. Here again hereditary damping results in an effective stiffness matrix whose elements are functions of the frequency. The natural undamped modes are then best defined as those corresponding to phase resonance. Any other definition would be of artificial character and would prevent the experimental determination of such modes, thus defeating one of the essential purposes of resonance testing.

A convenient concept for discussing the forced response in general and resonance in particular is that of a 'characteristic phase lag'. Originally devised for the special case of purely viscous damping [1], it is extended to cover all types of damping and provides and expansion of the response in orthogonal 'response modes'. At each phase resonance one of the orthogonal 'response modes' becomes identical with a natural undamped mode as previously defined. It is shown that at phase resonance the total reactive energy is zero and stationary with respect to small arbitrary variations of the excitation forces. The most difficult problem is to obtain reasonably pure natural modes while using only a limited number of shaking points. In this connection several principles can be applied: local phase resonance, pseudo-resonance in neighbouring modes, and the stationary property of the partial reactive energy. These are discussed in terms of the orthogonal response modes.

Under certain restrictive assumptions the inertia matrix, the effective stiffness matrix, and the damping matrix can be made diagonal simultaneously. The system is then effectively split into sybsystems with a single degree of freedom.

These assumptions, already considered by Lord Rayleigh [2] in the case of pure viscous damping, are easily extended to cover all types of damping. One is the basis of a method of shaker force adjustment previously evolved by Lewis and Wrisley [3]. The more general one was reformulated in France by Basile [4]. It is the writer's opinion that such assumptions are unduly restrictive and stand no change of being corroborated by experimental evidence.

Terminology

Square matrices are generally denoted by a capital Roman letter as follows:

M	matrix of inertia coefficients
K	matrix of stiffness coefficients
R	matrix of equivalent viscous damping coefficients
A-iB	matrix of dynamic flexibilities.

Column matrices are denoted by a lower case Roman letter, thus:

q	column matrix of coordinates
f	column matrix of excitations
y	column matrix of excitation amplitudes
x=a-ib	column matrix of response amplitudes (complex)
r	column matrix of response amplitudes
z	column matrix of natural mode amplitudes.

The transpose of a matrix is denoted by a 'prime', and the derivative with respect to time by a dot.

The elements of a matrix are indicated by subscripts, e.g.,

q_i is the ith element of column q

A_{ij} is the element of the ith line and j^{th} column of A.

Modes are indicated by a subscript between brackets, e.g.,

$z_{(r)}$ is the rth natural mode (a column matrix).

The convolution or Duhamel integral between two time-functions is denoted by an asterisk, e.g.,

$$\phi * q_i = \int_0^t \phi(t-\tau) q_i(\tau) d\tau \ .$$

2. MECHANICAL AND MATHEMATICAL MODELS FOR THE STUDY OF ELASTIC SPRING BEHAVIOUR

2.1 *Simple Viscous Internal Damping*

The simplest mechanical model of a structural unit where spring action is accompanied by energy dissipation is shown in Figure 1. It consists of a purely elastic spring, with stiffness k_1, and a dashpot in parallel. The force developed by the dashpot is proportional to the velocity $\dot{q}_1(t)$ of the displacement. The force-displacement relation of such a unit is therefore of the type

$$f_1 = k_1 q_1 + c_1 \dot{q}_1 \ . \qquad (1)$$

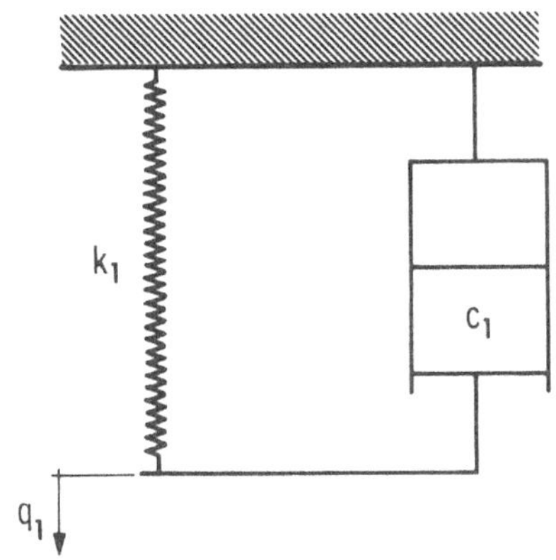

Figure 1 - Structural Unit with Simple Viscous Damping

This equation furnishes immediately the reaction force of the unit when the displacement is known as a function of time.

Conversely, when the force is known as a function of time, the solution of the differential equation (1) yields

$$q_1(t) = q_1(o) + \frac{1}{c_1} \int_o^t e^{-\tau/t_1} f_1(t-\tau)d\tau \qquad (2)$$

where $t_1 = c_1/k_1$ is the 'time constant' of the unit.

For instance, when a constant force f_1 is suddenly applied at $t = o$ to the unstrained unit $(q_1(o) = o)$,

$$q_1 = \frac{f_1}{k_1} (1 - e^{-t/t_1})$$

and the displacement tends ultimately to the static displacement f_1/k_1. When the force suddenly applied is a harmonic function of time

$$f_1 = \text{Re} \left\{ y_1 e^{i\omega t} \right\}$$

we find

$$q_1 = \text{Re}\left\{\frac{y_1}{k_1 + i\omega c_1}\left(e^{i\omega t} - e^{-t/t_1}\right)\right\} .$$

Here again the motion consists of a transient associated with the time constant t_1 of the unit and a steady state part which is the forced vibration or frequency response of the unit. The amplitude and phase relationship between force and motion in the forces response is also obtained by the reverse procedure of imposing on the unit a harmonic displacement

$$q_1 = \text{Re}\{x_1 e^{i\omega t}\} \tag{3}$$

and determining the force directly from equation (1)

$$f_1 = \text{Re}\left\{(k_1 + i\omega c_1)x_1 e^{i\omega t}\right\} . \tag{4}$$

Equations (3) and (4) are actually simpler to discuss; they also make comparison with the behaviour of other structural units easier. From them we obtain

$$q_1 = x_1 \cos\omega t$$

$$f_1 = x_1(k_1 \cos\omega t - \omega c_1 \sin\omega t) .$$

The elimination of the time between these equations leads to the hysteresis curves in an (f_1, q_1) diagram. It is convenient to make these diagrams non dimensional by using the reduced variables

$$\frac{q_1}{x_1} = \cos\omega t$$

$$\frac{f_1}{k_1 x_1} = \cos\omega t - \omega\frac{c_1}{k_1} \sin\omega t .$$

This amounts to taking the amplitude of vibration as the unit of displacement and the static force necessary to produce this displacement as the unit of force. The family of hysteresis curves (Figure 2) is then dependent on a single parameter $(\omega c_1/k_1)$. All curves have a common vertical tangent at the points (1,1) and (-1,-1).

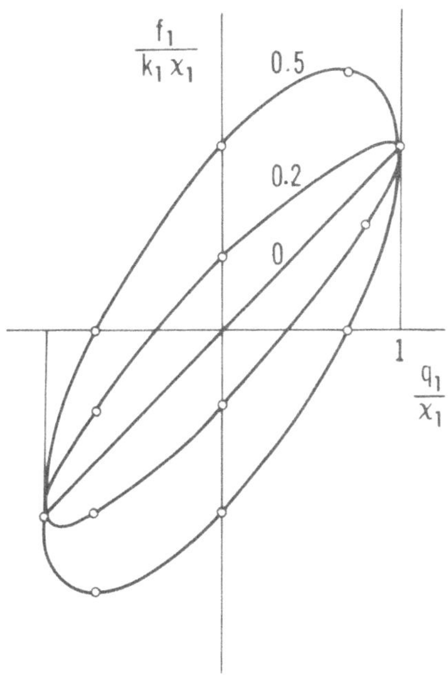

Figure 2 - Hysteresis Curves of Structural Unit with Simple Viscous Damping for Different Values of the Parameter $\omega c_1/k_1$

The energy dissipated during a cycle is

$$\int_0^{\frac{2\pi}{\omega}} f_1 \dot{q}_1 dt = \pi x_1^2 c_1 \omega \ .$$

For a given amplitude of vibration it is seen to increase proportionally to the frequency.

2.2 *Hereditary Internal Damping*

Viscous damping may be retained as a source of energy dissipation in more complicated models. The Maxwell model of Figure 3 incorporates a second spring in series with the dashpot and forms really a system having two degrees of freedom. Its behaviour is fully described by the two equations:

$$f_1 = k_1 q_1 + c_2(\dot{q}_1 - \dot{q}_2) \tag{6}$$

$$k_2 q_2 = c_2(\dot{q}_1 - \dot{q}_2) \; . \tag{7}$$

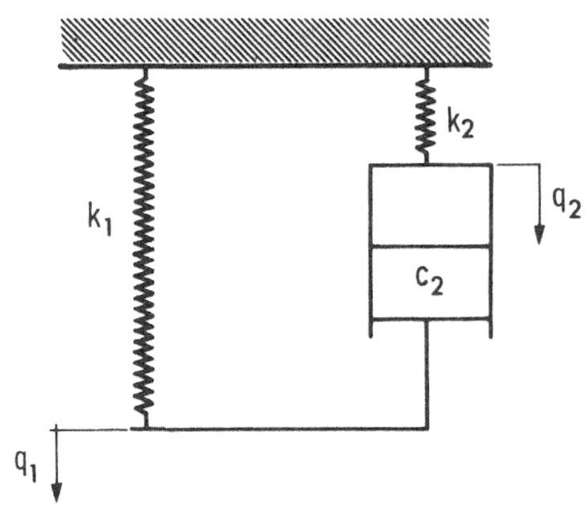

Figure 3 - Maxwell Type of Structural Unit

Since we are not interested in the internal degree of freedom q_2, we solve equation (7) for q_2 and substitute the result in equation (6) to obtain

$$f_1 = k_1 q_1 + \int_0^t \phi(t-\tau)\dot{q}_1(\tau)d\tau \tag{8}$$

where

$$\phi(t) = k_2 \exp\left(-\frac{k_2}{c_2}t\right) , \tag{9}$$

with the underlying assumption that for t = o the unit is unstrained.

In equation (1) the damping term is directly proportional to the instantaneous velocity of displacement. In the present case it is seen to depend on the past history of this velocity. For this reason it is suitably called 'hereditary damping' and the function $\phi(t)$ is known as the 'hereditary function' of the damping.

In solving problems of transient behaviour for such units the operational form of the force-displacement relationship is useful.

With the usual definition of the Laplace transform of a function x(t)

$$\bar{x}(s) = \int_{o}^{\infty} e^{-st} x(t)dt$$

and assuming $q_1(o) = o$ and $q_2(o) = o$, the basic equations (6) and (7) have the following Laplace transforms

$$\bar{f}_1(s) = (k_1 + sc_2)\bar{q}_1 - sc_2\bar{q}_2$$

$$(k_2 + sc_2)\bar{q}_2 = sc_2\bar{q}_1$$

whence, after elimination of $\bar{q}_2$

$$\bar{f}_1 = \left(k_1 + \frac{c_2k_2s}{k_2 + sc_2} \right)\bar{q}_1 \ . \tag{10}$$

In this operational relation the coefficient of $\bar{q}_1$, namely

$$z_1(s) = k_1 + \frac{c_2k_2s}{k_2 + sc_2} \tag{11}$$

is known as the 'operational modulus' of the unit and its reciprocal the 'operational flexibility'. (The names operational impedance and operational admittance are misleading, since the electrical analogies, from which they are derived, are actually violated.)

Solving equation (10) for $\bar{q}_1$ and re-interpreting we obtain

$$q_1(t) = \frac{f_1(t)}{k_1+k_2} + \frac{1}{c_2}\left(\frac{k_2}{k_1+k_2}\right)^2 \int_0^t e^{-\tau/t_2} f_1(t-\tau)d\tau \qquad (12)$$

where

$$t_2 = c_2\left(\frac{1}{k_1} + \frac{1}{k_2}\right) .$$

A comparison of this general solution with equation (2) shows a significant difference in behaviour due to the presence of an instantaneous displacement associated with a spring modulus (k_1+k_2). Such a behaviour is physically evident from a simple inspection of the structure of the unit.

From equation (12) the response of the unit to a suddenly applied constant force is readily found to be

$$q_1(t) = \frac{f_1}{k_1}\left(1 - \frac{k_2}{k_1+k_2} e^{-t/t_2}\right)$$

and tends ultimately to the same static displacement as in the previous model.

The forced response to harmonic excitation is of greater extent. Using the reverse procedure we put

$$q_1 = \mathrm{Re}\{x_1 e^{i\omega t}\} \text{ and } q_2 = \mathrm{Re}\{x_2 e^{i\omega t}\}$$

in equations (6) and (7), and eliminating x , we obtain

$$f_1 = \mathrm{Re}\left\{\left(k_1 + i\,\frac{\omega c_2 k_2}{k_2+i\omega c_2}\right) x_1 e^{i\omega t}\right\} .$$

In accordance with general theory, the complex modulus of the forced response

$$z_1(\omega) = k_1 + i\,\frac{\omega c_2 k_2}{k_2+i\omega c_2} \qquad (13)$$

is also obtained from the operational modulus (11) by substitution of $i\omega$ for s.

From equation (4) the complex modulus in the case of simple viscous damping is found to be

$$z_1(\omega) = k_1 + i\omega c_1 \; . \tag{14}$$

When equation (13) is split into its real and imaginary parts it may be given a similar form, namely

$$z_1(\omega) = k_e + i\omega c_e \tag{15}$$

where

$$k_e = k_1 + k_2 \frac{\omega^2 c_2^2}{k_2^2 + \omega^2 c_2^2} \tag{16}$$

$$c_e = c_2 \frac{k_2^2}{k_2^2 + \omega^2 c_2^2} \; . \tag{17}$$

The behaviour of the hereditary damping unit is thus reduced to that of a simple viscous damping unit with charactersitics depending on the frequency. This procedure is convenient for purposes of comparison and since moreover it corresponds to a decomposition of the complex modulus into real and imaginary parts it is also physically significant.

It is seen (Figure 4) that the equivalent spring stiffness k_e increases with frequency to the asymptotic value (k_1+k_2), whilst the equivalent coefficient of viscous damping c_e decreases from c_2 to zero. The energy dissipated per cycle, $\pi x_1^2 \omega c_e$, also tends to zero for very high frequencies. Per unit time, however, it would tend ultimately to the finite value $\frac{1}{2}\, x_1^2 k_2^2 / c_2$.

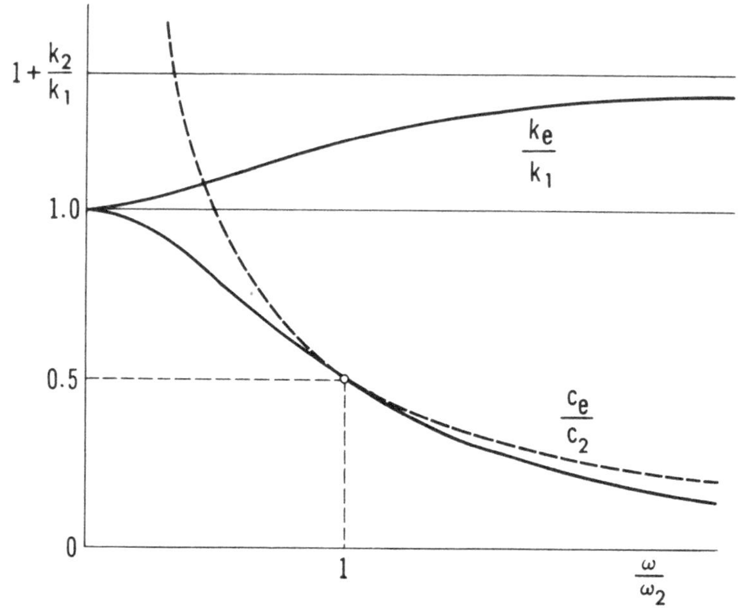

Figure 4 - Effective Spring Stiffness Re and Equivalent Viscous Damping Coefficient c_e for Maxwell Unit

There is a characteristic circular frequency of the unit for which the energy dissipated per cycle passes through a maximum. The maximum of the product ωc_e is easily found to occur for the value

$$\omega_2 = \frac{k_2}{c_2}$$

where

$$c_e = \frac{1}{2} c_2 \ .$$

This value is conveniently taken as a reference in the non-dimensional diagram of Figure 4, so that the curve

$$\frac{c_e}{c_2} = \frac{1}{1 + \left(\frac{\omega}{\omega_2}\right)^2} \tag{18}$$

becomes valid for all Maxwell units.

The behaviour of c_e as a function of the frequency is also reflected in the hysteresis curves of Figure 5. At zero frequency we have a straight line of pure spring behaviour; for $\omega = \omega_2$ an elliptical curve of maximum area; and, when the frequency tends to infinity, we have again a pure spring behaviour associated with the modulus (k_1+k_2) and resulting in a straight line of greater slope.

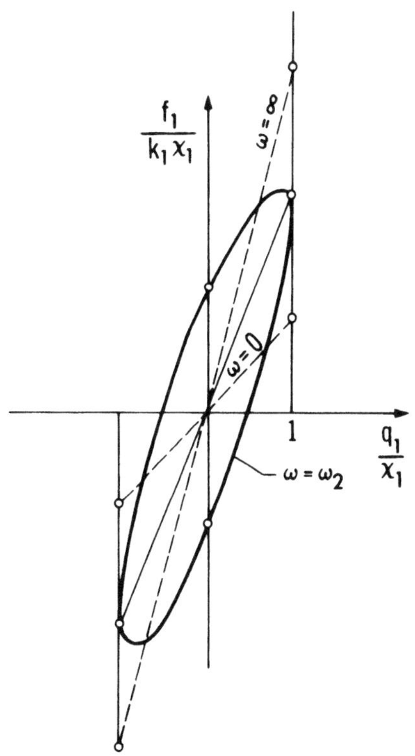

Figure 5 - Hysteresis Curves for Maxwell Unit

The Maxwell model is only a simple example of a unit with hereditary damping. More complicated models are easily imagined. For example, several Maxwell units with different time constants can be connected in parallel. A large number of hidden coordinates will facilitate the correlation between the theoretical behaviour of k_e and c_e as functions of the frequency and experimental evidence. The transient behaviour of such units will still depend on a relation of the form of equation (8) with a suitable hereditary function $\phi(t)$. On the basis of equation (9) we can write as a generalized form of the hereditary function, involving a continuous relaxation spectrum:

$$\phi(t) = \int_0^\infty F(r)\, e^{-rt} dr$$

or, in operational form,

$$\bar{\phi}(s) = \int_0^\infty \frac{F(r)}{r+s}\, dr \ .$$

Then, from the operational form of equation (8), namely

$$\bar{f}_1 = z_1(s)\bar{q}_1 \tag{19}$$

where

$$z_1(s) = k_1 + s\bar{\phi}(s) \tag{20}$$

we can deduce the corresponding operational modulus

$$z_1(s) = k_1 + s \int_0^\infty \frac{F(r)}{r+s}\, dr \ .$$

The following example is given by Biot:

$$F(r) = \frac{2}{\pi} k_1 g_1 \frac{1}{r} \qquad \text{for } r > \varepsilon$$

$$= 0 \qquad \text{for } r < \varepsilon \ .$$

Then,

$$z_1(s) = k_1 - \frac{2}{\pi} k_1 g_1 \ln \frac{\varepsilon}{\varepsilon + s} .$$

After replacing s by $i\omega$, the complex modulus of the frequency response appears to be

$$z_1(\omega) = k_1 \left(1 + \frac{2}{\pi} g_1 \ln \sqrt{1 + \frac{\omega^2}{\varepsilon^2}} \right) + i k_1 g_1 \frac{2}{\pi} \tan^{-1}$$

For small ε the imaginary part tends very rapidly to the asymptotic value $k_1 g_1$. Since g_1 would be small, the real part, or effective spring stiffness, is also practically constant in a wide frequency band, though it ultimately grows to infinity. This behaviour is approximately in accord with the so-called 'structural damping' law considered in more detail in the next section. The corresponding hereditary function is

$$\phi(t) = \frac{2}{\pi} k_1 g_1 \int_{\varepsilon}^{\infty} \frac{e^{-rt}}{r} dr = - \frac{2}{\pi} k_1 g_1 \, \mathrm{Ei}(-\varepsilon t) .$$

2.3 *Structural Damping*

If the complex impedance is constant in the whole frequency range

$$z_1(\omega) = k_1(1 + i g_1) \tag{21}$$

and the hysteresis curve (Figure 6) is also independent of frequency. This 'structural damping law' is widely used in flutter calculations. However, it cannot be followed in all its implications because, as will now be shown, it leads to a physically inacceptable transient behaviour. This behaviour can be deduced from the frequency response by the Fourier integral technique. In doing this it must be remembered that the imaginary part of the complex impedance is necessarily an odd function of the frequency. Consequently equation (21) holds for positive values of ω and must be extended to negative values by changing the sign of the imaginary part.

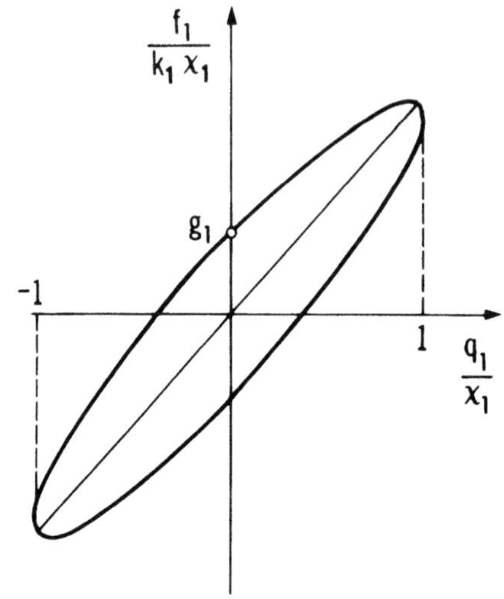

Figure 6 - Hysteresis curve of structural damping

Let us begin by assuming that $q_1(t)$ follows some periodic law expressed by a convergent Fourier series of period $T = 2\pi/\omega$ in the form

$$q_1(t) = \sum_{-\infty}^{\infty} C_n e^{-in\omega t}$$

where

$$C_n = A_n + iB_n \text{ for positive values of } n$$

$$C_n = A_n + iB_n \text{ for negative values of } n$$

so that also

$$q_1(t) = C_o + 2 \sum_{1}^{\infty} (A_n \cos n\omega t + B_n \sin n\omega t) \quad .$$

In order to find the corresponding force exerted by the unit we must multiply each term of the complex Fourier series by $k_1(1+ig_1)$ if n is positive, by $k_1(1-ig_1)$ if n is negative and simply by k_1 for the term corresponding to $n = 0$. The force is then given by

$$f_1(t) = k_1g_1(t)+k_1g_1 \sum_1^\infty \left\{i(A_n+iB_n)e^{-in\omega t} - i(A_n - iB_n)e^{-in\omega t}\right\}$$

$$= k_1g_1(t)-k_1g_1 \sum_1^\infty (2A_n\sin n\omega t-2B_n\cos n\omega t) .$$

The last series is known as the 'allied Fourier series' to the series for $q_1(t)$ and, according to the theory for the summation of such series [5], the result can be expressed in the form

$$f_1(t) = k_1g_1(t) + \frac{k_1g_1}{2\pi} \int_{-\pi/\omega}^{\pi/\omega} \omega q_1(\tau)\cot\frac{\omega(\tau-t)}{2}\, d\tau .$$

The integral must naturally be understood as a 'principle Cauchy value'. Passing to the limit for $\omega \to o$ $(T \to \infty)$, or using the corresponding analysis for the summation of 'allied Fourier integrals', we obtain

$$f_1(t) = k_1q_1(t) + \frac{1}{\pi} k_1g_1 \int_{-\infty}^{\infty} q_1(\tau) \frac{d\tau}{\tau-t} . \qquad (22)$$

It is now necessary to verify this general law of 'structural damping' by recomputing the frequency response.

Putting $q_1 = x_1e^{i\omega\tau}$ and $f_1 = y_1e^{i\omega t}$ we have

$$z_1(\omega) = k_1 + \frac{1}{\pi} k_1g_1 \int_{-\infty}^{\infty} e^{i\omega(\tau-t)} \frac{d\tau}{\tau-t}$$

or, with the change of variable $u = \omega(\tau-t)$

$$z_1(\omega) = k_1 + \frac{1}{\pi} k_1g_1 \left[\int_{-\infty}^{\infty} \frac{\cos u\, du}{u} + i \int_{-\infty}^{\infty} \frac{\sin u\, du}{u}\right] .$$

The first integral is zero by reason of symmetry. The second is well known to be equal to π. We are thus led back to equation (21) and we may note that in the case when ω is negative, the limits of the second integral will be reversed, confirming that the imaginary part of z_1 will have to change sign.

Besides the severe restrictions put on the behaviour of $q_1(t)$ to ensure the existence of the integral, the general form obtained for structural damping is physically absurd. It would imply that the force exerted by the unit does not only depend on the past history of the motion but on its future as well. Though structural damping does not appear to enjoy a sound physical basis, it is an acceptable approximation in some definite frequency band. This was already apparent from the example of the generalized hereditary function. It is also true when compared with the simple Maxwell model. To show this we deduce from equation (21) the equivalent coefficient of viscous damping

$$c_e = k_1 g_1 \frac{1}{\omega}$$

and compare it with equation (18). This comparison is easier if we write

$$\frac{c_e}{c_2} = g_1 \frac{k_1}{k_2}\left(\frac{\omega}{\omega_2}\right)^{-1} .$$

This curve can be made to touch the curve defined by equation (18), provided

$$g_1 = \frac{1}{2}\frac{k_2}{k_1} \tag{23}$$

and the contact point is then at $\omega = \omega_2$ (Figure 4).

If the simple Maxwell model is acceptable, equation (23) furnishes a correlation between the observed increase in spring stiffness with frequency and the value which should be taken for the coefficient g when using a structural damping law. The difference between isothermal and adiabatic moduli of metallic materials is quoted to be of the order of 1%. This would give

$$g_1 = 0.005 .$$

This figure is probably of the correct order of magnitude. It must however be remembered that other sources of damping, like slip in riveted joints, are present, so that the value of g , which integrates the other effects, may well rise to about 0.02. The increasing use of bonded or integrally stiffened structures will tend to reduce the value of this damping coefficient to its purely internal damping contribution.

2.4 *Internal Damping in Continuous Media*

Enrico Volterra, who considered the vibrations of discrete systems with hereditary characteristics, has also extended the hereditary damping concept to continuous media [6]. The analysis shows that for an isotropic medium the stress tensor $\{\sigma_{ij}\}$ and the strain tensor $\{e_{ij}\}$ are related by two heredity functions $\phi(t)$ and $\psi(t)$, so that

$$\sigma_{ij} = \delta_{ij}\left(\lambda e + \int_0^t \phi(t-\tau)\,\frac{\partial \ell}{\partial \tau}\,d\tau\right) + 2\mu e_{ij} + 2\int_0^t \psi(t-\tau)\,\frac{\partial e_{ij}}{\partial \tau}\,d\tau . \quad (24)$$

In this relation δ_{ij} is Kronecker's delta, $e = \Sigma e_{ii}$, and λ and μ are the usual Lamé constants.

By using a single relaxation constant for each type of heredity, the operational form of the relations is found to be

$$\bar{\sigma}_{ij} = \delta_{ij}\left(\lambda + \frac{sk_\phi c_\phi}{k_\phi + sc_\phi}\right)\bar{e} + 2\left(\mu + \frac{sk_\psi c_\psi}{k_\psi + sc_\psi}\right)\bar{e}_{ij} \quad .$$

They agree, as a particular form, with more general stress-strain relations justified by M.A. Biot on the basis of thermodynamical considerations [7, 8].

Letting k_ϕ and k_ψ tend to infinity, we obtain the extension of the simple viscous damping model of Figure 1 for which

$$\bar{\sigma}_{ij} = \delta_{ij}(\lambda + sc_\phi)\bar{e} + 2(\mu + sc_\psi)\bar{e}_{ij} \quad .$$

For harmonic motion

$$e_{ij} = \mathrm{Re}\{Y_{ij}e^{i\omega t}\}$$

and this yields

$$\sigma_{ij} = \mathrm{Re}\{\delta_{ij}(\lambda+i\omega c_\phi)\Sigma\gamma_{ii}e^{i\omega t}+2(\mu+i\omega c_\psi)\gamma_{ij}e^{i\omega t}\} \ . \qquad (25)$$

The natural extension of the structural damping law would then be

$$\sigma_{ij} = \mathrm{Re}\{\delta_{ij}\lambda(1+ig_\phi)\Sigma\gamma_{ii}e^{i\omega t}+2\mu(1+ig_\psi)\gamma_{ij}e^{i\omega t}\} \ . \qquad (26)$$

When these relations are solved for the amplitudes of strain γ_{pq}, it appears that Poisson's ratio has to be replaced by a complex quantity

$$\nu(1 + ig_\nu) \qquad (27)$$

where g_ν depends on g_ϕ and g_ψ and the usual shear modulus G and traction modulus E may then be respectively replaced by

$$G(1 + ig_\psi) \qquad (28)$$

and

$$2G(1 + ig_\psi)[1 + \nu(1 + ig_\nu)] \ . \qquad (29)$$

In the special case $g_\phi = g_\psi = g$, it is found that $g_\nu = o$. Poisson's ratio then remains real, whilst the same complex factor affects both moduli, namely

$$\begin{cases} G(1 + ig) \ , & \text{and} \\ E(1 + ig) \ . \end{cases} \qquad (30)$$

As will be seen later, this leads to an important simplification in the study of the forced response of a continuous isotropic medium. It has the essential consequence that elastically uncoupled modes are also automatically uncoupled with respect to internal damping.

3. DYNAMIC AMPLIFICATION FACTORS FOR A SINGLE DEGREE OF FREEDOM

Before attempting a discussion of the influence of internal damping on aircraft resonance it will be useful to examine the resonance phenomenon for a single degree of freedom. The addition of the effects of inertia is the essential new feature we must introduce.

3.1 *Resonance in the Presence of Simple Viscous Damping*

With the addition of a mass m_1 to the model of Figure 1, equation (1) becomes

$$f_1 = k_1 q_1 + c_1 \dot{q}_1 + m_1 \ddot{q}_1 \qquad (31)$$

and, under the assumption of simple harmonic displacements, when

$$q_1 = \mathrm{Re}\{x_1 e^{i\omega t}\} . \qquad (32)$$

The external force required becomes

$$f_1 = \mathrm{Re}\{y_1 e^{i\omega t}\} \qquad (33)$$

where

$$y_1 = (k_1 + i\omega c_1 - \omega^2 m_1) x_1 . \qquad (34)$$

Whilst varying the frequency, let us keep the amplitude of excitation constant and always referred to the same phase. This is the case if y_1 is taken as a real constant. By solving equation (34) for x_1 and substituting in equation (32), the forced response is found to be

$$q_1 = \mathrm{Re}\left\{\frac{y_1}{k_1 + i\omega c_1 - \omega^2 m_1} e^{i\omega t}\right\} . \qquad (35)$$

Noting that the low frequency or quasi-static response is

$$q_1 = \mathrm{Re}\left\{\frac{y_1}{k_1} e^{i\omega t}\right\}$$

we may rewrite equation (35) in the form

$$q_1 = \mathrm{Re}\left\{\frac{y_1}{k_1} D\, e^{i\omega t}\right\} \tag{36}$$

where

$$D = \frac{1}{1 + i\omega \dfrac{c_1}{k_1} - \omega^2 \dfrac{m_1}{k_1}} \tag{37}$$

is the 'dynamic amplification factor'.

It is convenient at this stage to introduce the quantities

$$\omega_o = \sqrt{k_1/m_1} \quad \text{and} \quad \varepsilon = \frac{c_1}{2\omega_o m_1} = \frac{c_1}{2\sqrt{m_1 k_1}} \tag{38}$$

which are the most important characteristics of the free vibrations of the system. The free vibrations are governed by equations (31) with $f_1 = o$. With the aid of definitions (38) this is transformed to

$$\omega_o^2 q_1 + 2\varepsilon\omega_o \dot{q}_1 + \ddot{q}_1 = o$$

and the roots of its characteristic equation are

$$\omega_o[-\varepsilon \pm \sqrt{\varepsilon^2 - 1}] \ .$$

In the absence of damping (when $c_1 = o$ or $\varepsilon = o$) these roots are $\pm\, i\omega_o$, so that ω_o is the natural circular frequency of the undamped system. The dimensionless quantity ε along governs the complex nature of the roots and is the true measure of the dynamic effects of damping. Since the critical value, above which the roots become real, is $\varepsilon = 1$, ε may be considered as the ratio of the actual damping coefficient c_1 to its critical value.

Definitions (38) also simplify the expression (37) and we obtain

$$D = \left[1 - \left(\frac{\omega}{\omega_o}\right)^2 + 2i\varepsilon\frac{\omega}{\omega_o}\right]^{-1} \ . \tag{39}$$

The significance of D as a complex quantity is best revealed by its polar form

$$D = Ae^{-i\phi} \tag{40}$$

where

$$A = \left[\left(1 - \frac{\omega^2}{\omega_o^2}\right) + 4\varepsilon^2 \frac{\omega^2}{\omega_o^2}\right]^{-1/2} \tag{41}$$

and

$$\tan\phi = \frac{2\varepsilon \frac{\omega}{\omega_o}}{1 - \left(\frac{\omega}{\omega_o}\right)^2} . \tag{42}$$

Substitution of equation (40) in equation (36) yields

$$q_1 = \frac{y_1}{k_1} A \cos(\omega t - \phi) \tag{43}$$

so that A appears as the dynamic amplification factor for the amplitude of vibration and ϕ as the phase lag angle between the motion and the excitation.

Figure 7 is a representation of D as a complex quantity in the Argand plane. Figure 8 is the classical representation of A and ϕ as functions of the frequency.

Phase resonance is defined as the condition for which the motion is in quadrature with the excitation. The condition $\phi = \pi/2$ is precisely seen to occur when $\omega = \omega_o$. Since (at least for small values of ε) the phase is very sensitive to frequency alterations in the neighbourhood of resonance, the measurement of phase is an ideal way to ascertain resonance conditions.

Amplitude resonance is defined as the condition for which A is a maximum. This occurs for a slightly lower frequency $\omega = \omega_o \sqrt{1-2\varepsilon^2}$, and the maximum of A, which is $[2\varepsilon\sqrt{1-\varepsilon^2}]^{-1}$, is not much greater than the value $[2\varepsilon]^{-1}$ it has at phase resonance.

Except for large damping, amplitude resonance, though less pure, is a good substitute for phase resonance.

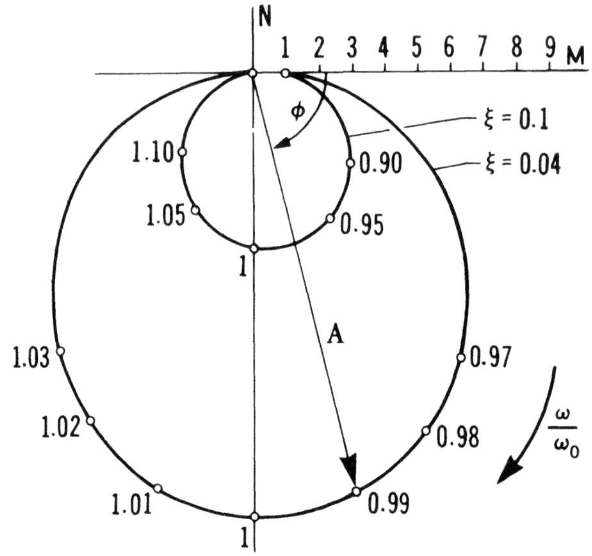

Figure 7 - Polar Representation of Dynamic Amplification Factor. Simple Viscous Damping

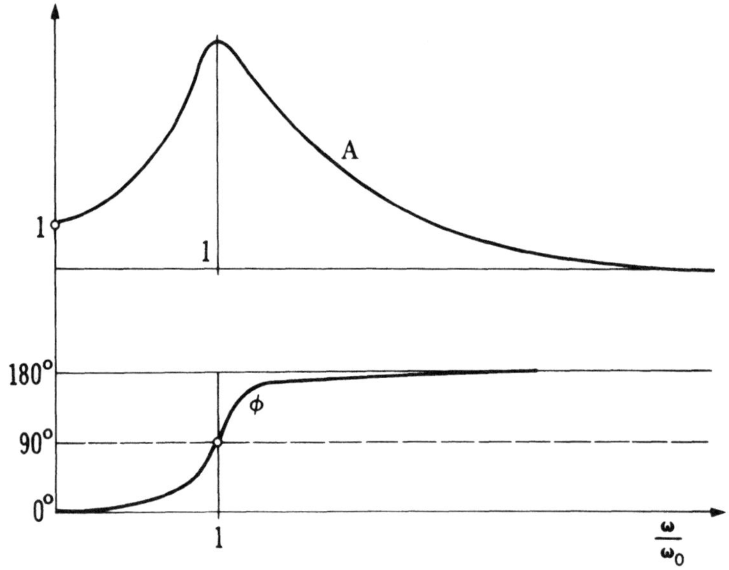

Figure 8 - Amplitude and Phase of Dynamic Amplification Factor. Simple Viscous Damping

3.2 *Resonance in the Presence of Hereditary Damping*

The addition of a mass m_1 to the simple Maxwell model of Figure 3 produces, after similar calculations, a dynamic amplification factor

$$D = \left(1 + i\,\frac{\omega c_2 k_2}{k_1(k_2 + i\omega c_2)} - \omega^2\,\frac{m_1}{k_1}\right)^{-1} . \tag{44}$$

Transformation and examination of this expression reveals two difficulties. The first is how to define the natural undamped frequency of the system. If one puts $c_2 = 0$ the natural frequency is

$$\omega_0 = \sqrt{k_1/m_1} . \tag{45}$$

But, when c_2 tends to infinity, the system is also undamped and its natural frequency is given by

$$\sqrt{(k_1 + k_2)/m_1} = \omega_0\sqrt{1 + \kappa} \tag{46}$$

where

$$\kappa = k_2/k_1 . \tag{47}$$

Since the ratio κ is fairly small, these two definitions of a natural undamped frequency are not very different. However, as will appear later, there will be reason to choose a third definition, giving an intermediate value.

The second difficulty is the definition of a reduced damping coefficient. The characteristic equation of the free vibrating system is of the third degree and the nature of the roots is consequently more difficult to ascertain. It will be sufficient for our purpose to introduce the dimensionless coefficient

$$\varepsilon = \frac{c_2}{\sqrt{m_1 k_1}} . \tag{48}$$

By taking (45), (46) and (48) into account, equation (44) may be written

$$D = \left(1 - \alpha^2 + i\,\frac{\varepsilon\kappa\alpha}{\kappa + i\varepsilon\alpha}\right)^{-1} \tag{49}$$

with

$$\alpha = \frac{\omega}{\omega_o} .$$

The resulting phase lag angle ϕ is given by

$$\tan\phi = \frac{\kappa^2\varepsilon\alpha}{(\kappa^2+\varepsilon^2\alpha^2)(1-\alpha^2)+\kappa\varepsilon^2\alpha^2} . \tag{50}$$

The denominator of expression (50) is a quadratic in α^2. One of the roots is real and negative and has consequently no special significance. The other root is real and positive and it may be concluded that there is one value of the forced frequency corresponding to phase resonance. The position of the root with respect to the values $\alpha = 1$ and $\alpha = \sqrt{1+\kappa}$, which correspond to undamped natural frequencies, is easily found. Insertion of the first value gives to the quadratic the positive value $\kappa\varepsilon^2$. Insertion of the second gives the negative value $-\kappa^3$. Since the coefficient of α is negative this proves that phase resonance occurs between the undamped natural frequencies which were previously considered.

Since, from a physical standpoint, the definition of the undamped natural frequency is necessarily somewhat artificial, there is some virtue in adopting a precise definition based on the principle of phase resonance. In other words, the natural undamped frequencies of linear visco-elastic systems will be considered as best defined by their phase resonant frequencies. In other respects the behaviour of the phase angle with increasing frequency is qualitatively similar to that of a system with simple viscous damping. Because of the reduction in equivalent damping at high frequencies it simply approaches sooner its asymptotic value of 180°.

Consideration of amplitude resonance does not reveal any remarkable features.

3.3 *Resonance in the Presence of Structural Damping*

The dynamic amplification factor due to structural damping as defined by equation (21) is

$$D = \left(1 + ig_1 - \frac{\omega^2}{\omega_o^2}\right)^{-1} \tag{51}$$

where

$$\omega_o = \sqrt{k_1/m_1} \ .$$

The behaviour of the phase lag angle

$$\tan\phi = \frac{g_1}{1 - \left(\frac{\omega}{\omega_o}\right)^2}$$

shows phase resonance for $\omega = \omega_o$, which is thus rightly defined as the natural undamped frequency of the system. A significant difference is the existence of a small, but non-zero, phase angle at zero frequency.

The representation of D in the Argand plane is exceptionally simple. The reciprocal of D is clearly a straight line parallel to the real axis and, from inversion theory, the locus of D itself is a circle with its centre on the imaginary axis (Figure 9).

It is obvious from this circle diagram that for this type of damping amplitude resonance and phase resonance are coincident.

Kennedy and Pancu [9] have shown that in the case of structural damping $d\phi/d\omega^2$ also reaches its maximum at resonance. The method of determining resonance by measurements of the rate of change of phase with ω^2 has the advantage of avoiding the measurement of absolute phase angles, which may present difficulties for certain types of excitation. In cases other than those with structural damping, the maximum rate of phase change is not exactly coincident with phase resonance but very close to it, provided damping is small.

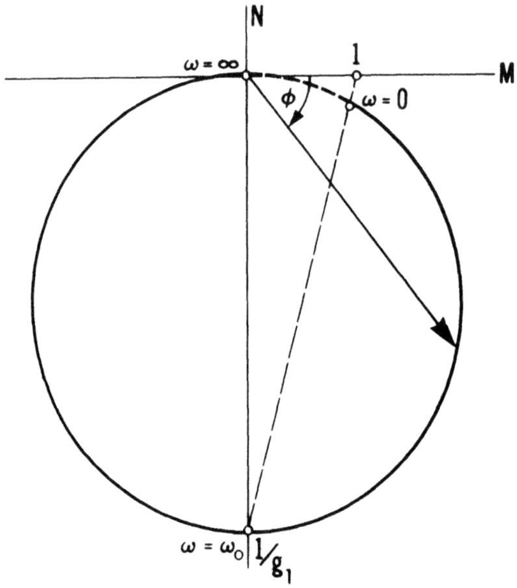

Figure 9 - Circular Locus of Dyanmic Amplification Factor for Structural Damping

4. INFLUENCE OF DAMPING ON RESONANCE IN MANY DEGREES OF FREEDOM

4.1 *Matrix Formulation of the Problem*

To discuss frequency response problems with internal damping on systems with a large but finite number of degrees of freedom it is most convenient to use matrices because of the simplicity offered by matrix formulation. The behaviour of continuous structures may be approximated closely by such methods and most of the conclusions so reached may be extended to continuous structures without further justification.

Let us return to the equation of motion of the single degree of freedom system with hereditary damping in the form

$$m_1\ddot{q}_1 + k_1 q_1 + \phi * \dot{q}_1 = f_1 \tag{53}$$

which is derived from equation (8) by the simple addition of inertia effects. The notation used for the convolution or Duhamel integral

$$\phi * \dot{q}_1 = \int_0^t \phi(t-\tau)\,\dot{q}_1(\tau)d\tau \tag{54}$$

is convenient for the extension required by the matrix formulation. The extension of equation (53) to many degrees of freedom is simply

$$M\ddot{q} + Kq + \Phi * \dot{q} = f \tag{55}$$

where q denotes the column matrix of the coordinates and $\dot{q}$ and $\ddot{q}$ are the column matrices of their first and second time derivatives. M is a square symmetric and positive definite matrix of inertia coefficients, K a square symmetric and positive semi-definite matrix of stiffness coefficients. Φ is a square symmetric matrix of heredity functions, whose product with $\dot{q}$ involves a convolution integral of type (54) for each product of elements. Finally, f is a column vector of external exciting forces.

If each heredity function is of type (9) we have a simple 'Maxwell damping'. Simple viscous damping may be considered a special case where each stiffness k_2 is taken to be infinite. When k_2 tends to infinity the function

$$\phi(t) = \begin{cases} o & \text{for} \quad t < o \\ k_e \exp\left(-\frac{k_2}{c_2}t\right) & \text{for} \quad t > o \end{cases}$$

which has the property

$$\int_o^\infty \phi(t)dt = c_2 \quad \text{(independent of } k_2\text{)}$$

behaves like the Dirac operator $c_2\delta(t)$ and the convolution integral reduces to

$$c_2 \int_0^t \delta(t-\tau)\ \dot{q}_1(\tau)d\tau = c_2\dot{q}_1(t)$$

in conformity with the simple viscous damping law.

The matrix Φ is then reduced to $\delta(t)C$, where C is a matrix of constant viscosity coefficients and equation (55) becomes

$$M\ddot{q} + Kq + C\dot{q} = f \ . \tag{56}$$

'Structural damping' cannot be included in equation (55) since it does not admit of a representation by a Duhamel integral.

The operational stiffness matrix corresponding to (55) is

$$s^2M + K + s\Phi(s)$$

and the frequency response of the system when

$$\left.\begin{aligned} q &= e^{i\omega t}x \\ \text{and} \qquad & \\ f &= e^{i\omega t}y \end{aligned}\right\} \tag{57}$$

derived from it by the substitution $s = i\omega$, is given by

$$[K - \omega^2M + i\omega\Phi(i\omega)]_x = y \ . \tag{58}$$

Since the elements of Φ are real functions of $i\omega$, a change of sign in ω is equivalent to changing i into -i. Consequently, when Φ is split into its real and imaginary parts, the result is of the form

$$\Phi(i\omega) = R(\omega^2) + i\omega I(\omega^2) \ . \tag{59}$$

The matrix

$$\tilde{K} \equiv K - \omega^2 I(\omega^2) \tag{60}$$

may be considered as a matrix of effective stiffnesses which are functions of the frequency, and the matrix $R(\omega^2)$ as the matrix of equivalent viscous damping coefficients. This reduces the structure of the equations to that which would result from consideration of equation (56).

4.2 *Natural Undamped Modes*

With the use of (59), let us rewrite equation (58) in the form

$$[K - \omega^2 M + i\omega R]x = y \tag{61}$$

where K now stands for the matrix $\tilde{K}$ defined by (60), and where functional dependence of the elements of K and R on frequency is understood. In view of the discussion in Section 3.2, and since it is the purpose to obtain the natural mode shapes by resonance testing, these modes are best defined by the principle of phase resonance. Phase resonance occurs, y being a matrix of real elements (all excitations in phase), when the matrix of response amplitudes is of the type $x = iz$, where z has real elements.

By splitting equation (61) into real and imaginary parts, we obtain

$$-\omega Rz = y \tag{62}$$

$$(K - \omega^2 M)z = o \ . \tag{63}$$

The last equation, which is homogeneous, has then by definition the natural frequencies ω_K as eigenvalues, and the corresponding natural modes $z_{(K)}$ as eigensolutions. The required excitations follow from equation (62). From (63) the square of a natural frequency is expressed by the ratio

$$\omega_K^2 = \frac{z'_{(K)} K(\omega_K^2) z_{(K)}}{z'_{(K)} M z_{(K)}} \ . \tag{64}$$

It is perhaps important to observe that, because $K(\omega_r^2)$ and $K(\omega_s^2)$ are not necessarily proprotional matrices, the modes $z_{(r)}$ and $z_{(s)}$ associated with two distinct natural frequencies ω_r and ω_s need not satisfy the usual orthogonality relation

$$\left.\begin{aligned} z'_{(r)} M z_{(s)} &= o \\ \omega_r &\neq \omega_s \end{aligned}\right\} . \tag{65}$$

The loss of this property may lead to doubts about the adequacy of our definition of natural modes. Other assumptions, however, like the ones based on quasi-static stiffnesses K(o) or on on $K(\infty)$, whilst saving equations (65), are artificial in character and would produce definitions of modes unattainable experimentally. From a practical point of view, the variations in stiffnesses with frequency are small enough to yield negligible differences between these definitions. Hence modes obtained from resonance definition should satisfy approximately equations (65).

For pure viscous damping, where $I(\omega^2) \equiv o$ and $R = C$ is a matrix of constants like K, and also in the case of 'structural damping', the definitions coincide and orthogonality should be satisfied exactly. These cases represent, however, an idealization of the physical situation.

4.3 *The Characteristic Phase Lags*

An analysis of the frequency response based on equation (61) is conveniently based on a type of orthogonal expansion using the concept of 'characteristic phase lag'. In its original form [1] it was developed for the pure viscous damping case. Since it does not use equations (65) it is easily extended to the general case where K and R are functions of the frequency.

At any given frequency the existence of a set of normal excitation modes and corresponding normal frequence response modes will be assumed as follows

$$\left.\begin{aligned} y &= y_{(K)} \quad \text{real} \\ x &= e^{-i\phi_K} r_{(K)} \quad \text{with real } r_{(K)} \end{aligned}\right\} . \tag{66}$$

It is therefore postulated that in a natural mode all coordinates vibrate with the same phase lag angle ϕ_K with respect to a mode of excitation forces which are in phase.

Substitution of these expressions in equation (61) yields after separation of terms in phase and terms in quadrature:

$$[\omega\cos\phi_K R - \sin\phi_K(K - \omega^2 M)]r_{(K)} = o \qquad (67)$$

and

$$[\cos\phi_K(K - \omega^2 M) + \omega\sin\phi_K R]r_{(K)} = y_{(K)} \ . \qquad (68)$$

Equation (67) is homogeneous and poses an eigenvalue problem for the 'characteristic phase lags'. The eigenvalue $\tan\phi_K$ are the roots of the algebraic equation in $\tan\phi$, namely

$$t|[\omega R - \tan\phi(K - \omega^2 M)]| = 0 \ .$$

They are all real since

$$\tan\phi_K = \frac{\omega r'_{(K)} R\ r_{(K)}}{r'_{(K)}(K - \omega^2 M)r_{(K)}} \qquad (69)$$

and, if $r'_{(K)}$ denotes the transpose conjugate of $r_{(K)}$, these are ratios of two Hermitian forms. When $\tan\phi_K$ is known the values ϕ_K are defined without ambiguity by the condition

$$0 \leq \phi_K \leq \pi \ . \qquad (70)$$

It will be shown that this determination is consistent with a continuous variation of ϕ_K with the frequency.

Once a ϕ_K is determined, the corresponding mode r_{iK} follows from equation (67), except for an undetermined multiplier. The corresponding excitation mode follows from equation (69) or one of the following combinations of equations (67) and (68), namely

$$\cos\phi_K y_{(K)} = (K - \omega^2 M)r_{(K)} \qquad (71)$$

and

$$\sin\phi_K y_{(K)} = \omega R r_{(K)} \ . \qquad (72)$$

Obviously both $r_{(K)}$ and $y_{(K)}$ may be taken as real. They then satisfy the initial assumptions of (66); and moreover, in equation (69) we may take $r'_{(K)}$ to represent simply the transpose of $r_{(K)}$.

The energy dissipated per unit cycle in such a type of excitation is

$$E_K = \pi \sin\phi_K \, r'_{(K)} y_{(K)} = \pi\omega r'_{(K)} R \, r_{(K)} \ . \tag{73}$$

It should be small compared to the maximum deformation energy, so that the quantities

$$\gamma_K = \frac{\omega r'_{(K)} R \, r_{(K)}}{r'_{(K)} R \, r_{(K)}} \tag{74}$$

should be small compared to unity.

Now the fact that energy is really dissipated per unit cycle shows that the equation

$$\omega R \, r_{(K)} = 0$$

cannot be true, except perhaps for $\omega = o$ or $\omega = \infty$. It follows from equation (67), that the value $\tan\phi_K = o$ cannot occur in the frequency range, except at the limits $\omega = o$ and $\omega = \infty$. As a consequence the continuous variation of a root $\tan\phi_K$ will yield a continuous variation of ϕ_K consistent with equation (70).

Let us now introduce the frequencies λ_K defined by

$$\lambda_K^2 = \frac{r'_{(K)} R \, r_{(K)}}{r'_{(K)} M \, r_{(K)}} \ . \tag{75}$$

In view of this definition and (74), equation (69) can be written as

$$\tan\phi_K = \frac{\gamma_K}{1 - \left(\frac{\omega}{\lambda_K}\right)^2} \ .$$

Note that, since K, R and $r_{(K)}$ are dependent on frequency, λ_K and γ_K are themselves functions of the forcing frequency.

Whilst λ_K^2 always remains finite and positive (the limits K(o) and K(∞) exist for the stiffnesses) as a ratio of two positive definite quadratic forms, γ_K behaves in a different manner for $\omega = o$ and $\omega = \infty$ according to the assumptions made for damping.

For hereditary damping. Finite, positive definite limits exist for R when $\omega = o$ and for $\omega^2 R$, when $\omega = \infty$. Then γ_K and $\tan\phi_K$ behave like ω near $\omega = o$ and like ω^{-1} and ω^{-3} respectively near $\omega = \infty$.

For pure viscous damping. K and R are matrices of constants and nothing is changed in their behaviours near $\omega = o$. However, near $\omega = \infty$ γ_K now tends to infinity as ω (these values are thus not small any more at high frequency) and $\tan\phi_K$ still tend to zero but only as ω^{-1}.

For structural damping. K and ωR are matrices of constant elements. Both γ_K and $\tan\phi_K$ tend to small positive quantities near $\omega = o$. Near $\omega = \infty$ γ_K remains finite, whilst $\tan\phi_K$ tends to zero as ω^{-2}.

From equation (67) again, it appears that $\tan\phi_K = \infty$ can be an eigenvalue if $r_{(K)}$ satisfies

$$(K - \omega^2 M) r_{(K)} = o \ . \tag{76}$$

This is recognized as the problem of equation (63) and justifies the assumption that

$$r_{(K)} = z_{(K)} \ , \tag{77}$$

$$\phi_K = \pi/2 \ , \tag{78}$$

$$\lambda_K^2 = \omega_K^2 \quad \text{for} \quad \omega = \omega_K \ , \tag{79}$$

$$y_{(K)} = \omega_K R z_{(K)} \ . \tag{80}$$

To sum up:

There are as many excitation modes and normal response modes as degrees of freedom in the system.

Each characteristic phase lag is zero at zero frequency (or small and positive for the assumption of structural damping), passes through $\pi/2$ for one of the natural frequencies of the system (the response mode being then identical with the natural mode) and tends to π at high frequency. It thus behaves qualitatively like the phase lag in a system of one degree of freedom.

For pure viscous damping [1] and structural damping [10] it may even be shown that $\omega^{-1}\tan\phi_K$ and $\tan\phi_K$ respectively increase monotonically with ω^2.

4.4 *Orthogonality of the Normal Response Modes*

From equation (67) it follows that a normal response mode satisfies the equation

$$\omega R r_{(K)} = \tan\phi_K (K-\omega^2 M) r_{(K)} \ .$$

Let $r_{(j)}$ be another response mode at the same frequency. Then, by pre-multiplication by $r'_{(j)}$ we obtain

$$\omega r'_{(j)} R r_{(K)} = \tan\phi_K \ r'_{(j)} (K - \omega^2 M) r_{(K)} \ .$$

Similarly, using the equation satisfied by $r_{(j)}$, transposing and post-multiplying by $r_{(K)}$, we derive

$$\omega r'_{(j)} R r_{(K)} = \tan\phi_j \ r'_{(j)} (K - \omega^2 M) r_{(K)} \ .$$

Hence, if $\tan\phi_K \neq \tan\phi_j$, the following orthogonal properties are satisfied:

$$r'_{(j)} R r_{(K)} = o \ , \tag{81}$$

$$r'_{(j)}(K - \omega^2 M)r_{(K)} = o \ . \tag{82}$$

These can still be shown to hold true if a multiple root occurs for a cheracteristic phase-lag by orthogonalizing the associated independent modes. (From matrix theory the number of independent modes is known to correspond to the degree of multiplicity of the root).

Pre-multiplying equation (68) by $r'_{(j)}$ and using the orthogonality relations we obtain

$$r'_{(j)}y_{(K)} = o \qquad j \neq K \ . \tag{83}$$

This proves a most essential property, namely that the energy input from a given excitation goes only into its associated response mode.

4.5 *Expansion of the Frequency Response to Arbitrary Excitation*

At any given frequency the $y_{(K)}$'s form a complete set, or base, of the vectorial space. Indeed, a dependency relation

$$\Sigma \alpha_K y_{(K)} = o$$

would imply, after multiplication by $r'_{(j)}$ and use of equations (83), that

$$\alpha_j [r'_{(j)}y_{(j)}] = o \quad \text{for all j's} \ .$$

Now, from equations (71) and (72)

$$[r'_{(j)}y_{(j)}]^2 = [r'_{(j)}(K-\omega^2 M)r_{(j)}]^2 + \omega^2 [r'_{(j)}Rr_{(j)}]^2 > 0$$

and consequently all α_j's are necessarily zero.

Any excitation mode may then be represented in a unique manner as an expansion

$$y = \Sigma\ \beta_K y_{(K)}\ . \tag{84}$$

It will be assumed that all exciters are kept in phase so that only real coefficients β_K are to be considered; they may be regarded as generalized coordinates for the arbitrary excitation.

The corresponding expansion of the frequency response will be

$$\kappa = \Sigma\ \beta_K e^{-i\phi_K} r_{(K)} = a - ib$$

where

$$a = \Sigma\ \beta_K \cos\phi_K r_{(K)} \tag{85}$$

is the part in phase with the excitation and

$$b = \Sigma\ \beta_K \sin\phi_K r_{(K)} \tag{86}$$

is the part lagging 90° behind.

From equations (84) and (83), and then from equation (72),

$$\beta = \frac{r'_{(K)}y}{r'_{(K)}y_{(K)}} = \sin\phi_K \frac{r'_{(K)}y}{\omega r'_{(K)}Rr_{(K)}} \tag{87}$$

or, using definitions (74) and (75),

$$\beta_K = \frac{\sin\phi_K}{\gamma_K\lambda_K^2} \frac{r'_{(K)}y}{r'_{(K)}Mr_{(K)}}\ . \tag{88}$$

It will be convenient to normalize the response modes by assigning to each the same generalized mass

$$\mu = r'_{(K)}Mr_{(K)}\ . \tag{89}$$

The phase and quadrature responses to arbitrary excitation as given by (85) and (86) will then appear in the form

$$a = \frac{1}{\mu}\ \Sigma\ \frac{\sin\phi_K \cos\phi_K}{\gamma_K\lambda_K^2}\ [r'_{(K)}y]r_{(K)} \tag{90}$$

$$b = \frac{1}{\mu} \Sigma \frac{\sin^2\phi_K}{\gamma_K \lambda_K^2} [r'_{(K)} y] r_{(K)} \quad . \tag{91}$$

4.6 *Dynamic Flexibility Matrices*

If the energy input terms shown in brackets $[r'_{(K)} y]$, are written as the last factor in each term of the expansions (they are scalar quantities) and if use is made of the associative properties of matrix multiplication, the excitation y may be factored out to the right. We then find

$$a = Ay \quad \text{and} \quad b = By \tag{92}$$

with

$$A = \frac{1}{\mu} \Sigma \frac{\sin\phi_K \cos\phi_K}{\gamma_K \lambda_K^2} r_{(K)} r'_{(K)} \tag{93}$$

$$B = \frac{1}{\mu} \Sigma \frac{\sin^2\phi_K}{\gamma_K \lambda_K^2} r_{(K)} r'_{(K)} \quad . \tag{94}$$

They are dynamic flexibility matrices and they appear as expansions in dyadic products of the response modes. An element $A_{ij}(\omega)$ of A is the amplitude of response in phase for the coordinate q_i, due to unit amplitude of excitation in the coordinate q_j; $B_{ij}(\omega)$ is the corresponding response in quadrature. Since

$$x = (A - iB)y$$

we have actually inverted relation (61) and found the reciprocal

$$[K - \omega^2 M + i\omega R]^{-1} = A - iB \quad . \tag{95}$$

4.7 *Stationary Character of the Reactive Energy in Pure Mode Excitation* [10]

Consider the energy input per cycle under arbitrary excitation, namely the 'active energy'

$$E_a = \pi y'b = \pi b'y \tag{96}$$

and define the 'reactive energy' by

$$E_r = \pi y'a = \pi a'y \ . \tag{97}$$

By substituting equation (84) for y, equations (85) and (86) for a and b, using equation (83) and then substituting equation (71) for $y_{(K)}$, we obtain

$$E_a = \pi\omega\Sigma\beta_K^2[r'_{(K)}Rr_{(K)}]$$

$$E_b = \pi\omega\Sigma\beta_K^2\cot\phi_K[r'_{(K)}Rr_{(K)}] \ .$$

Clearly E_a is a positive definite quadratic form, which corresponds to its character of really dissipated energy. On the other hand E_r is generally not positive definite. Each time a phase resonance is passed, one of the $\cot\phi_K$ coefficients passes from a positive to a negative value, thereby changing the sign of the expression.

The conditions for the reactive energy to be stationary with respect to arbitrary small variations of the generalized coordinates of the excitation are

$$\frac{\partial E_r}{\partial \beta_K} = 2\pi\omega\beta_K\cot\phi_K[r'_{(K)}Rr_{(K)}] = o \quad \text{for all values of K.}$$

Except for the trivial solution of zero excitation (all β_K's zero) the conditions can only be satisfied if

(1) All β_K's are zero except one, say β_j,

(2) Cot $\phi_j = o$ or $\phi_j = \pi/2$.

It follows then from equations (77) to (80) that

$\omega = \omega_j$ (a natural frequency),

$$x = -i\beta_j z_{(i)} \quad \text{(a natural mode in quadrature)}$$

$$y = \beta_j \omega_j R z_{(j)} \quad \text{(a pure mode excitation).}$$

Under these conditions the stationary value of E_r is zero. Hence the stationary property of the reactive energy provides a control for the ideal excitation cases and allows the experimental determination of a pure natural mode.

An alternative interesting proof is as follows. From equations (97) and (92)

$$E_r = \pi y' A y$$

and if we now consider the elements of y as the excitation coordinates, the conditions for stationary values are

$$\delta E_r = \pi(\delta y' A y + y' A \delta y) = 2\pi \delta y'(Ay) = o$$

for arbitrary variations δy. Hence

$$a = Ay = o \ . \tag{100}$$

Indeed, the vanishing of the response in phase implies a natural response mode in quadrature which can only occur at a resonant frequency. The natural frequencies of the system are therefore also solutions of the determinant

$$|A(\omega)| = 0 \tag{101}$$

and the pure excitation modes $Rz_{(j)}$ are the corresponding eigensolutions of equation (100).

4.8 *Pure Modes from Partial Excitation*

A pure excitation mode for a continuous structure would involve distributed volume forces and is clearly impossible to achieve. For a finite number of shaking points there arises the problem of optimizing the distribution of force amplitudes between the shakers to obtain a response as close as possible to a natural mode.

The principles involved may be illustrated on a system with a finite number of degrees of freedom, provided only some of the coordinates be excited.

4.8.1 *The resonance principle*

Let $z_{(r)}$ and ω_r be the natural mode and the natural frequency aimed at in a resonance test. Then, provided (1) ω_r is well separated from the other natural frequencies, so that for $\omega = \omega_r$ we have $\sin\phi_r = 1$ and the other $\sin\phi_K$'s are small quantities of $O(\gamma)$ like the γ_K's; (2) the energy input bracket $[z'_{(r)}y]$ be of the same order (at least) as the other input brackets, we get the following situation for $\omega = \omega_r$:

(a) The dominant term in the response in quadrature is

$$b_{(r)} = \frac{[z'_{(r)}y]}{\mu\omega_r^2\gamma_r(\omega_r)} z_{(r)} \tag{102}$$

all other terms being small of $O(\gamma^2)$ compared to it.

(b) The term in $z_{(r)}$ disappears in the response in phase ($\cos\phi_r = o$) and the other terms are small of $O(\gamma)$ compared to the dominant term in quadrature. Consequently we may determine experimentally the natural mode $z_{(r)}$, with errors of $O(\gamma^2)$ only, by measuring the response in quadrature at resonance. It is necessary for this purpose to use pick-ups capable of discriminating between the phase and quadrature components. If simple pick-ups are used, giving only peak response amplitude, caution must be exercised in the neighbourhood of nodes of $b_{(r)}$. For if a local amplitude of $b_{(r)}$ is large, the addition of terms of $O(\gamma^2)$ in phase with it and of $O(\gamma)$ in quadrature still results in a peak amplitude correct to $O(\gamma^2)$. But in the vicinity of nodes of $b_{(r)}$ the error may be of $O(\gamma)$.

Numerical results for torsional and flexural vibrations of a continuous beam with uniform damping, excited at one end, show that the relative importance of the response in phase grows significantly with the order of resonance.

The situation which has just been described is an ideal one and, provided we do not apply it to a high order of resonance, excellent results may be obtained from tests with a single shaker.

To achieve the energy bracket requirement the shaker should be located at a station where large vibration amplitudes occur for each particular resonance condition.

However, the following question arises; how are we going to decide that we have a resonance condition? An answer to this is provided by the use of the local phase resonance principle.

4.8.2 *The local phase resonance principle*

Let us compute the variation in the response x for a small alteration of the forcing frequency, y being kept constant. More precisely let us calculate the column matrix:

$$v = \frac{dx}{d\omega^2} .$$

This will be done to a first approximation by assuming ωR and K not to vary appreciably with frequency. Then, by differentiating equation (61) with respect to ω^2, we obtain

$$-Mx + [K - \omega^2 M + i\omega R]v = o$$

and from equation (95)

$$v = (A - iB)Mx = (AMa - BMb) - i(AMb + BMa) . \qquad (103)$$

In the ideal case described earlier, the dominant term in the real part of v is easily determined from equations (90),

(91), (93) and (94) and found to be

$$\frac{da}{d\omega^2} = -\frac{[z'_{(r)}y]}{\mu\omega_r^4\gamma_r^2(\omega_r)} z_{(r)} \ . \tag{104}$$

The other terms are of $O(\gamma^2)$ compared to it and the terms of the imaginary part are of $O(\gamma)$.

Suppose now that the resonance condition is decided by picking out the response in phase in one of the coordinates, say the ith coordinate, and by adjusting the forced frequency so as to make it vanish. This procedure may be called the local phase resonance principle and is justified if, besides condition (2) of the resonance principle, the amplitude $z_{(r)i}$ in the ith coordinate is of the same order as the other $r_{(K)i}$. For then a comparison of the correction term

$$\Delta\omega^2 \frac{d}{d\omega^2} a_i = -\frac{\Delta\omega^2}{\omega_r^2} \frac{[z'_{(r)}y]}{\mu\omega_r^2\gamma_r^2(\omega_r)} z_{(r)i}$$

with the term $a_i(\omega_r^2)$ which it must cancel (as may be found from equation (90)) shows that

$$\frac{\Delta\omega^2}{\omega_r^2} = O(\gamma^2) \quad \text{only} \ .$$

The natural frequency will thus be measured within a relative error of $O(\gamma^2)$ and the corrections due to $\Delta\omega^2$ in the response in quadrature consists of terms of $O(\gamma^2)$ with respect to the dominant term given by equation (102).

In conclusion it may be expected that, when a natural frequency is well separated and single excitation is applied at a point of large response amplitude, local phase resonance at any other (or the same) point of large response amplitude provides a measurement of the natural frequency and the mode shape both with relative errors of $O(\gamma^2)$.

It may be observed that this principle amounts to replacing the frequency equation (101) by the approximate one

$$A_{ij}(\omega) = 0$$

where j is the index of the coordinate being excited and i the index of the coordinate where the phase response is made to vanish.

4.8.3 *The pseudo-resonance principle*

In complex structures the natural frequencies are not always well separated. In theory two or several distinct mode shapes may even belong to the same natural frequency [11, 12]. In such cases and more generally in the higher frequency spectrum, where there is a tendency for natural frequencies to be closer to each other, it is impossible to obtain pure response modes without having to resort to a multiplicity of shaking points. By using several shakers it becomes theoretically possible to achieve pseudo-resonance in the response modes which are too close in frequency with the one it is wished to measure. In other words the energy brackets of these response modes may be made to vanish, thereby cancelling in the expansion (91) terms which would otherwise be relatively large because their $\sin\phi_K$ values are no longer of $O(\gamma)$.

There remains, therefore, the problem of deciding from experimental evidence whether these pseudo-resonance conditions are present.

For the sake of simplicity we will begin with the problem of two close frequencies which will be denoted by ω_1 and ω_2. Taking only the dominant terms, when the forces frequency is close to ω_1 and ω_2, we have

$$b = \frac{\sin^2\phi_1}{\mu\gamma_1\lambda_1^2}\,[r'_{(1)}y]r_{(1)} + \frac{\sin^2\phi_2}{\mu\gamma_2\lambda_2^2}\,[r'_{(2)}y]r_{(2)} \quad ,$$

$$a = \frac{\sin\phi_1\cos\phi_1}{\mu\gamma_1\lambda_1^2}[r'_{(1)}y]r_{(1)} + \frac{\sin\phi_2\cos\phi_2}{\mu\gamma_2\lambda_2^2}[r'_{(2)}y]r_{(2)} \ .$$

Supposing $\omega = \omega_1$, exactly, then $\cos\phi_1 = o$, $\sin\phi_1 = 1$, $r_{(1)} = z_{(1)}$ and the second term in b is the parasitic term to be eliminated by an excitation vector y such that $[r'_{(2)}y] = o$ and $[r'_{(1)}y]$ is of the same order as the energy brackets. It is seen that the remaining dominant term in a disappears at pseudo-resonance and that we then have almost pure phase resonance in every coordinate.

In fact, if we have two shakers we may adjust their relative force amplitude to have exact phase resonance at a point of large amplitude. Again, by a small adjustment $\Delta\omega^2/\omega_1^2$ of $O(\gamma^2)$ we may, when taking into account the terms neglected in a, provide for exact phase resonance at some other point of large amplitude. It may therefore be concluded that all requirements for accurate measurements, and in particular for pseudo-resonance in the parasitic mode, are met by using two shakers, preferably located at points of large response amplitude, and adjusting frequency and relative shaker forces to provide for phase resonance in two other (or eventually the same) points of large response amplitudes.

If we distinguish by the subscripts j_1 and j_2 the coordinates excited and by i_1 and i_2 the coordinates measured, this method amounts to replacing the pure excitation condition of equation (100) by the approximate one

$$a_{i_1} = A_{i_1 j_1}(\omega)y_{j_1} + A_{i_1 j_2}(\omega)y_{j_2} \ ,$$

$$a_{i_2} = A_{i_2 j_1}(\omega)y_{j_1} + A_{i_2 j_2}(\omega)y_{j_2} \ .$$

The extension of this principle to any number of shakers is straightforward and leads to the conditions

$$A_{(s)}y_{(s)} = o \tag{105}$$

where $A_{(s)}$ is the submatrix of A obtained by selecting the elements common to thc lines $(i_1, i_2, \ldots, i_m)$ and the columns $(j_1, j_2, \ldots, j_m)$, and $y_{(s)}$ is the column matrix of the excitation amplitudes used in the test.

We may note that in the special case where

$$i_t = j_t \qquad (t = 1, 2, \ldots, m)$$

i.e., when phase resonance is imposed at the locations of the shakers, the principle is an expression of the stationary property of the reactive energy of the partial excitation.

The practical imposition of conditions (105) becomes increasingly difficult with the number of shakers in action. More research is needed to examine the convergence of possible methods of successive approximations. One may for instance think about the possibility of extending the Gauss-Seidel process to equation (105). This would consist in practice of adjusting in succession each shaker force amplitude y_{j_t} to provide for phase resonance in coordinate q_{i_t} with a cyclic frequency adjustment as well.

Another method would be to measure experimentally (by bringing only one of the exciters to life) each column of $A_{(s)}$ and the phase responses. The frequency would then be found by numerical evaluations of $|A_{(s)}|$ until it vanishes, and the required excitations by solving numerically the compatible system fo equations (105). This procedure has actually been proposed by Traill-Nash [13].

A useful control of the pseudo-resonance condition in the neighbouring modes is provided by the purity of the response loops (quadrature versus phase response) in each coordinate when the frequency is varied in the vicinity of the natural frequency under examination. A good response loop would have to resemble that a single degree of freedom system, which is almost a pure circle, a point specially emphasized by Kennedy and Pancu [9].

4.9 *Forced Response under Special Assumptions*

If we make the assumption that

$$K(\omega^2) = [1 + \theta(\omega)]K(o) \quad \text{with} \quad \theta(o) = o \tag{106}$$

the natural modes are orthogonal with respect to both the matrices M and K(o) and

$$\left. \begin{aligned} z'_{(r)} Mz &= o \\ z'_{(r)} K(o) z_{(s)} &= o \quad r \neq s \end{aligned} \right\} . \tag{107}$$

If we now make the further assumption that $R(\omega^2)$ is defined by

$$R = \gamma(\omega) K(o) \tag{108}$$

we have the extension of a case considered by Rayleigh [2], which leads to very simple results. Indeed the substitution of the assumptions in equation (67) and comparison with the equation satisfied by a natural mode shows that at any frequency we may write

$$r_{(K)} = z_{(K)}$$

and that

$$\tan\phi_K = \frac{\omega\gamma(\omega)}{1 + \theta(\omega) - [1 + \theta(\omega)_K] \dfrac{\omega^2}{\omega_K^2}}$$

whilst from equation (72) and the previous results

$$y_{(K)} = \frac{\omega\gamma(\omega)}{1 + \theta(\omega_K)} Mz_{(K)}$$

Hence the shapes of the forced response modes become independent of the frequency, the behaviour of a characteristic phase lag is obtained explicitly and the shape of the normal excitation mode depends only on the inertial matrix M.

This last feature is important because from the knowledge of the mass distribution and the measurements of the response amplitudes the optimum distribution of the shaker forces may be progressively adjusted. This is the basis of a method proposed by Lewis and Wrisley [3].

Turning back to the more general equations of motion (55), let us introduce the assumption

$$\Phi = \phi(t)K(o) \tag{109}$$

which affects both equations (106) and (108) when applied specifically to the case of forced response. (Note that in equation (55) K stands really for what is now called K(o)). Then, introducing normal coordinates $\eta_k(t)$, we obtain

$$q = \sum_1^n \eta_k(t) z_{(K)}$$

and pre-multiplying by $z'_{(r)}$ and making use of the orthogonality relations (107), we obtain the normal equations

$$[z'_{(r)} M z_{(r)}]\ddot{\eta}_r + [z'_{(r)} K(o) z_{(r)}](\eta_r + \phi^* \dot{\eta}_r) = z'_{(r)} f \quad (r = 1,2,\ldots,n) .$$

Or, normalizing the natural modes as in equation (89) and considering equation (64) we derive

$$\ddot{\eta}_r + \frac{\omega_r^2}{1 + \theta(\omega_r)} (\eta_r + \phi^* \dot{\eta}_r) = \frac{1}{\mu} (z'_{(r)} f) . \tag{110}$$

Because of assumption (109), the normal coordinates are entirely uncoupled. In such a case a further test of pure excitation becomes possible, for when the natural mode $z_{(r)}$ is the only one excited and the excitation is suddenly cut off, $z_{(r)}$ is also a mode of decay. Moreover the analysis of the decay in any local amplitude provides information as to the exact nature of the heredity function $\phi(t)$.

In the general case where equation (109) is not valid, $z_{(r)}$ is not a mode of dacay; the coupling due to damping gradually introduces the other modes in the decay response.

The assumption made in (109) is not the most general that will lead to entirely uncoupled modes. As pointed out by Rayleigh [2] (in the particular case of simple viscous damping) the same type of reduction is possible when is a linear function of K and M, namely

$$\Phi = \phi(t)K(o) + \psi(t)M \ . \qquad (111)$$

Furthermore, from matrix theory, it is possible to show that this is the most general assumption under which

$$z'_{(r)}\phi z_{(s)} = o \qquad r \neq s$$

holds together with equations (107). Then K(o), M and Φ are reducible together to diagonal forms, a property known in France under the name of Basile's assumption [4]. The normal equations corresponding to assumption (111) are

$$\ddot{\eta}_r + \left(\psi + \frac{\omega_r^2}{1 + \theta(\omega_r)}\,\phi\right) * \dot{\eta}_r + \frac{\omega_r^2}{1 + \theta(\omega_r)}\,\eta_r = \frac{1}{\mu}\,(z'_{(r)}f)$$

and all the properties of the forces response are easily deduced from them.

REFERENCES

[1] FRAEIJS DE VEUBEKE, B.M., "Dephasages caractéristiques et vibrations forcées d'un système amorti", Acad. R. de Belgique, *Bull. de la Classe des Sc.* XXXIV, 1948, pp. 626-641.

[2] RAYLEIGH, Lord, *The Theory of Sound*, Dover, N. Y., Vol. 1, Chapter V., Section 97, 1950.

[3] LEWIS, A.C. and WRISLEY, D.L., "A System for the Excitation of Pure Modes of Complex Structures", *Journal of the Aeronautical Sciences*, Vol. 17, Nov. 1950, pp. 705-722.

[4] BASILE, B., "Recherche des Caractéristiques Dynamiques des Systèmes Continus", Actes du Colloque International de Mecanique, Poitiers, P.S.T. No. 261, 1950.

[5] JEFFREYS, H., *Mathematical Physics*, Cambridge University Press, 1946, Sections 14.04 and 14.113.

[6] VOLTERRA, Enrico, "On Elastic Continua with Hereditary Characteristics", *Journal of Applied Mechanics*, Vol. 18, No. 3, Sept., 1951, p. 273.

[7] BIOT, M.A., "Variational Principles in Irreversible Thermodynamics with Application to Viscoelasticity", *The Physical Review*, Vol. 97, No. 6, March, 1955, pp. 1463-1469.

[8] BIOT, M.A., "Theory of Stress-Strain Relations in Anisotropic Visco-Elasticity and Relaxation Phenomena", *Journal of Applied Physics*, Vol. 25, Nov., 1955, pp. 1385-1391.

[9] KENNEDY, C.C. and PANCU, C.D.P., "Use of Vectors in Vibration Measurement and Analysis", *Journal of the Aeronautical Sciences*, Vol. 14, Nov., 1947, pp. 603-630.

[10] FRAEIJS DE VEUBEKE, B.M., "A Variational Approach to Pure Mode Excitation Based on Characteristic Phase Lag Theory", AGARD Report No. 39, April, 1956.

[11] GAUZY, H. and PIRONNEAU, Y., "Un Avion Peut-il Avoir Deux Fréquences Propres Egales?", La Recherche Aeronautique No. 64, Mai-Juin, 1958, pp. 43-48.

[12] FRAEIJS DE VEUBEKE, B.M., "Quelques Propriétés des Structures Ayant Deux ou Plusieurs Fréquences Propres Confondues", Actes de 3eme Congrès Aéronautique Européen, Bruxelles, September, 1958.

[13] TRAILL-NASH, R.W., "On the Excitation of Pure Natural Modes in Aircraft Resonance Testing", Aeronautical Research Laboratories of Australia, Report S.M. 254, July, 1957.

[14] BIOT, M.A., "Linear Thermodynamics and the Mechanics of Solids", *Proc. of the Third U.S. National Congress of Applied Mechanics*, American Society of Mechanical Engineers, June, 1958.

UPPER AND LOWER BOUNDS IN MATRIX STRUCTURAL ANALYSIS

B.M. Fraeijs de Veubeke

University of Liège, Belgium

1. INTRODUCTION

The numerical calculation of static influence coefficients for a complex structure is necessarily based on simplifying assumptions. The results of such calculations are of value if the accuracy of the approximation is known. Within some measure of probability a quantitative estimate of the accuracy can be gained through previous experience with a similar structure. A theoretically more satisfactory situation arises if the calculation procedures are such that both upper and lower bounds are obtained for the coefficients. The maximum possible errors are then known quantitatively and no further investigation is needed should they lie inside the margin acceptable to the designer. Should this margin be exceeded, further refinements are required in the structural idealization.

The analytical procedures leading to upper and lower bounds are well established for the small deflexions case. It is not mandatory, though preferable for obvious reasons of economy, that the two procedures be applicable to the same idealization of

---

Original publication, *AGARDograph 72*, by the Advisory Group for Aerospace Research and Development, North Atlantic Treaty Organization (AGARD/OTAN), 1964.

the given structure. For the sake of completeness a short account of these procedures is given in the Appendix A of the present paper. Their chief characteristic is that they should be "pure". By "pure" is meant that no assumptions are allowed that would violate *both* the compatibility conditions and the equilibrium equations.

The approach yielding lower bounds must be purely compatible and, to this purpose, built on a continuous single-valued field of small displacements. The approach yielding upper bounds must satisfy everywhere the equilibrium conditions.

One must here distinguish carefully between:

(1) the structural decomposition into elements: plates, beams ... assumed to behave in accordance with simplified displacement or stress fields,

(2) the assumptions involved in the procedures for connecting the elements together, especially with regard to continuity of the displacements or stresses,

(3) the procedures adopted for solving the redundancies.

The terminology "displacement method" and "force method" used in the literature too often refers only to the last aspect of the overall problem and is not entirely relevant to the distinction made here between a "purely compatible approach" and "a purely equilibrium approach".

A displacement method of solution, or stiffness method, can be used to solve the connexion problem between elements and still violate some continuity of displacement between panels. A force method of solution can be applied and still violate local equilibrium conditions.

The important consideration to apply, if one wishes to be certain that the coefficients are upper or lower bounds, is to ensure that the geometry of the elements, their assumed displacement or stress field and the geometry of their connexions produce either a purely compatible overall field or a pure overall equilibrium field. Once this is ascertained it will be seen that, in either case, both the force method of resolution or the stiffness method are generally applicable, though perhaps with different degrees of efficiency.

2. DUAL TREATMENT OF THE TRIANGULAR PANEL

The subdivision of panels in triangular elements is an attractive proposition for complicated goemetries (taper, sweepback, skew ribs, ...). The triangular panel is also the natural shape for the purely compatible or purely equilibrium connexions between *uniform* stress-strain fields.

2.1 *Uniform Field*

A linear two-dimensional displacement field

$$\begin{aligned} u &= a + x\varepsilon_x + y\left(\frac{1}{2}\gamma_{xy} - \omega\right) \\ v &= b + x\left(\frac{1}{2}\gamma_{xy} + \omega\right) + y\varepsilon_y \end{aligned} \tag{2.1}$$

depends on six parameters: two translations, a, b, a rotation ω accounting for the rigid body displacements, and three uniform strain components (ε_x, γ_{xy}, ε_y).

The corresponding uniform stress field is given by

$$\begin{pmatrix} \sigma_x \\ \sigma_y \\ \tau_{xy} \end{pmatrix} = M \begin{pmatrix} \varepsilon_x \\ \varepsilon_y \\ \gamma_{xy} \end{pmatrix} \tag{2.2}$$

with a stress-strain matrix M, which for isotropic panels is

$$M = \frac{E}{1 - \nu^2} \begin{pmatrix} 1 & \nu & 0 \\ \nu & 1 & 0 \\ 0 & 0 & \frac{1-\nu}{2} \end{pmatrix} . \tag{2.3}$$

The reciprocal relations are

$$\begin{pmatrix} \varepsilon_x \\ \varepsilon_y \\ \gamma_{xy} \end{pmatrix} = M^{-1} \begin{pmatrix} \sigma_x \\ \sigma_y \\ \tau_{xy} \end{pmatrix} , \quad M^{-1} = \begin{pmatrix} 1 & -\nu & 0 \\ -\nu & 1 & 0 \\ 0 & 0 & 2(1+\nu) \end{pmatrix} \frac{1}{E} \quad . \qquad \begin{matrix} (2.4) \\ (2.5) \end{matrix}$$

Retaining the assumption that there are no body forces, the uniform stress field satisfies the internal equilibrium equations [see equations (A.6)]. So, with regard to its own interior domain, the uniform stress-strain field is both purely compatible and a purely equilibrium field.

The duality between compatibility and equilibrium arises solely when such uniform fields are edged and pieced together.

2.2 *Compatible Net of Triangular Fields*

Along any straight boundary line both the u and v components vary linearly and are entirely determined by the values taken in two different points. Consequently, if two linear displacement fields of type (2.1) are made to coincide in two points, they are coincident along the whole straight line joining these points. This property leads to the construction of a compatible net of triangular fields by enforcing the coincidence of the displacements at each common vertex. A natural requirement for such a construction is then that the parameters of each field be expressed in terms of the six displacement components at the vertices. Let the vertices be numbered (p,q,r) in a counter-clockwise sense. Then

$$A = \begin{vmatrix} 1 & x_p & y_p \\ 1 & x_q & y_q \\ 1 & x_r & y_r \end{vmatrix} = y_{rp}x_{qr} - x_{rp}y_{qr} \qquad (2.6)$$

is equal to twice the area of the triangle, where

$$x_{pq} = x_p - x_q \ , \quad \text{etc.}$$

An elementary calculation yields for the strain components

$$\begin{pmatrix} \varepsilon_x \\ \varepsilon_y \\ \gamma_{xy} \end{pmatrix} = \frac{1}{A} Nu \tag{2.7}$$

where N is the rectangular (3 x 6) matrix

$$N = \begin{pmatrix} y_{qr} & y_{rp} & y_{pq} & 0 & 0 & 0 \\ 0 & 0 & 0 & -x_{qr} & -x_{rp} & -x_{pq} \\ -x_{qr} & -x_{rp} & -x_{pq} & y_{qr} & y_{rp} & y_{pq} \end{pmatrix} \tag{2.8}$$

and u the column matrix of displacements, whose transpose or row matrix will be written

$$u' = (u_p \ u_q \ u_r \ v_p \ v_q \ v_r) \ . \tag{2.9}$$

The strain energy of the panel of uniform thickness t is conveniently obtained from Clapeyron's theorem (see equation (A.5) for the energy density)

$$U = \frac{1}{2} \, t \, \frac{A}{2} \, (\varepsilon_x \ \varepsilon_y \ \gamma_{xy}) \begin{pmatrix} \sigma_x \\ \sigma_y \\ \tau_{xy} \end{pmatrix} .$$

Substitution of equations (2.2) and (2.7) gives

$$U = \frac{t}{4A} u'N'MNu = \frac{1}{2} u'Ku \tag{2.10}$$

with

$$K = (t/2A)N'MN \text{ a symmetrical } (6 \times 6) \text{ matrix} \ . \tag{2.11}$$

The elements of the u matrix are to be considered as generalized coordinates for the field. The corresponding generalized forces (or the so-called corner forces) are obtained from Castigliano's formulae. If f denotes their column matrix, the variational procedure equivalent to Castigliano's theorem gives

$$\delta U = \frac{1}{2}\,\delta u' K u + \frac{1}{2}\,u' K \delta u = \delta u' K u = \delta u' f \ .$$

Hence,

$$f = Ku \tag{2.12}$$

and K is the so-called "stiffness matrix". When formula (2.11) is expanded the identity of K with the forms given by Turner or Argyris can easily be checked. The elements of the f matrix will be denoted as follows

$$f' = (H_p \ H_q \ H_r \ V_p \ V_q \ V_r) \ .$$

The relations between stresses and displacements follows from equations (2.2) and (2.7)

$$\begin{pmatrix} \sigma_x \\ \sigma_y \\ \tau_{xy} \end{pmatrix} = \frac{1}{A}\,MNu \tag{2.13}$$

and from this can be derived the equation

$$f = Ku = \frac{t}{2A}\,N'MNu = \frac{t}{2}\,N' \begin{pmatrix} \sigma_x \\ \sigma_y \\ \tau_{xy} \end{pmatrix} \tag{2.14}$$

yielding an interpretation of the generalized forces. For example, it is found that

$$H_p = \frac{t}{2}\,(\sigma_x y_{qr} + \tau_{xy} x_{rq}) \ .$$

It can then be stated [5] that each generalized force at a vertex is equal to half the resultant of stresses acting on the opposite edge or to the resultant of stresses acting on half of the adjacent edges (Figure 1).

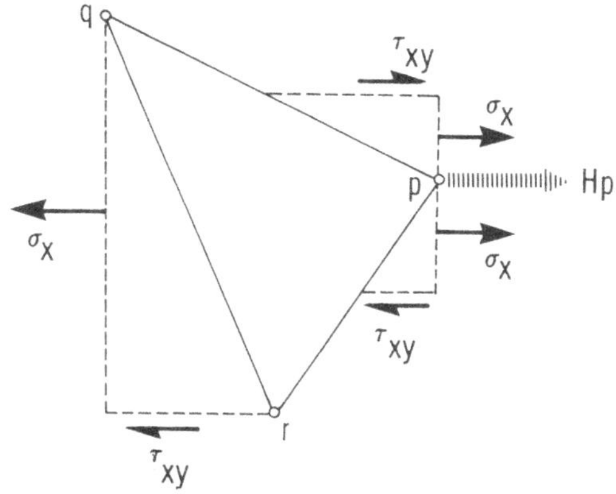

Figure 1

Obviously a compatible net of triangular fields is not an equilibrium field. The equilibrium equations between adjacent fields are of a global character; they state that the vectorial sum of all generalized loads meeting in a common vertex is equal to the external load applied there. They do not prevent the stresses to vary discontinuously across a common edge.

2.3 *Equilibrium Net of Triangular Fields*

In this case it is the continuous transmission of stresses that is required and this can only be achieved by relaxing compatibility requirements. The equilibrium requirements are met by taking as generalized displacement coordinates the displacements at the middle of each edge of the triangle. The mid-points define the vertices of a "skeleton triangle" (hatched in Figure 2), whose area is four times smaller than that of the real panel:

$$\frac{1}{4} A = y_{rp} x_{qr} - x_{rp} y_{qr} \; . \tag{2.15}$$

Otherwise the theory of par. 2.2 applies without modifications and, keeping the definition (2.8) of the N matrix, one obtains the stiffness matrix

$$K^* = \frac{8t}{A} N'MN \tag{2.16}$$

and the relationships

$$\begin{pmatrix} \varepsilon_x \\ \varepsilon_y \\ \gamma_{xy} \end{pmatrix} = \frac{4}{A} Nu \text{ and } f = 2tN' \begin{pmatrix} \sigma_x \\ \sigma_y \\ \tau_{xy} \end{pmatrix} . \tag{2.17}$$

From this last result follows that each generalized load applied to a mid-point is the resultant of the stresses acting along the corresponding side (Figure 2).

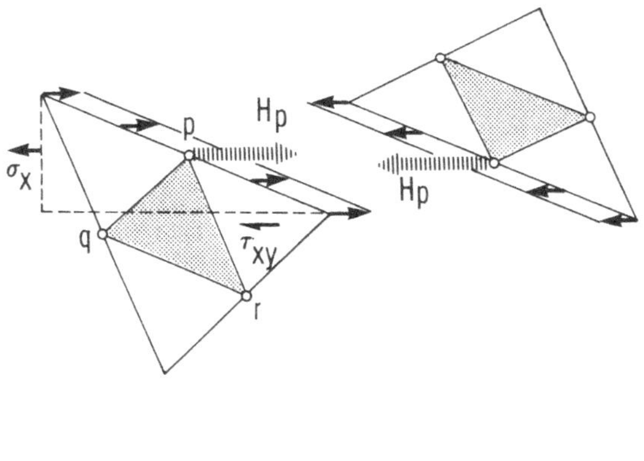

Figure 2

Whereas in the compatible net a vertex or node is common to several fields, a node of an equilibrium net is only common to two panels. The global equilibrium condition is, barring external loading, equivalent to stating that the generalized forces at the node are equal and opposite (Figure 2). In view of the fact that the stresses are constant along each edge this implies that the stresses are transmitted continuously as was required. Any external load, represented by a generalized resultant applied at a node, will really have to consist of a uniformly distributed load along the edge.

If local equilibrium is satisfied in an equilibrium net, compatibility is now generally violated for two reasons:

(1) the strain along a common edge can have different values in each panel;

(2) the orientation of the edge can be different.

This possibility of rotation of one panel with respect to the adjacent one leads to peculiar difficulties to be discussed later.

3. THE STIFFNESS METHOD OF RESOLUTION

The essential steps of the stiffness method of resolution are briefly recalled:

(1) Addition of the stiffnesses of the individual elements in a stiffness matrix for the complete structure. This step expresses the single-value of the displacement vector at a node where the elements are assembled and the additivity of the corresponding internal generalized loads, balancing the external load;

(2) Elimination of externally unloaded nodes, where deflexions are not to be determined. Suppose equation (2.12) has been extended by step 1 to represent the relations between the external loads matrix f and the displacement matrix u for the complete structure. If the ith element of f is zero, we have

$$\sum_j K_{ij} u_j = 0 \ .$$

This equation is used to express u_i as a linear combination of the other displacements. The jth column of the original stiffness matrix is thereby modified by the addition of the ith column multiplied by the factor $-K_{ij}/K_i$. The ith column itself becomes a column of zeroes and is deleted together with the ith row from which equation (2.18) was derived.

Steps 1 and 2 can be taken progressively as an intertwined process of growth and reduction until the complete stiffness matrix $K_{(f)}$ (free-structure matrix) is obtained for the loaded nodes.

(3) Elimination of all rigid body modes to produce a final nonsingular stiffness matrix $K_{(s)}$ (supported structure matrix). This step consists in prescribing certain displacements to be zero in order that the structure be at least isostatically, or even hyperstatically, supported. Rows corresponding to the suppressed displacements are deleted in the equation $f = K_{(f)}u$. They are used afterwards to know the reaction loads due to the remaining independent loads.

(4) Inversion of the nonsingular matrix $K_{(s)}$ to produce the matrix of influence coefficients.

No difficulties are encountered in implementing those steps in the case of compatible triangular fields associated to beam elements or, more generally, in the case of fields where the displacements are defined at the corners. The treatment of equilibrium fields is by no means restricted to the use of the dual "force" method and there are some obvious advantages to be gained from attempting to solve them also by the stiffness procedure. The same basic programme can be used in the computer and the sometimes delicate choice of convenient self-strainings is avoided. To achieve this, however, some care must be exercised with respect to the addition of beam elements and the elimination of rigid-body modes:

(a) *Addition of a Beam Element* (Figure 3). The total load g_p applied by the fields (and possibly by an external source) to the isolated beam element is in the form of a uniform shear flow

g_p/c_p. There are also end loads g_m and g_n with conventional positive sense identical to the sense chosen for the end displacements v_m and v_n. The internal tension T in a cross-section positioned by the nondimensional coordinate $-\frac{1}{2} \leq \xi \leq \frac{1}{2}$ is to be calculated by equilibrium considerations

$$T = -g_m - \frac{1}{2} g_p - \xi g_p \; . \tag{3.1}$$

The (complementary) strain-energy of the beam, assumed here for simplicity to have a uniform cross-sectional area, is

$$U = \frac{c_p}{2ES} \int_{-1/2}^{1/2} T^2 d\xi = \frac{c_p}{2ES} \left[(g_m + \frac{1}{2} g_p)^2 + \frac{1}{12} g_p^2 \right] \; . \tag{3.2}$$

A simple application of Castigliano's formulae yields

$$v_m - v_n = \frac{\partial U}{\partial g_m} = \frac{c_p}{ES} (g_m + \frac{1}{2} g_p) \tag{3.3}$$

as relative displacement of the end-sections, and

$$v_p - v_n = \frac{\partial U}{\partial g_p} = \frac{c_p}{2ES} (g_m + \frac{2}{3} g_p) \; . \tag{3.4}$$

It should be noted that v_p is not the displacement of the mid-span cross-section; it is a generalized displacement corresponding to the uniform shear flow. It is equal to the ordinary average of all cross-section displacements as calculated by integration of the strain. A best fit in compatibility is reached when this v_p is identified with that belonging to the field edges which are loading the beam. In that respect it would have been more appropriate, when discussing the equilibrium nets of triangular fields, to define the displacements as averages along each edge. However, since the field is uniform the average is identical with the displacement of the middle of the edge.

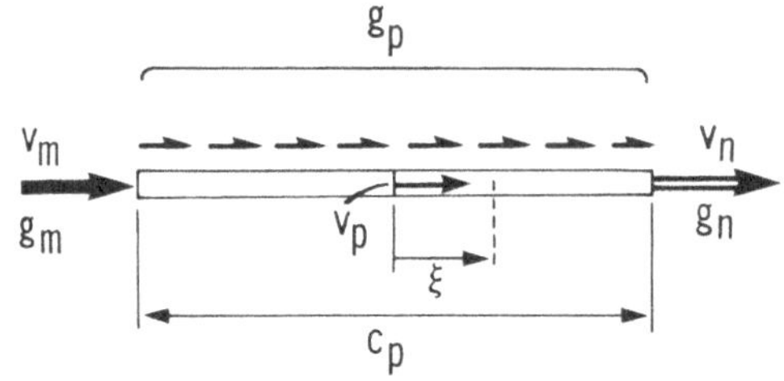

Figure 3

While the addition of a beam segment to a compatible net introduces no additional nodes, in the present case the end-sections of the beam are new points where displacements must be introduced and where concentrated external loads can be applied. To obtain the component stiffness matrix due to the beam, equations (3.3) and (3.4) are solved for g_m and g_p; the result substituted in equation (3.2) yields the strain-energy

$$U = \frac{2ES}{c_p} (v_m^2 + v_n^2 + 3v_p^2 + v_m v_n - 3v_p v_m - 3v_p v_n) \ . \qquad (3.5)$$

The other Castigliano's formulae give the load-displacement relations expressed by means of the required stiffness matrix in the form

$$\begin{pmatrix} g_m \\ g_p \\ g_n \end{pmatrix} = \frac{2ES}{c_p} \begin{pmatrix} 2 & -3 & 1 \\ -3 & 6 & -3 \\ 1 & -3 & 2 \end{pmatrix} \cdot \begin{pmatrix} v_m \\ v_p \\ v_n \end{pmatrix} \ . \qquad (3.6)$$

These equations are also directly obtainable from equations (3.3) and (3.4) together with (3.1) for $\xi = \frac{1}{2}$ and $T = g_n$. It should be noted that at the point where beam end sections meet together equilibrium will automatically be satisfied by the stiffness addition process.

(b) *Elimination of Rigid Body Modes.* The possibility of relative panel rotation already mentioned is generally checked by the presence of connected beam segments. The case can, however, arise of structural deformation modes due to panel rotations without strain energy. They will be called rigid deformation modes and are best determined by inspection of the kinematics of the skeleton triangles.

Two cases must be distinguished. In the first, passing from the free to the supported structure, the nature of the support prevents the rigid deformation modes as well as the rigid body displacements and no special treatment is required.

In the second case some rigid deformation modes remain even after the structure is supported. They are characterized by non-zero displacement vectors satisfying the matrix equations

$$K_{(s)}w_{(r)} = 0 \ , \ r = 1, 2, \ldots, t \ .$$

The theory of linear equations systems gives as necessary and sufficient conditions for solving $f = K_{(s)}u$ that the loading f verifies the virtual work equations

$$w'_{(r)}f = 0 \ . \tag{3.7}$$

Suppose now that the rigid deformation modes are just suppressed by preventing the first t displacements (this will generally require a renumbering of the displacements and is only advocated for simplicity of exposition). The first t components of f and u are now isolated in submatrices $f_{(1)}$ and $u_{(1)}$, their complements denoted by $f_{(2)}$ and $u_{(2)}$ and the matrix $K_{(s)}$ subdivided in four component blocks so that

$$f_{(1)} = K_{11}u_{(1)} + K_{12}u_{(2)} \quad ,$$

$$f_{(2)} = K_{21}u_{(1)} + K_{22}u_{(2)} \quad .$$

By making $u_{(1)} = 0$ we have, since K_{22} is by assumption non-singular

$$u_{(1)} = 0 \qquad u_{(2)} = K_{22}^{-1} f_{(2)} \ , \tag{3.8}$$

$$f_{(1)} = K_{12} K_{22}^{-1} f_{(2)} \ . \tag{3.9}$$

The last equation furnishes the reactions in the rigid links suppressing the $u_{(1)}$ displacements as linear functions of the independent loads of the $f_{(2)}$ matrix.

A unit load on the structure is now considered as a group formed by a unit vector of the $f_{(2)}$ matrix associated with the corresponding set of $f_{(1)}$ loads so that the reactions in the links disappear and the $u_{(1)}$ displacements can be freed. Under this circumstance the general expression for the displacements contains arbitrary contributions from the rigid deformation modes

$$u_{(1)} = \sum_1^t a_r w_{(r)(1)}$$

$$u_{(2)} = K_{22}^{-1} f_{(2)} + \sum_1^t a_r w_{(r)(2)} \ .$$

The analysis shows that the situation is unsatisfactory except for those influence coefficients that would happen to be independent of the arbitrary coefficients a_r.

4. THE QUADRILATERAL PANEL SUB-DIVIDED BY DIAGONALS

A remarkable case where the relative rotations between panels do not affect the loading possibilities and constitutes a convenient new building block is the quadrilateral panel sub-divided into four triangular fields by the internal diagonals (Figure 4). The four skeleton triangles articulated in points I, J, K and L, form a kinematically deformable chain. Yet the 2 x 4 load components applied on the outside vertices E, F, G and H are only restricted by three conditions expressing their statical equivalence to zero (rigid-body equilibrium).

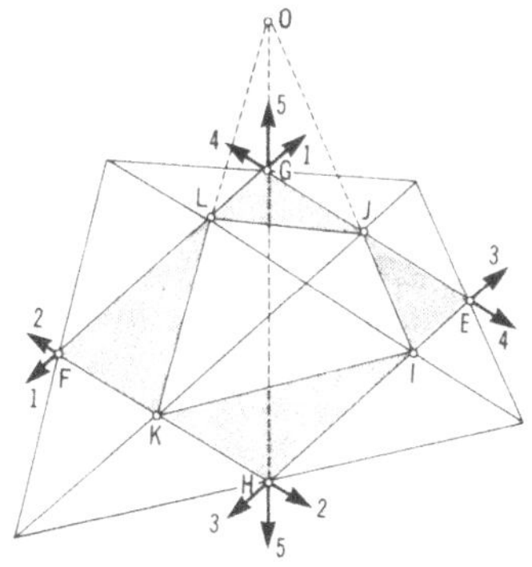

Figure 4

One proof consists in obtaining 2 x 4 - 3 = 5 independent states of loading which can be transmitted through the kinematical chain. Four of them are obvious and consist in the equal and opposite force pairs numbered from 1 to 4. The transmission is here based on the fact that the sides of the skeleton triangles are aligned two by two to form a skeleton parallelogram.

A fifth independent state of loading has been analysed for clarity on Figure 5. The pair of forces numbered 5 can be transmitted through the sides IJ and KL, keeping each skeleton triangle in equilibrium, provided the extensions of the lines IJ, KL and HG intersect in a common point O. Suppose this to be true; then by Ceva's theorem applied to the triangle FGH cut by the transversla KLO:

$$\frac{GO}{GH} \cdot \frac{LF}{LG} \cdot \frac{KH}{KF} = 1 .$$

Again from Ceva's theorem applied to the triangle GHE cut by the transversal IJO:

$$\frac{GO}{GH} \cdot \frac{JE}{JG} \cdot \frac{IH}{IE} = 1 .$$

Whence a necessary and sufficient condition that 0 be a common intersection is that we should have

$$\frac{LF}{LG} \cdot \frac{KH}{KF} = \frac{JE}{JG} \cdot \frac{IH}{IE} .$$

This, however, is true since LF = IH, LG = IE, KH = JE and KF = JG.

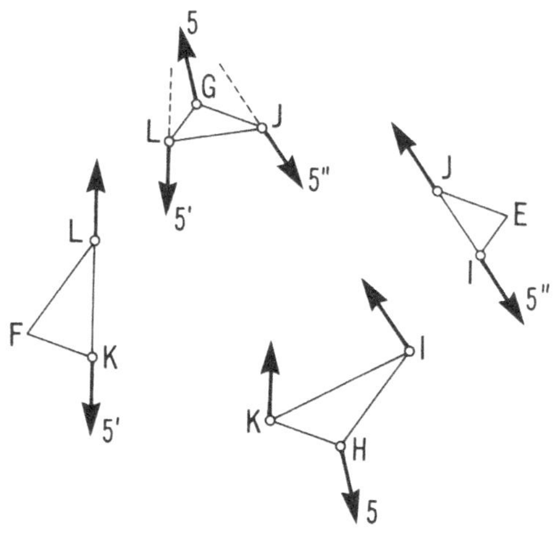

Figure 5

Another way to look at this property of the quadrilateral is to prove that the kinematical deformability of the chain can take place with each skeleton rotating about the outside vertex. The geometrical proof is similar. There follows that no virtual work of the loads applied to the outside vertices is ever involved and no other restrictions placed on them than static equilibrium with respect to rigid body motion. For the purpose of building up the structure the skeleton parallelogram of the quadrilateral can be considered as rigid.

Obviously a sixth state of loading is provided by a pair of opposite forces along EF. This is, however, easily seen to be a linear combination of the preceding states.

5. THE FORCE METHOD OF RESOLUTION

A drawback of the stiffness method, as applied to equilibrium fields, is a relatively large increase in the number of nodes requiring a correspondingly large number of eliminations. Experience should tell whether the accuracy of the computations is thereby seriously affected.

The opposite situation occurs when one wishes to apply the force method of resolution simultaneously to compatible fields and equilibrium fields. We shall here assimilate the force method to the complementary energy method and, since the prescribed displacements are usually zero, it will further reduce to the minimum principle of the strain energy expressed in terms of stresses.

The stress field is built up by adding a convenient particular equilibrium field with a unit load, ignoring the compatibility restrictions, and a complete independent set of self-strainings. The most convenient self-strainings are those confined to a minimum number of elements.

In that respect the compatible triangular nets are very simply dealt with. A complete set of independent self-strainings is obtained by considering the interactions between pairs of triangles (Figure 6) or between a triangle and a beam segment (Figure 7). The effect of such a self-straining on a triangular field is to produce a state of simple traction in a direction parallel to the interacted edge. This is easily verified by equation (2.14) with the relation (Figure 8)

$$\frac{1}{2}\, t h_p \sigma = F_p \ .$$

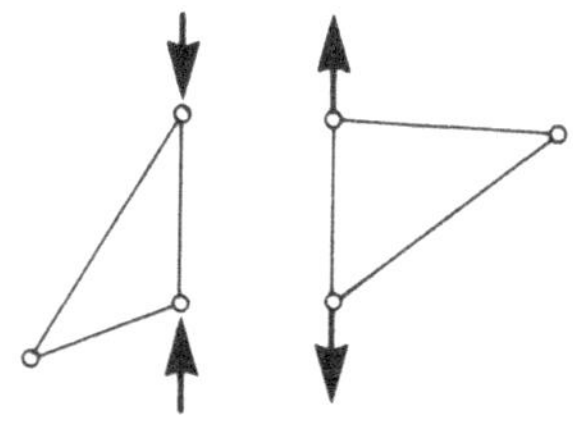

Figure 6

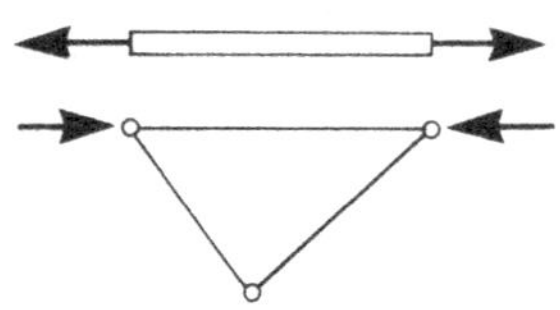

Figure 7

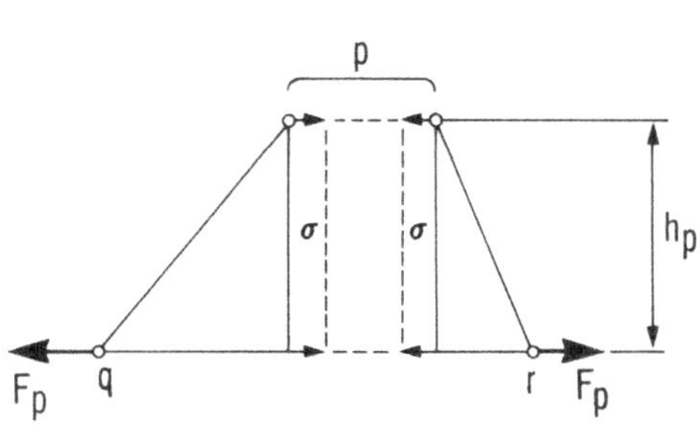

Figure 8

Turning to the cartesian axes of reference, we find

$$\sigma_x = \frac{2F_p}{th_p}\left(\frac{x_{qr}}{c_p}\right)^2 ,$$

$$\sigma_y = \frac{2F_p}{th_p}\left(\frac{y_{qr}}{c_p}\right)^2 ,$$

$$\tau_{xy} = \frac{2F_p}{th_p}\left(\frac{x_{qr}}{c_p}\frac{y_{qr}}{c_p}\right) .$$

Adding the effect of the self-strainings F_q and F_r along the other edges and substituting into

$$U = \frac{1}{2}\frac{tA}{2E}\,(\sigma_x^2 + \sigma_y^2 + 2\tau_{xy}^2 - 2\nu(\sigma_x\sigma_y - \tau_{xy}^2)) ,$$

the following standard form of the strain energy is obtained

$$U = \frac{1}{Et}\left\{\frac{(F_pc_p+F_qc_q+F_rc_r)^2}{A} - 2A(1+\nu)\left(\frac{F_pF_q}{c_pc_q} + \frac{F_qF_r}{c_qc_r} + \frac{F_rF_p}{c_rc_p}\right)\right\} .$$

The use of this standard form in adding together all the energy contributions implies that the particular equilibrium field be locally resolved in a (F_p, F_q, F_r) system. In that case each force of this system is the sum of a component from the particular equilibrium field and one hyperstatic unknown.

The beam energies are of the type $(c_pN^2)/(2ES)$ where the traction load N is the sum of a component fixed by the particular equilibrium state and unknowns from the interaction with adjacent fields.

The disadvantage of a large number of self-strainings is partially offset by their simplicity and the possibility of eliminating them gradually before the structure is completed. Indeed a self-straining intensity X can be eliminated by a Menabrea type of equation $\partial U/\partial X = 0$ as soon as the elements concerned by this self-straining have been added to the structure.

The equilibrium net of triangular fields contains fewer self-strainings but of a more complex character. One example is shown on Figure 9. As already observed [6] the setting up of the simplest independent self-strainings by the computer itself represents a major step in the automation of the computations.

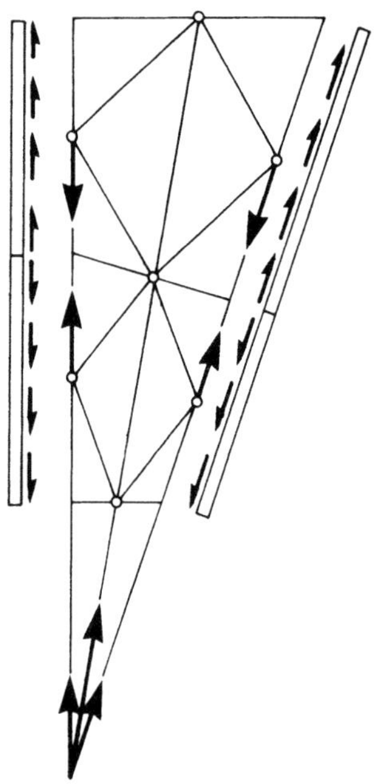

Figure 9

The state of stress in a triangle can this time be regarded as a superposition of three F states (Figure 10) similar to those of Figure 10 used in the compatible case. By virtue of equation (2.17) we have here $th_p\sigma = F_p$. Hence, with the exception of a numerical factor the strain energy of the superposed fields is the same as before

$$U = \frac{1}{4Et}\left\{\frac{(F_pc_p+F_qc_q+F_rc_r)^2}{A} - 2A(1+\nu)\left(\frac{F_pF_q}{c_pc_q} + \frac{F_qF_r}{c_qc_r} + \frac{F_rF_p}{c_rc_p}\right)\right\} .$$

To make use of this standard form, both the particular equilibrium and the effects of the various self-strainings to which a triangle would happen to be subjected to must be resolved in an F system. Note incidentally that the partial derivatives of this expression would yield the elongations of the skeleton triangle sides under the F forces. The strain energy of a beam segment as a function of loads was already dealt with. The loads appear again as linear combinations of a term due to the particular equilibrium condition considered and the self-straining terms.

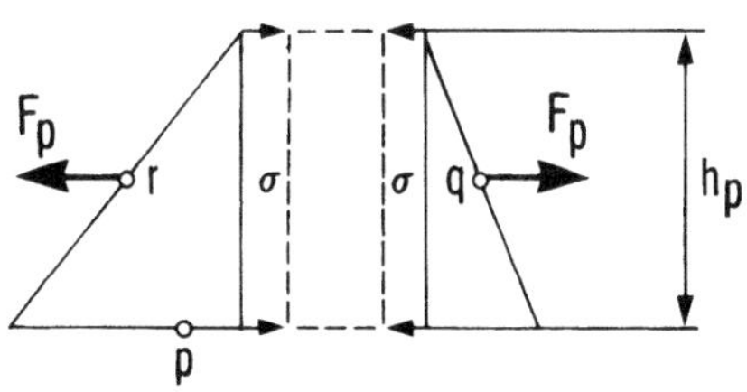

Figure 10

REFERENCES

[1] TREFFTZ, E., "Ein Gegenstück zum Ritzschen Verfahren", *Proc. of the 2nd Int. Cong. for Appl. Mech.*, Zürich, 1926.

[2] WEBER, C., "Eingrenzung von Verschiebungen und Zerrungen mit Hilfe der Minimälsatze", *Z.A.M.M.*, Vol. 22, 1942, p. 130.

[3] PRAGER, W. and SYNGE, J.L., "Approximations in Elasticity Based on the Concept of Function Space", *Quart. Appl. Math.*, Vol. 5, 1947.

[4] FRAEIJS DE VEUBEKE, B.M., "Sur Certaines inégalités fondamentales et leur généralisation dans la théorie des bornes supérieures et inférieures en élasticité", *Rev. Universelle des Mines*, XVII, No. 5, 1961.

[5] TURNER, M.J., "The Direct Stiffness Method of Structural Analysis", Structures and Materials Panel, AGARD, Aachen, 1959.

[6] DENKE, P.H., "A General Digital Computer Analysis of Statically Indeterminate Structures", Structures and Materials Panel, AGARD, Aachen, 1959.

[7] GALLAGHER, R.H., "A Correlation Study of Methods of Matrix Structural Analysis", Oral progress report to the Structures and Materials Panel of AGARD, Paris, 1961.

[8] ARGYRIS, J.H. and KELSEY, S., "Note on the Theory of Aircraft Structural Analysis", *Zeitschrift für Flugwissenschaften*, Vol. 7, No. 3, 1959.

APPENDIX A

*Principles for the Determination of Upper and Lower Bounds to Influence Coefficients**

1. LINEARIZED THEORY

The principles apply only to the fully linearized theory of elasticity. This means that

(a) Rotations and strains are small enough to justify the linear relationships

$$\varepsilon_x = \frac{\partial u}{\partial x} , \quad \varepsilon_y = \frac{\partial v}{\partial y} , \quad \gamma_{xy} = \frac{\partial u}{\partial y} + \frac{\partial v}{\partial x} \tag{A.1}$$

between strain components $(\varepsilon_x, \varepsilon_y, \gamma_{xy})$ and displacement vector (u, v). No distinction need be made between Lagrangian and Eulerian coordinates: the cartesian coordinates (x, y) are either those of the initial reference state or those of the deformed state.

(b) The strain energy density $W(\varepsilon_x, \gamma_{xy}, \varepsilon_y)$ relating stresses and strains

$$\sigma_x = \frac{\partial W}{\partial \varepsilon_x} , \quad \tau_{xy} = \frac{\partial W}{\partial \gamma_{xy}} , \quad \sigma_y = \frac{\partial W}{\partial \varepsilon_y} \tag{A.2}$$

is taken to be a quadratic, positive definite, homogeneous form. The stress-strain relations are thus of the homogeneous linear type

*For simplicity the two-dimensional case is considered. The extension to three dimensions is trivial.

(generalized Hooke's law).

(c) The complementary energy density Φ $(\sigma_x, \tau_{xy}, \sigma_y)$, whose general definition is through the Legendre contact transformation

$$\Phi = \sigma_x \varepsilon_x + \tau_{xy} \gamma_{xy} + \sigma_y \varepsilon_y - W \tag{A.3}$$

is also a quadratic homogeneous form (in the stress components). This is an obvious consequence of the assumption of a generalized Hooke's law.

Differentiating equation (A.3)

$$d\Phi = \left(\sigma_x - \frac{\partial W}{\partial \varepsilon_x}\right) d\varepsilon_x + \left(\tau_{xy} - \frac{\partial W}{\partial \gamma_{xy}}\right) d\gamma_{xy} + \left(\sigma_y - \frac{\partial W}{\partial \varepsilon_y}\right) d\varepsilon_y$$

$$+ \varepsilon_x d\sigma_x + \gamma_{xy} d\tau_{xy} + \varepsilon_y d\sigma_y$$

and using equations (A.2) there follows

$$\varepsilon_x = \frac{\partial \Phi}{\partial \sigma_x} \quad , \quad \gamma_{xy} = \frac{\partial \Phi}{\partial \tau_{xy}} \quad , \quad \varepsilon_y = \frac{\partial \Phi}{\partial \sigma_y} \quad . \tag{A.4}$$

Equations (A.4) are dual to equations (A.2); they state again the generalized Hooke's law, resolved this time with respect to strains. Euler's theorem on homogeneous quadratic forms is applicable both to W and to Φ. In each case, by virtue of equations (A.2) or (A.4), the same result is obtained:

$$W = \Phi = \frac{1}{2} (\sigma_x \varepsilon_x + \tau_{xy} \gamma_{xy} + \sigma_y \varepsilon_y) \quad . \tag{A.5}$$

This is conveniently referred to as the local form of Clapeyron's theorem: the energy densities are both numerically equal to half the sum of the products of stresses by corresponding strains.

(d) The internal equilibrium equations

$$\frac{\partial \sigma_x}{\partial x} + \frac{\partial \tau_{xy}}{\partial y} + X = 0 \quad , \quad \frac{\partial \tau_{xy}}{\partial x} + \frac{\partial \sigma_y}{\partial y} + Y = 0 \tag{A.6}$$

and the surface equilibrium equations

$$\ell\sigma_x + m\tau_{xy} = p \quad , \quad \ell\tau_{xy} + m\sigma_y = q \tag{A.7}$$

are linear. Here (X, Y) and (p, q) denote cartesian components of body forces and surface forces, while (ℓ, m) are the direction cosines of the outward normal at the surface.

(e) As a consequence of the linear character of all equations the principle of superposition applies.

2. SCALAR PRODUCT OF TWO FIELDS

By "field" is implied an arbitrary field of stresses and corresponding strains, related by the generalized Hooke's law. This field need not satisfy either the compatibility conditions equations (A.1) or the equilibrium equations (A.6) and (A.7). It should, however, be integrable in the sense that the total strain energy:

$$\frac{1}{2} \iint (\sigma_x \varepsilon_x + \tau_{xy}\gamma_{xy} + \sigma_y \varepsilon_y) dS$$

should exist.

If two fields are distinguished by the subscripts A and B, consider the following mixed expression

$$(A, B) = \iint (\sigma_{x,A}\varepsilon_{x,B} + \tau_{xy,A}\gamma_{xy,B} + \sigma_{y,A}\varepsilon_{y,B}) dS \ . \tag{A.8}$$

The integrand can be transformed as follows:

$$I = \sigma_{x,A} \frac{\partial \Phi}{\partial \sigma_{x,B}} + \tau_{xy,A} \frac{\partial \Phi}{\partial \tau_{xy,B}} + \sigma_{y,A} \frac{\partial \Phi}{\partial \sigma_{y,B}} \ .$$

The last member is clearly the bilinear form attached to Φ and is equivalent to

$$I = \sigma_{x,B} \frac{\partial \Phi}{\partial \sigma_{x,A}} + \tau_{xy,B} \frac{\partial \Phi}{\partial \tau_{xy,A}} + \sigma_{y,B} \frac{\partial \Phi}{\partial \sigma_{y,A}} .$$

Hence we obtain the commutativity property

$$(A, B) = (B, A) \tag{A.9}$$

which is recognized as a statement of the Betti-Rayleigh reciprocity principle. On the other hand if α and β denote scalar multipliers we obviously have:

$$(\alpha A, \beta B) = \alpha\beta (A, B)$$

and also

$$(A, B + C) = (A, C) + (B, C) .$$

Expression (A.8) has the properties of a scalar product, it is conveniently referred to as the scalar product of the two fields.

By virtue of Clapeyron's theorem the norm of a field

$$(A, A) \geq 0$$

is equal to twice the strain energy; it vanishes if and only if the field itself vanishes.

3. POTENTIAL ENERGY AND COMPLEMENTARY POTENTIAL ENERGY OF THE EXTERNAL LOADS

The following notations will prove to be useful by their concision:

C will denote a particular compatible field. This implies that the strains can be derived from continuous single-valued displacement functions (u, v) according to equations (A.1) and that those displacements take values

$$u = \bar{u} \quad , \quad v = \bar{v} \quad \text{on} \quad c_2 \tag{A.10}$$

prescribed on parts of the boundary.

H will denote a general homogeneous compatible field. Here the values prescribed to the displacements on c_2 are zero. That this field is "general" means either that it is the most general one, or, if approximate solutions are sought, that it contains unknown functions or unknown parameters to be determined by the application of the energy theorems.

According to these definitions and to the superposition principle the general compatible field of a problem can be denoted by

$$C + H .$$

E will denote a particular equilibrium field. This implies that the stresses satisfy the equilibrium equations (A.6) and (A.7) on the complementary part c_1 of the boundary.

A will denote a general self-straining field. It is an equilibrium field with the prescribed loads set equal to zero. It also contains unknown functions or parameters.

According to these definitions the general equilibrium field of a problem can be denoted by

$$E + A .$$

S will denote the exact field of the problem, or the solution. It is both a compatible and an equilibrium field.

In a scalar product, if one field is compatible and the other an equilibrium field, the value of the product is expressible in terms of the displacements of the compatible field and the external loads with which the other field is in equilibrium. Take the case of the product (C, E). Since C is a compatible field, the product can be written as

$$(C,\ E) = \iint \left\{ \sigma_{x,E} \frac{\partial u_c}{\partial x} + \tau_{xy,E} \left(\frac{\partial u_c}{\partial y} + \frac{\partial v_c}{\partial x} \right) + \sigma_{y,E} \frac{\partial v_c}{\partial y} \right\} dx\ dy .$$

If we now integrate by parts

$$(C,E) = \int_{c_1+c_2} \left\{ (\ell\sigma_{x,E} + m\tau_{xy,E})u_c + (\ell\tau_{xy,E} + m\sigma_{y,E})v_c \right\} ds$$

$$- \iint \left\{ u_c \left(\frac{\partial\sigma_{x,E}}{\partial x} + \frac{\partial\tau_{xy,E}}{\partial y} \right) + v_c \left(\frac{\partial\tau_{xy,E}}{\partial x} + \frac{\partial\sigma_{y,E}}{\partial y} \right) \right\} dx\ dy.$$

However, the stresses of the equilibrium field verify equations (A.6) and also equations (A.7) on the part c_1 of the boundary, while the displacements of C verify equation (A.10) on c_2. Hence

$$(C,\ E) = -P_c - Q_e \tag{A.11}$$

where

$$P_c = -\iint (u_c X + v_c Y)\ dx\ dy - \int_{c_1} (pu_c + qv_c) dx$$

is the potential energy of the prescribed loads associated with the displacements of the field C, and

$$Q_e = -\int_{c_2} \left\{ \bar{u}(\ell\sigma_{x,E} + m\tau_{xy,E}) + \bar{v}(\ell\tau_{xy,E} + m\sigma_{y,E}) \right\} ds$$

is the complementary potential energy of the reactions of field E associated with the prescribed displacements. From this result, which has a general character, one finds immediately the following similar properties:

$$(H,\ E) = -P_h\ , \tag{A.12}$$

$$(C,\ A) = -Q_a\ , \tag{A.13}$$

$$(H,\ A) = 0\ . \tag{A.14}$$

This last result is specially useful and can be stated as the following theorem: "The scalar product of a homogeneous compatible field and a self-straining field vanishes".

In most problems the prescribed boundary displacements are zero and the particular displacement field associated with C does not appear. This simplification will be used henceforward. Should it arise, the problem involving non-zero prescribed displacements can be treated by the same techniques as presented below.

4. PRINCIPLE OF MINIMUM TOTAL ENERGY

Under the assumption that C = 0, the minimum total energy principle can be stated as follows:

$$\mu = \frac{1}{2}\ (H,\ H)\ -\ (H,\ E)$$

takes its minimum value for H = S.

Proof: Since the solution S is also a compatible field, the difference between S and H is a compatible and homogeneous field. This is written

$$H = S + \delta H\ .$$

Substitution of this into the expression of μ, expansion of the product and rearrangement of the terms produces

$$\mu = \frac{1}{2}\ (S,S)\ -\ (S,E)\ +\ \frac{1}{2}\ (\delta H,\delta H)\ +\ (S-E,\delta H)\ .$$

Now S-E is the difference between two particular equilibrium fields and is consequently a self-straining field. Its product with a compatible homogeneous field vanishes, so the last term vanishes. The two first terms are the value taken by μ when H = S. The third term is positive unless δH vanishes. This completes the proof.

5. PRINCIPLE OF MINIMUM COMPLEMENTARY ENERGY

Since the prescribed boundary displacements are assumed to be zero there is no complementary potential energy. The principle states that the true solution makes the complementary strain energy (the strain energy expressed in terms of stresses) a minimum. In terms of a scalar produce

$$\lambda = \frac{1}{2} (E + A, E + A) \text{ is minimum for } E + A = S .$$

If we subtract from E + A the particular equilibrium field S we obtain a self-straining field; this can be written

$$E + A = S + \delta A .$$

Substitution of this into the scalar product and rearrangement produces

$$\lambda = \frac{1}{2} (S,S) + \frac{1}{2} (\delta A, \delta A) + (S, \delta A) .$$

Again the last term vanishes as the product of a self-straining field and a homogeneous compatible field. The first term is the value taken by λ for E + A = S. The second term is positive unless δA vanishes. This completes the proof.

6. CLAPEYRON'S EXTERIOR THEOREM

Since S is a homogeneous compatible field and S - E a self-straining field we have

$$(S, S - E) = 0 .$$

This can also be written in the form

$$\frac{1}{2} (S,S) = \frac{1}{2} (S,E) = - \frac{1}{2} P \qquad (A.15)$$

the last equality following from property (A.12) so that P denotes the potential energy of the prescribed loads under the displacements of the true solution. This is also the usual theorem of Clapeyron according to which the strain energy is equal to half the virtual work done by the loads under the true displacements.

7. UPPER AND LOWER BOUNDS TO DIRECT INFLUENCE COEFFICIENTS

From the principle of minimum total energy and Clapeyron's theorem (A.15) follows

$$\frac{1}{2}(H,H) - (H,E) \geq \frac{1}{2}(S,S) - (S,E) = -\frac{1}{2}(S,S) = \frac{1}{2}P .$$

From the principle of minimum complementary energy follows

$$\frac{1}{2}(E + A,E + A) \geq \frac{1}{2}(S,S) = -\frac{1}{2}P .$$

Hence by changing signs and the sense of the last inequality

$$\frac{1}{2}(H,H) - (H,E) \geq \frac{1}{2}P \geq -\frac{1}{2}(E + A,E + A) . \qquad (A.16)$$

Under restrictive assumptions used for displacements in some approximate but compatible approach, the left-hand side of the inequality will be the value estimated for half the potential energy; denote it by $\frac{1}{2}\bar{P}$. Denote similarly by $\frac{1}{2}\underline{P}$ the right-hand side which will be the value estimated under restrictive assumptions for the stresses in some equilibrium approximation to the same problem. The free parameters or functions contained in the approximations are naturally determined by the minimum principles themselves in order that the bracket for P

$$\bar{P} \geq P \geq \underline{P}$$

be as tight as possible.

If the load consists of a single force F with corresponding displacement w, the influence coefficient c is defined by

$$w = cF$$

and the potential energy can be written

$$P = -Fw = -cF^2 \quad \text{(exact solution)}.$$

In a compatible approximation we would find

$$\bar{P} = -\underline{c}F^2 \quad \text{(compatible approximation)}.$$

In an equilibrium approximation

$$\underline{P} = -\bar{c}F^2 \quad \text{(equilibrium approximation)} .$$

Insertion of these expressions in the bracket for the potential energy produces after cancellation of the common factor F^2 and a change of sign altering the sense of the inequalities

$$\bar{c} \geq c \geq \underline{c} . \tag{A.17}$$

Thus an approximate compatible approach produces a lower bound and an approximate equilibrium approach an upper bound to a direct influence coefficient.

8. UPPER AND LOWER BOUNDS TO CROSS INFLUENCE COEFFICIENTS

Let

$$\begin{aligned} w_1 &= c_{11}F_1 + c_{12}F_2 \\ w_2 &= c_{21}F_1 + c_{22}F_2 \end{aligned} \qquad c_{21} = c_{12}$$

be the (exact) displacements associated to the two loads F_1 and F_2. If $(\underline{c}_{11}, \underline{c}_{12}, \underline{c}_{22})$ are the approximate influence coefficients resulting from a compatible approach and $(\bar{c}_{11}, \bar{c}_{12}, \bar{c}_{22})$ those resulting from an equilibrium approach, previous results enable us to write

$$\bar{c}_{11} \geq c_{11} \geq \underline{c}_{11} \quad , \quad \bar{c}_{22} \geq c_{22} \geq \underline{c}_{22} \ . \tag{A.18}$$

The problem is to establish a similar connexion for the cross influence coefficient c_{12}. To this purpose we set

$$F_2 = \lambda F_1 \ ,$$

and consider the potential energy P given by

$$-P = w_1F_1 + w_2F_2 = F_1^2(c_{11} + 2\lambda c_{12} + \lambda^2 c_{22}) \ .$$

Again from the general bracket applicable to the potential energy there is found that

$$\bar{c}_{11}+2\lambda\bar{c}_{12}+\lambda^2\bar{c}_{22} \geq c_{11}+2\lambda c_{12}+\lambda^2 c_{22} \geq \underline{c}_{11}+2\lambda\underline{c}_{12}+\lambda^2\underline{c}_{22} \ .$$

Take the first inequality and solve it for c_{12} *assuming* λ *to be a positive quantity:*

$$2c_{12} \leq \frac{1}{\lambda}\,(\bar{c}_{11}-c_{11}) + 2\bar{c}_{12}+\lambda(\bar{c}_{22}-c_{22}) \ .$$

It follows *a fortiori* by virtue of equations (A.18) that

$$2c_{12} \leq \frac{1}{\lambda}\,(\bar{c}_{11}-\underline{c}_{11}) + 2\bar{c}_{12} + \lambda(\bar{c}_{22}-\underline{c}_{22}) \ .$$

The positive λ giving the smallest upper bound is readily found to be

$$\lambda = \sqrt{(\bar{c}_{11}-\underline{c}_{11})/(\bar{c}_{22}-\underline{c}_{22})}$$

from which follows

$$c_{12} \leq \bar{c}_{12} + \sqrt{(\bar{c}_{11}-\underline{c}_{11})(\bar{c}_{22}-\underline{c}_{22})} \ . \tag{A.19}$$

Had we solved the same inequality assuming a negative λ value we would have

$$2c_{12} \geq \frac{1}{\lambda}\,(\bar{c}_{11}-c_{11}) + 2\bar{c}_{12} + \lambda(\bar{c}_{22}-c_{22}) \ .$$

It follows again *a fortiori* by virtue of equations (A.18) that

$$2c_{12} \geq \frac{1}{\lambda}\,(\bar{c}_{11}-\underline{c}_{11}) + 2\bar{c}_{12} + \lambda(\bar{c}_{22}-\underline{c}_{22}) \ .$$

A highest lower bound is obtained for the negative λ value

$$\lambda = -\sqrt{(\bar{c}_{11}-\underline{c}_{11})/(\bar{c}_{22}-\underline{c}_{22})}$$

from which follows

$$c_{12} \geq \bar{c}_{12} - \sqrt{(\bar{c}_{11}-\underline{c}_{11})(\bar{c}_{22}-\underline{c}_{22})} \ . \tag{A.20}$$

The second inequality can be treated by the same technique and produces the new bounds

$$c_{12} \leq \underline{c}_{12} + \sqrt{(\bar{c}_{11}-\underline{c}_{11})(\bar{c}_{22}-\underline{c}_{22})} \ , \tag{A.21}$$

$$c_{12} \geq \underline{c}_{12} - \sqrt{(\bar{c}_{11}-\underline{c}_{11})(\bar{c}_{22}-\underline{c}_{22})} \ . \tag{A.22}$$

If, as usually the case, it is found numerically that $\bar{c}_{12} > \underline{c}_{12}$ the best bracket is made of inequalities (A.20) and (A.21).

APPENDIX B

Examples and Numerical Results

B.M. Fraeijs de Veubeke and Guy Sander

1. STRUCTURAL MODEL

The ultimate step is the analysis of a complete and complicated structure. Very instructive results are, however, obtained concerning the validity of the dual approach by the consideration of a rather simple structural element.

A rectangular element, or panel, of length 2a and height 2b, edged by beams of identical cross-section S, was analysed by various methods and the influence coefficients compared. The beams are assumed to be devoid of flexural rigidity. The two structural parameters of the structure are then

$$r = a/b$$

$$R = S/(bt)$$

where t is the constant thickness of the panel. Exploratory values of 1 and 5 for r, and 0.4-1.0-2.0 and 4.0 for R were adopted.

The reference to Model 1 or Model 2 relates to the type of support of the structure and is explained in Figure 11.

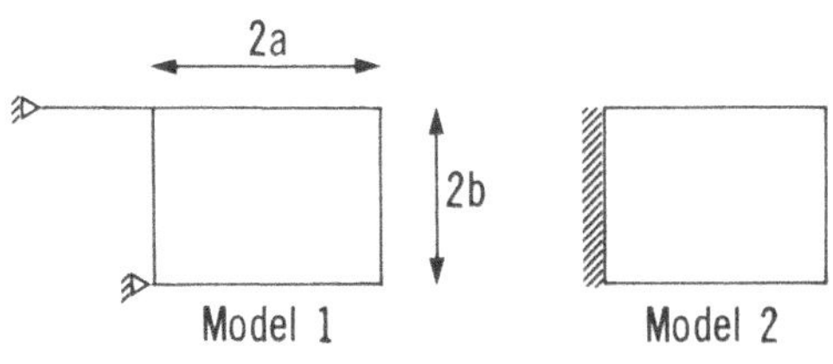

Figure 11

The loading cases investigated are shown in Figure 12 and referred to as Cases I, II, III or IV.

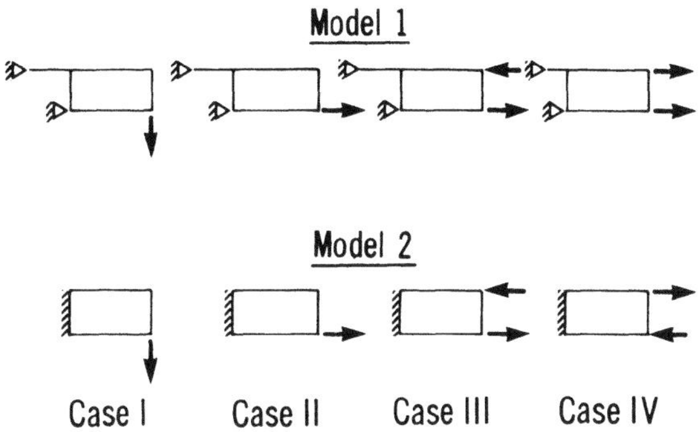

Figure 12

2. METHODS OF ANALYSIS

The methods of analysis that were compared are the following:

(1) *Tension 4.* This denomination refers to a subdivision of the panel in four triangles by the diagonals and construction of a pure equilibrium field as explained in Sections 2.3 and 3a of the paper. It is found that the state of stress in the panel is either zero or a state of pure shear.

(2) *Tension 16.* The panel is first subdivided into four equal rectangular fields of dimensions a and b. Each rectangle is then subdivided in four triangles by the diagonals. Again a pure equilibrium field is constructed.

In contrast to the "Tension 4" case there are redundancies in the form of self-strainings. Figure 13 shows the self-strainings involved in the Model 1 type of support. In the case of Model 2 there are additional self-strainings depicted in Figure 14.

"Tension 4" and "Tension 16" are the only methods used to provide upper bounds to the influence of coefficients.

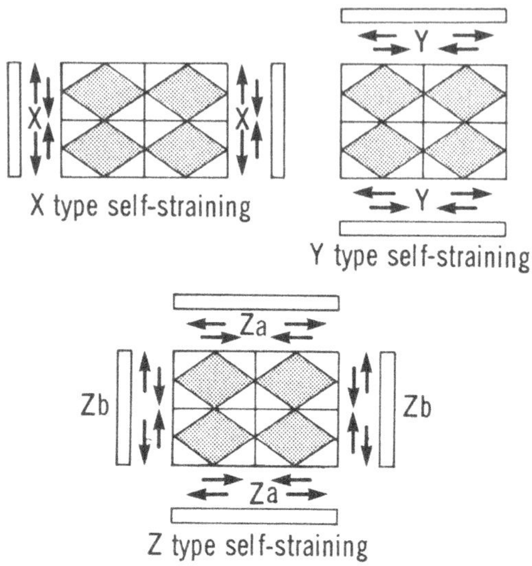

Figure 13

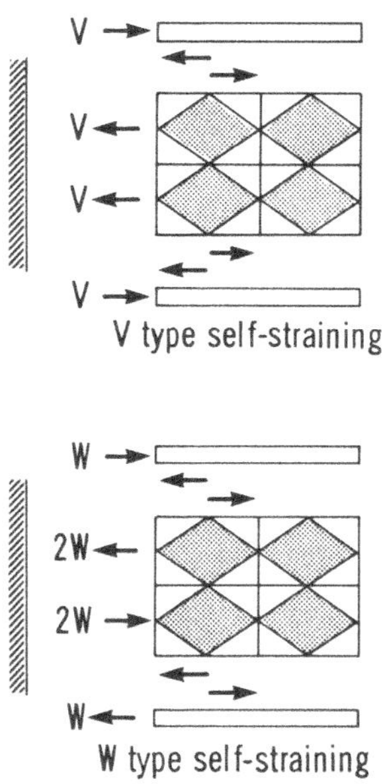

Figure 14

(3) *Turner 16.* The subdivision of the panel in triangles is the same as for "Tension 16". However, the field of displacements is purely compatible. It is identical to the Argyris-Turner triangularization method and the calculations were essentially made according to the Turner stiffness method [5].

(4) *Turner 4.* It is the triangularization and stiffness method applied to the subdivision used in Tension 4.

(5) *Argyris 4.* The panel is first subdivided into the four equal rectangular fields. In each field the displacement assumption is used

$$u = u_o + px + qy + r\,xy \quad ,$$

$$v = v_o + \ell x + my + n\,xy \quad .$$

The eight parameters can be expressed in terms of the corner displacements. As in the case of triangular fields continuity of the displacements for the whole panel is obtained by simply requiring identical displacements at the common corners. Corner loads and stiffness matrices are obtained by the same procedures as used for the stiffness method applied to triangular fields [8]. The major difference with respect to a subdivision into triangles is that the stress field associated with the displacement assumption is not an equilibrium field. The equilibrium conditions are violated continuously but smaller discontinuities in the stresses are recorded across the common edges. This method, together with Turner 4 and 16, must yield lower bounds. Whenever used, it gave results between the Turner 4 and 16 methods, generally closer to the last one.

(6) *Beam Theory.* This refers to a purely compatible Bernoulli type of bending theory, applicable to loading case I. It consists in assuming for the panel a field of the type

$$u = yB(x) \quad ,$$
$$v = V(x) \quad .$$

The unknown functions B(x) and V(x) are determined from the principle of variations for displacements. The value obtained for the influence coefficient is classical

$$\frac{d}{P} = \frac{(2a)^3}{3EI} + \frac{2a}{2btG}$$

where

$$I = 2Sb^2 + \frac{2tb^3}{3(1-\nu^2)}$$

is the moment of inertia of the cross-section with an increased web contribution due to the assumption of inextensibility in the y direction. Because of this assumption the result is poor when r and R are small. Because the sections normal to x are assumed to remain plane, the method applies as well to Model 1 as to Model 2.

The method provides an interesting lower bound for large values of r. It is seen to be very close to the Tension 16 method and consequently both are then very accurate.

(7) *Fictitious Spar Theory*. This is the only hybrid assumption investigated that does not with certainty provide an upper or lower bound. It consists in replacing the direct stress-carrying capacity of the plate by fictitious edge and central spars. When one-sixth of a transverse area of the plate is concentrated as an edge spar and two-thirds as a central spar both the total cross-sectional area and the total moment of inertia are conserved. The stresses transmitted by the plate are then reduced to a system of four shear fields. In practice the assumption works well. The influence coefficients it provides are located between the Turner and Tension 16 methods, which are generally the best upper and lower bounds.

It is however, recognized that in this idealization both the Poisson's ratio effect and sweep-back effects are lost.

It would therefore be interesting to pursue the investigations on a model incorporating sweep-back and perhaps taper.

Note: The values recorded in the tables and compared in Figures 17, 18, 19 and 20 are displacements corresponding to specific values:

$$\text{load } P = 1000 \text{ kilos}$$
$$E = 2.2\ 10^4 \text{ kilo/mm}^2$$
$$b = 1000 \text{ mm}$$
$$t = 2 \text{ mm}$$

To obtain the non-dimensional quantities

$$d\,\frac{Et}{P} = f(r,R)$$

where d is the displacement corresponding to the load, the table values should be multiplied by the factor 44. In the figures, heavy continuous lines are for lower bounds, dotted lines for upper bounds. The thin continuous line is for the fictitious spar theory.

3. REDUCTION OF SELF-STRAINING CONDITIONS FROM SYMMETRY CONSIDERATIONS

Sometimes advantage can be gained from the symmetries in reducing the number of self-strainings to consider. Take for example Model 1 with loading case II. Figure 15 shows that this can be considered as the superposition of loading cases IV and III.

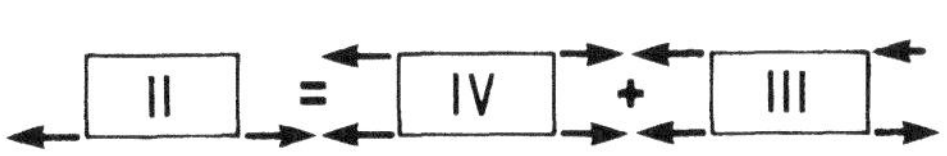

Figure 15

(a) *Case IV.* It is symmetrical with respect to both the x and y axes. Hence we need only consider a self-straining of type Z when dealing with the Tension 16 method. It is readily found by investigation of the equilibrium conditions of the triangles (Figure 16) that in triangles 1 and 2 the state of stress is one of pure shear:

$$\sigma_x = 0 \quad , \quad \sigma_y = 0 \quad , \quad \tau_{xy} = Z \ ,$$

while in triangles 3 and 4 it is the state

$$\sigma_x = -Z\,\frac{a}{b} \quad , \quad \sigma_y = -Z\,\frac{b}{a} \quad , \quad \tau_{xy} = 0 \ .$$

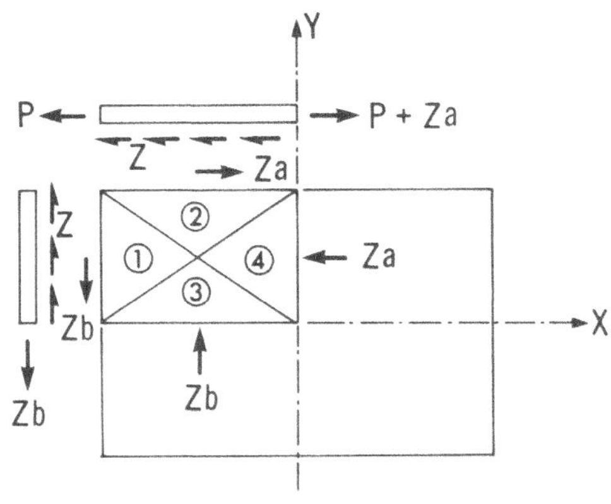

Figure 16

The corresponding strain energies (see Section 1) are

$$\frac{1}{2}\,\frac{ab}{2}\,\frac{t}{E}\,\{2(1+\nu)\tau_{xy}^2\} = \frac{abt}{2E}\,(1+\nu)Z^2$$

$$\frac{1}{2}\,\frac{ab}{2}\,\frac{t}{E}\left\{Z^2\,\frac{b^2}{a^2} + Z^2\,\frac{a^2}{b^2} - 2\nu Z^2\right\} = \frac{abt}{4E}\left(\frac{a^2}{b^2} + \frac{b^2}{a^2} - 2\nu\right) Z^2 \ .$$

Adding and multiplying by 4 (the other three rectangular fields have symmetrical states of stress) the panel energy is

$$U_p = \frac{abt}{E}\left\{2 + \frac{a^2}{b^2} + \frac{b^2}{a^2}\right\} Z^2 .$$

In half of a vertical beam the tension, computed from equilibrium considerations, is

$$T = Zt(b-y)$$

and the corresponding strain energy for both complete verticals

$$U_v = \frac{2}{3}\,\frac{t^2 b^3}{ES}\, Z^2 .$$

In half of a horizontal beam the tension is the superposition of the self-straining effect and a particular equilibrium state under the external horizontal load P applied at the end:

$$T = P + Zt(a+x) .$$

The corresponding strain energy for both complete horizontals

$$U_h = \frac{2}{ES}\left\{aP^2 + a^2PZt + \frac{1}{3}\,a^3t^2Z^2\right\} .$$

The total energy of the structure is thus found to be

$$U = \frac{Z^2}{E}\left\{abt\left(2 + r^2 + \frac{1}{r^2}\right) + \frac{2tb^2r^3}{3R}\right\} + 2\,\frac{ZP}{E}\,\frac{r^2b}{R} + \frac{P^2}{E}\,\frac{2r}{Rt} .$$

From Menabrea's theorem $\partial U/\partial Z = 0$ follows

$$Z = -\,\frac{P}{bt}\left\{\frac{3r}{3R(2+r^2+r^{-2})+2r^2}\right\}$$

and finally from Castigliano's formula

$$d = \frac{1}{2}\,\frac{\partial U}{\partial P} = \frac{1}{E}\left\{\frac{2r^2b}{R}\,Z + \frac{4r}{Rt}\,P\right\}$$

where d is the deflexion under each P load of Model 1. After substitution of the Z value there comes

$$d = \frac{P}{Et}\frac{r}{R}\left\{2 - \frac{3r^2}{3R(2+r^2+r^{-2})+2r^2}\right\} \quad .$$

Note that in the tension 4 approach no self-strainings exist and we would have Z = 0 with a corresponding deflexion

$$\frac{P}{Et} 2 \frac{r}{R} \; .$$

This is the first term of the previous formula which represents the elongation of a horizontal beam under the end load when the plate is under no stress and does not participate to the general stiffness. The second term represents the plate contribution in reducing the deflexion.

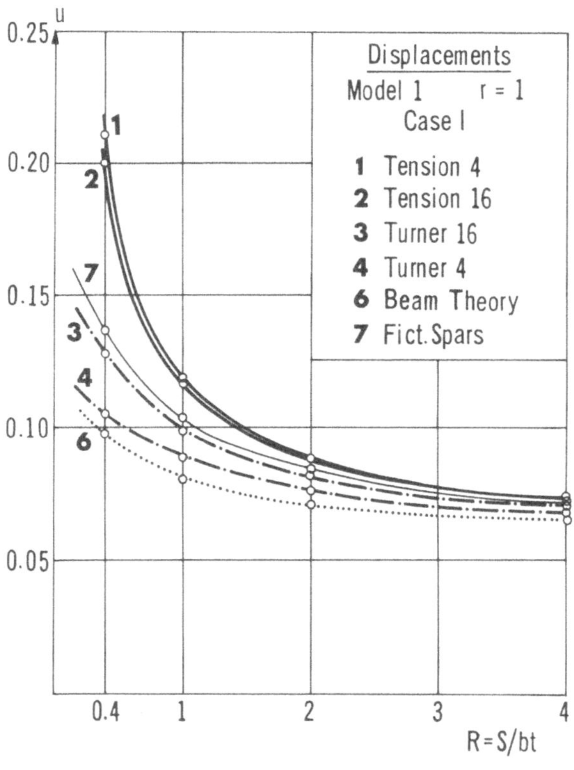

Figure 17

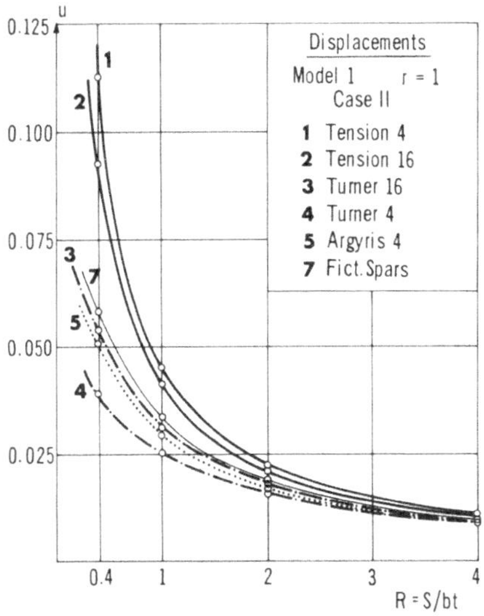

Figure 18

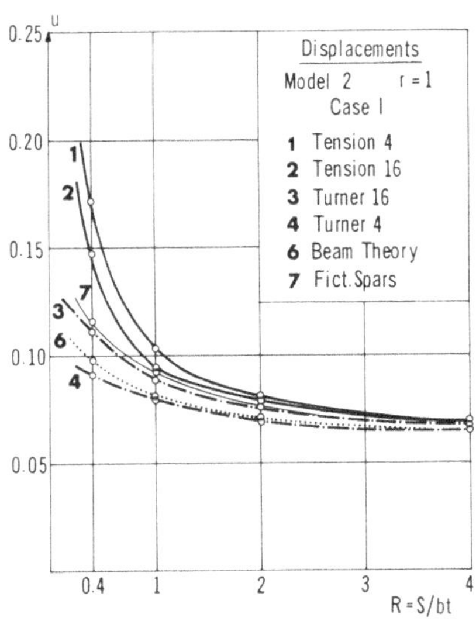

Figure 19

(b) *Case III.* It is symmetrical with respect to the y axis, antisymmetrical with respect to the x axis. It can be dealt with by considering only the Y type of self-straining. By similar calculations there is found

$$Y = \frac{P}{tR} \frac{r}{\frac{2}{3}\frac{r}{R} + \frac{1}{r^3} + 3r + \frac{2(1+2\nu)}{r}}$$

$$d = \frac{P}{Et}\frac{r}{R}\left\{2 - \frac{r^4}{\frac{2}{3}r^4 + R + 3Rr^4 + 2(1+2\nu)r^2 R}\right\} .$$

Again the first term represents the beam elongation in the absence of the place, as would be found in the tension 4 approach. The second term is the plate effect given by the tension 16 approach.

(c) *Case II.* Returning to case II, it is obvious from the principle of superposition that the deflexion under each load is the average between those of case III and case IV.

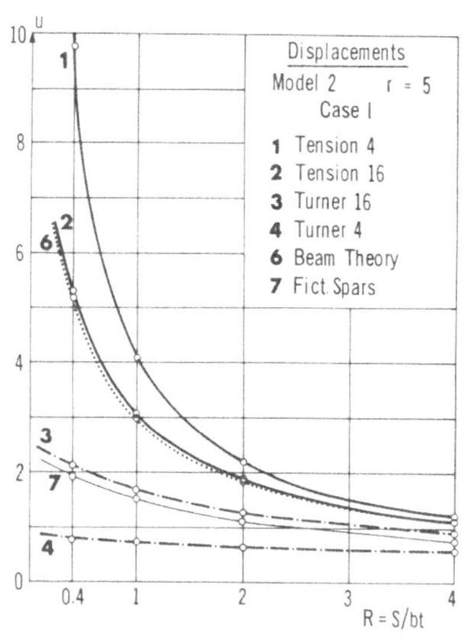

Figure 20

Table 1

$r = \frac{a}{b}$	$r = \frac{a}{b} = 1$ 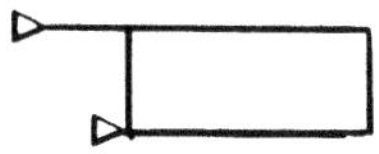				$r = \frac{a}{b} = 5$ 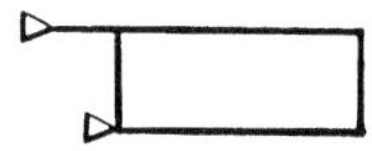			
$R = \frac{S}{bt}$	0.4	1.0	2.0	4.0	0.4	1.0	2.0	4.0
Methods	*Case I*  *P*							
Tension 4	0.2106	0.1196	0.0893	0.0742	9.840	4.113	2.204	1.249
Tension 16	0.2038	0.1185	0.0890	0.0741	9.182	3.989	2.171	1.241
Turner 16	0.1287	0.0991	0.0825	0.0719	2.139	1.699	1.303	0.939
Turner 4	0.1059	0.0894	0.0767	0.0690	0.810	0.765	0.561	0.606
Argyris 1	0.1099	0.0895	0.0773	0.0692				
Argyris 4	0.1238	0.0948	0.0807	0.06945				
Beam. theory	0.0986	0.0813	0.0719	0.06603	5.239	3.068	1.896	1.1630
Fict. spars	0.1374	0.1045	0.0850	0.0730	1.985	1.556	1.159	0.788
	Case II  *P*							
Tension 4	0.1136	0.0454	0.0227	0.01136	0.5681	0.227	0.113	0.0568
Tension 16	0.0930	0.0415	0.0217	0.0111	0.365	0.178	0.0999	0.0529
Turner 16	0.0545	0.0315	0.0185	0.0102	0.143	0.103	0.0705	0.0434
Turner 4	0.0393	0.0256	0.0162	0.0094	0.101	0.0749	0.0534	0.0349
Argyris 1	0.0413	0.0263	0.0165	0.00953				
Argyris 4	0.0468	0.0299	0.0177	0.00982				
Fict. spars	0.0588	0.0340	0.0194	0.01048	0.109	0.085	0.0619	0.0401

Table 2

Methods	*Case III* 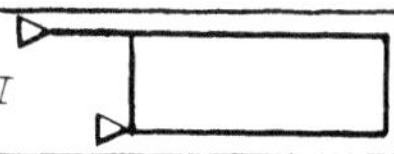 ← *P* → *P*							
Tension 4	0.1136	0.0454	0.0227	0.0113	0.5681	0.227	0.113	0.0568
Tension 16	0.0976	0.0425	0.0220	0.01117	0.4201	0.1973	0.1054	0.05466
Turner 16	0.0570	0.0325	0.0189	0.0103	0.117	0.0895	0.0642	0.0410
Turner 4	0.0468	0.0291	0.0176	0.0099	0.0409	0.0370	0.0317	0.0248
Argyris 1	0.0508	0.0304	0.0182	0.0101				
Argyris 4	0.0632	0.0321	0.0186	0.0102				
Fict. Spars	0.0588	0.0340	0.0194	0.0104	0.109	0.0852	0.0619	0.0401
	Case IV 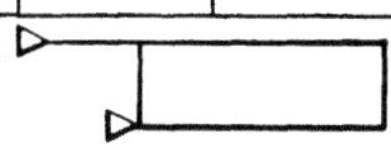 → *P* → *P*							
Tension 4	0.113	0.0454	0.0227	0.0113	0.568	0.227	0.113	0.0568
Tension 16	0.0885	0.0406	0.0214	0.0110	0.310	0.162	0.0935	0.0511
Turner 16	0.0520	0.0305	0.0182	0.0100	0.169	0.116	0.0768	0.0458
Turner 4	0.0318	0.0222	0.0148	0.0089	0.161	0.112	0.0750	0.0450
Argyris 1	0.0318	0.0222	0.0148	0.00895				
Argyris 4	0.0399	0.0267	0.0166	0.00953				
Fict. Spars	0.0588	0.0340	0.0194	0.0104	0.109	0.0852	0.0619	0.0401

Table 3

$r = \frac{a}{b}$	r = 1 				r = 5 			
$R = \frac{S}{bt}$	0.4	1.0	2.0	4.0	0.4	1.0	2.0	4.0
Methods	*Case I*  *P*							
Tension 4	0.172	0.104	0.0818	0.0704	9.803	4.098	2.196	1.246
Tension 16	0.148	0.0953	0.0803	0.0701	5.345	3.089	1.903	1.166
Turner 16	0.112	0.0893	0.0767	0.0687	2.131	1.700	1.300	0.937
Turner 4	0.0920	0.0805	0.0753	0.0664	0.800	0.756	0.687	0.601
Argyris 1	0.0942	0.0807	0.0753	0.0665				
Argyris 4	0.0949	0.0823	0.0759	0.0678				
Beam. Theory	0.0986	0.0813	0.0719	0.06603	5.239	3.068	1.896	1.1630
Fict. Spars	0.1179	0.0931	0.0758	0.0695	1.965	1.544	1.152	0.795
	Case II  → *P*							
Tension 4	0.1136	0.0454	0.0227	0.0113	0.568	0.227	0.113	0.0568
Tension 16	0.0877	0.0404	0.0211	0.0110	0.365	0.0178	0.0999	0.0529
Turner 16	0.0496	0.0296	0.0178	0.00998	0.140	0.101	0.0700	0.0432
Turner 4	0.0390	0.0255	0.0170	0.00944	0.099	0.0740	0.0530	0.0348
Argyris 1	0.0410	0.0261	0.0171	0.00952				
Argyris 4	0.0476	0.0290	0.0175	0.00963				
Fict. Spars	0.0588	0.0340	0.0194	0.01048	0.109	0.085	0.0619	0.0401

Table 4

Methods	*Case III* 							
Tension 4	0.1136	0.0454	0.0227	0.0113	0.568	0.227	0.113	0.0568
Tension 16	0.0971	0.0418	0.0216	0.0110	0.420	0.197	0.105	0.0544
Turner 16	0.0569	0.0324	0.0189	0.0103	0.117	0.0896	0.0641	0.0410
Turner 4	0.0469	0.0259	0.0150	0.0099	0.0409	0.0370	0.0317	0.0248
Argyris 1	0.0508	0.0304	0.0182	0.0101				
Argyris 4	0.0631	0.0321	0.0186	0.0102				
Fict. Spars	0.0588	0.0340	0.0194	0.01048	0.109	0.085	0.0619	0.0401
	Case IV 							
Tension 4	0.1136	0.0454	0.0227	0.0113	0.568	0.227	0.113	0.0568
Tension 16	0.0783	0.0390	0.0206	0.0110	0.310	0.162	0.0935	0.0511
Turner 16	0.0424	0.0268	0.0167	0.00963	0.163	0.114	0.0759	0.0454
Turner 4	0.0312	0.0219	0.0146	0.0089	0.158	0.111	0.0744	0.0448
Argyris 1	0.0312	0.0219	0.0147	0.00893				
Argyris 4	0.0390	0.0260	0.0165	0.00950				
Fict. Spars	0.0588	0.0340	0.0194	0.01048	0.109	0.085	0.0619	0.0401

DISPLACEMENT AND EQUILIBRIUM MODELS IN THE FINITE ELEMENT METHOD

B.M. Fraeijs de Veubeke

University of Liège, Belgium

1. INTRODUCTION

The purpose of the finite element method in matrix structural analysis is: to define generalized loads and generalized displacements in terms of which the connection between elements is expressible by simple identification of displacements and transmission of loads; to relate an internal stress field and, if possible, an internal displacement field to the generalized quantities; to establish a stiffness matrix or any other convenient form of relations between generalized loads and generalized displacements describing the elastic properties of the element; to solve for the unknown quantities of the connected structure, taking into account the external loading and the appropriate boundary conditions.

The discretization, that necessarily occurs at some stage of a structural analysis when numerical results are to be produced, is here incorporated at the outset in the limited number of degrees of freedom allocated to the element deformation. This,

Original publication, *Stress Analysis*, edited by Zienkiewicz and Holister, John Wiley & Sons Limited, Baffins Lane, Chichester, Sussex, U.K., 1965.

as a rule, implies approximations either to the compatibility conditions or to the equilibrium conditions or to both. The variety of models that can be set up to approximate the element behaviour can therefore be subdivided into three classes according to their properties of:

- satisfying compatibility but not equilibrium;
- satisfying equilibrium but not compatibility;
- violating both equilibrium and compatibility.

Confronted with so many possibilities, it is natural that the question of accuracy of the overall analysis be raised with respect to any particular choice of model. In general the accuracy is established on an experimental basis by checking the analysis of simple structures against a known exact solution or by examining the convergence trends of analyses repeated with a finer network of elements. Much theoretical work remains to be done in order to establish the model requirements ensuring convergence to the exact solution when the size of the elements tends to zero.

There are however two particular types of models that allow a direct estimate of the convergence from energy considerations. For convenience they will be referred to as the 'displacement models' and the 'equilibrium models'.

Displacement models are the best known. They are based on parametric displacement fields ensuring compatibility of deformations both internal to the elements and across their boundaries. Under those conditions the whole displacement field of the connected structure is continuous and piecewise differentiable. The examples given are presented with a particular emphasis on the role of the displacement modes attached to the generalized variables as weight functions for the external loads.

Equilibrium models are based on parametric stress fields securing internal equilibrium and continuous stress transmission between elements. The whole stress field is then in equilibrium.

(This paper presents later a general theory for constructing such equilibrium models.) The examples illustrate the possibilities of false kinematical deformation modes appearing either in a single element or in a combination of them and methods for dealing with such situations.

Arguments based on two-field variational principles are presented to justify the superiority of the minimal principles in deriving the stiffness properties of the elements. Still more powerful arguments are to be found in the upper and lower bound character of the approximate influence coefficients so obtained. Short proofs of the upper and lower bound properties are presented to make this manuscript self contained.

A dual analysis of a given structure, based on the alternate use of displacement and equilibrium models, and submitted to the same external loads and boundary conditions allows a direct quantitative estimate of the convergence to the true solution by comparison of the upper and lower bounds obtained.

Both the direct stiffness method and the self-straining method for solving the connected fields of elements are given a brief review. The last one is the more promising for dealing with equilibrium models.

2. VARIATIONAL PRINCIPLES APPLICABLE TO SIMPLIFY THE ANALYSIS OF FINITE ELEMENTS

The use of selected deformation modes or stress-transmission modes to approximate the elastic behaviour of a finite element must in the end result in a set of equations relating generalized loads to generalized displacements. This set is conveniently derived from energy theorems or variational principles, the most widely used being the principle of variation of displacements and the principle of variation of stresses, also called the complementary energy principle.

More general principles exist which, in theory, allow simultaneous approximations on displacements and stresses. It seems worthwhile to discuss them briefly, if only to show their limited usefulness.

For brevity, only the plane stress case will be considered; the extensions to three dimensional problems being obvious. By the term 'field' will be understood a field of stresses and strains related by the energy equations

$$\sigma_x = \frac{\partial W}{\partial \varepsilon_x} \quad , \quad \tau_{xy} = \frac{\partial W}{\partial \gamma_{xy}} \quad , \quad \sigma_y = \frac{\partial W}{\partial \varepsilon_y} \quad , \tag{1}$$

where $W(\varepsilon_x, \gamma_{xy}, \varepsilon_y)$, more briefly denoted by $W(\varepsilon)$, is the strain energy density, or energy per unit surface area in the reference condition. Hence $(\sigma_x, \tau_{xy}, \sigma_y)$ are not properly speaking stresses but rather normal loads and shear flows defined per unit thickness of slice. The same stress-strain relations are expressed by

$$\varepsilon_x = \frac{\partial \phi}{\partial \sigma_x} \quad , \quad \gamma_{xy} = \frac{\partial \phi}{\partial \tau_{xy}} \quad , \quad \varepsilon_y = \frac{\partial \phi}{\partial \sigma_y} \quad , \tag{2}$$

where $\phi(\sigma_x, \tau_{xy}, \sigma_y)$, more briefly denoted by $\phi(\sigma)$, is the complementary energy density defined by the Legendre transformation

$$\phi = \sigma_x \varepsilon_x + \tau_{xy}\gamma_{xy} + \sigma_y \varepsilon_y - W \ . \tag{3}$$

A field is 'integrable' if single-valued, continuously differentiable displacement functions $u(x, y)$ and $v(x, y)$ exist such that

$$\varepsilon_x = \frac{\partial u}{\partial x} = u_x \ , \quad \gamma_{xy} = \frac{\partial u}{\partial y} + \frac{\partial v}{\partial x} = u_y + v_x \quad , \quad \varepsilon_y = \frac{\partial v}{\partial y} = v_y \ . \tag{4}$$

The use of those equations implies that the analysis is also restricted to small strains and small rotations.

A field is 'compatible' if, in addition to equations (4) the displacement functions satisfy prescribed boundary displacements on parts of the boundary denoted by Γ_u

$$u = \bar{u} , \quad v = \bar{v} \quad \text{on} \quad \Gamma_u . \tag{5}$$

A field is an 'equilibrium field' if the stresses satisfy the following equilibrium equations

$$\frac{\partial \sigma_x}{\partial x} + \frac{\partial \tau_{xy}}{\partial y} + \bar{X} = 0 \quad , \quad \frac{\partial \tau_{xy}}{\partial x} + \frac{\partial \sigma_y}{\partial y} + \bar{Y} = 0 , \tag{6}$$

$$p_x = l\sigma_x + m\tau_{xy} = \bar{p}_x , \quad p_y = l\tau_{xy} + m\sigma_y = \bar{p}_y \text{ on } \Gamma_\sigma \tag{7}$$

where (l, m) are the direction cosines of the outward normal to the boundary, $(\bar{X}, \bar{Y})$ are prescribed internal loads and $(\bar{p}_x, \bar{p}_y)$ prescribed boundary loads. The Γ_σ part of the boundary is complementary to Γ_u.

2.1 *The General Variational Principle*

The general variational principle of elasticity can be stated as follows:

$$\delta(U + P + D) = 0 \tag{8}$$

where,

$$U = \iint_\Delta W(\varepsilon)\, dxdy$$

is the total strain energy, calculated from a field represented by the strains $(\varepsilon_x, \gamma_{xy}, \varepsilon_y)$

$$P = - \iint_\Delta (\bar{X}u + \bar{Y}v)dxdy - \int_{\Gamma_\sigma} (\bar{p}_x u + \bar{p}_y v)ds$$

is the potential energy of prescribed loads, calculated from an integrable field represented by the displacements (u, v);

$$D = \iint_\Delta [\sigma'_x(u_x-\varepsilon_x)+\tau'_{xy}(u_y+v_x-\gamma_{xy})+\sigma'_y(v_y-\varepsilon_y)]dxdy$$

$$+ \int_{\Gamma_u} [p'_x(\bar{u}-u)+p'_y(\bar{v}-v)]ds \ ,$$

is a dislocation potential introducing a third field represented by the stresses $(\sigma'_x, \tau'_{xy}, \sigma'_y)$. These stresses can be considered as Lagrangean multipliers removing the compatibility constraints (4) and (5) that must be satisfied *a priori* in the simpler principle of variation of displacements [1, 2, 3]

$$\delta(U + P) = 0 .$$

In this simpler principle only displacements (u, v) are involved and subject to variations. In the general principle all three fields are subject to independent variations. Those on the Lagrangean multipliers restore the compatibility conditions (4) and (5). Those on the $(\varepsilon_x, \gamma_{xy}, \varepsilon_y)$ field require that

$$\sigma_x = \frac{\partial W}{\partial \varepsilon_x} = \sigma'_x \quad , \quad \tau_{xy} = \frac{\partial W}{\partial \gamma_{xy}} = \tau'_{xy} \quad , \quad \sigma_y = \frac{\partial W}{\partial \varepsilon_y} = \sigma'_y \ . \tag{9}$$

In other words they require that the Lagrangean multipliers become identical with the stresses of the $(\varepsilon_x, \gamma_{xy}, \varepsilon_y)$ field. Finally the variations on (u, v) require that the Lagrangean multipliers be an equilibrium field; the stresses $(\sigma'_x, \tau'_{xy}, \sigma'_y)$ should satisfy the equilibrium equations (6) and (7). A functional diagram of this general principle is illustrated on Figure 1 and is self explanatory. In the end there is complete identification between the three fields and the solution must as it should, be compatible and in equilibrium.

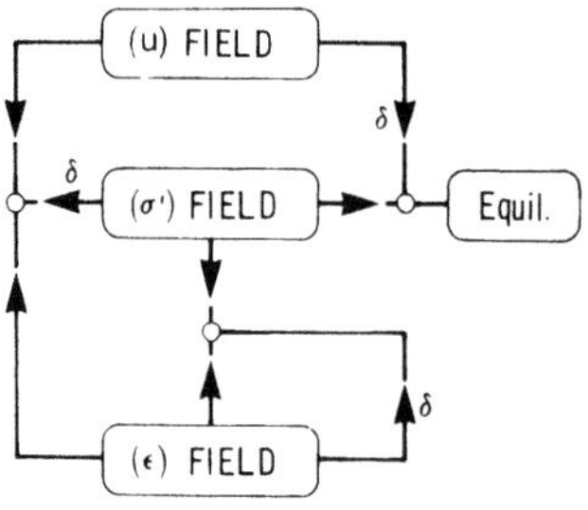

Figure 1

2.2 *E. Reissner's Principle*

By making *a priori* assumptions on the multipliers the general principle can be simplified, still leaving us with two independent fields. The assumption leading to Reissner's principle [3, 4] is that equations (9) are satisfied *a priori*. Then, in view of definition (3), the general theorem reduces to

$$\delta \iint_{\Delta} (\sigma_x u_x + \tau_{xy}(u_y + v_x) + \sigma_y v_y - \phi(\sigma))\,dxdy + P + \int_{\Gamma_u} (p_x(\bar{u}-u) + p_y(\bar{v}-v))\,ds = 0. \tag{10}$$

The functional diagram of this is illustrated on Figure 2.

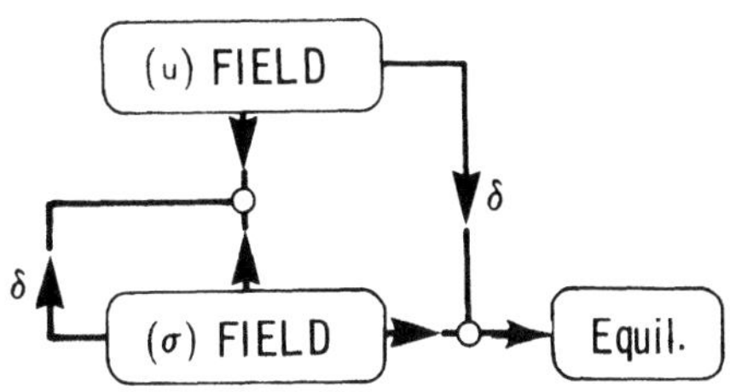

Figure 2

The theoretical possibility of using separate approximations on the displacements (u, v) and the stresses (σ_x, τ_{xy}, σ_y), with this theorem, raises the following points:

(a) If no restrictive assumptions are made on the stresses, their variation requires

$$u_x = \frac{\partial\phi}{\partial\sigma_x} = \varepsilon_x \quad , \quad u_y + v_x = \frac{\partial\phi}{\partial\tau_{xy}} = \gamma_{xy} \quad , \quad v_y = \frac{\partial\phi}{\partial\sigma_y} = \varepsilon_y \tag{11}$$

and

$$u = \bar{u} \qquad v = \bar{v} \quad \text{on } \Gamma_u . \tag{12}$$

Equations (11) show that $(\sigma_x, \tau_{xy}, \sigma_y)$ are the stresses related to the strains $(u_x, u_y + v_x, v_y)$ so that, in view of (3), Reissner's principle (10) reduces to the ordinary displacement principle

$$\delta \left[\iint_\Delta W(u_x, u_y + v_x, v_y) dxdy + P \right] = 0 . \tag{13}$$

This result can be stated in the form of a *limitation principle:* If a net of finite elements is analyzed by compatible displacement modes and the stresses left free to be determined by energy considerations, the best stresses are those associated with the strains derived from the displacements and the degrees of freedom in the displacement modes are governed by the ordinary principle of variation of displacements. In other words it is useless to look for a better solution by injecting additional degrees of freedom in the stresses, although the stresses obtained will not, as a rule, satisfy the detailed equilibrium conditions.

(b) Conversely, if no restrictive assumptions are made on the displacements, variations on those will require the stresses to be solutions of the equilibrium equations. This is precisely the type of approach one wishes to use in the stress model analysis of finite elements. If we then assume a stress field in equilibrium and consider the following integration by parts

$$\iint_\Delta (\sigma_x u_x + \tau_{xy}(u_y+v_x)+\sigma_y v_y) \, dxdy$$

$$= \int_{\Gamma_u+\Gamma_\sigma} (p_x u+p_y v) dx - \iint_\Delta \left[u\left(\frac{\partial\sigma_x}{\partial x} + \frac{\partial\tau_{xy}}{\partial y}\right) + v\left(\frac{\partial\tau_{xy}}{\partial x} + \frac{\partial\sigma_y}{\partial y}\right)\right] dxdy$$

we obtain, in view of equations (6) and (7)

$$\iint_\Delta (\sigma_x u_x+\tau_{xy}(u_y+v_x)+\sigma_y v_y)dxdy = - P + \int_{\Gamma_u} (p_x u+p_y v)ds .$$

This result, substituted into (10) reduces Reissner's principle to the complementary energy principle

$$\delta \left[\iint_{\Delta} \phi(\sigma)\,dxdy - \int_{\Gamma_u} (p_x\bar{u}+p_y\bar{v})ds \right] = 0 \ . \tag{14}$$

This can again be stated as another limitation principle: If the stresses (σ_x, τ_{xy}, σ_y) form an equilibrium field *a priori*, their degrees of freedom are governed by the ordinary complementary energy principle. Inasmuch as the resulting strains are not, as a rule, integrable the principle gives no indication concerning a best associated displacement field.

It would be wrong to conclude from the limitation principles that Reissner's principle is worthless. It can be used to advantage in developing a consistent equilibrium analysis based on *partial* assumptions on the stresses. A good example is provided by Reissner's theory of plate bending [5, 6].

2.3 *Another Two-Field Variational Principle*

Another way to simplify the general principle is to endow the multipliers with the property of equilibrium. Then, through an integration by parts similar to that achieved before, the general principle reduces to:

$$\delta \left\{ \iint_{\Delta} [W(\varepsilon)-(\sigma'_x\varepsilon_x+\tau'_{xy}\gamma_{xy}+\sigma'_y\varepsilon_y)]\ dxdy + \int_{\Gamma_u} (p'_x\bar{u}+p'_y\bar{v})ds \right\} = 0 \ . \tag{15}$$

It contains the equilibrium field (σ'_x, τ'_{xy}, σ'_y) and the arbitrary field (ε_x, γ_{xy}, ε_y). Its functional diagram is represented on Figure 3. The limitations involved in the use of simultaneous approximations are very similar to those of Reissner's principle;

(a) A strain field free of restrictive assumptions will result in Euler-Lagrange equations

$$\sigma_x = \frac{\partial W}{\partial \varepsilon_x} = \sigma'_x , \quad \tau_{xy} = \frac{\partial W}{\partial \gamma_{xy}} = \tau'_{xy} , \quad \sigma_y = \frac{\partial W}{\partial \varepsilon_y} = \sigma'_y ,$$

identifying the best strain field with the one associated to the equilibrium field. As a result (15) again degenerates into the complementary energy principle (14) that will govern the degrees of freedom in the equilibrium field. Unless the strain field associated to this equilibrium field is integrable, no indications are given concerning displacements.

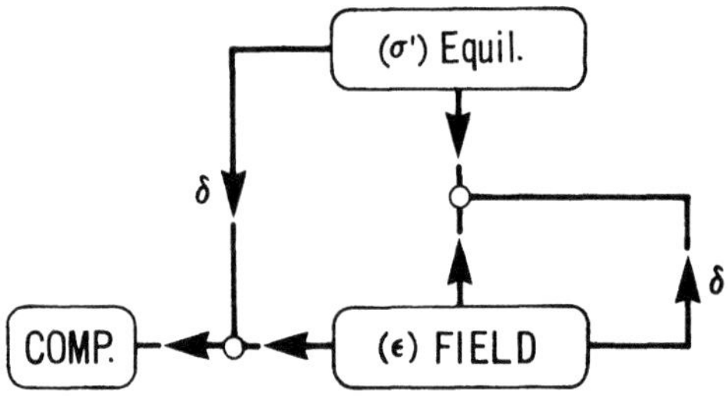

Figure 3

(b) If, in order to obtain displacements, the strain field is assumed *a priori* to be integrable, substitution of equations (4) followed by an integration by parts and consideration of the equilibrium equations satisfied by the (σ') field, transform the principle back to the displacement principle (13).

2.4 *Orthogonality of Equilibrium Solutions and Compatible Solutions in Function Space*

Up to now the stress-strain relations expressed by equations (1) or (2) were kept of general nature; in particular the conclusions reached before are valid for nonlinear relations.

In the case of linear stress-strain relations the abscence of interaction between compatible fields and equilibrium fields can be presented in a more striking form. Both the strain energy density and the complementary energy density are quadratic homogeneous positive definite forms so that, by Euler's theorem, the following numerical equivalence exists

$$W(\varepsilon) = \frac{1}{2}(\sigma_x\varepsilon_x + \tau_{xy}\gamma_{xy} + \sigma_y\varepsilon_y) = \phi(\sigma) \ . \tag{16}$$

It is a local expression of Clapeyron's theorem. Furthermore the expansion of the energy density of the difference between two fields can be written

$$W(\varepsilon-\varepsilon') = W(\varepsilon) + W(\varepsilon') - I(\varepsilon, \sigma') \tag{17}$$

with

$$W(\varepsilon-\varepsilon') = \phi(\sigma-\sigma') \quad , \quad W(\varepsilon) = \phi(\sigma) \ , \quad W(\varepsilon') = \phi(\sigma')$$

where

$$I(\varepsilon, \sigma') = (\varepsilon_x\sigma'_x+\gamma_{xy}\tau'_{xy}+\varepsilon_y\sigma'_y)$$

$$= (\varepsilon'_x\sigma_x + \gamma'_{xy}\tau_{xy}+\varepsilon'_y\sigma_y) = I(\sigma, \varepsilon') \ . \tag{18}$$

This is a local form of Betti-Rayleigh's reciprocal theorem. If one considers a field in the domain Δ as a point F in function space, the distance d(F, F') between two fields is conveniently defined by $d(F, F') = \sqrt{d^2(F, F')}$,

$$d^2(F, F') = 2\iint_\Delta W(\varepsilon-\varepsilon')dxdy = 2\iint_\Delta \phi(\sigma-\sigma')dxdy \ . \tag{19}$$

This definition satisfies the required axioms of distance (see for instance reference 5)

$$d(F, F') = d(F', F)$$

$$d(F, F') + d(F', F'') \geq d(F , F'')$$

$$d(F, F') = 0 \leftrightarrow F \equiv F' \ .$$

The function space is thereby provided with a metric [6, 7]; in particular the scalar product (F, F') between two fields is thereby defined. From

$$d^2(F, F') = d^2(F, 0) + d^2(F', 0) - 2(F, F') ,$$

where 0 denotes the origin (stresses and strains identically zero), applying the definition of distance

$$(F, F') = \iint_\Delta [W(\varepsilon) + W(\varepsilon') - W(\varepsilon-\varepsilon')]\, dxdy$$

and finally, in view of equations (17) and (18)

$$(F, F') = \iint_\Delta (\varepsilon_x\sigma'_x+\gamma_{xy}\tau'_{xy}+\varepsilon_y\sigma'_y)dxdy$$

$$= \iint_\Delta (\varepsilon'_x\sigma_x+\gamma'_{xy}\tau_{xy}+\varepsilon'_y\sigma_y)dxdy = (F', F). \tag{20}$$

Consequently the scalar product is the total interference energy between the two fields.

Of special interest will be the study of the distance between a compatible field, for which equations (4) and (5) are valid and an equilibrium field in which the stresses satisfy equations (6) and (7). Let this be the case respectively for the fields F and F'. Their scalar product is then expressible in terms of the prescribed loads and boundary displacements:

$$(F, F') = \iint_\Delta (u_x\sigma'_x + (u_y+v_x)\tau'_{xy} + v_y\sigma'_y)dxdy$$

becomes after integration by parts and use of the equilibrium equations satisfied by the field F' and equations (5)

$$(F, F') = \iint_\Delta (\bar{X}u+\bar{Y}v)dxdy + \int_{\Gamma_\sigma} (\bar{p}_xu+\bar{p}_yv)ds$$

$$+ \int_{\Gamma_u} (p'_x\bar{u}+p'_y\bar{v})ds . \tag{21}$$

It follows then from (17) that the distance between the two fields can be placed into the form

$$\frac{1}{2}\, d^2(F,\, F') = \iint_\Delta W(\varepsilon)dxdy - \iint_\Delta (\bar{X}u+\bar{Y}v)dxdy - \int_{\Gamma_\sigma} (\bar{p}_x u+\bar{p}_y v)ds$$

$$+ \iint_\Delta \phi(\sigma')dxdy - \int_{\Gamma_u} (p'_x\bar{u}+p'_y\bar{v})ds \;. \tag{22}$$

Since the exact solution must enjoy both qualities of equilibrium and compatibility, the distance between a compatible approach and an equilibrium approach must in the end vanish. Conversely, when the distance vanishes we know that both fields are identical and must then represent the exact solution. Hence a suitable energy principle will be

$$d^2(F,\, F')\text{minimum} \;. \tag{23}$$

Inspection of equation (22) reveals however that this principle spontaneously degenerates in two other principles:

$$\iint_\Delta W(u_x,u_y + v_x,v_y)dxdy + P \quad \text{minimum} \quad , \tag{24}$$

$$\iint_\Delta \phi(\sigma')dxdy - \int_{\Gamma_u} (p'_x\bar{u}+p'_y\bar{v})ds \quad \text{minimum} \quad . \tag{25}$$

The first is the principle of minimum total potential, the second the principle of minimum complementary energy. They are both more precise statements of principles (13) and (14) under the linear stress-strain assumption.

It is clear that approximations on displacements will not interact with approximations on equilibrium stresses and conversely, each will be governed by its own independent principle. It is one aspect of the orthogonality in function space of the subspace of equilibrium solutions and the subspace of compatible solutions.

In the applications, the minimum of total potential will be used to establish the properties of displacement models, the minimum complementary energy for stress models. By securing the continuity of displacements between elements, or the continuity of stress transmission, both principles will also apply to the structure as a whole. It can then be established [9, 10] that a lower bound to a direct influence coefficient is produced by the displacement analysis, an upper bound by the stress analysis. Furthermore the results of both analyses allow a similar enclosure to be built for the mutual influence coefficients [11, 12, 13, 14]. This already provides a powerful control on the overall accuracy of the analysis in finite elements. Bounds on local stress values can also be obtained in principle through the use of Green functions [8]; the amount of analytical and numerical work involved seems however too high to be of real practical utility.

3. DISPLACEMENT MODEL ANALYSIS

3.1 *General Theory*

Let

$$u = \sum_{i=1}^{k} \alpha_i U_i(x, y) \quad , \quad v = \sum_{j=k+1}^{n} \alpha_j V_j(x, y) \quad , \qquad (26)$$

be a displacement field expressed in terms of assumed displacement modes $(U_i(x, y), V_i(x, y))$ with unknown amplitudes (α_i). The column matrix of the amplitudes will be denoted by α; its transpose α^T is the row matrix

$$\alpha^T = (\alpha_1, \alpha_2, \ldots, \alpha_n) \quad .$$

The following matrix notations are also introduced

$$\varepsilon^T = (\varepsilon_x, \varepsilon_y, \gamma_{xy})$$

$$\sigma^T = (\sigma_x, \sigma_y, \sigma_{xy}) \ .$$

From the displacement field follows by taking partial derivatives

$$\varepsilon = B\alpha \ , \quad \varepsilon^T = \alpha^T B^T \tag{27}$$

with the (n x 3) matrix

$$B^T = \left(\frac{\partial U_i}{\partial x} , \frac{\partial V_i}{\partial y} , \frac{\partial U_i}{\partial y} + \frac{\partial V_i}{\partial x} \right), \quad \text{row index } i = 1,2,3,\ldots,n.$$

The linear stress-strain relations are given in matrix form by

$$\sigma = D\varepsilon \tag{28}$$

where, taking as example an isotropic plate in plane stress, of thickness $t(x, y)$,

$$D = \frac{Et(x, y)}{1-\nu^2} \begin{bmatrix} 1 & \nu & 0 \\ \nu & 1 & 0 \\ 0 & 0 & \frac{1-\nu}{2} \end{bmatrix} = D^T \ . \tag{29}$$

The strain energy density can now be expressed as a quadratic form in the (α_i) starting from Clapeyron's theorem

$$W = \frac{1}{2} \sigma^T \varepsilon = \frac{1}{2} \varepsilon^T D \varepsilon = \frac{1}{2} \alpha^T (B^T D B) \alpha \ . \tag{30}$$

Generalized displacements (q_j) are next defined according to certain rules:

- connection of the element with another element across a common boundary is to be expressed by equating, two by two, the generalized displacements pertaining to the boundary,

- the whole displacement field along this boundary must be determined uniquely by the generalized displacements pertaining to the boundary and conversely. In this manner, equating the generalized displacements secures the continuity of displacements along the boundary,

- the number of parameters (α_i) must be equal or superior to the number of required generalized displacements. If superior, convenient additional generalized displacements can be chosen so that the matrices α and q can be connected by a nonsingular linear transformation of matrix $T = (T_{ij})$

$$\alpha = Tq \ , \qquad q = T^{-1}\alpha \ . \tag{31}$$

The unrequired components of q can be eliminated later (see example of Section 3.4). From (30) and (31) the total strain energy of the element becomes a quadratic function of the generalized displacements

$$U = \iint_{\Delta} W \, dxdy = \frac{1}{2} q^T K \, q \ , \tag{32}$$

with

$$K = T^T \left\{ \iint_{\Delta} (B^T D B) dxdy \right\} T = K^T \ . \tag{33}$$

No prescribed boundary displacements are used; to apply the minimum total energy principle there remains to express the potential energy of the prescribed loads. From (26)

$$P = - \sum_i \alpha_i P_i \ , \tag{34}$$

$$P_i = \iint_{\Delta} (\bar{X}U_i + \bar{Y}V_i) dxdy + \int_{\Gamma_\sigma} (\bar{p}_x U_i + \bar{p}_y V_i) ds \ . \tag{35}$$

With the matrix notation

$$p^T = (P_1, P_2, \ldots, P_n) \ ,$$

equation (34) assumes the form of a scalar product

$$P = -p^T\alpha = -g^Tq \ , \tag{36}$$

the last result following from transformation (31) and introduction of the row matrix

$$g^T = (Q_1,\ Q_2,\ \ldots,\ Q_n) = p^TT \ . \tag{37}$$

As implied by the scalar product form of the potential energy, the elements of the row matrix g^T are the generalized loads conjugate to the generalized displacements. They finally derive from the actual loading of the element by the use of weight functions

$$X_j(x,\ y) = \sum_i U_i(x,\ y)T_{ij} \quad , \quad Y_j(x,\ y) = \sum_i V_i(x,\ y)T_{ij} \tag{38}$$

$$Q_j = \iint_\Delta (\bar{X}X_j+\bar{Y}Y_j)dxdy + \int_{\Gamma_\sigma} (\bar{p}_xX_j+\bar{p}_yY_j)ds \ . \tag{39}$$

The weight functions depend solely on the assumed displacement modes and the choice of the generalized displacements. Once calculated they can also be used directly to establish the stiffness matrix, for one has

$$u = \sum_j q_jX_j(x,\ y) \quad , \qquad v = \sum_j q_jY_j(x,\ y) \ , \tag{40}$$

and

$$\varepsilon = Nq \ , \qquad \text{with} \qquad N = BT \ . \tag{41}$$

The elements of the matrix N^T are calculated from partial derivatives of the weight functions as those of B^T were from the originally assumed modes.

The matrix K can then be calculated from the formula

$$K = \iint_\Delta (N^TDN)dxdy \ . \tag{42}$$

Stated in terms of the generalized quantities, the minimum total energy principle is

$$\frac{1}{2}\, q^T K q - q^T g \quad \text{should be a minimum .}$$

It furnishes the condition $\delta q^T (Kq-g) = 0$ for an arbitrary row matrix δq^T and consequently the matrix equation

$$g = Kq \; . \tag{43}$$

This set of equations gives the generalized loads in terms of the generalized displacements; the matrix K is known as the "stiffness matrix" of the element. Observe that under a change of generalized displacements $q = J\bar{q}$, the potential energy preserves its scalar product form $-\bar{g}^T\bar{q} = -g^T q$ if g transforms according to the rule $\bar{g} = J^T g$ and equation (43) turns into

$$\bar{g} = \bar{K}\bar{q} \quad \text{with} \quad \bar{K} = J^T K J \; ,$$

q is of course the matrix representation of a contravariant tensor, g that of a covariant tensor and K that of a twice covariant tensor.

3.2 *Stiffness of Triangular Panels and Beam Segments. Linear Displacement Analysis*

Those are early examples of displacement models introduced in the literature by Turner, Clough, Martin and Topp [15]. The assumed displacement field is linear and contains the three rigid body modes

$$u = \alpha_1 + \alpha_2 x + \alpha_3 y \; ,$$

$$v = \alpha_4 + \alpha_5 x + \alpha_6 y \; .$$

The generalized coordinates are chosen to be the local displacements at the vertices of the triangle (Figure 4)

$$q^T = (u_1,\, u_2,\, u_3,\, v_1,\, v_2,\, v_3) \; .$$

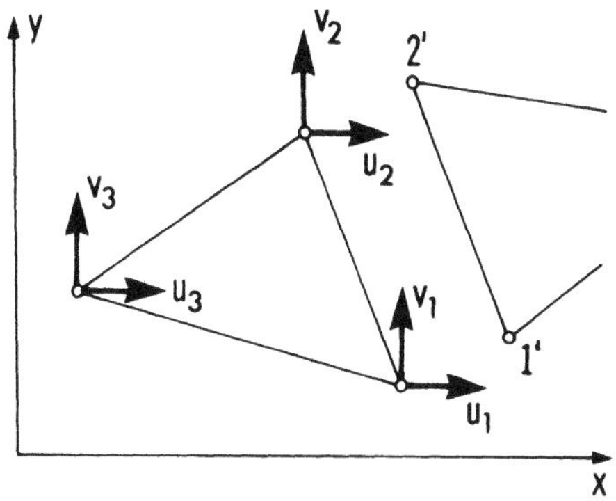

Figure 4

This choice is according to the rules, for along a side of the triangle the displacements vary linearly and are uniquely determined by the displacements at both of its ends. With

$$\begin{aligned} u_i &= \alpha_1 + \alpha_2 x_i + \alpha_3 y_i \,, \\ v_i &= \alpha_4 + \alpha_5 x_i + \alpha_6 y_i \,, \end{aligned} \qquad i = 1,\ 2,\ 3\ .$$

the inverse of the transformation matrix is deduced

$$T^{-1} = \begin{bmatrix} 1 & x_1 & y_1 & 0 & 0 & 0 \\ 1 & x_2 & y_2 & 0 & 0 & 0 \\ 1 & x_3 & y_3 & 0 & 0 & 0 \\ 0 & 0 & 0 & 1 & x_1 & y_1 \\ 0 & 0 & 0 & 1 & x_2 & y_2 \\ 0 & 0 & 0 & 1 & x_3 & y_3 \end{bmatrix}$$

It can be observed that, provided the vertices are numbered in the anticlockwise sense, the determinant

$$\begin{vmatrix} 1 & x_1 & y_1 \\ 1 & x_2 & y_2 \\ 1 & x_3 & y_3 \end{vmatrix} = 2A \ ,$$

is equal to twice the area of the triangle. After an elementary inversion to obtain the matrix T, the weight functions are found to be

$$\left.\begin{aligned} u &= u_1W_1(x, y) + u_2W_2(x, y) + u_3W_3(x, y) \ , \\ v &= v_1W_1(x, y) + v_2W_2(x, y) + v_3W_3(x, y) \ , \end{aligned}\right\} \qquad (44)$$

$$W_1(x, y) = \frac{1}{2A}\,[x_2y_3 - x_3y_2 + x(y_2 - y_3) + y(x_3 - x_2)] \ , \qquad (45)$$

the other two following from a cyclic index permutation.

The weight function $W_1(x, y)$ is illustrated on Figure 5. The generalized load conjugate to u_i is

$$\iint_\Delta \bar{X} W_i(x, y)dxdy + \int_{\Gamma_\sigma} W_i(x, y)\bar{p}_x ds \ .$$

The loads conjugate to v_i follows by replacing $\bar{X}$ with $\bar{Y}$ and $\bar{p}_x$ with $\bar{p}_y$.

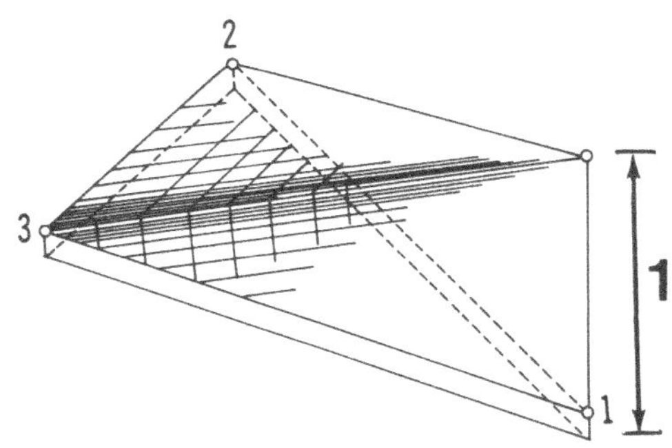

Figure 5

From equations (44) and (45) the matrix N of equation (41) turns out to be

$$N = \frac{1}{2A}\begin{pmatrix} y_2-y_3 & y_3-y_1 & y_1-y_2 & 0 & 0 & 0 \\ 0 & 0 & 0 & x_3-x_2 & x_1-x_3 & x_2-x_1 \\ x_3-x_2 & x_1-x_3 & x_2-x_1 & y_2-y_3 & y_3-y_1 & y_1-y_2 \end{pmatrix} . \tag{46}$$

Finally, denoting by $\tilde{t}$ the average thickness of the plate

$$\bar{t} = \frac{1}{A} \iint_{\Delta} t(x, y)dxdy$$

the stiffness matrix of the triangular plate turns out to be

$$K = \frac{EA\bar{t}}{1-\nu^2} N^T \begin{pmatrix} 1 & \nu & 0 \\ \nu & 1 & 0 \\ 0 & 0 & \frac{1-\nu}{2} \end{pmatrix} N . \tag{47}$$

In a structure that can be subdivided into such triangular panels and beam segments the only additional information necessary is the stiffness matrix of a beam segment (Figure 6). It is assumed that this element undergoes only uniaxial strain and, for simplicity of calculation, the reference axis 0x is taken parallel to the element. For displacement compatibility with the edges of the triangular panels to which the beam segment is attached, the axial displacement is again taken to be a linear function

$$u = \alpha_1 + \alpha_2 x .$$

In terms of the end displacements taken as generalized coordinates

$$u = u_1 W_1(x) + u_2 W_2(x) ,$$

$$W_1(x) = \frac{x_2 - x}{x_2 - x_1} , \quad W_2(x) = \frac{x - x_1}{x_2 - x_1} .$$

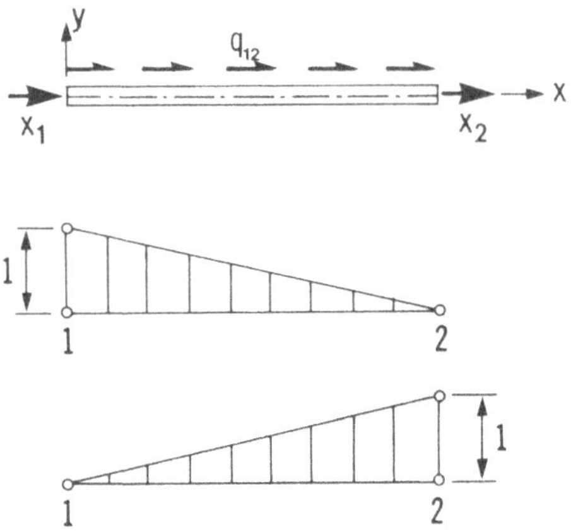

Figure 6

If the loads on the segment consist of end loads X_1 and X_2 and a shear flow distribution $q_{12}(x)$, the potential energy

$$-P = \int_{x_1}^{x_2} q_{12}(x)u(x)dx + X_1u_1 + X_2u_2 ,$$

becomes, after substitution of u(x)

$$-P = u_1Q_1 + u_2Q_2$$

and the generalized loads conjugate to u_1 and u_2 are

$$Q_i = X_i + \int_{x_1}^{x_2} q_{12}(x)W_i(x)dx \quad (i = 1, 2).$$

The weight functions $W_i(x)$ are illustrated on Figure 6. The strain in the segment is

$$\varepsilon = u_1 \frac{dW_1}{dx} + u_2 \frac{dW_2}{dx} = \frac{u_2-u_1}{x_2-x_1} = n^Tq ,$$

$$n^T = \frac{1}{x_2-x_1} (-1, 1) .$$

The normal load in the segment is $L = ES(x)\varepsilon$, where $S(x)$ is the cross sectional area of the beam. The strain energy is

$$\frac{1}{2}\int_{x_1}^{x_2} \varepsilon L dx = \frac{1}{2}\left(\int_{x_1}^{x_2} ES(x)dx\right) q^T nn^T q \quad .$$

The stiffness matrix is

$$K = \left(\int_{x_1}^{x_2} ES(x)dx\right) nn^T = \frac{\int_{x_1}^{x_2} ES(x)dx}{(x_2-x_1)^2} \begin{pmatrix} 1 & -1 \\ -1 & 1 \end{pmatrix} .$$

3.3 *Stiffness of Triangular Panels and Beam Segments. Quadratic Displacement Analysis*

Both displacements are taken to be general quadratic expressions in the cartesian coordinates. Along an edge they will both vary according to a parabolic law and will be determined by their values in three points along the edge. Accordingly a suitable choice for the generalized displacement coordinates are the twelve local displacement components at the vertices and at the mid-points of the edges, Figure 7.

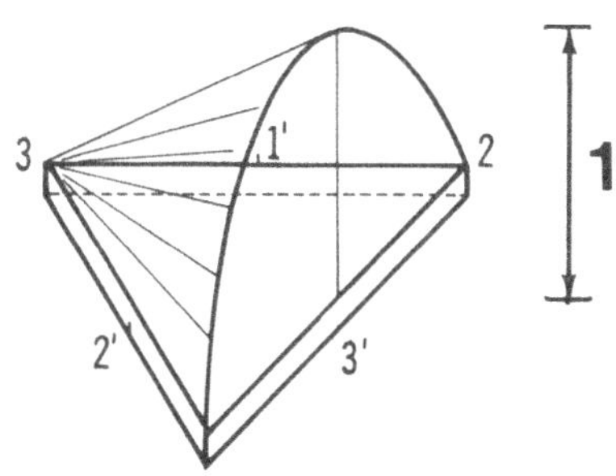

Figure 7

The inverse of the matrix T is obtained from the equations

$$u_i = \alpha_1+\alpha_2 x_i+\alpha_3 y_i+\alpha_4 x_i+\alpha_5 x_i y_i+\alpha_6 y_i^2 \qquad i = 1,2,3,1',2',3'.$$

$$v_i = \alpha_7+\alpha_8 x_i+\alpha_9 y_i+\alpha_{10} x_i^2+\alpha_{11} x_i y_i+\alpha_{12} y_i^2 \ .$$

Instead of inverting this matrix analytically, it is much simpler to establish directly the weight functions. Only two different types of weight function need to be calculated: $W_{3'}(x, y)$ sketched on Figure 7 and $W_3(x, y)$ sketched on Figure 8. The other ones follow from a cyclic index permutation. Both must be of the type

$$W = A + Bx + Cy + Dx^2 + Exy + Fy^2 \ .$$

In the case of $W_{3'}$ we can express that it is identically zero along the straight lines

$$y = y_3 + \frac{y_{13}}{x_{13}}(x-x_3) \quad \text{and} \quad y = y_3 + \frac{y_{23}}{x_{23}}(x-x_3)$$

where, in general,

$$x_{mn} = x_m - x_n \quad \text{and} \quad y_{mn} = y_m - y_n \ .$$

This yields six equations between the constants (A, B, C, D, E, F) only five of which are independent. The sixth equation expresses that the weight function is equal to unity in the point $x = \frac{1}{2}(x_1+x_2)$, $y = \frac{1}{2}(y_1+y_2)$. These equations are extremely simple to solve and one obtains

$$A^2 W_{3'}(x, y) = -a_{13}a_{23} - x(y_{13}a_{23}+y_{23}a_{13}) + y(x_{13}a_{23}+x_{23}a_{13})$$

$$- x^2 y_{13}y_{23} + xy(x_{13}y_{23}+x_{23}y_{13}) - y^2 x_{13}x_{23} \qquad (48)$$

with the additional notation $a_{mn} = x_m y_n - x_n y_m$.

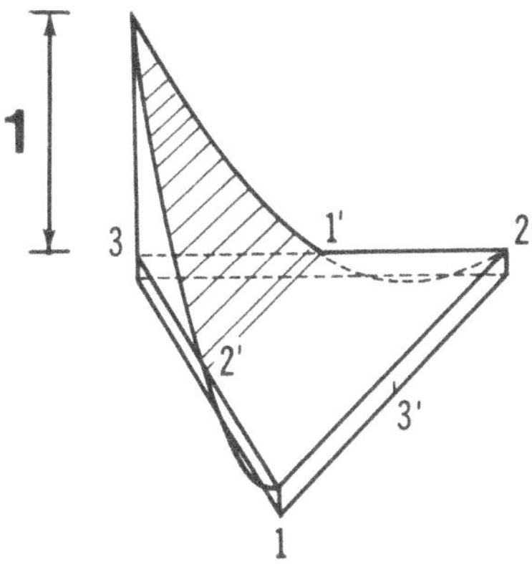

Figure 8

In the case of W_3 one expresses its nullity along the lines

$$y = y_{1'} + \frac{y_{21}}{x_{21}}(x - x_{1'}) = \frac{1}{2}(y_2 + y_3) + \frac{y_{21}}{x_{21}}\left[x - \frac{1}{2}(x_2 + x_3)\right]$$

and

$$y = y_1 + \frac{y_{21}}{x_{21}}(x - x_1) \; ,$$

and its value of unity at the point $x = x_3$, $y = y_3$. Thus

$$2A^2 W_3(x, y) = a_{21}(a_{21} + A) + xy_{21}(2a_{21} + A) - yx_{21}(2a_{21} + A)$$
$$+ x^2 y_{21}^2 - 2xyy_{21}x_{21} + y^2 x_{21}^2 \; . \tag{49}$$

As before the symbol A denotes the area of the triangle. The displacement field can now be written as

$$u = u_1W_1 + u_2W_2 + u_3W_3 + u_{1'}W_{1'} + u_{2'}W_{2'} + u_{3'}W_{3'} \; ,$$

$$v = v_1W_1 + v_2W_2 + v_3W_3 + v_{1'}W_{1'} + v_{2'}W_{2'} + v_{3'}W_{3'} \; ,$$

and the matrix N is set up by partial differentiation of the weight functions. For the purpose of numerical calculation it is convenient to expand the N matrix in the form

$$2A^2N = N_0 + xN_x + yN_y ,$$

where the matrices N_0, N_x and N_y have constant elements. The stiffness matrix calculated from equation (42) is then expanded in the form

$$K = \frac{1}{4A^2}\left(N_0^T D_0 N_0 + 2N_0^T D_x N_x + 2N_0^T D_y N_y + N_x^T D_{xx} N_x + 2N_x^T D_{xy} N_y + N_y^T D_{yy} N_y\right)$$

where

$$D_0 = \iint_\Delta D\, dxdy \quad , \quad D_x = \iint_\Delta Dx\, dxdy \quad , \quad D_y = \iint_\Delta Dydxdy \quad ,$$

$$D_{xx} = \iint_\Delta Dx^2dxdy \; , \quad D_{xy} = \iint_\Delta Dxydxdy \; , \; D_{yy} = \iint_\Delta Dy^2dxdy .$$

In the case of an isotropic plate of constant thickness, D is a matrix of constants and

$$D_0 = DA \; , \quad D_x = D\iint_\Delta xdxdy \; , \quad D_y = D\iint_\Delta ydxdy \; ,$$

$$D_{xx} = D\iint_\Delta x^2dxdy \; , \quad D_{xy} = D\iint_\Delta xydxdy \; , \quad D_{yy} = D\iint_\Delta y^2dxdy \; .$$

An appreciable saving of the computational time is then obtained by a preliminary shift of the coordinates to the centre of area; then $(x_1+x_2+x_3) = 0$, $(y_1+y_2+y_3) = 0$, D_x and D_y vanish, while

$$\iint_\Delta x^2dxdy = \frac{A}{12}\,(x_1^2+x_2^2+x_3^2) \; ,$$

$$\iint_\Delta xydxdy = \frac{A}{12}\,(x_1y_1+x_2y_2+x_3y_3) \; ,$$

$$\iint_\Delta y^2dxdy = \frac{A}{12}\,(y_1^2+y_2^2+y_3^2) \; .$$

A beam segment attached to an edge of such a triangular plate will have a compatible deformation provided the axial displacement be a parabolic function determined by the local displacements $(u_1,\ u_2,\ u_3)$ at the ends and at the central station

(Figure 9). Thus

$$u = u_1W_1(x) + u_2W_2(x) + u_3W_3(x) ,$$

where

$$W_1(x) = \frac{x_3(x_1+x_3)-x(x_1+3x_3)+2x^2}{(x_3-x_1)^2} ,$$

$$W_2(x) = 4\,\frac{-x_1x_3+x(x_1+x_3)-x^2}{(x_3-x_1)^2} ,$$

$$W_3(x) = \frac{x_1(x_1+x_3)-x(3x_1+x_3)+2x^2}{(x_3-x_1)^2} .$$

The axial strain is $\varepsilon = n^T q$ with

$$n^T = \frac{1}{(x_3-x_1)^2}\left(-(x_1+3x_3)+4x\ ,\ 4(x_1+x_3-2x)\ ,\ -(x_3+3x_1)+4x\right) .$$

The stiffness matrix can then be calculated from

$$K = \int_{x_1}^{x_3} ES(x)nn^T dx .$$

In the case when the cross section S is a constant:

$$K = \frac{ES}{3\ell}\begin{pmatrix} 7 & -8 & 1 \\ -8 & 16 & -8 \\ 1 & -8 & 7 \end{pmatrix}$$ ℓ being the length of the segment.

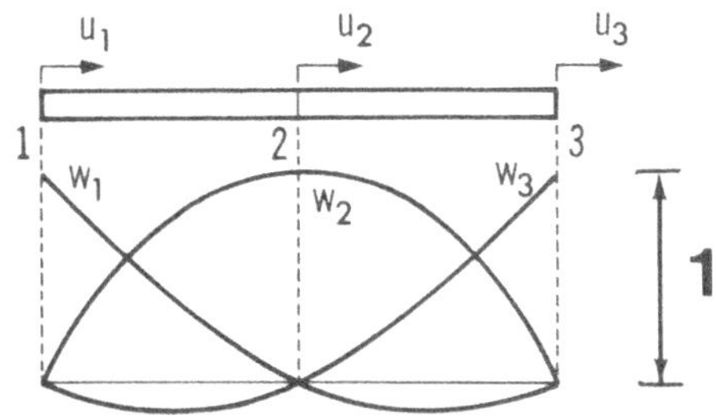

Figure 9

3.4 *Stiffness of a Spar Segment. Displacement Model*

To avoid an exaggerated stiffness of a segment of a spar of the type used in aircraft construction, elementary considerations should first be applied to establish the minimum order of polynomial expansions involved. The representation should be able to provide an almost exact solution to the bending of the spar under constant transverse shear or linearly increasing bending moment. This implies that the axial strain be at a linear function of x, the axial displacement a parabolic function of x. From integration of the elementary curvature equation of the beam it is then also true that the vertical displacement should be a function of the third degree. Accordingly the following assumptions are introduced for spar bending (see Figure 10)

$$u(x,\ z) = z\alpha(x)\ , \tag{50}$$

where the cross section rotation is

$$\alpha(x) = \alpha_1 + \alpha_2 x + \alpha_3 x^2\ , \tag{51}$$

and

$$w(x,\ z) = W(x) = \alpha_4+\alpha_5 x+\alpha_6 x^2+\alpha_7 x^3\ . \tag{52}$$

Assumption (50) is equivalent to stating that the cross sections remain plane. It also implies that the spar is symmetrical with respect to the Ox axis and, if it is part of a box beam, that the cover sheets are symmetrical. Assumption (52) implies that the vertical fibers of the web are inextensible; in addition to the reasonable simplification it brings into the calculation of the strain energy, it has also the important advantage of introducing transverse loading of the spar only through the resultant transverse loads in each cross section. Hence one needs only to consider a transverse distributed load p(x) and the transverse shears T_1 and T_3 at the end sections. The choice of the generalized coordinates is guided by the following considerations.

For full compatibility in the attachment of the spar to cover sheet elements the axial displacements at the ends and in the middle of the upper spar cap are selected:

$$u_1 = h\alpha(a) \quad , \quad u_2 = h\alpha(0) \ , \ u_3 = h\alpha(-a) \ . \tag{53}$$

They determine completely the parabolic distribution of the axial displacement along the spar cap; an identical distribution will result along the edge of a triangular panel as analyzed in Section 3.3, when the corresponding local displacements are made identical.

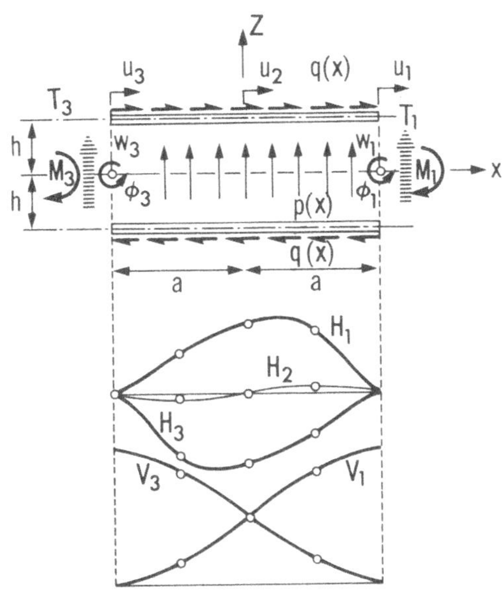

Figure 10

The displacement of an end section of the spar is entirely determined by its vertical displacement and its rotation. Since the end rotations are already represented by u_1 and u_3 it is only necessary to add the coordinates

$$w_1 = W(a) \quad , \quad w_3 = W(-a) \ , \tag{54}$$

to have a complete representation of the boundary displacements and to secure full compatibility with respect to the spar connection to neighbouring elements. Since the original field contains seven parameters we are left with two internal degrees of freedom. For those we select the end slopes of the W(x) function

$$\phi_1 = W'(a) \quad , \quad \phi_3 = W'(-a) \ . \tag{55}$$

In terms of the coordinates introduced by (53 - 55) the original assumptions (51 - 52) can now be written

$$h\alpha(x) = u_2 + \frac{1}{2}(u_1-u_3)\frac{x}{a} + \frac{1}{2}(u_1+u_3-2u_2)\left(\frac{x}{a}\right)^2 , \tag{56}$$

$$4W(x) = 2(w_1+w_3)-a(\phi_1-\phi_3)+[3(w_1-w_3)-a(\phi_1+\phi_3)]\frac{x}{a}$$

$$+ a(\phi_1-\phi_3)\left(\frac{x}{a}\right)^2 + [w_3-w_1+a(\phi_1+\phi_3)]\left(\frac{x}{a}\right)^3 . \tag{57}$$

For an isotropic web, taking into account the inextensibility in the Oz direction

$$\sigma_z = \nu\sigma_x \ , \quad \sigma_x = \frac{Et}{1-\nu^2} z\alpha' \ , \ \tau_{xy} = Gt(\alpha+W') \ .$$

The normal load in the upper spar cap will be

$$N = ESh\alpha' \ .$$

For simplicity both the thickness t of the web and the area S of the spar caps will be assumed constant. The bending moment and the transverse shear in a cross section becomes

$$M = 2Nh + \int_{-h}^{h} \sigma_x z dz = EI\alpha' \quad , \quad I = \frac{2th^3}{3(1-\nu^2)} + 2Sh^2$$

$$T = \int_{-h}^{h} \tau_{xz} dz = 2Gth(\alpha+W') \ .$$

The total strain energy can be calculated by the expression

$$\frac{1}{2}\int_{-a}^{a}[M\alpha'+T(\alpha+W')]dx = \frac{1}{2}EI\int_{-a}^{a}\alpha'^2\,dx+2Gth\int_{-a}^{a}(\alpha+W')^2dx\ .$$

The assumptions (56 - 57) allow the integration to be carried out explicitly, whereby the strain energy becomes a quadratic form in the generalized coordinates whose partial derivatives are the conjugate generalized loads. We denote those respectively by $2U_i$ (there are two spar caps), W_i and Φ_i. The resulting stiffness relations (43) are given by the following

$$\begin{Bmatrix}2U_1\\2U_2\\2U_3\end{Bmatrix} = K_{uu}\begin{Bmatrix}u_1\\u_2\\u_3\end{Bmatrix} + K_{uw}\begin{Bmatrix}w_1\\w_3\end{Bmatrix} + K_{u\phi}\begin{Bmatrix}\phi_1\\\phi_3\end{Bmatrix}\ , \qquad (58)$$

$$\begin{Bmatrix}W_1\\W_3\end{Bmatrix} = K^T_{uw}\begin{Bmatrix}u_1\\u_2\\u_3\end{Bmatrix} + K_{ww}\begin{Bmatrix}w_1\\w_3\end{Bmatrix} + K_{w\phi}\begin{Bmatrix}\phi_1\\\phi_3\end{Bmatrix}\ , \qquad (59)$$

$$\begin{Bmatrix}\Phi_1\\\Phi_2\end{Bmatrix} = K^T_{u\phi}\begin{Bmatrix}u_1\\u_2\\u_3\end{Bmatrix} + K^T_{w\phi}\begin{Bmatrix}w_1\\w_3\end{Bmatrix} + K_{\phi\phi}\begin{Bmatrix}\phi_1\\\phi_3\end{Bmatrix}\ , \qquad (60)$$

where the various stiffness matrices are

$$K_{uu} = \frac{EI}{6h^2a}\begin{Bmatrix}7 & -8 & +1\\-8 & 16 & -8\\+7 & -8 & 7\end{Bmatrix} + \frac{Gta}{15h}\begin{Bmatrix}8 & 4 & -2\\4 & 32 & 4\\-2 & 4 & 8\end{Bmatrix}\ ,$$

$$K_{uw} = \frac{Gt}{5}\begin{Bmatrix}1 & -1\\8 & -8\\1 & -1\end{Bmatrix}\ , \qquad K_{ww} = \frac{6Gth}{5a}\begin{Bmatrix}1 & -1\\-1 & 1\end{Bmatrix}\ ,$$

$$K_{w\phi} = \frac{Gth}{5}\begin{Bmatrix}-1 & -1\\1 & 1\end{Bmatrix}\ , \qquad K_{u\phi} = \frac{Gta}{15}\begin{Bmatrix}7 & -3\\-4 & -4\\-3 & 7\end{Bmatrix}\ ,$$

$$K_{\phi\phi} = \frac{2Gtah}{15}\begin{Bmatrix}4 & -1\\-1 & 4\end{Bmatrix}.$$

Turning now to the task of interpreting the generalized loads, we compare the scalar product form of the potential energy and its expression in terms of the applied loads:

$$-P = 2(u_1U_1+u_2U_2+u_3U_3)+w_1W_1+w_3W_3+\phi_1\Phi_1+\phi_3\Phi_3$$

$$= 2\int_{-a}^{a} q(x)h\alpha(x)dx + \int_{-a}^{a} p(x)W(x)dx+w_1T_1+w_3T_3$$

$$+ \frac{u_1}{h} M_1 + \frac{u_3}{h} M_3 .$$

Substitution of equations (56) and (57) and identification of the coefficients of the generalized displacements results in

$$U_1 = \frac{1}{2}\int_{-a}^{a} q(x)\,\frac{x}{a}\left(\frac{x}{a} + 1\right) dx + \frac{M_1}{h} \quad ,$$

$$U_2 = \int_{-a}^{a} q(x)\left[1 - \left(\frac{x}{a}\right)^2\right] dx \quad ,$$

$$U_3 = \frac{1}{2}\int_{-a}^{a} q(x)\,\frac{x}{a}\left(\frac{x}{a} - 1\right) dx + \frac{M_3}{h} \quad ,$$

$$W_1 = \int_{-a}^{a} p(x)\,\frac{1}{4}\left[2 + 3\,\frac{x}{a} - \left(\frac{x}{a}\right)^3\right] dx + T_1 \quad ,$$

$$W_3 = \int_{-a}^{a} p(x)\,\frac{1}{4}\left[2 - 3\,\frac{x}{a} + \left(\frac{x}{a}\right)^3\right] dx + T_3 \quad ,$$

$$\Phi_1 = \frac{a}{4}\int_{-a}^{a} p(x)\left(\frac{x}{a} - 1\right)\left(\frac{x}{a} + 1\right)^2 dx \quad ,$$

$$\Phi_3 = \frac{a}{4}\int_{-a}^{a} p(x)\left(\frac{x}{a} + 1\right)\left(\frac{x}{a} - 1\right)^2 dx \quad .$$

The auxiliary variables ϕ_1 and ϕ_3 can now be taken from equations (60) and substituted into equations (58) and (59) to give the final stiffness relations:

$$\begin{pmatrix} 2H_1 \\ 2H_2 \\ 2H_3 \end{pmatrix} = K_{hh} \begin{pmatrix} u_1 \\ u_2 \\ u_3 \end{pmatrix} + K_{hv} \begin{pmatrix} w_1 \\ w_3 \end{pmatrix} , \tag{61}$$

$$\begin{pmatrix} V_1 \\ V_3 \end{pmatrix} = K_{hv}^T \begin{pmatrix} u_1 \\ u_2 \\ u_3 \end{pmatrix} + K_{vv} \begin{pmatrix} w_1 \\ w_3 \end{pmatrix} , \tag{62}$$

$$K_{hh} = \frac{EI}{6h^2a} \begin{pmatrix} 7 & -8 & +1 \\ -8 & 16 & -8 \\ +1 & -8 & 7 \end{pmatrix} + \frac{Gta}{9h} \begin{pmatrix} 1 & 4 & 1 \\ 4 & 16 & 4 \\ 1 & 4 & 1 \end{pmatrix} , \tag{63}$$

$$K_{hv} = \frac{Gt}{3} \begin{pmatrix} 1 & -1 \\ 4 & -4 \\ 1 & -1 \end{pmatrix} , \quad K_{vv} = \frac{Gth}{a} \begin{pmatrix} 1 & -1 \\ -1 & 1 \end{pmatrix} . \tag{64}$$

The new generalized loads are

$$H_1 = U_1 - \frac{1}{12h} (5\Phi_1 - \Phi_3) ,$$

$$H_2 = U_2 + \frac{1}{13h} (\Phi_1 + \Phi_3) ,$$

$$H_3 = U_3 - \frac{1}{12h} (5\Phi_3 - \Phi_1) ,$$

$$V_1 = W_1 + \frac{1}{12a} (\Phi_1 + \Phi_3) ,$$

$$V_3 = W_3 - \frac{1}{10a} (\Phi_1 + \Phi_3)$$

or, explicitly:

$$H_1 = \frac{M_1}{h} + \frac{1}{2} \int_{-a}^{a} q(x) \frac{x}{a} \left(1 + \frac{x}{a}\right) dx + \frac{a}{24h} \int_{-a}^{a} p(x) \left(1 - \frac{x}{a}\right)\left(1 + \frac{x}{a}\right) \times \left(3 + 2\frac{x}{a}\right) dx ,$$

$$H_2 = \int_{-a}^{a} q(x) \left[1 - \left(\frac{x}{a}\right)^2\right] dx + \frac{a}{36h} \int_{-a}^{a} p(x) \frac{x}{a} \left[\left(\frac{x}{a}\right)^2 - 1\right] dx ,$$

$$H_3 = \frac{M_3}{h} + \frac{1}{2}\int_{-a}^{a} q(x)\frac{x}{a}\left(\frac{x}{a} - 1\right) dx + \frac{a}{24h}\int_{-a}^{a} p(x)\left(1 - \frac{x}{a}\right)$$

$$\times\left(1 + \frac{x}{a}\right)\left(-3 + 2\frac{x}{a}\right) dx \ ,$$

$$V_1 = T_1 + \frac{1}{12}\int_{-a}^{a} p(x)\left(1 + \frac{x}{a}\right) dx \ ,$$

$$V_3 = T_3 + \frac{1}{12}\int_{-a}^{a} p(x)\left(1 - \frac{x}{a}\right) dx \ .$$

The essential feature of the elimination of the auxiliary variables is this modification of the generalized loads and in particular the appearance of contributions from the external loading $p(x)$ in the loads conjugate to the spar cap displacements are illustrated in Figure 10. The weight functions attached to $q(x)$ are the same as those encountered for the beam segment in Section 3.3.

If the shearing deformation of the spar is neglected, the stiffness problem appears in a different light. The shearing deformation can be made to vanish by letting the shear modulus G go to infinity. Then in equation (61) the terms proportional to G must cancel and this is immediately seen to occur if the following relation is satisfied

$$\frac{a}{3h}(u_1 + 4u_2 + u_3) + (w_1 - w_3) = 0 \ . \qquad (65)$$

This purely geometrical condition is a consequence of the fact that the cross sections remain normal to the parabolic curve of the neutral axis. The general statement of this property is of course the disappearance of the shearing strain:

$$\alpha + W' = 0 \ .$$

After substitution of equation (56 - 57) and identification of the powers of x/a this results into

$$u_1 = -h\phi_1 \ , \quad u_3 = -h\phi_3 \ , \ u_2 = \frac{3h}{4a}(-w_1+w_3) + \frac{h}{4}(\phi_1+\phi_3) \ ,$$

and the elimination of the slopes again results in equation (65). The only stiffness relations that remain are now

$$\begin{pmatrix} 2H_1 \\ 2H_2 \\ 2H_3 \end{pmatrix} = K_{hh} \begin{pmatrix} u_1 \\ u_2 \\ u_3 \end{pmatrix}$$

between the equivalent bending moments $2hH_i$ and the spar cap displacements, where K_{hh} is reduced to its part proportional to the bending rigidity. This matrix is singular. In fact if one sets $u_1 = u_2 = u_3$ no equivalent bending moments are induced, an obvious result since we then have a rigid body (rotation) mode. Corresponding to this we have the equilibrium relation

$$H_1 + H_2 + H_3 = 0 \ .$$

The procedure for finding the spar deflexions are then as follows. The spar must be supported against rotation, indeed u_1 or u_3 are generally known from boundary conditions or previous calculations. The stiffness relations determine the two other u values. Inserting those in equation (65) the deflexion increment can be calculated and again either w_1 or w_2 are known from boundary conditions or previous calculations. As was indicated previously the present stiffness matrix was devised to give a correct behaviour under constant transverse shear load or linearly increasing bending moment. Incorporated in a box beam it requires a parabolic displacement analysis of the cover sheet elements in the direction of the spar flanges. Previous attempts at setting up spar stiffnesses have generally aimed at a linear displacement along the flanges to be used in conjunction with linear cover sheet elements.

Our analysis can be modified to incorporate this simplification *a posteriori* by the change (and reduction) in coordinates

$$\begin{pmatrix} u_1 \\ u_2 \\ u_3 \end{pmatrix} = T \begin{pmatrix} u_1 \\ u_3 \end{pmatrix} , \qquad T = \begin{pmatrix} 1 & 0 \\ \frac{1}{2} & \frac{1}{2} \\ 0 & 1 \end{pmatrix} .$$

This states that the middle displacement u_2 is now constrained by the assumption of linearity to become the average of the end displacements. Remembering the types of changes that generalized loads and stiffness matrices undergo as a result of their tensorial character the simplified representation will read

$$T^T \begin{pmatrix} 2H_1 \\ 2H_2 \\ 2H_3 \end{pmatrix} = \begin{pmatrix} 2H_1+H_2 \\ H_2+2H_3 \end{pmatrix} = T^T K_{hh} T \begin{pmatrix} u_1 \\ u_3 \end{pmatrix} + T^T K_{hv} \begin{pmatrix} w_1 \\ w_3 \end{pmatrix} ,$$

$$\begin{pmatrix} V_1 \\ V_3 \end{pmatrix} = K_{hv}^T T \begin{pmatrix} u_1 \\ u_3 \end{pmatrix} + K_{vv} \begin{pmatrix} w_1 \\ w_3 \end{pmatrix} .$$

The element is now adapted only to the transmission of a constant bending moment and very much stiffened thereby. It will need a much finer spanwise subdivision to avoid the poor results reported [16, 17] for displacement models of similar linear characteristics.

It is perhaps of interest to note that similar simplifications can be introduced in the quadratic displacement analysis of triangular panels to keep them flexible in a preferential direction and linearize them transversally.

4. EQUILIBRIUM ANALYSIS

4.1 *General Theory*

While the general theory of displacement models is reasonably well understood [18, 19] this does not seem to be the case for equilibrium analysis. The reason is probably that it appears less intuitive from a geometric standpoint. The emphasis is on stress transmission modes rather than deformation modes and the first step is to assume a parametric stress field. In two dimensional analysis, for example

$$\sigma_x = \Sigma\beta_i S^{(i)}_{xx}(x, y) + \Sigma P_r T^{(r)}_{xx}(x, y) ,$$

$$\sigma_y = \Sigma\beta_i S^{(i)}_{yy}(x, y) + \Sigma P_r T^{(r)}_{yy}(x, y) ,$$

$$\tau_{xy} = \Sigma\beta_i S^{(i)}_{xy}(x, y) + \Sigma P_r T^{(r)}_{xy}(x, y) , \tag{66}$$

$$\frac{\partial S^{(i)}_{xx}}{\partial x} + \frac{\partial S^{(i)}_{xy}}{\partial y} = 0 , \qquad \frac{\partial S^{(i)}_{xy}}{\partial x} + \frac{\partial S^{(i)}_{yy}}{\partial y} = 0 , \tag{67}$$

$$\frac{\partial T^{(r)}_{xx}}{\partial x} + \frac{\partial T^{(r)}_{xy}}{\partial y} = -X_r(x, y) , \qquad \frac{\partial T^{(r)}_{xy}}{\partial x} + \frac{\partial T^{(r)}_{yy}}{\partial y} = -Y_r(x, y) . \tag{68}$$

The stress modes of unknown amplitudes (β_i) are in equilibrium without internal forces, while the modes of amplitudes (P_r) are in equilibrium with assumed body force distributions (P_rX_r, P_rY_r). In order to extract a maximum of information from the complementary energy principle no loads are prescribed. Instead, displacements are specified along the whole boundary and also the displacements conjugate to the body force distributions. This requires a slight extension in the presentation of the principle: from virtual work considerations the displacement conjugate to the loading mode (X_r, Y_r) is the weighed average

$$v_r = \iint_\Delta (X_r u + Y_r v)\, dxdy . \tag{69}$$

If this is specified, the complementary energy principle takes the extended form

$$\iint_\Delta \phi(\sigma)dxdy - \int_\Gamma (p_x\bar{u}+p_y\bar{v})dx - \Sigma\bar{v}_r P_r \quad \text{minimum} . \tag{70}$$

Exactly like (p_x, p_y) are the reactions along the boundary against the specified boundary displacements, the P_r are the reaction amplitudes of stress modes against the prescribed generalized displacements $\bar{v}_r$. In this manner the parameters (β_i) and (P_r) of the equilibrium stress field are arbitrary and independent in the principle.

Generalized boundary loads (Q_j) are next defined according to the following rule: along each connection boundary the stress distribution is uniquely determined by the generalized loads pertaining to this boundary and conversely. In this manner a reciprocity between generalized loads across the boundary entails complete continuity of stress transmission.

The conjugate boundary displacements (q_j) are obtained from the virtual work equation

$$\int (\bar{u}p_x + \bar{v}p_y)dx = \Sigma q_j Q_j \quad \textit{along each boundary.} \tag{71}$$

Let us now express all this in matrix form with the notations

$$b^T = (\beta_1 \cdots \beta_m) \qquad q^T = (\bar{q}_1 \cdots \bar{q}_n)$$
$$p^T = (P_1 \cdots P_t) \qquad v^T = (\bar{v}_1 \cdots \bar{v}_t)$$
$$g^T = (Q_1 \cdots Q_n) \; .$$

The stress field (66) can be written as

$$\sigma = Sb + Tp \; , \tag{72}$$

with the stress modes entered as columns of the (3 x m) matrix S and the (3 x t) matrix T. The complementary strain energy will be

$$\iint_\Delta \phi(\sigma)dxdy = \frac{1}{2}\iint_\Delta \sigma^T \varepsilon dxdy = \frac{1}{2}\iint_\Delta \sigma^T D^{-1}\sigma dxdy$$
$$= \frac{1}{2} b^T F_{bb} b + b^T F_{bp} p + \frac{1}{2} p^T F_{pp} p \; , \tag{73}$$

with flexibility matrices,

$$F_{bb} = \iint\limits_{\Delta} S^T D^{-1} S dxdy \quad , \qquad F_{bp} = \iint\limits_{\Delta} S^T D^{-1} T dxdy \quad ,$$
$$F_{pp} = \iint\limits_{\Delta} T^T D^{-1} T dxdy \ . \tag{74}$$

From their definitions the generalized loads will be expressed in terms of the field parameters by a matrix equation

$$g = Cb + Ap \ . \tag{75}$$

The (n x m) matric C and the (n x a) matrix A are load connection matrices. The virtual work of the boundary loads against the prescribed conjugate displacements will be

$$q^T g = q^T Cb + q^T Ap$$

and is now expressed in terms of the field parameters. The principle of minimum complementary energy

$$\frac{1}{2} b^T F_{bb} b + b^T F_{bp} p + \frac{1}{2} p^T F_{pp} p - q^T Cb - q^T Ap - v^T p \quad \text{minimum} \ ,$$

yields the stationary conditions with respect to variations on b and p,

$$F_{bb} b + F_{bp} p = C^T q \ ,$$
$$F_{bp}^T b + F_{pp} p = A^T q + v \ .$$

Introducing the matrices

$$K = CF_{bb}^{-1} C^T \quad , \tag{76}$$

$$\left.\begin{aligned} & A - CF_{bb}^{-1} F_{bp} \ , \\ ' &= F_{pp} - F_{bp}^T F_{bb}^{-1} F_{bp} \ , \end{aligned}\right\} \tag{77}$$

the stationary conditions can be rewritten in the form

$$g = Kq + Rp \; , \tag{78}$$

$$v = Lp - R^T q \; . \tag{79}$$

The first expresses the boundary loads as the sum of a part due to the boundary displacements and a part due to a reaction against the body loads. The second expresses the displacements conjugate to the body loads as the sum of a fixed boundary term and a term due to boundary displacement. This form of the equations is the more natural since, instead of considering the p loads as unknown reactions for the purpose of applying the complementary energy principle, these loads are generally given. In many cases there are no body forces and equation (78) reduces to the form (43) with the stiffness matrix (76).

It should be observed that the knowledge of the displacements is reduced to that of a certain number of weighed averages. Sometimes the strain field is integrable and a complete internal displacement field available.

4.2 *Triangular Panels. Equilibrium Analysis*

The assumed stress field is the simplest conceivable

$$\sigma_x = \beta_1 \; , \quad \sigma_y = \beta_2 \; , \quad \tau_{xy} = \beta_3 \; . \tag{80}$$

There are no body forces. Along the edge 1-2 (Figure 11) the generalized loads

$$\begin{aligned} f_{12} &= \int_1^2 p_x ds = \int_1^2 \left(\sigma_x dy - \tau_{xy} dx \right) = \beta_1 y_{21} - \beta_3 x_{21} \; , \\ g_{12} &= \int_1^2 p_y ds = \int_1^2 \left(\tau_{xy} dy - \sigma_y dx \right) = \beta_3 y_{21} - \beta_2 x_{21} \; , \end{aligned} \tag{81}$$

are defined by the edge stresses and conversely the edge tractions are

$p_x = f_{12}/c_{12}$, $p_y = g_{12}/c_{12}$, c_{12} = length of the edge.

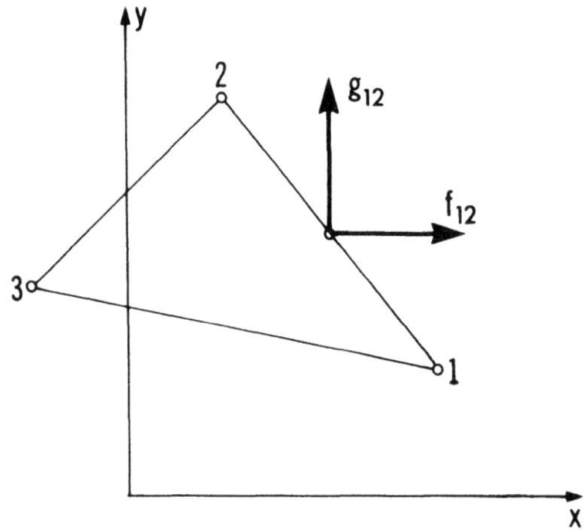

Figure 11

Conjugate edge displacements are obtained from the virtual work equations along this edge

$$\int_1^2 (p_x\bar{u}+p_y\bar{v})ds = \frac{f_{12}}{c_{12}}\int_1^2 \bar{u}ds + \frac{g_{12}}{c_{12}}\int_1^2 \bar{v}ds = f_{12}u_{12}+g_{12}v_{12}.$$

Hence

$$u_{12} = \frac{1}{c_{12}}\int_1^2 \bar{u}ds \ , \qquad v_{12} = \frac{1}{c_{12}}\int_1^2 \bar{v}ds \ . \tag{82}$$

They are the ordinary averages of the displacements along the edge. Results for the other edges follow by cyclic index permutations. Adopting the order

$$g^T = (f_{23},f_{31},f_{12},g_{23},g_{31},g_{12}) \ ,$$

$$q^T = (u_{23},u_{31},u_{12},v_{23},v_{31},v_{12}) \ ,$$

the connexion matric C condensing in the equation g = Cb all relations of type (81) is

$$C = \begin{bmatrix} y_{32} & 0 & x_{23} \\ y_{13} & 0 & x_{31} \\ y_{21} & 0 & x_{12} \\ 0 & x_{23} & y_{32} \\ 0 & x_{31} & y_{13} \\ 0 & x_{12} & y_{21} \end{bmatrix} .$$

Since in the present case we have simply

$$\sigma = b ,$$

as equation (72), the flexibility matrix is

$$F = \iint_{\Delta} D^{-1} dxdy .$$

For an isotropic panel

$$F = \frac{A}{E\tilde{t}} \begin{pmatrix} 1 & -\nu & 0 \\ -\nu & 1 & 0 \\ 0 & 0 & 2(1+\nu) \end{pmatrix} , \qquad \frac{1}{\tilde{t}} = \frac{1}{A} \iint_{\Delta} \frac{dxdy}{t(x,y)} .$$

The stiffness matrix is

$$CF^{-1}C^T = \frac{E\tilde{t}}{A(1-\nu^2)} C \begin{pmatrix} 1 & \nu & 0 \\ \nu & 1 & 0 \\ 0 & 0 & \frac{1-\nu}{2} \end{pmatrix} C^T .$$

When this is compared to the stiffness matrix of the same panel calculated in Section 3.2 we note that since $C = 2AN^T$ the present matrix can be deduced from that of Section 3.2 by multiplication with a factor $4(\tilde{t}/\bar{t})$.

Under constant thickness conditions the strain field is integrable and the displacements can be expressed by equations of the same type as in Section 3.2, i.e., by linear forms. The coefficients (α_i) of the forms are completely determined by the present generalized displacements. In fact the integration of formulae of type (82) yields

$$u_{ij} = \alpha_1+\alpha_2 \frac{1}{2} (x_i+x_j) + \alpha_3 \frac{1}{2} (y_i+y_j) ,$$

$$v_{ij} = \alpha_4+\alpha_5 \frac{1}{2} (x_i+x_j) + \alpha_6 \frac{1}{2} (y_i+y_j) ,$$

and these equations can be solved for the coefficients. The displacements at the vertices can be deduced and the following relations, obvious in view of the linearity of the field, are valid

$$u_{ij} = \frac{1}{2} (u_i+u_j) , \quad v_{ij} = \frac{1}{2} (v_i+v_j) .$$

This was used as starting point for the original derivation of the equilibrium model of the panel [14]. The present derivation is a more correct application of the basic principles. It also applies to the variable thickness case for which the strain field might not be integrable.

The major difference between the direct stiffness analysis and the equilibrium analysis is in the connection properties of the models. For the equilibrium model the stresses are transmitted continuously from one panel to the other but only the average displacements of the edges are coincident. In the other model the edge displacements are everywhere coincident but the stresses suffer discontinuities from one panel to the other.

4.3 *Beam Segment Equilibrium Analysis*

The equilibrium equation governing the axial load L in the beam is (Figure 12)

$$\frac{dL}{dx} = - \tau . \tag{83}$$

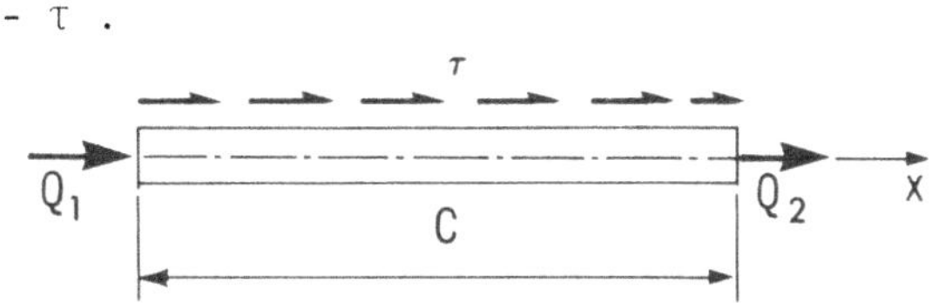

Figure 12

We require a solution for a constant shear flow τ. The general solution of equation (83) is then

$$L = \beta - \tau x \ . \tag{84}$$

In accordance with the philosophy of the general theory, β is the only parameter of the stress field in equilibrium without body forces, τ the only amplitude parameter of the field in equilibrium with the assumed body force distribution $\tau = 1$. Let v be the conjugate displacement to this distribution

$$v = \int_0^c u dx.$$

This is supposed to be specified together with the boundary displacements u(0) and u(c), leading to a virtual work of the reactions equal to

$$-\bar{u}(0)L(0) + \bar{u}(c)L(c) + \bar{v}\tau \ .$$

The generalized loads defined for boundary connections are

$$Q_1 = -L(0) = -\beta \ , \qquad Q_2 = L(c) = \beta - c\tau \ . \tag{85}$$

In this example the body force τ can be generated partially or totally by attachment of the beam to shear panels. It is for this reason that a constant τ solution was indicated; it can produce a continuous stress transmission between the beam and our triangular shear panels of Section 4.2. For the same reason the generalized load

$$Q_3 = c\tau \ , \tag{86}$$

that will correspond to a generalized load of the panels must be added to the set of connection loads. Equations (85 - 86) result in the following connexion matrix

$$g = \begin{pmatrix} Q_1 \\ Q_3 \\ Q_2 \end{pmatrix} = C \begin{pmatrix} \beta \\ \tau \end{pmatrix} \qquad C = \begin{pmatrix} -1 & 0 \\ 0 & c \\ 1 & -c \end{pmatrix} . \tag{87}$$

The significance of the conjugate generalized displacements is deduced from the virtual work equations for each boundary separately

$$q_1 Q_1 = -\bar{u}(0)L(0) = -\bar{u}(0)\beta ,$$

$$q_3 Q_3 = \tau \int_0^c u dx ,$$

$$q_2 Q_2 = \bar{u}(c)L(c) = \bar{u}(c)(\beta - c\tau) .$$

Substitution of equations (85 - 86) and identification produces

$$q_1 = \bar{u}(0) , \qquad q_3 = \frac{1}{c}\int_0^c u dx , \qquad q_2 = \bar{u}(c) . \tag{88}$$

The flexibility matrix F results from the expansion of the complementary strain energy, using equation (84)

$$\frac{1}{2}\int_0^c \frac{L^2}{ES} dx = \frac{1}{2} (\beta \ \tau) F \begin{pmatrix} \beta \\ \tau \end{pmatrix}$$

$$F = \frac{1}{E} \begin{pmatrix} \int_0^c \frac{dx}{S} & -\int_0^c \frac{xdx}{S} \\ -\int_0^c \frac{xdx}{S} & \int_0^c \frac{x^2 dx}{S} \end{pmatrix} ,$$

and we have finally $g = Kq$ with $K = CF^{-1}C^T$.

For the particular case of a beam of constant cross section S, the flexibility matrix turns out to be

$$F = \frac{c}{6ES} \begin{pmatrix} 6 & -3c \\ -3c & 2c^2 \end{pmatrix}$$

and the stiffness matrix

$$K = \frac{2ES}{c}\begin{pmatrix} 2 & -3 & 1 \\ -3 & 6 & -3 \\ 1 & -3 & 2 \end{pmatrix} .$$

The same result was established before [14] in another way.

4.4 *Spar Bending Equilibrium Theory*

Although, as in the case of displacement models, spars could be treated as combinations of triangular plate elements and beam segments, an independent analysis taking into account the symmetry properties of the problem, appears to be more efficient. The following three parameter field is assumed for the web stresses

$$\sigma_x = \frac{6z}{h^2}(\beta_1 + x\beta_2) \quad , \quad \sigma_z = 0 \ , \quad \tau_{xz} = \beta_3 - 3\frac{z^2}{h^2}\beta_2 \ . \tag{89}$$

It is an equilibrium field without body forces and can easily be derived from an Airy stress function expanded in odd powers of z. The expansion in the x direction is limited so as to give a constant shear flow τ_{xz} along the connection with the spar flanges. In this manner the spar flanges can be added as beam segments treated according to the theory of Section 4.3 and implementing the equilibrium conditions with the web. Hence it proves adequate to treat the web separately. It is also worth noting that, because the stress transverse to the fibers is zero, the external vertical loading on the spar will have to take place through shears in the sections $x = \pm a$.

Generalized loads giving a complete description of the stress distributions along the connection boundaries are:

(a) $$Q_3 = \int_{-a}^{a} (\tau_{xz})_{z=h} dx = -6a\beta_2 + 2a\beta_3 \ , \tag{90}$$

the total load transmitted to the flanges; the shear flow between web and spar flanges is then $Q_3/(2a)$ and the conjugate displacement u_3 can be deduced from the virtual work equation

$$\left.\begin{aligned} &\int_{-a}^{a} (\tau_{xz}u)_{z=h}dx = \frac{Q_3}{2a}\int_{-a}^{a} udx = Q_3u_3, \\ &\text{whence} \\ &u_3 = \frac{1}{2a}\int_{-a}^{a} udx , \end{aligned}\right\} \tag{91}$$

(b)
$$M_2 = \int_{-h}^{h} (\sigma_x)_{x=a}zdz = 4h(\beta_1+a\beta_2) \tag{92}$$

the web bending moment in the x = a section. The normal load distribution in this section is then $\sigma_x = 3zM_2/(2h^3)$ and the conjugate displacement ϕ_2 follows from

$$\int_{-h}^{h} (\sigma_x u)_{x=a}dz = \frac{3M_2}{2h^3}\int_{-h}^{h} (u)_{x=a}zdz = M_2\phi_2$$

whence

$$\phi_2 = \frac{3}{2h^3}\int_{-h}^{h} (u)_{x=a}zdz . \tag{93}$$

Similar definitions

$$M_1 = -\int_{-h}^{h} (\sigma_x)_{x=-a}zdz = 4h(-\beta_1+a\beta_2) , \tag{94}$$

$$\phi_1 = \frac{3}{2h^3}\int_{-h}^{h} (u)_{x=-a}zdz , \tag{95}$$

are used in the section x = -a.

(c)
$$T_2 = \int_{-h}^{h} (\tau_{xz})_{x=a}dz = 2h(-\beta_2+\beta_3) , \tag{96}$$

$$V_2 = \int_{-h}^{h} \left(1 - 3\frac{z^2}{h^2}\right) (\tau_{xz})_{x=a}dz = \frac{8}{5}h\beta_2 . \tag{97}$$

These generalized loads describe the shear flow along x = a

$$(\tau_{xz})_{x=a} = \frac{1}{2h} T_2 + \frac{5}{8h} V_2 \left(1 - 3 \frac{z^2}{h^2}\right) ,$$

in terms of a constant distribution of total shear T_2 and a parabolic distribution of amplitude V_2 which is statically equivalent to zero. The conjugate displacements follow from the virtual work equation

$$\int_{-h}^{h} (\tau_{xz} w)_{x=a} dz = \frac{T_2}{2h} \int_{-h}^{h} (w)_{x=a} dz + \frac{5V_2}{8h} \int_{-h}^{h} \left(1 - 3 \frac{z^2}{h^2}\right) (w)_{x=a} dz ,$$

whence

$$w_2 = \frac{1}{2h} \int_{-h}^{h} (w)_{x=a} dz , \quad v_2 = \frac{5}{8h} \int_{-h}^{h} \left(1 - 3 \frac{z^2}{h^2}\right) (w)_{x=a} dz . \tag{98}$$

Similar definitions are used in the x = -a section.

$$T_1 = - \int_{-h}^{h} (\tau_{xz})_{x=-a} dz = 2h(\beta_2 - \beta_3) , \tag{99}$$

$$V_1 = - \int_{-h}^{h} \left(1 - 3 \frac{z^2}{h^2}\right) (\tau_{xz})_{x=-a} dz = - \frac{8}{5} h\beta_2 ,$$

$$w_1 = \frac{1}{2h} \int_{-h}^{h} (w)_{x=-a} dz , \quad v_1 = \frac{5}{8h} \int_{-h}^{h} \left(1 - 3 \frac{z^2}{h^2}\right) (w)_{z=-a} dz . \tag{100}$$

Adopting the following sequence of generalized quantities

$$g^T = (T_1, V_1, M_1, T_2, V_2, M_2, 2Q_3) ,$$

$$q^T = (w_1, v_1, \phi_1, w_2, v_2, \phi_2, u_3) ,$$

the connexion matrix, defined by equation (75), follows from equations (90 - 99) as

$$C^T = \begin{pmatrix} 0 & 0 & -4h & 0 & 0 & 4h & 0 \\ 2h & -(8/5)h & 4ah & -2h & (8/5)h & 4ah & -12a \\ -2h & 0 & 0 & 2h & 0 & 0 & 4a \end{pmatrix} . \tag{101}$$

For an isotropic web, the strain energy can be calculated from

$$\frac{1}{2}\int_{-a}^{a}\int_{-h}^{h}\left(\frac{\sigma_x^2}{Et}+\frac{\tau_{xz}^2}{Gt}\right)dxdy\ .$$

After substitution of assumptions (89) the F matrix can be obtained and, in case of a constant thickness t, turns out to be

$$F = \frac{4a}{ht}\begin{pmatrix} \frac{12}{E} & 0 & 0 \\ 0 & \frac{4a^2}{E}+\frac{9h^2}{5G} & -\frac{h^2}{G} \\ 0 & -\frac{h^2}{G} & \frac{h^2}{G} \end{pmatrix} . \qquad (102)$$

The web stiffness matrix follows from inversion and matrix multiplications as $K = CF^{-1}C^T$. These operations can be carried out numerically; there is also no difficulty in carrying them out analytically.

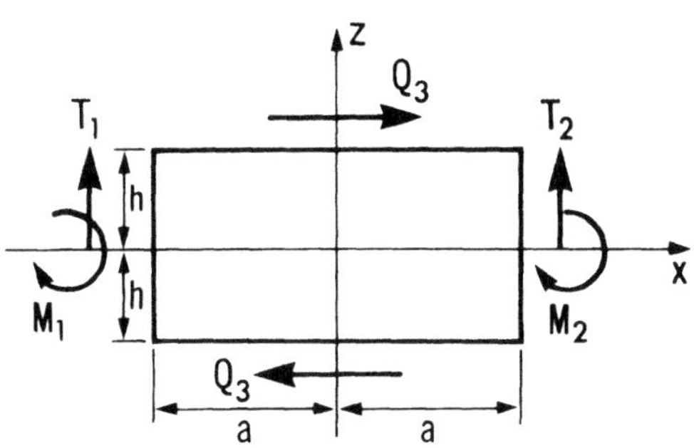

Figure 13

5. PROPERTIES OF ELEMENT STIFFNESS MATRICES

The complete system of equations

$$g = Kq \quad , \quad K = K^T \; , \tag{103}$$

between the generalized loads and displacements of an element is always singular, since the corresponding homogeneous system

$$Ku = 0 \; , \tag{104}$$

possesses amongst its non vanishing solutions at least the rigid body displacement modes of the element:

$$Ku_{(i)} = 0 \qquad (i = 1,2, \ldots, r) \; . \tag{105}$$

In displacement models this is a consequence of the inclusion *a priori* of the rigid modes in the original parametric field. In the case of equilibrium models it is a consequence of equilibrium properties of the generalized loads, calculated from an original equilibrium field. Indeed, considering the homogeneous system

$$C^T z = 0 \; , \tag{106}$$

and in view of the structure (76) of the stiffness matrix, it is observed that the non trivial solutions

$$C^T z_{(j)} = 0 \; , \quad (j = 1, 2, \ldots, s) \; , \tag{107}$$

are also solutions of equation (104). Moreover, denoting by

$$v_{(j)} = -R^T z_{(j)} \; , \tag{108}$$

the body load displacements due to the boundary displacements $z_{(j)}$, it is found that, by virtue of equation (78)

$$z^T_{(j)}g + v^T_{(j)}p = 0 \ . \tag{109}$$

These equations express overall equilibrium properties of the generalized loads in the form of vanishing virtual work equations.

The difference between element stiffness matrices derived from displacement models and from equilibrium models is that, in the latter case, there may exist $z_{(j)}$ modes that are not of the rigid body type. It will be remembered that for displacement models it is exactly the opposite: there may be more field parameters than generalized coordinates required for connection purposes. A good example of an equilibrium model where generalized loads exceed in number the stress field parameters by more than the rigid body freedoms is the spar bending model of Section 4.4. An independent set of $z_{(j)}$ modes derived from equation (107) is as follows:

$$z^T_{(1)} = (1 \quad 0 \quad 0 \quad 1 \quad 0 \quad 0 \quad 0)$$
$$z^T_{(2)} = (a \quad 0 \quad 1 \quad -a \quad 0 \quad 1 \quad h)$$
$$z^T_{(3)} = (0 \quad 1 \quad 0 \quad 0 \quad 1 \quad 0 \quad 0)$$
$$z^T_{(4)} = (0 \quad 5a \quad 1 \quad 0 \quad 0 \quad 1 \quad 0) \ .$$

The first is easily recognized as a vertical translation mode, the second as a rotation mode about the origin and this exhausts the rigid body freedoms if due account is taken of the symmetry of the problem. The other two modes are kinematical deformation freedoms that leave the element unstressed. They may seem to be undesirable characteristics of the element representation but represent, admittedly in an extreme form, deficiencies in deformation compatibility that must necessarily be expected from an approximate equilibrium analysis. The equilibrium equations of type (109) associated with these $z_{(j)}$ modes are respectively:

$$T_1 + T_2 = 0 \ , \quad \text{or vertical equilibrium}$$

$a(T_1 - T_2) + M_1 + M_2 + 2hQ_3 = 0$, or rotational equilibrium

$$V_1 + V_2 = 0 \ , \tag{110}$$

$$M_1 + M_2 + 5aV_1 = 0 \ . \tag{111}$$

We conclude this section by showing how, in this particular case, the kinematical deformation freedoms can be removed at the expense of some stress continuity violation at the boundaries but retaining the major benefit of the equilibrium analysis: the proviso of upper bounds to the remaining influence coefficients. The method is generally applicable whenever the strain field is integrable. This is precisely the case in our example, provided the web thickness is constant. For then

$$\begin{aligned} Etu &= -\alpha_1 z + 6\beta_1 \frac{xz}{h^2} + \beta_2 \left[3\frac{zx^2}{h^2} - (2+\nu)\frac{z^3}{h^3} \right] + 2(1+\nu)\beta_3 z \ , \\ Etw &= \alpha_2 + \alpha_1 x - 3\beta_1 \left[\frac{x^2}{h^2} + \nu \frac{z^2}{h^2} \right] - \beta_2 \left[\frac{x^3}{h^2} + 3\nu \frac{xz^2}{h^2} \right], \end{aligned} \tag{112}$$

is a displacement field associated to (89); α_1 and α_2 are integration constants representing the rigid body modes. When this field is substituted into the definitions (91 - 100), the seven generalized displacements are correlated with the five parameters (α_1, α_2, β_1, β_2, β_3). By eliminating the parameters it turns out that we should have

$$\begin{aligned} v_2 + v_1 &= \nu \frac{h^2}{6a} (\phi_2 - \phi_1) \\ v_2 - v_1 &= \nu \frac{h(\phi_1+\phi_2) - 2u_3}{2\left[\frac{a}{h} + \frac{2+\nu}{5}\frac{h}{a}\right]} \ . \end{aligned} \tag{113}$$

Hence v_1 and v_2 are dependent Poisson ratio effects and, conversely, as can be seen from equations (110 - 111) the conjugate loads are determined by the bending moments.

Inasmuch as the equilibrium theory considers the seven generalized displacements to be independent, equations (113) are generally unsatisfied. This shows more explicitly how the kinematical deformation freedoms are of the nature of compatibility violations. Compatibility can be restored as follows: the original minimum principle can be written

$$\frac{1}{2}\, b^T F b - (w_1T_1+w_2T_2+M_1\phi_1+M_2\phi_2+2Q_3u_3) - (v_1V_1+v_2V_2)\ , \quad \text{minimum}$$

where the F matrix is given by expression (102), the generalized loads are to be expressed in terms of the stress parameters and the generalized displacements are to be prescribed. This is now modified by calculating the last bracket in terms of stress parameters; from equations (97 - 99) one has

$$v_1V_1 + v_2V_2 = \frac{8}{5}\, h\beta_2(v_2-v_1)\ .$$

But also from the displacement field (112) and definitions (98 - 100)

$$v_2 - v_1 = \frac{2\nu a}{Et}\, \beta_2\ ,$$

so that finally

$$v_1V_1+v_2V_2 = \frac{1}{2}\left(\frac{32}{5}\,\nu\,\frac{ah}{Et}\right)\beta_2^2\ .$$

Substitution of this into the original principle gives

$$\frac{1}{2}\, b^T H b - (w_1T_1+w_2T_2+\phi_1M_1+\phi_2M_2+2u_3Q_3)\ , \quad \text{minimum}$$

with a modified flexibility matrix

$$H = F - \frac{32\nu ah}{5Et}\begin{pmatrix} 0 & 0 & 0 \\ 0 & 1 & 0 \\ 0 & 0 & 0 \end{pmatrix} = \frac{8a}{Eth}\begin{pmatrix} 6 & 0 & 0 \\ 0 & \left(2a^2 + \frac{9+5\nu}{5}h^2\right) & -(1+\nu)h^2 \\ 0 & -(1+\nu)h^2 & (1+\nu)h^2 \end{pmatrix}\ . \tag{114}$$

The generalized loads matrix is now reduced to

$$g^T = (T_1,\ M_1,\ T_2,\ M_2,\ 2Q_3)\ ,$$

with a connection matrix B^T, deduced from (101) by suppression of the second and fifth columns leading to the new stiffness matrix $BH^{-1}B^T$. As the representation still follows from a minimum complementary energy principle, the upper bound character of the influence coefficients is preserved. The advantages of the new representation are: the reduction in size of the stiffness matrix, the elimination of the kinematical deformation freedoms and the existence of a correlated internal displacement field. In counterpart there can exist a violation in stress continuity in the form of different amplitudes of the parabolic shear distributions $(1 - 3z^2/h^2)$ between the edges of vertical web junctions. However equilibrium is restored if one conceives this difference to be artificially absorbed by an external load distribution. Since this local load is statically equivalent to zero, it can be expected, on the basis of de Saint-Venant's principle, that its effect on the deflections and slopes will be small.

6. PROPERTIES OF CONNECTED ELEMENTS STIFFNESS MATRICES

From a kinematic standpoint the connections between elements are defined by relating their generalized displacements to the nodal displacements of the structure. Ordering the various elements by means of a subscript (k), so that $q_{(k)}$ denotes the displacements matrix of the kth element, and denoting simply by q the column matrix of nodal displacements, the kinematic connections can be written in the form

$$q_{(k)} = L_k q \tag{115}$$

In this equation L_k is the localizing matrix of element k. Often the element displacements are all oriented with respect to a common axes system; the localizing matrix is then simply composed of zero or unit elements and serves to identify each element displacement component with some, similarly defined, nodal displacement

However, in the more general case, the localizing matrix can incorporate the changes in orientation required for conformity with the definitions of the nodal displacements.

The force relationships due to the connections follow directly from consideration of the virtual work of the external loads acting at the level of the nodes. If g denotes the column matrix of these loads, conjugate to the nodal displacements, and $g_{(k)}$ the generalized loads on the isolated kth element, the following virtual work equality must hold for an arbitrary displacement matrix q,

$$g^T q = \sum_k g^T_{(k)} q_{(k)} = \sum_k g^T_{(k)} L_k q \ . \tag{116}$$

Whence

$$g^T = \sum_k g^T_{(k)} L_k \quad \text{or}$$

$$g = \sum_k L^T_k g_{(k)} \ . \tag{117}$$

This equation shows how the loads on the isolated elements add up to balance the external loads on the nodes. Because of the rule adopted in equilibrium models to attach each generalized quantity to a single boundary, the summation for any row in equation (117) really extends only to two subscript values. If the corresponding external load happens to be zero, we then have a statement of reciprocity of action between two elements. In view of the identity of stress distributions represented by the reciprocal element loads, this is equivalent to complete continuity in the stress transmission. When the external load component is present the stress continuity is enforced only if the external load possesses the same type of distribution.

In displacement models, while displacement continuity is built in *a priori*, the summations generally extend to more than two subscript values and stress continuity is not enforced.

In this case equations (117) are easily interpreted in terms of weight functions applicable to the arbitrary external loading of the structure for conversion into equivalent nodal forces.

Referring back to equation (78) and writing it for each element

$$g_{(k)} = K_k q_{(k)} + R_k p_{(k)} , \tag{118}$$

we obtain, after substitution into equation (117)

$$g = Kq - p \tag{119}$$

where

$$K = \sum_k L_k^T K_k L_k , \tag{120}$$

$$-p = \sum_k L_k^T R_k p_{(k)} . \tag{121}$$

In these equations p represents a column matrix of external body loads converted into equivalent nodal forces, K is the complete stiffness matrix of the unsupported structure. It is seen to be built up by properly addressing the element matrices by means of the localizing matrices.

In building the complete stiffness matrix by connecting displacement models, no kinematic deformation modes are generated that do not properly belong to the structure. With equilibrium models however artificial kinematic freedoms can appear. This was already shown to happen within a single element; it can also result from relative motions between elements. Special care must therefore be exercised with respect to the pattern of subdivision into equilibrium models.

It should be observed that a structural representation containing such unreal kinematic freedoms is not necessarily to be rejected. The restrictions put on the external loads to avoid production of virtual work in the kinematic modes can be acceptable.

It can happen, for instance, that these modes only displace unloaded nodes. A simple example is provided by the quadrilateral panel when subdivided into triangular elements by the diagonals, each triangle being treated as an equilibrium model according to the analysis of Section 4.2.

For the purpose of visualizing the kinematic mode generated by this subdivision (Figure 14), the triangular panels can be assumed to be pin-jointed at the middle of their edges, since for unstressed panels the displacements of these points are the averages of the edge displacements and comply with the definitions of the generalized displacements. The unstressed quadrilateral can obviously behave as a four bar linkage but it can be proved by geometry [14] that the points, E, F, G and H can be taken as instantaneous rotation centers. Consequently the quadrilateral can transmit any self equilibrated load system applied to the nodes E, F, G and H. The internal nodes I, J, K and L should either remain unloaded or their load system restricted by a virtual work condition. The impact of this on the mathematical process of elimination of the unloaded nodes is worth investigating. By splitting the nodal displacements and nodal loads matrices in two parts: $q_{(1)}$ and $g_{(1)}$ for the fixed nodes and $q_{(2)}$ and $g_{(2)}$ for the others, the following stiffness equations can be written

$$g_{(1)} = K_{11}q_{(1)} + K_{12}q_{(2)} \quad , \tag{122}$$

$$g_{(2)} = K_{21}q_{(1)} + K_{22}q_{(2)} \quad . \tag{123}$$

By assumption the kinematic mode is of the type

$$q_{(1)} = 0 \quad , \qquad q_{(2)} = m$$

and, since it exists without external loads,

$$K_{12}m = 0 \quad , \tag{124}$$

$$K_{22}m = 0 \quad . \tag{125}$$

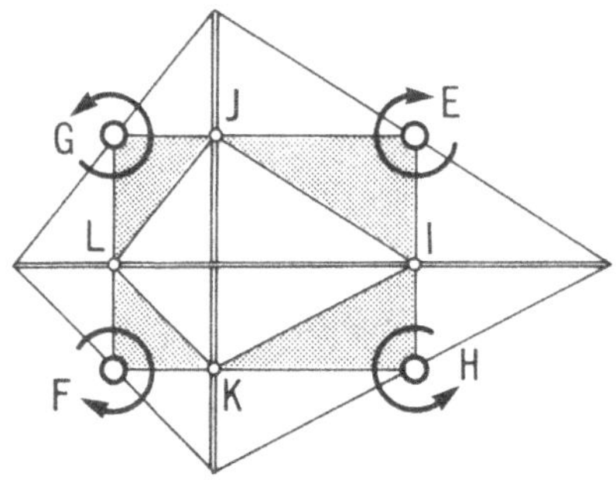

Figure 14

If the $g_{(2)}$ loads exist

$$g^T_{(2)}m = q^T_{(1)}K^T_{21}m + q^T_{(2)}K^T_{22}m \ .$$

But, since $K^T_{21} = K_{12}$ and $K^T_{22} = K_{22}$, it follows from equations (124 - 125) that

$$g^T_{(2)}m = 0 \ . \tag{126}$$

This is the virtual work equation restricting these loads. If however the nodes affected by the kinematic mode are unloaded, that is $g_{(2)} = 0$, we shall generally wish to eliminate the conjugate displacements by solving the system

$$K_{22}q_{(2)} = -K_{21}q_{(1)} \tag{127}$$

and substituting for $q_{(2)}$ in equation (122). Although the homogeneous adjoint to equation (127) has the non trivial solution $q_{(2)} = m$, the second member has the orthogonality property

$$q^T_{(1)}K^T_{21}m = 0 \ ,$$

required for the existence of a solution. The general solution

$$q_{(2)} = Pq_{(1)} + \alpha m$$

depends on the arbitrary scalar α. However, in view of equation (124) the final result does not depend on it

$$g_{(1)} = (K_{11}+K_{12}P)q_{(1)} , \tag{128}$$

and holds for any particular matrix P satisfying

$$K_{22}P = -K_{21} . \tag{129}$$

A particular matrix can be found by preventing the kinematic deformation to take place, e.g., in the case of the quadrilateral by adding a rigid rod between JK or LI. In this matter a displacement component of node J (or L) is related to one of node K (or I) and system (127) reduced to one with a non singular matrix; the discarded equation being automatically satisfied.

7. SOLUTION BY SELF-STRAININGS

The theory of equilibrium models developed in Section 4.1 furnishes stiffness matrices and allows in principle a solution of the overall structural problem by the same stiffness methods and the same computer programmes as for displacement models. It is however true that the requirements of stress continuity and the resulting definitions of generalized variables increase considerably the number of nodal displacements. As a consequence the size of the complete stiffness matrix can be much larger than in the case of a structure analyzed in the same number of displacement models. Problems of conditioning of the stiffness equations and accuracy of the numerical procedures are raised thereby, which only experience can help to evaluate.

As a matter of fact the topology of the connections between elements is generally such that displacements are better choices for the unknowns in a displacement model analysis and forces in an equilibrium model analysis. When the internal forces themselves are chosen it is better to look for self-equilibrating combinations, whose intensities constitute the generalized unknowns. This section is devoted to a short mathematical description of this procedure, using the matrices previously defined.

We start by expressing the external loads on the nodes in terms of the stress parameters of the elements. This is achieved by substituting the relations (75)

$$g_{(k)} = C_k b_{(k)} + A_k p_{(k)} , \tag{130}$$

into the equilibrium equation (5.3)

$$g = \sum_k L_k^T C_k b_{(k)} + \sum_k L_k^T A_k p_{(k)} . \tag{131}$$

From equation (121) we also obtain equivalent nodal forces due to the external body loads and transform the expression by means of the definition (77) of the R matrices

$$p = \sum_k L_k^T C_k F_{bbk}^{-1} F_{bpk} p_{(k)} - \sum_k L_k^T A_k p_{(k)} . \tag{132}$$

Adding (131) and (133) and denoting by

$$f = g + p , \tag{133}$$

the total nodal external forces, we place the result in the form

$$\sum_k L_k^T C_k b_{(k)} = f - \sum_k L_k^T C_k F_{bbk}^{-1} F_{bpk} p_{(k)} \tag{134}$$

of a linear system to determine the unknowns $b_{(k)}$. The general solution is of the form

$$b_{(k)} = b^0_{(k)} + \sum_i x_i b^i_{(k)} \ . \tag{135}$$

The particular solution is of the type

$$b^0_{(k)} = M_k (f - \sum_k L^T_k C_k F^{-1}_{bbk} F_{pbk} p_{(k)}) \ . \tag{136}$$

The other terms are the general solution without second member

$$\sum_k L^T_k C_k b^i_{(k)} = 0 \qquad i = 1, 2, \ldots, s \ . \tag{137}$$

They represent the stress parameters of self-equilibrating states of stress, whose unknown intensities x_i are taken to be the generalized quantities to determine in order to satisfy the compatibility requirements. Introducing the matrices

$$B_k = (b^1_{(k)}, b^2_{(k)}, \ldots, b^s_{(k)}) \ ,$$

$$x^T = (x_1, x_2, \ldots, x_s) \ ,$$

the solution (135) takes the form

$$b_{(k)} = M_k (f - \sum_k L^T_k C_k F^{-1}_{bbk} F_{bpk} p_{(k)}) + B_k x \ , \tag{138}$$

it is a complete description of the possible equilibrium states of the structure. We substitute it into the strain energy (73) of an element

$$\frac{1}{2} b^T_{(k)} F_{bbk} b_{(k)} + b^T_{(k)} F_{bpk} p_{(k)} + \frac{1}{2} p^T_{(k)} F_{ppk} p_{(k)} \ ,$$

add over all the elements and obtain, after cancellation of terms, the following expression of the total strain energy

$$\psi = \frac{1}{2} f^T [\delta_{00}] f + x^T [\delta_{10}] f + \frac{1}{2} x^T [\delta_{11}] x + \frac{1}{2} \sum_k p^T_{(k)} F_{ppk} p_{(k)} \ , \tag{139}$$

where

$$[\delta_{00}] = \sum_k M_k^T F_{bbk} M_k \tag{140}$$

$$[\delta_{10}] = \sum_k B_k^T F_{bbk} M_k \tag{141}$$

$$[\delta_{11}] = \sum_k B_k^T F_{bbk} B_k \ . \tag{142}$$

From equations (119) that can be written $f = Kq$, the nodal displacements matrix q is conjugate to the loads matrix f and the virtual work of external loads if $f^T q$. We can apply the complementary energy principle in the form

$$\delta(\psi - f^T q) = 0 \ , \tag{143}$$

where the matrix q is prescribed and f considered as a reaction loads matrix.

Taking variations on the selfstraining intensities matrix x produces the stationarity condition

$$[\delta_{11}]x + [\delta_{10}]f = 0 \quad \text{or}$$

$$x = -[\delta_{11}]^{-1}[\delta_{10}]f \ . \tag{144}$$

The stresses are then determined in the whole structure in terms of the external forces. Taking variations on the reaction loads f

$$[\delta_{10}]^T x + [\delta_{00}]f = q$$

or, in view of equation (144)

$$q = If \tag{145}$$

with the influence coefficients matrix

$$I = [\delta_{00}] - [\delta_{10}]^T[\delta_{11}]^{-1}[\delta_{10}]. \tag{146}$$

Our last step implies that the elements of f can be varied independently; this of course is only the case when the complete structure has been supported against rigid body modes by preventing some of the nodal displacements and when the f matrix does not include the reactions against the supports. Those reactions can be determined from overall equilibrium considerations. Should the structure be subjected to additional prescribed displacements (externally hyperstatic) the corresponding unknown reactions can be obtained from the set of equations (145).

As already mentioned the size of the column matrix x is generally much smaller than that of q. The advantage of the self-strainings method is then that it requires only the inversion of the square matrix $[\delta_{11}]$ of the same size as x. The difficulty lies in the determination of the simplest matrices $b^{i}_{(k)}$ or B_k. A great step forward in the automation of the calculations would be achieved if the computer itself could be taught to investigate the topology of the matrices $L^T_k C_k$ and deduce the self-strainings confined to the smallest number of elements.

8. CONVERGENCE

Because the solutions based on displacement models depend on a minimum principle, the lower bounds on the influence coefficients can only be raised by further subdivision of the structure, provided the original system of displacements be contained as a particular case of the new one. This point has been stressed by Melosh [18]. The requirement is however automatically satisfied when the type of element analysis is not altered. For along a new subdividion line the original displacement field behaves with the same polynomial approximation (linear, quadratic or higher) as it will be endowed with in the new analysis. The only difference is that some of the coefficients of the polynomial, which were originally dependent, become independent.

The same considerations apply to the analyses with equilibrium models where further subdivision can only lower the upper bounds on influence coefficients.

This monotonic convergence is however insufficient to guarantee convergence to the true solution and it is obvious that the subdivision pattern must play an important role in the convergence process. The interest of the dual method of analysis is precisely that by comparison between upper and lower bounds the degree of true convergence can be ascertained.

9. BOUNDS TO THE DIRECT INFLUENCE COEFFICIENTS

9.1 *Lower Bounds by the Minimum Total Energy Principle*

The displacements are assumed to verify some homogeneous boundary conditions

$$u = 0 \text{ and/or } v = 0 \quad \text{on} \quad \Gamma_u \tag{147}$$

so that the structure is at least isostatically supported and can resist the application of unit loads.

Since the first variation of the total energy must vanish

$$\iint (\sigma_x \delta\varepsilon_x + \tau_{xy}\delta\gamma_{xy} + \sigma_y \delta\varepsilon_y)dxdy - \iint (\bar{X}\delta u + \bar{Y}\delta v)dxdy$$

$$- \int_{\Gamma_\sigma} (\bar{p}_x \delta u + \bar{p}_y \delta v)ds = 0 \; . \tag{148}$$

Consider the special variation ($\delta\varepsilon$ a constant)

$$\delta u = u\delta\varepsilon \; , \quad \delta v = v\delta\varepsilon \; , \tag{149}$$

where u and v are the final displacement components given by application of the principle. This special variation is compatible with (147) and gives also

$$\delta\varepsilon_x = \delta \frac{\partial u}{\partial x} = \frac{\partial}{\partial x}(\delta u) = \delta\varepsilon\, \varepsilon_x$$

and similarly

$$\delta\varepsilon_y = \delta\varepsilon\, \varepsilon_y \quad , \quad \delta\gamma_{xy} = \delta\varepsilon\, \gamma_{xy} \; .$$

Substitution into (148) produces

$$\iint_\Delta (\varepsilon_x\sigma_x+\gamma_{xy}\tau_{xy}+\varepsilon_y\sigma_y)dxdy - \iint_\Delta (\bar{X}u+\bar{Y}v)dxdy - \int_{\Gamma_\sigma} (\bar{p}_x u+\bar{p}_y v)ds = 0 \; .$$

In view of equation (16) this is a form of Clapeyron's theorem

$$U = -\frac{1}{2} P = \frac{1}{2} \iint_\Delta (\bar{X}u+\bar{Y}v)dxdy + \frac{1}{2} \int_{\Gamma_\sigma} (\bar{p}_x u+\bar{p}_y v)ds \; . \tag{150}$$

This special variation certainly exists if no approximations are made; hence for the exact solution

$$U_e = -\frac{1}{2} P_e \; . \tag{151}$$

The special variation also exists for approximate fields of the assumed type (26) as it is possible to take $\delta\alpha_i = \delta\varepsilon\, \alpha_i$. Hence, for the approximate solution

$$U_a = -\frac{1}{2} P_a \; . \tag{152}$$

Because the total energy of the exact solution must be an absolute minimum

$$U_a + P_a \geq U_e + P_e$$

and then, by virtue of equation (151 - 152)

$$P_a \geq P_e \; . \tag{153}$$

Let F be a prescribed generalized load conjugate to the displacement q. Then $P_a = -\bar{F}q_a$ and $P_e = -\bar{F}q_e$, whereby (153) gives

$$\bar{F}q_a \leq \bar{F}q_e \ .$$

The proof of the lower bound character of the approximate influence coefficient follows after division by $\bar{F}$

$$\frac{q_a}{\bar{F}} \leq \frac{q_e}{\bar{F}} \ . \tag{154}$$

9.2 *Upper Bounds by the Minimum Complementary Energy Principle*

Since the prescribed displacements on the boundary, where the structure is supported are zero, the principle reduces to

$$\iint_{\Delta} \phi(\sigma)dxdy - \bar{q}F = \psi - \bar{q}F \quad \text{minimum} \ . \tag{155}$$

The generalized displacement is now prescribed, the conjugate load a reaction. Vanishing of the first variation yields

$$\iint_{\Delta} (\varepsilon_x\delta\sigma_x+\gamma_{xy}\delta\tau_{xy}+\varepsilon_y\delta\sigma_y)dxdy - \bar{q}\delta F = 0 \ . \tag{156}$$

Consider the special variations

$$\delta\sigma_x = \sigma_x \frac{\delta F}{F} \ , \quad \delta\tau_{xy} = \tau_{xy}\frac{\delta F}{F} \ , \quad \delta\sigma_y = \sigma_y \frac{\delta F}{F} \ ,$$

that certainly exist for the exact solution and also for the approximate stress fields of type (66), where one can set

$$\delta\beta_i = \beta_i \frac{\delta F}{F} \ , \quad \delta P_r = P_r \frac{\delta F}{F} \ .$$

For such special variations, condition (156) gives

$$\frac{1}{F} \iint_{\Delta} (\sigma_x \varepsilon_x + \tau_{xy} \gamma_{xy} + \delta_y \varepsilon_y) dxdy - \bar{q} = 0$$

or, in view of equation (16), again a form of Clapeyron's theorem

$$\psi = \frac{1}{2} \bar{q} F .$$

From the statement (155) it follows that

$$\psi_a - \bar{q} F_a \geq \psi_e - \bar{q} F_e$$

and since

$$\psi_a = \frac{1}{2} \bar{q} F_a \quad \text{and} \quad \psi_e = \frac{1}{2} \bar{q} F_e$$

$$\bar{q} F_a \leq \bar{q} F_e$$

ψ_a and ψ_e being strain energies are both positive, so that F_a and F_e have both the sign of $\bar{q}$. Division by the positive quantity $F_a F_e$ proves the upper bound character of the approximate influence coefficient:

$$\frac{\bar{q}}{F_a} \geq \frac{\bar{q}}{F_e} . \tag{157}$$

The derivation of bounds to the cross influence coefficients is a straight-forward algebraic process [11, 12, 13, 14].

REFERENCES

[1] COURANT, R. and HILBERT, D., *Methoden der Mathematischen Physik*, Vol. 1, Chap. IV, Springer, Berlin, 1937, p. 199.

[2] BOUCKAERT, L.P., *Théorie de L'Elasticité*, Chap. X, Wouters, Louvain, 1960.

[3] FRAEIJS DE VEUBEKE, B.M., "Diffusion des inconnues Hyperstatiques dans les voilures à longerons couplés", *Bull. 24 Serv. Tech. Aeron.*, Bruxelles, 1951.

[4] REISSNER, E., "On a Variational Theorem in Elasticity", *J. Math. Phys.*, Vol. 29, 1950, p. 90.

[5] REISSNER, E., *J. Math. Phys.*, Vol. 23, 1944, p. 184.

[6] FRAEIJS DE VEUBEKE, B.M., "Flexion et extension des plaques d'épaisseur modérée", *Acad. Roy. Belg. Classe. Sci. Mem.*, 1959, 1698 p.

[7] KELLEY, J.L., *General Topology*, Van Nostrand, 1955, 118 p.

[8] SYNGE, J.L., *The Hypercircle in Mathematical Physics*, Cambridge University Press, 1957, 336 p.

[9] TREFFTZ, E., "Ein gegenstuck zum Ritzchen verfahren", *Proc. 2nd Int. Congress Appl. Mech.*, Zurich, 1926.

[10] PRAGER, W. and SYNGE, J.L., "Approximations in Elasticity Based on the Concept of Function Space", *Quart. Appl. Math.*, Vol. 5, 1947.

[11] WEBER, C., "Eingrenzung von verschiebungen und zerrungen mit hilfe der minimalsatze", *Z.A.M.M.*, Vol. 20, 1952, p. 130.

[12] BENTHEM, J.P., "On the Analysis of Swept Wing Structures", *Tech. Rept., S.578, Nat. Aero. Astron. Res. Inst.*, Amsterdam, 1961.

[13] FRAEIJS DE VEUBEKE, B.M., "Sur certaines inégalités fondamentales et leur généralisation dans la théorie des bornes supérieures et inférieures en élasticité", *Rev. Universelle Mines*, 1961.

[14] FRAEIJS DE VEUBEKE, B.M., "Upper and Lower Bounds in Matrix Structural Analysis", *AGARDograph*, Vol. 72, Pergamon Press, 1964, 165 p.

[15] TURNER, M.J., CLOUGH, R. W., MARTIN, H.C. and TOPP, L.J., "Stiffness and Deflexion Analysis of Complex Structures", *J. Aeron. Sci.*, 805, September, 1956.

[16] GALLAGHER, R.H. and RATTINGER, I., "The Experimental and Theoretical Determination of the Elastic Characteristics of Modern Airframes", *AGARD Rept.* 400, 1960.

[17] GALLAGHER, R.H., "A Correction Study of Methods of Matrix Structural Analysis", *AGARDograph*, Vol. 69, Pergamon Press, 1963.

[18] MELOSH, R.J., "Basis for Derivation of Matrices for the Direct Stiffness Method", *AIAA Journal*, Vol. 1, No. 7, 1963.

[19] BESSELING, J.F., "The Complete Analogy Between the Matrix Equations and the Continuous Field Equations of Structural Analysis", *Int. Sym. on Analogue and Digital Techniques Applied to Aeron.*, Liège, 1963.

VARIATIONAL PRINCIPLES IN FLUID MECHANICS

B.M. Fraeijs de Veubeke

University of Liège, Belgium

INTRODUCTION

There has been in recent years renewed interest for the powerful variational methods in fluid mechanics. Surveys of known applications to transonic and supersonic flow have been published by W. Fiszdon [10, 11].

The use of a lagrangean description of the flow, coupled with lagrangean variations of particle displacements, provides the most natural formalism for a variational principle. It is a natural extension of the powerful methods introduced in mechanics by Lagrange, Hamilton and Jacobi. The only difficulty is the introduction of the side condition of conservation of particle mass when passing from a discrete system to the continuous medium. Most of the older work on variational principles is in this direction [2, 4, 5]; an excellent account of this approach is to be found in a more recent work of Eckart [8].

For practical applications however variational principles based on the eulerian description of fields, coupled with eulerian,

Original publication, 7ème Conférence Scientifique sur la "Mécanique des Fluides", Académie Polonaise Des Sciences, Jurata, 1-8 September, 1965.

or local, variations are more useful. Such principles have first been constructed, rather artificially, to yield the known equations of conservation in eulerian form [3]. The first attempt to derive them logically from Hamilton's principle seems to be due to Herivel [6]. As reported by Serrin [7], Herivel's principle, which yields only irrotational flow, was completed by Lin.

The present paper attempts a completely logical derivation from Hamilton's principle to a purely eulerian principle. It is an extension of an earlier work [9] where the same difficulty of removing the restriction to irrotational flow was dealt with rather artificially. A logical step, which departs from the procedure adopted by Lin, has been taken. Since independent variations are to be taken on the velocity field, it is only natural to remove the constraints implying that the velocity is the derivative of the particle's position vector. This introduces a new vector multiplier enjoying a property of constant circulation along any segment fixed in the flow. This procedure yields almost immediately a general variational principle than can easily be specialized to the principles given by Bateman.

1. LAGRANGEAN AND EULERIAN VARIATIONALS

Let

$$x_i = X_i(a_j, t, \varepsilon) , \quad i = 1,2,3, \quad j = 1,2,3, \tag{1.1}$$

denote the cartesian coordinates of a particle at time t; each particle is identified by its parameters, or lagrangean coordinates, a_j. Equations (1.1) describe a family of flows; the true flow corresponding to $\varepsilon = 0$, varied flows to $\varepsilon \neq 0$.

The displacement during the interval dt in true motion is

$$d_a x_i = \left(\frac{\partial X_i}{\partial t}\right)_{\varepsilon=0} dt , \quad i = 1,2,3. \tag{1.2}$$

Similarly, the lagrangean variation of displacement, is defined as

$$\delta_a x_i = \left(\frac{\partial X_i}{\partial \varepsilon}\right)_{\varepsilon=0} d\varepsilon \ , \qquad i = 1,2,3. \tag{1.3}$$

Both notations suggest that the lagrangean coordinates are kept fixed so that both displacements refer to the same particle.

More generally, let

$$f = F(a_j, t, \varepsilon) \tag{1.4}$$

denote an intensive variable of the field, then

$$d_a f = \left(\frac{\partial F}{\partial t}\right)_{\varepsilon=0} dt \tag{1.5}$$

is the material or total change due to true motion:

$$\delta_a f = \left(\frac{\partial F}{\partial \varepsilon}\right)_{\varepsilon=0} d\varepsilon \tag{1.6}$$

is the *lagrangean variation.*

Obviously the operators d_a and δ_a commute:

$$\delta_a d_a f = d_a \delta_a f = \left(\frac{\partial^2 F}{\partial \varepsilon \partial t}\right)_{\varepsilon=0} d\varepsilon dt \tag{1.7}$$

and commute with partial derivatives with respect to the lagrangean coordinates. Expression (1.4) is the lagrangean description of the field of the intensive variable f. By solving equations (1.1) for the lagrangean coordinates

$$a_j = a_j(x_i, t, \varepsilon) \tag{1.8}$$

and substituting in (1.4) one obtains in principle the eulerian description of the field:

$$f = F(a_j(x_i, t, \varepsilon), t, \varepsilon) = f(x_i, t, \varepsilon) \ . \tag{1.9}$$

This naturally suggests corresponding definitions for the changes in f when the cartesian coordinates are kept fixed:

$$d_x f = \left(\frac{\partial f}{\partial t}\right)_{\varepsilon=0} dt \tag{1.10}$$

is the local change in f due to true motion,

$$\delta_x f = \left(\frac{\partial f}{\partial \varepsilon}\right)_{\varepsilon=0} d\varepsilon \tag{1.11}$$

is the local or *eulerian variation* of the field.

The operators d_x and δ_x commute with each other and with the partial derivatives with respect to the cartesian coordinates.

In subsequent calculations general use will be made of the eulerian description of fields. The connection between the lagrangean and eulerian changes will therefore be explicited as follows:

$$d_a f = d_x f + \sum_1^3 \frac{\partial f}{\partial x_i} d_a x_i \ , \tag{1.12}$$

$$\delta_a f = \delta_x f + \sum_1^3 \frac{\partial f}{\partial x_i} \delta_a x_i \, . \tag{1.13}$$

When the more common notations

$$\frac{df}{dt} = \frac{d_a f}{dt} \ , \qquad \frac{\partial f}{\partial t} = \frac{d_x f}{dt}$$

are used for the material and local time-derivatives, (1.12) reduces to the known expression

$$\frac{df}{dt} = \frac{\partial f}{\partial t} + \underline{u} \cdot \text{grad } f \ , \tag{1.14}$$

where $u_i = \dfrac{d_a x_i}{dt}$ are the velocity-field components.

Similarly, (1.13) can be written

$$\delta_a f = \delta_x f + \delta_a \underline{x} \cdot \text{grad } f \ . \tag{1.15}$$

Observe that, by their very definitions

$$\frac{da_j}{dt} = \frac{\partial a_j}{\partial t} + \underline{u} \cdot \text{grad } a_j = 0 \, ,$$

$$j = 1,2,3.$$

$$\delta_a a_j = \delta_x a_j + \delta_a \underline{x} \cdot \text{grad } a_j = 0 \, ,$$

$$\frac{\partial x_i}{\partial t} = u_i - \underline{u} \cdot \text{grad } x_i = 0 \, ,$$

$$i = 1,2,3.$$

$$\delta_x x_i = \delta_a x_i - \delta_a \underline{x} \cdot \text{grad } x_i = 0 \, ,$$

We need corresponding results for the volume element

$$dv = J da_1 da_2 da_3 \, , \quad J = \frac{D(X_1, X_2, X_3)}{D(a_1, a_2, a_3)} \, .$$

The change in volume for the same aggregate of particles will be

$$\frac{d}{dt} dv = \left(\frac{d}{dt} J\right) da_1 da_2 da_3 \, .$$

In calculating the derivative of the jacobian determinant use is made of the following expansions

$$\frac{d_a}{dt} \frac{\partial X_j}{\partial a_i} = \frac{\partial}{\partial a_i} \frac{d_a x_j}{dt} = \frac{\partial u_j}{\partial a_i} = \sum_{m=1}^{3} \frac{\partial u_j}{\partial x_m} \frac{\partial X_m}{\partial a_i} \, .$$

Rearrangement of the terms produces the well known result

$$\frac{d}{dt} J = J \text{ div } \underline{u}$$

so that

$$\frac{d}{dt} dv = dv \text{ div } \underline{u} \, . \tag{1.16}$$

Similarly for the lagrangean variation

$$\delta_a dv = dv \text{ div } \delta_a \underline{x} \, . \tag{1.17}$$

Again, by definition,

$$\frac{\partial}{\partial t} dv = 0 , \quad \delta_x dv = 0 . \tag{1.18}$$

We are now in a position to evaluate the changes occuring in any extensive quantity fdv:

$$\frac{d}{dt}(fdv) = \frac{df}{dt} dv + f \frac{d}{dt} dv = \left(\frac{df}{dt} + f \operatorname{div} \underline{u}\right) dv . \tag{1.19}$$

But also, in view of (1.14)

$$\frac{d}{dt}(fdv) = \left(\frac{\partial f}{\partial t} + \operatorname{div}(f\underline{u})\right) dv . \tag{1.20}$$

Hence, integrating in a volume D, bounded by a surface S of outward normal $\underline{n}$,

$$\frac{d}{dt} \int_D fdv = \frac{\partial}{\partial t} \int_D fdv + \int_S f(\underline{u}\cdot\underline{n})dS . \tag{1.21}$$

This gives the rate of change of the volume integral when the particles are followed in their true motion.

The perfectly similar calculation for variations gives

$$\begin{aligned} \delta_a(fdv) &= (\delta_a f + f\operatorname{div}\delta_a\underline{x})dv = (\delta_x f + \operatorname{div}(f\delta_a\underline{x}))dv \\ \delta_a \int_D fdv &= \delta_x \int_D fdv + \int_S f(\delta_a\underline{x}\cdot\underline{n})dS . \end{aligned} \tag{1.22}$$

From this basic formula, a variational principle involving lagrangean variations can be transformed in one involving only local variations, provided the additional surface integral can be cancelled or suitably interpreted. The variational principles of dynamics also involve time integrals. A useful transformation is then obtained from (1.21) by integration

$$\int_{t_1}^{t_2}\int_D \frac{\partial f}{\partial t} dvdt = - \int_{t_1}^{t_2} \int_S f(\underline{u}\cdot\underline{n})dSdt + \left| \int_D fdv \right|_{t_1}^{t_2} . \tag{1.23}$$

2. HAMILTON'S PRINCIPLE

We consider the case of isentropic flow of a perfect fluid.

The specific internal energy E is a function of specific mass

$$E = E(\rho) \quad \text{with} \quad \frac{dE}{d\rho} = \frac{p}{\rho^2} . \tag{2.1}$$

The specific potential energy Ω of external forces is a function of cartesian coordinates only, so that

$$\delta_x \Omega = 0 , \quad \delta_a \Omega = \text{grad}\Omega \cdot \delta_a \underline{x} . \tag{2.2}$$

The Lagrangian per unit mass will be

$$L = \frac{1}{2} (\underline{u} \cdot \underline{u}) - (E + \Omega) . \tag{2.3}$$

Hamilton's principle applies to a system that is closed with respect to mass exchange; the particles must be followed in their motion, so that the basic formulation is lagrangean in nature:

$$\delta_a \int_{t_1}^{t_2} \int_D \rho L dv dt - \int_{t_1}^{t_2} \int_{S^*} p^* (\delta_a \underline{x} \cdot \underline{n}) dS dt = 0 . \tag{2.4}$$

The boundary conditions envisaged are as follows:

(1) on a portion S* of the bounding surface the pressure is given ($p = p^*$) and the corresponding virtual work due to variation of particle displacements is included in the principle,

(2) on the remainder S-S* of the surface the fluid is gliding along flexible moving walls. At any time the variation of particle displacement must verify the geometrical constraint

$$\delta_a \underline{x} \cdot \underline{n} = 0 \quad \text{on} \quad S\text{-}S^* . \tag{2.5}$$

The only independent variation in the principle is $\delta_a \underline{x}$.

The corresponding variation on the specific mass ρ must be obtained from a separate statement of conservation of mass of a particle:

$$\delta_a(\rho dv) = 0 \,. \tag{2.6}$$

As a result of this the variational principle (2.4) assumes the equivalent form

$$\int_{t_1}^{t_2}\int_D \rho\delta_a L dv dt - \int_{t_1}^{t_2}\int_{S^*} p^*(\delta_a \underline{x}\cdot\underline{n}) dS dt = 0 \,. \tag{2.7}$$

Expanding the variation of the Lagrangian:

$$\delta_a L = \underline{u}\cdot\delta_a\underline{u} - \frac{p}{\rho^2}\,\delta_a\rho - \text{grad}\Omega\cdot\delta_a\underline{x} \,, \tag{2.8}$$

where

$$\delta_a\underline{u} = \delta_a \frac{d\underline{x}}{dt} = \frac{d}{dt}\,\delta_a\underline{x} \tag{2.9}$$

and, by virtue of (2.6) and (1.17)

$$\delta_a\rho = -\rho\,\text{div}(\delta_a\underline{x}) \,. \tag{2.10}$$

In preparation for the required integrations by parts we write finally

$$\rho\delta_a L dv = \frac{d}{dt}\,(\delta_a\underline{x}\cdot\rho\underline{u}dv) + \text{div}(p\delta_a\underline{x})dv - \delta_a\underline{x}\cdot\underline{M}dv \,,$$

where

$$\underline{M}dv = \frac{d}{dt}\,(\rho\underline{u}dv) + (\text{grad}p + \rho\,\text{grad}\Omega)dv \,. \tag{2.11}$$

Consequently, the Euler equation of the variational principle is $\underline{M}dv = 0$, a statement of conservation of momentum of a particle. It will be observed that it assumes the newtonian form of the equations of motion

$$\rho\,\frac{d\underline{u}}{dt} + \text{grad}p + \rho\,\text{grad}\Omega = 0 \tag{2.12}$$

after an independent statement of conservation of mass during true motion

$$\frac{d}{dt}(\rho dv) = 0 . \tag{2.13}$$

The boundary conditions obtained are

$$\int_{t_1}^{t_2}\int_S p(\delta_a \underline{x}\cdot\underline{n})dSdt - \int_{t_1}^{t_2}\int_{S^*} p^*(\delta_a \underline{x}\cdot\underline{n})dSdt = 0 .$$

By virtue of the geometrical constraint (2.5), they reduce correctly to

$$p - p^* = 0 \quad \text{on} \quad S^* . \tag{2.14}$$

Finally there are initial conditions

$$\int_D \delta_a \underline{x}\cdot\rho u dv = 0 \quad \text{for} \quad t = t_1 \text{ and } t = t_2 .$$

They are satisfied by the usual rule stating that the varied configurations must coincide with the true configuration in the initial and final state

$$\delta_a \underline{x} = 0 \quad \text{for } t = t_1 \text{ and } t = t_2 . \tag{2.15}$$

3. EULERIAN FORM OF HAMILTON'S PRINCIPLE

A direct application of formula (1.22) brings the Hamilton principle into the eulerian form

$$\delta_x \int_{t_1}^{t_2}\int_D \rho L dv dt + \int_{t_1}^{t_2}\int_S \rho L(\delta_a \underline{x}\cdot\underline{n})dSdt - \int_{t_1}^{t_2}\int_{S^*} p^*(\delta_a \underline{x}\cdot\underline{n})dSdt = 0 . \tag{3.1}$$

Introducing

$$E + \rho\frac{dE}{d\rho} = E + \frac{p}{\rho} = I \tag{3.2}$$

the thermodynamical enthalpy, and remembering (1.18) and (2.2), we calculate

$$\delta_x(\rho L) = \left(\frac{1}{2}\,(\underline{u}\cdot\underline{u}) - I - \Omega\right)\delta_x\rho + \rho\underline{u}\cdot\delta_x\underline{u}\ , \tag{3.3}$$

wherein we must substitute

$$\delta_x\rho = -\mathrm{div}(\rho\delta_a\underline{x})$$

$$\delta_x\underline{u} = \delta_a\underline{u} - (\delta_a\underline{x}\cdot\mathrm{grad})\underline{u} = \frac{d}{dt}\,\delta_a\underline{x} - (\delta_a\underline{x}\cdot\mathrm{grad})\underline{u}\ .$$

Hence

$$\delta_x(\rho L) = (I+\Omega)\mathrm{div}(\rho\delta_a\underline{x}) + \rho\underline{u}\cdot\frac{d}{dt}\,\delta_a\underline{x} - \frac{1}{2}\,(\underline{u}\cdot\underline{u})\mathrm{div}(\rho\delta_a\underline{x}) - \rho\underline{u}\cdot$$

$$\cdot\ (\delta_a\underline{x}\cdot\mathrm{grad})\underline{u}\ .$$

The two last terms on the last line combine into

$$-\ \mathrm{div}\left(\rho\,\frac{\underline{u}\cdot\underline{u}}{2}\,\delta_a\underline{x}\right)$$

and finally

$$\delta_x(\rho L)dv = \mathrm{div}\left[\rho\left(I+\Omega - \frac{\underline{u}\cdot\underline{u}}{2}\right)\delta_a\underline{x}\right]dv + \frac{d}{dt}\,(\rho\underline{u}\cdot\delta_a\underline{x}dv) - \delta_a\underline{x}\cdot\underline{M}dv\ ,$$

where

$$\underline{M}dv = \frac{d}{dt}\,(\rho\underline{u}dv) + \rho\mathrm{grad}(I+\Omega)dv\ .$$

After use of (2.13), the Euler equation appears in the equivalent form

$$\frac{d\underline{u}}{dt} + \mathrm{grad}(I+\Omega) = 0 \tag{3.4}$$

showing the existence of an acceleration potential.

Because

$$\rho\left(I+\Omega - \frac{\underline{u}\cdot\underline{u}}{2}\right) = p - \rho L$$

the boundary conditions of the eulerian form are identical to those of the lagrangean form.

4. GENERALIZATION OF THE EULERIAN PRINCIPLE

By incorporating the constraint on the conservation of mass:

$$\frac{\partial\rho}{\partial t} + \operatorname{div}(\rho\underline{u}) = 0 \tag{4.1}$$

with a lagrangean multiplier, the principle can be made self-supporting. This step alone, however, is known to be insufficient. In transfering to variations on the velocity field it restricts the flow to the irrotational case. If variations on the velocity field are to replace variations on particle coordinates, it is necessary to remove also the constraints

$$\frac{d\underline{x}}{dt} - \underline{u} = 0 \,. \tag{4.2}$$

This second step departs from the procedure adopted by Lin to amplify the eulerian principle formulated by Herivel [6]. It seems more satisfactory from the viewpoint of proceeding without preconceived knowledge of the results to obtain. The generalized principle is then

$$\delta_x \int_{t_1}^{t_2} \int_D \left[\rho L + \phi\left(\frac{\partial\rho}{\partial t} + \operatorname{div}(\rho\underline{u})\right) + \rho\underline{\psi}\cdot\left(\frac{d\underline{x}}{dt} - u\right)\right] dvdt$$

$$+ \int_{t_1}^{t_2} \int_S \rho L(\delta_a\underline{x}\cdot\underline{n})dSdt - \int_{t_1}^{t_2} \int_{S^*} p^*(\delta_a\underline{x}\cdot\underline{n})dSdt = 0 \,. \tag{4.3}$$

Variation of the scalar multiplier ϕ yields (4.1) as Euler-Lagrange equation, variation of the vector multiplier $\underline{\psi}$ yields the constraints (4.2).

The independent variations $\delta_x\underline{u}$ produce as Euler equation

$$\underline{u} - \text{grad}\phi - \underline{\psi} = 0 \tag{4.4}$$

while the independent variation $\delta_x \rho$ yields

$$\frac{\underline{u}\,\underline{u}}{2} - I - \Omega - \frac{d\phi}{dt} = 0 \,. \tag{4.5}$$

To establish the Euler equation for $\delta_a \underline{x}$ we must manipulate

$$\delta_x \frac{d\underline{x}}{dt} = \delta_a \frac{d\underline{x}}{dt} - (\delta_a \underline{x}\cdot\text{grad})\underline{u} = \frac{d}{dt}\,\delta_a \underline{x} - (\delta_a \underline{x}\cdot\text{grad})\underline{u} \,.$$

And consequently

$$\rho\underline{\psi}\cdot\delta_x \frac{d\underline{x}}{dt}\,dv = \frac{d}{dt}\,(\delta_a \underline{x}\cdot\rho\underline{\psi}dv) - \delta_a \underline{x}\cdot\frac{d}{dt}\,(\rho\underline{\psi}dv) - \rho dv\underline{\psi}\cdot(\delta_a \underline{x}\cdot\text{grad})\underline{u} \,.$$

Using the property (2.13), furnished by the principle in the equivalent form (4.1), it turns out that the components of the vector multiplier verify, according to (4.6),

$$\frac{d\psi_i}{dt} + \sum_{j=1}^{3} \psi_j \frac{\partial u_j}{\partial x_i} = 0 \,, \qquad i = 1,2,3. \tag{4.7}$$

An equivalent vectorial equation is

$$\frac{\partial\underline{\psi}}{\partial t} + \text{grad}(\underline{\psi}\cdot\underline{u}) + \text{rot}\,\underline{\psi} \times \underline{u} = 0 \,. \tag{4.8}$$

The independent boundary conditions turn out to be:

$$\int_{t_1}^{t_2} \int_S \phi\rho(\delta_x \underline{u}\cdot\underline{n})\,dSdt = 0 \tag{4.9}$$

and, because of (4.5)

$$\rho L = p + \rho\,\frac{d\phi}{dt} \,,$$

$$\int_{t_1}^{t_2} \int_S \left(p + \rho\,\frac{d\phi}{dt}\right) (\delta_a \underline{x}\cdot\underline{n})\,dSdt - \int_{t_1}^{t_2} \int_{S^*} p^*(\delta_a \underline{x}\cdot\underline{n})\,dSdt = 0 \,. \tag{4.10}$$

A necessary constraint on the velocity field is

$$(\delta_x \underline{u}\ \underline{n}) = 0 \quad \text{on} \quad S - S^* , \tag{4.11}$$

if the compared motions are to remain tangent to the moving walls. On the surface S^*, the variations $\delta_x \underline{u}$ are free and consequently (4.9) imposes a natural condition

$$\phi = 0 \quad \text{on} \quad S^* . \tag{4.12}$$

Since particles on the free surface will remain on it, (4.12) will entail

$$\frac{d\phi}{dt} = 0 \quad \text{on} \quad S^* . \tag{4.13}$$

Thus, while along the moving walls the constraint (2.5) prevails, (4.10) is again satisfied along S^* by the pressure condition (2.14).

Initial conditions for the principle are

$$\int_D (\rho\underline{\psi}\cdot\delta_a \underline{x})dv = 0 \quad \text{for} \quad t = t_1 \text{ and } t = t_2 , \tag{4.14}$$

which is satisfied by the rule (2.15), and

$$\int_D \phi\delta_x \rho dv = 0 \quad \text{for} \quad t = t_1 \text{ and } t = t_2 . \tag{4.15}$$

Indeed, to have (2.6) satisfied at all times, we need to have it satisfied initially. After that, equation (2.13) which is now catered for by the principle, will keep it true at later times. Thus we need to have

$$\delta_x \rho = - \rho \operatorname{div}(\delta_a \underline{x}) \quad \text{for} \quad t = t_1$$

and by virtue of (2.15) this reduces to the condition

$$\delta_x \rho = 0 \quad \text{for} \quad t = t_1 . \tag{4.16}$$

The similar condition for $t = t_2$ will then be automatically satisfied. To achieve the consistency proof of the generalized principle it remains to show that it furnishes the newtonian equations of motion. Indeed, taking $\underline{\psi}$ from (4.4) and $d\phi/dt$ from (4.5), (4.8) becomes

$$\frac{\partial \underline{u}}{\partial t} + \text{grad}\left(\frac{\underline{u}\cdot\underline{u}}{2} + I + \Omega\right) + \text{rot}\ \underline{u}\times\underline{u} = 0\ . \qquad (4.17)$$

5. $\underline{\psi}$ AS A CONSTANT CIRCULATION FIELD

Since the final aim is to remove entirely the necessity for variations of particle displacements, the corresponding Euler-Lagrange equation (4.8) must somehow be solved and the solution substituted into the principle.

The nature of the solution can be recognized by the property of the $\underline{\psi}$ field to keep a constant circulation along any segment carried with the particles in their motion.

If $d\underline{x}(dx_1, dx_2, dx_3)$ is such a segment

$$\frac{d}{dt}\left(\sum_1^3 \psi_i dx_i\right) = \sum_1^3\left(\frac{d\psi_i}{dt}\, dx_i + \psi_j du_j\right) = 0 \qquad (5.1)$$

by virtue of the differential equations (4.7). Hence the differential form

$$\sum_1^3 \psi_i dx_i = \sum_1^3 \psi_i(a_j,t)\,dX_i(a_j,t) = \sum_{j=1}^3\left(\sum_{i=1}^3 \psi_i \frac{\partial X_i}{\partial a_j}\right) da_j$$

expressed in lagrangean coordinates, is explicitly independent of time and can be written

$$\sum_1^3 \psi_i dx_i = \sum_1^3 A_j(a_k)\,da_j\ , \qquad (5.2)$$

where the functions A_j depend only on the chosen set of lagrangean variables. It will be convenient to think of the Pfaffian form

(5.2) in the lagrangean variables as reduced to any canonical form:

$$\sum_1^3 A_j(a_k)da_j = d\Gamma + AdB \ .$$

Then, returning to the eulerian representation,

$$\sum_1^3 \psi_i dx_i = d\gamma(x_i,t)+\alpha(x_i,t)d\beta(x_i,t) \ , \qquad (5.3)$$

where

$$\frac{d\alpha}{dt} = 0 \ , \quad \frac{d\beta}{dt} = 0 \ , \quad \frac{d\gamma}{dt} = 0 \ . \qquad (5.4)$$

It appears finally from (5.3) that the general solution of (4.8) is

$$\underline{\psi} = \text{grad}\gamma+\alpha\text{grad}\beta \ ,$$

where α, β and γ are lagrangean variables.

6. ELIMINATION OF THE VARIATIONS $\delta_a \underline{x}$

The solution (5.4) and (5.5) of the vector multipliers is substituted into the generalized principle as follows.

From (5.4)

$$\frac{d\beta}{dt} = \frac{\partial\beta}{\mu t} + \frac{d\underline{x}}{dt} \cdot \text{grad}\beta = 0 \ , \quad \frac{d\gamma}{dt} = \frac{\partial\gamma}{\partial t} + \frac{d\underline{x}}{dt} \cdot \text{grad}\gamma = 0 \ .$$

This allows us to write

$$\underline{\psi} \cdot \frac{d\underline{x}}{dt} = (\text{grad}\gamma+\alpha\text{grad}\beta) \cdot \frac{d\underline{x}}{dt} = -\left(\frac{\partial\gamma}{\partial t} + \alpha\,\frac{\partial\beta}{\partial t}\right)$$

and the kernel of the variational principle becomes

$$K = \rho L + \phi\left(\frac{\partial\rho}{\partial t} + \text{div}(\rho\underline{u})\right) - \left(\frac{\partial\gamma}{\partial t} + \underline{u}\cdot\text{grad}\gamma\right) - \rho\alpha\left(\frac{\partial\beta}{\partial t} + \underline{u}\cdot\text{grad}\beta\right) \ .$$

There are no $\delta_a \underline{x}$ terms stemming any more from variation of the kernel. Accordingly all boundary terms in $\delta_a \underline{x}$ in the variational principle are also removed, which will entail that any pressure condition will have to be satisfied independently. In expanding the principle

$$\delta_x \int_{t_1}^{t_2} \int_D K \, dvdt = 0 \tag{6.1}$$

it is easily seen that $d\gamma/dt = 0$ is lost as an Euler equation but, simultaneously, that variations on ϕ and γ play a similar role. As a matter of fact we can write the second term of K dv as follows:

$$\phi\left(\frac{\partial\rho}{\partial t} + \operatorname{div}(\rho\underline{u})\right) dv = \frac{d}{dt}(\phi\rho dv) - \rho\left(\frac{\partial\phi}{\partial t} + \underline{u}\cdot\operatorname{grad}\phi\right) dv .$$

Removal of the total time derivative will only modify the conditions at $t = t_1$ and $t = t_2$, while it is seen that ϕ and γ combine into a single variable $\theta = \phi+\gamma$. Hence we consider the principle (6.1) with the simplified kernel

$$K = \rho\left\{\frac{\underline{u}\,\underline{u}}{2} - E - \Omega - \left(\frac{\partial\theta}{\partial t} + \underline{u}\cdot\operatorname{grad}\theta\right) - \alpha\left(\frac{\partial\beta}{\partial t} + \underline{u}\cdot\operatorname{grad}\beta\right)\right\} .$$

The following Euler equations are obtained

$$\delta_x\underline{u} \quad \underline{u} = \operatorname{grad}\theta + \alpha\operatorname{grad}\beta , \tag{6.2}$$

$$\delta_x\rho \quad \frac{\underline{u}\cdot\underline{u}}{2} = I + \Omega + \frac{d\theta}{dt} + \alpha\frac{d\beta}{dt} , \tag{6.3}$$

$$\delta_x\theta \quad \frac{\partial\rho}{\partial t} + \operatorname{div}(\rho\underline{u}) = 0 , \tag{6.4}$$

$$\delta_x\alpha \quad \frac{d\beta}{dt} = 0 , \tag{6.5}$$

$$\delta_x\beta \quad \frac{\partial}{\partial t}(\rho\alpha) + \operatorname{div}(\rho\alpha\underline{u}) = 0 \tag{6.6}$$

together with initial conditions requiring $\delta_x\theta$ and $\delta_x\beta$ to vanish for $t = t_1$ and $t = t_2$.

From (6.4) and (6.6) the result $d\alpha/dt = 0$ is retrieved and (6.2) is then the formulation of the velocity field originally given by Clebsch [1].

The vorticity is represented by

$$\operatorname{rot} \underline{u} = \operatorname{grad}\alpha \times \operatorname{grad}\beta \tag{6.7}$$

so that the vortex lines are intersections of the surfaces α = constant and β = constant which move with the particles.

Combining the result

$$\underline{u} \cdot \underline{u} = \underline{u} \cdot (\operatorname{grad}\theta + \alpha \operatorname{grad}\beta) = \frac{d\theta}{dt} + \alpha \frac{d\beta}{dt} - \left(\frac{\partial\theta}{\partial t} + \alpha \frac{\partial\beta}{\partial t} \right)$$

obtained from (6.2) with (6.3), we find the energy integral

$$I + \Omega + \frac{\underline{u} \cdot \underline{u}}{2} + \frac{\partial\theta}{\partial t} + \alpha \frac{\partial\beta}{\partial t} = 0 \ . \tag{6.8}$$

The newtonian equations of motion can be derived by taking the material derivative of (6.2).

The absence of natural surface boundary conditions is a matter of closer scrutiny of the technique of integration by parts. In the stationary case there are surface boundary terms; in the nonstationary case those terms are cancelled by application of formula (1.23). A natural way to correct this situation is to replace (1.21) by the statement

$$\frac{d}{dt} \int_D f dv = \frac{\partial}{\partial t} \int_D f dv + \int_S f(\underline{w} \cdot \underline{n}) dS \ , \tag{6.9}$$

where $\underline{w}$ is equal to the fluid velocity $\underline{u}$ on the free surface S^*, but elsewhere equal to the local velocity of the wall. If the fluid is gliding along the wall there is no flux across it and (6.9) holds true as well as (1.21). Replacing (1.23) by the new formula

$$\int_{t_1}^{t_2} \int_D \frac{\partial f}{\partial t} dv dt = - \int_{t_1}^{t_2} \int_S f(\underline{w} \cdot \underline{n}) dS dt + \left| \int_D f dv \right|_{t_1}^{t_2} \tag{6.10}$$

the surface boundary terms obtained are

$$\int_{t_1}^{t_2} \int_S \rho(\delta_x \theta + \alpha \delta_x \beta)(\underline{w} - \underline{u} \cdot n) dS dt = 0 \ .$$

They require correctly that

$$\underline{u}\cdot\underline{n} = \underline{w}\,\underline{n} \quad \text{on} \quad S. \tag{6.11}$$

7. THE VARIATIONAL PRINCIPLES OF BATEMAN

There are two ways of modifying the presentation of the variational principle (6.1) with the simplified kernel.

When (6.3) is used *a priori*, the variations on ρ should disappear. This is achieved by substitution of (6.3) into the kernel, producing:

$$K = \rho(I-E) = p\ .$$

In this presentation p should be considered as a function of the enthalpy, the enthalpy itself being taken from (6.3). The procedure is very similar to the transformation of the total energy principle of elasticity into the complementary energy principle. One first looks after the variable conjugate to ρ with respect to the energy per unit volume, which turns out to be the enthalpy:

$$\frac{d}{d\rho}(\rho E) = E + \rho\frac{dE}{d\rho} = I\ .$$

The complementary energy is then built up by a contact transformation and turns out to be the pressure

$$\rho\frac{d}{d\rho}(\rho E) - \rho E = \rho(I-E) = p\ .$$

This complementary energy is to be considered as a function of the conjugate variable I. In fact, from the involutive property of contact transformations:

$$\frac{dp}{dI} = \rho\ . \tag{7.1}$$

Bateman's first principle is precisely

$$\delta_x \int_{t_1}^{t_2}\int_D p\,dv\,dt = 0\ , \tag{7.2}$$

where p = p(I) and one considers *a priori* that

$$I = \frac{\underline{u}\cdot\underline{u}}{2} - \Omega - \left(\frac{\partial\theta}{\partial t} + \underline{u}\cdot\text{grad}\theta\right) - \alpha\left(\frac{\partial\beta}{\partial t} + \underline{u}\cdot\text{grad}\beta\right) .$$

In view of (7.1) the Euler equations are again (6.2), (6.4), (6.5) and (6.6).

Suppose now that (6.2) is used *a priori*; then $\underline{u}$ and its variation should disappear. Replacing (6.2) into the kernel, it takes the form

$$K = -\rho\left(E+\Omega+ \frac{\partial\theta}{\partial t} + \alpha\frac{\partial\beta}{\partial t} + \frac{1}{2}(\text{grad}\theta+\alpha\text{grad}\beta)\cdot(\text{grad}\theta+\alpha\text{grad}\beta)\right) . \quad (7.4)$$

The Euler equations of this principle are (6.3), (6.4), (6.5) and (6.6), where $\underline{u}$ stands for the expression given by (6.2).

With this understanding, Bateman's second principle for the stationary case is sometimes losely written as

$$\delta_x \int_{t_1}^{t_2} \int_D \rho\left(E + \Omega + \frac{\underline{u}\cdot\underline{u}}{2}\right) dvdt = 0 . \quad (7.5)$$

In most practical applications both modifications are used. In applying (7.2), the pressure is expressed as a function of the enthalpy and the enthalpy directly expressed in terms of the functions (θ,α,β)

$$I = -\Omega - \frac{\partial\theta}{\partial t} - \alpha\frac{\partial\beta}{\partial t} - \frac{1}{2}(\text{grad}\theta+\alpha\text{grad}\beta)\cdot(\text{grad}\theta+\alpha\text{grad}\beta) .$$

8. ISOENERGETIC FLOWS

The restriction of isentropic flow is easily removed to treat the more general case of isoenergetic flow, characterized by

$$\frac{dS}{dt} = 0 . \quad (8.1)$$

This law of conservation of entropy for each particle, shows that S is a function of lagrangean variables only. Instead of introducing it as a side condition with a lagrangean multiplier, a procedure followed by Herivel [6] and Eckart [8], we prefer to keep the velocity field in the form (6.2) and consider the entropy as a function

$$S = S(\theta, \alpha, \beta) \ . \tag{8.2}$$

With this understanding, and remembering that:

$$\frac{\partial E(\rho, S)}{\partial \rho} = p/\rho^2 \quad \text{and} \quad \frac{\partial E(\rho, S)}{\partial S} = T \ , \tag{8.3}$$

where T is the absolute temperature, the Euler equations of principle (6.1) are modified as follows: (6.2) and (6.3) remain true, (6.4) remains true provided:

$$\frac{\partial S}{\partial \theta} = 0 \quad \text{i.e.,} \quad S = S(\alpha, \beta) \tag{8.4}$$

(6.5) and (6.6) are replaced by:

$$\frac{d\beta}{dt} = - T \frac{\partial S}{\partial \alpha} \ , \qquad \frac{d\alpha}{dt} = T \frac{\partial S}{\partial \beta} \ . \tag{8.5}$$

Observe that from these equations

$$\frac{dS}{dt} = \frac{\partial S}{\partial \alpha}\frac{d\alpha}{dt} + \frac{\partial S}{\partial \beta}\frac{d\beta}{dt} = 0$$

as required.

The vortex lines are still given by the intersections of the surfaces α = constant and β = constant but they are no more fixed in the fluid. Another consequence of (8.5) is

$$\frac{\partial \underline{u}}{\partial t} + \text{rot}\underline{u} \times \underline{u} = T \text{ grad } S \tag{8.6}$$

so that, by Crocco's theorem

$$\text{grad}\left(I + \Omega + \frac{\underline{u} \cdot \underline{u}}{2} \right) = 0 \ . \tag{8.7}$$

REFERENCES

[1] CLEBSCH, A., *J. reine angew. Math.*, Vol. 54, 1857, p. 293; I, Vol. 56, 1859.

[2] LICHTENSTEIN, L., *Grundlagen der Hydrodynamik*, Chapter 9, Springer, Berlin, 1929.

[3] BATEMAN, H., *Proc. Roy. Soc. London*, Series A, Vol. 125, 1929, p. 598.

[4] DE DONDER, Th. and VAN DEN DUNGEN, F.H., *Acad. R. Belg. Bull. Cl. Sci.*, Vol. 35, 1949, p. 841.

[5] DELVAL, J., *Acad. R. Belg. Bull. Cl. Sci.*, Vol. 37, 1951, p. 986.

[6] HERIVEL, J.W., *Proc. Cambridge Phil. Soc.*, Vol. 51, 1955, p. 344.

[7] SERRIN, J., *Handbuch der Physik*, Band VIII/I, Berlin/Göttingen/Heidelberg, Springer-Verlag, 1964.

[8] ECKART, C., *The Physics of Fluids*, Vol. 3, 1960, p. 421.

[9] FRAEIJS DE VEUBEKE, B.M., *Annales Soc. Sci.*, Bruxelles, Vol. 74 - III, 1960, p. 157.

[10] FISZDON, W., IUTAM Symposium Transsonicum, Aachen, 1962, 362, Berlin/Göttingen/Heidelberg, Springer-Verlag, 1964.

[11] FISZDON, W., *Z.A.M.M.*, Vol. 42, T134, 1962.

[12] DROBOT, S. and RYBARSKI, A., *Archive for Rational Mechanics and Analysis*, Vol. 2, 1959, p. 393.

STRAIN-ENERGY BOUNDS IN FINITE-ELEMENT ANALYSIS BY SLAB ANALOGY

B.M. Fraeijs de Veubeke
University of Liège, Belgium

O.C. Zienkiewicz
University of Wales, Swansea

It is well known that in approximate analysis of elasticity problems lower and upper bounds to the true strain-energy content can be obtained by the alternative assumptions of compatible displacement fields or of equilibrating stress fields. While formulations based on the first are relatively easy to achieve those based on the second present many difficulties. In this paper it is shown how by virtue of the analogy between plane elasticity and slab deflection problems compatible-displacement formulations in either one can be used to generage equilibrating formulations in the other. This should result in a direct application of existing programmes to a wider range of problems.

INTRODUCTION

When approximate solutions are formulated by the finite-element method it is possible to obtain 'bounding' values of the true strain-energy content. If the approximation involves the use of a compatible-displacement field, it will represent in general the lower bound. Alternatively, if an equilibrating stress field is used, an upper bound on the strain energy will be obtained [1].

Energy bounds can be translated into bounds for the structural deflections, providing a direct measure of convergence of the analysis.

Original publication, *Journal of Strain Analysis*, Vol. 2, n°4, (in collaboration with O.C. Zienkiewicz), Inst. of Mech. Engineers, 1967.

Generation of compatible solutions, while not always easy, presents fewer difficulties than the derivation of equilibrating solutions. In fact, many efficient solutions to both the plane-elasticity and plate-bending problems have been derived ([2], [3]) using compatible-displacement formulations. The object of this paper is to show how the slab analogy can help by using such solutions to generate equilibrating solutions and obtain reciprocal bounds.

In addition the slab analogy will always yield an alternative formulation of the problem which at times may be efficient from the computational point of view.

SLAB ANALOGY

The recognition of an analogy between the stress functions in plane problems and the lateral displacements of plates was evident early [4] through the identical biharmonic relations valid for homogeneous and isotropic situations. An extension of this to multiply connected regions and to non-homogeneous situations came later ([5],[6]). Southwell [7] extended the analogy concept to a direct relation between displacements in the plane problem and two new stress functions introduced for plate bending. Fung [8] derived Southwell's equations by the complementary-energy principle, extending them to plates of variable thickness and mixed boundary conditions.

In the notes given below it will be seen that a 'one to one' analogy is evident for all steps of the formulation of the two problems, irrespective of material properties assumed. The variables entering each problem will first be listed and then the analogy stated.

Plane Elasticity

The stress-dependent part of strain can be defined in terms of the displacements u and v in the direction of the x and y axes. Thus:

$$\{\varepsilon\} = \begin{Bmatrix} \varepsilon_x \\ \varepsilon_y \\ \varepsilon_{xy} \end{Bmatrix} = \begin{Bmatrix} \partial u/\partial x \\ \partial v/\partial y \\ \frac{1}{2}\,(\partial u/\partial y+\partial v/\partial x) \end{Bmatrix} - \begin{Bmatrix} C\alpha T \\ C\alpha T \\ 0 \end{Bmatrix} \quad . \qquad (1)$$

The last vector stands for thermal strains due to a temperature rise T and an expansion coefficient α. (The coefficient C is equal to unity for plane stress or $(1+\nu)$ for plane strain.)

In the absence of dislocations the displacements u and v are single-valued. With the help of equation (1) they can be determined by the variational principle of displacements, which requires the following 'functional' to be a minimum:

$$\int W(\varepsilon_x,\varepsilon_y,\varepsilon_{xy})dxdy - \int (\bar{X}u+\bar{Y}v)dxdy - \int (\bar{\sigma}_n u_n+\bar{\tau}_{tn}u_t)ds \quad . \qquad (2)$$

In equation (2), the strain-energy density W is a quadratic form containing the appropriate elastic constants to produce the linear stress-strain relations

$$\sigma_x = \frac{\partial W}{\partial \varepsilon_x} \quad , \quad \sigma_y = \frac{\partial W}{\partial \varepsilon_y} \quad , \quad \tau_{xy} = \frac{1}{2}\,\frac{\partial W}{\partial \varepsilon_{xy}} \quad . \qquad (3)$$

$\bar{X}$ and $\bar{Y}$ are specified body forces; $\bar{\sigma}_n$ and $\bar{\tau}_{tn}$ normal and tangential stresses specified along parts of the boundary. Along complementary parts of the boundary the normal and tangential displacements are specified:

$$u_n = \bar{u}_n \qquad u_t = \bar{u}_t \quad \ldots \quad . \qquad (4)$$

The variational derivatives of the principle (2) produce a pair of partial differential equations for the unknowns u and v, together with supplementing boundary-condition equation (4). They are respectively statements of equilibrium with the body forces

$$\frac{\partial \sigma_x}{\partial x} + \frac{\partial \tau_{xy}}{\partial y} + \bar{X} = 0 \qquad \frac{\partial \tau_{xy}}{\partial x} + \frac{\partial \sigma_y}{\partial y} + \bar{Y} = 0 \tag{5}$$

and equilibrium at the boundary with the specified stresses

$$\sigma_n = \bar{\sigma}_n \quad , \quad \tau_{nt} = \bar{\tau}_{nt} \quad \ldots \tag{6}$$

This defines uniquely the problem and permits its solution.

The finite-element method of approximation, relevant to this formulation of the problem, makes use of displacement models for the finite elements. Within each element the displacements are single-valued and able to be differentiated; the whole field can then be differentiated piece by piece. Furthermore, the elements are said to be 'conforming' if the displacements are single-valued at the interfaces. In such a case the variational principle (2) remains applicable to the structure as a whole and predicts a lower strain-energy bound if $\bar{u}_n = \bar{u}_s = 0$ are only specified displacements; an upper bound if $\bar{X} = \bar{Y} = 0$ and $\bar{\sigma}_n = \bar{\tau}_{sn} = 0$ but non-zero displacements or temperatures are given.

An alternative approach to the problem is through the use of a general solution to the equilibrium equations.

Such a solution is provided by setting

$$\{\sigma\} = \begin{Bmatrix} \sigma_x \\ \sigma_y \\ \tau_{xy} \end{Bmatrix} = \begin{Bmatrix} \partial^2\phi/\partial y^2 + F \\ \partial^2\phi/\partial x^2 + F \\ -\partial^2\phi/\partial x \partial y \end{Bmatrix} \quad \ldots \tag{7}$$

in which F is a body-force potential such that

$$\bar{X} = -\partial F/\partial x \quad , \quad \bar{Y} = -\partial F/\partial y \; .$$

Along those parts of the boundary where stresses are specified (Figure 1)

$$\left.\begin{aligned} \bar{\sigma}_n &= \frac{\partial^2\phi}{\partial t^2} + F = \frac{\partial^2\phi}{\partial s^2} + \frac{\partial\phi}{\partial n}\frac{d\psi}{ds} + F \\ -\bar{\tau}_{nt} &= \frac{\partial^2\phi}{\partial t\partial n} = \frac{\partial^2\phi}{\partial s\partial n} - \frac{\partial\phi}{\partial s}\frac{d\psi}{ds} \end{aligned}\right\} , \qquad (8)$$

where $d\psi/ds$ is the curvature of the boundary. The stress function $\phi(x, y)$ can be determined by the complementary energy principle requiring the minimum of

$$\int [(\Phi\ (\sigma_x, \sigma_y, \tau_{xy}) + \alpha CT(\sigma_x + \sigma_y)]\ dxdy$$

$$- \int \left(\bar{u}_n \left(\frac{\partial^2\phi}{\partial s^2} + \frac{\partial\phi}{\partial n}\frac{d\psi}{ds}\right) - \bar{u}_s \left(\frac{\partial^2\phi}{\partial n\partial s} - \frac{\partial\phi}{\partial s}\frac{d\psi}{ds}\right)\right) ds\ . \qquad (9)$$

The complementary energy density Φ (stress energy) provides the same stress-strain relations as equation (3), but is now solved for the strains as

$$\varepsilon_x = \frac{\partial\Phi}{\partial\sigma_x}\ , \quad \varepsilon_y = \frac{\partial\Phi}{\partial\sigma_y}\ , \quad \varepsilon_{xy} = \frac{1}{2}\frac{\partial\Phi}{\partial\tau_{xy}}\ . \qquad (10)$$

The variational derivative of this principle yields

$$\frac{\partial^2\varepsilon_x}{\partial y^2} + \frac{\partial^2\varepsilon_y}{\partial x^2} - 2\frac{\partial^2\varepsilon_{xy}}{\partial x\partial y} = -\nabla^2(\alpha CT)\ . \qquad (11)$$

This is also a necessary condition for the integrability of single-valued displacements u and v, defined according to equation (1). After substitution of equations (10) and (7), it yields the partial-differential equation governing the stress function, which reduces to the non-homogeneous biharmonic equation

$$\nabla^4\phi = -\frac{1}{mE}\nabla^2(\alpha CT)$$

in the isotropic case. The constant m is equal to unity for plane stress or $(1-\nu^2)$ for plain strain.

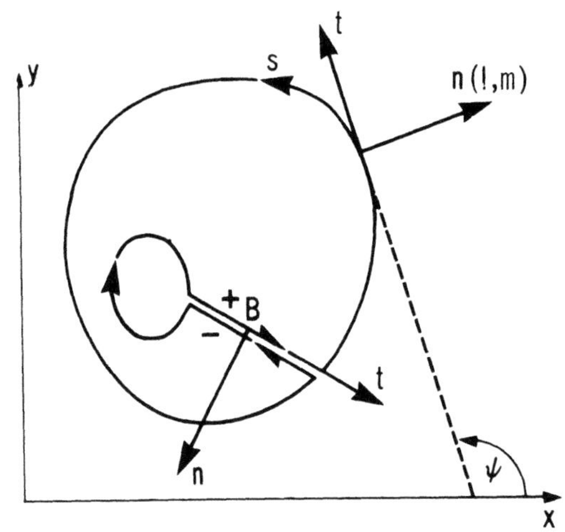

Figure 1 - A General Analysis Region

The variational boundary conditions, which supplement equations (8) when displacements are specified are: by continuity

$$\frac{\partial\bar{u}_t}{\partial s} + \bar{u}_n\frac{d\psi}{ds} = \varepsilon_t + \alpha CT \; \ldots \qquad (12)$$

and

$$\frac{\partial^2\bar{u}_n}{\partial s^2} - \frac{\partial}{\partial s}\left(\bar{u}_t\frac{d\psi}{ds}\right) = 2\frac{\partial\varepsilon_{nt}}{\partial s} + \frac{d\psi}{ds}(\varepsilon_n - \varepsilon_t) - \frac{\partial}{\partial n}(\varepsilon_t + \alpha CT) \; .$$

This permits the problem to be solved, provided the domain is simply connected. For simplicity we use straight 'barriers' to reduce the multi-connected case to the simply connected one

(Figure 1). Further statements of equilibrium are here necessary because stresses should be continuously transmitted across each 'barrier'. Thus

$$\left.\begin{aligned} \sigma_n^{\,+} &= \frac{\partial^2\phi^+}{\partial t^2} + F = \sigma_n^{\,-} = \frac{\partial^2\phi^-}{\partial t^2} + F \\ \tau_{nt}^{\,+} &= \frac{\partial^2\phi^+}{\partial n\partial t} = \tau_{nt}^{\,-} = -\frac{\partial^2\phi^-}{\partial n\partial t} \end{aligned}\right\} \qquad (13)$$

the signs referring to values at opposite sides of the barrier. Equivalent statements are

$$\left.\begin{aligned} \phi^+ &= \phi^- + tN_n + M_B \\ \frac{\partial\phi^+}{\partial n} &= \frac{\partial\phi^-}{\partial n} - N_t \end{aligned}\right\} \; . \qquad (14)$$

The significance of the constants (N_n, N_t, M_B) appears in Appendix 1. The variational transition equations based on equations (14) are obtained directly from equation (12). As there are no specified displacements on a barrier, the variation on ϕ_n^- gives

$$(\varepsilon_t + \alpha CT)^+ = (\varepsilon_t + \alpha CT)^-$$

and the variation on ϕ^-

$$2\,\frac{\partial\varepsilon_{nt}}{\partial t} - \frac{\partial}{\partial n}(\varepsilon_t + \alpha CT)\bigg|^+ = 2\,\frac{\partial\varepsilon_{nt}}{\partial t} - \frac{\partial}{\partial n}(\varepsilon_t + \alpha CT)\bigg|^- \, .$$

In view of definitions (1), these results are equivalent to

$$\left(\frac{\partial u_t}{\partial t}\right)^+ = \left(\frac{\partial u_t}{\partial t}\right)^- \qquad \left(\frac{\partial^2 u_n}{\partial t^2}\right)^+ = \left(\frac{\partial^2 u_n}{\partial t^2}\right)^-$$

and yield the classical property discovered by Weingarten [9] and Volterra [10] that both sides of the barrier can undergo a kinematic relative displacement:

$$\left.\begin{aligned} u_t^+ &= u_t^- + p \\ u_n^+ &= u_n^- + \omega t + q \end{aligned}\right\} \quad . \tag{15}$$

Hence equation (11) is not sufficient to ensure that the displacements have single values. Parameters in the solution for ϕ must be further adjusted to implement for each barrier the single-value conditions 'in the large'

$$p = 0 \ , \quad q = 0 \ , \quad \omega = 0 \ . \tag{16}$$

Those parameters are, for instance, the 'Michell' constants of the expressions $\alpha x + \beta y + \gamma$ that can be added to ϕ along each internal boundary of a cavity without disturbing satisfaction of the boundary conditions.

The finite-element approximation, relevant to this alternative formulation, makes use of equilibrium models for the finite elements. An equilibrium model can be generated by a single-valued, twice-differentiable stress function, defined in its interior. The elements will be 'stress diffusing' if the stresses defined at the interfaces are continuously transmitted between adjacent elements.

In view of equations (8), this property can be ensured if F, ϕ, and its normal slope $\partial\phi/\partial n$ remain single-valued at the interfaces. This condition is, however, not quite necessary, since the surface $z = \phi(x, y)$ of any element can be moved bodily (a vertical translation and two rotations about the x and y axes) without disturbing the stress field. This freedom, used along barriers, allows the treatment of multi-connected cases.

The variational principle (9) remains then applicable to the whole structure and predicts upper strain-energy bounds if $\bar{u}_n = \bar{u}_t = 0$ are the only prescribed displacements; lower bounds if $\bar{X} = \bar{Y} = 0$ and $\bar{\sigma}_n = \bar{\tau}_{nt} = 0$ and the stressing is due to the prescribed displacement.

Because, in the general case, the approximations will not allow ϕ to satisfy the partial-differential equation derived from equation (11), the displacement field within an element will not be integrable. The only knowledge provided about displacements will be in the form of weighted averages ([1],[11]).

Plate Flexure

In the Kirchhoff-Love theory of plate flexure, the curvatures produced by the internal moments are related to the lateral displacement w by

$$\{\kappa\} = \begin{Bmatrix} \kappa_x \\ \kappa_y \\ \kappa_{xy} \end{Bmatrix} = \begin{Bmatrix} \partial^2 w/\partial x^2 \\ \partial^2 w/\partial y^2 \\ \partial^2 w/\partial x \partial y \end{Bmatrix} - \begin{Bmatrix} \kappa_0 \\ \kappa_0 \\ 0 \end{Bmatrix} \tag{17}$$

in which κ_0 is an initial, isotropic curvature of the type resulting from a temperature change. The internal moments can be related to the curvatures through an energy $W(\kappa_x, \kappa_y, \kappa_{xy})$ per unit area containing appropriate elastic constants

$$M_x = -\frac{\partial W}{\partial \kappa_x}, \quad M_y = -\frac{\partial W}{\partial \kappa_y}, \quad M_{xy} = -\frac{1}{2}\frac{\partial W}{\partial \kappa_{xy}} . \tag{18}$$

The problem for the lateral displacement is governed by the variational principle requiring the minimum of

$$\int W\,dxdy - \int \bar{q}w\,dxdy + \int \left(\bar{M}_n \frac{\partial w}{\partial n} - \bar{K}_n w\right) ds - \sum_i \bar{Z}_i w_i . \tag{19}$$

The later displacement can be specified along parts of the boundary

$$w = \bar{w} . \tag{20}$$

Then, along complementary parts the shear distribution $\overline{K}_n$ is given with, possibly, some concentrated loads $\overline{Z}_i$.

In the same manner the normal bending moment $\overline{M}_n$ is given along parts of the boundary complementary to those where the normal slope is known

$$\frac{\partial w}{\partial n} = \left(\frac{\partial \bar{w}}{\partial n}\right) \quad . \tag{21}$$

Finally, a known transverse pressure $\bar{q}(x, y)$ is applied.

The variational derivative of the minimum total-energy principle, using equations (18) and (17), is

$$\frac{\partial^2 M_x}{\partial x^2} + 2\,\frac{\partial^2 M_{xy}}{\partial x \partial y} + \frac{\partial^2 M_y}{\partial y^2} = -\bar{q} \quad . \tag{22}$$

It is convenient, for later use, to introduce as auxiliary quantity the shear vector (Q_x, Q_y)

$$Q_x = \frac{\partial M_x}{\partial x} + \frac{\partial M_{xy}}{\partial y} \ ; \qquad Q_y = \frac{\partial M_{xy}}{\partial x} + \frac{\partial M_y}{\partial y} \quad . \tag{23}$$

Then, the variational derivative is more clearly a statement of vertical equilibrium

$$\frac{\partial Q_x}{\partial x} + \frac{\partial Q_y}{\partial y} = -\bar{q} \quad . \tag{24}$$

Expressed through equations (18) and (17) in terms of w, it becomes the partial-differential equation governing the lateral displacement. For the isotropic case it reduces to

$$\nabla^4 w = \frac{\bar{q}}{D} \qquad \left(D = \frac{Et^3}{12(1-\nu^2)} \text{ the bending rigidity} \right) \quad .$$

The variational boundary conditions supplementing equations (20) and (21) are the equilibrium requirements

$$K_n = Q_n + \frac{\partial M_{nt}}{\partial s} = \bar{K}_n \quad , \tag{25}$$

$$M_n = \bar{M}_n \quad , \tag{26}$$

$$M_{nt}(s_i+\varepsilon) - M_{nt}(s_i-\varepsilon) = \bar{Z}_i \qquad (\varepsilon \rightarrow 0) \quad . \tag{27}$$

Equation (25) introduces the general definition of the Kirchhoff equivalent shear distribution along a curved boundary.

A displacement model for a finite plate-flexure element will be defined by a parametric, single-valued, twice-differentiable lateral-displacement field. The elements will be conforming if w and $\partial w/\partial n$ are single-valued at the interfaces. The variational principle is applicable to a gridwork of conforming elements and predicts a lower bound to the strain energy if $\bar{w} = 0$, $(\partial \bar{w}/\partial n) = 0$ are the only prescribed displacement conditions; an upper bound if $\bar{q} = 0$, $\bar{K}_n = 0$, $\bar{M}_n = 0$; $\bar{Z}_i = 0$ and displacements are caused by boundary movements. The stress-diffusing properties of the elements, represented here by the single values of K_n and M_n at the interfaces, will only be averaged in the approximation, because the variations on w and $\partial w/\partial n$ at the interfaces are constrained by the finite number of degrees of freedom.

The alternative approach to plate flexure is again through a general solution to the equilibrium problem. If a stress-function vector (U, V) is introduced such that

$$\begin{Bmatrix} M_y \\ M_x \\ -M_{xy} \end{Bmatrix} = \begin{Bmatrix} \partial U/\partial x \\ \partial V/\partial y \\ \frac{1}{2}\,(\partial U/\partial y + \partial V/\partial x) \end{Bmatrix} - \begin{Bmatrix} P_0 \\ P_0 \\ 0 \end{Bmatrix} \quad . \tag{28}$$

Then we have for the shear load vector (23)

$$Q_x = \frac{\partial \Omega}{\partial y} - \frac{\partial P_0}{\partial x} \qquad Q_y = -\frac{\partial \Omega}{\partial x} - \frac{\partial P_0}{\partial y} \quad , \tag{29}$$

with
$$\Omega = \frac{1}{2}\left(\frac{\partial V}{\partial x} - \frac{\partial U}{\partial y}\right) , \tag{30}$$

and the equilibrium equation (24) is satisfied, provided

$$\nabla^2 P_0 = \bar{q} \quad . \tag{31}$$

The stress-function vector can be determined from the variational principle requiring the minimum of the 'functional'

$$\int \left\{ \Phi(M_x, M_y, M_{xy}) - \kappa_0 (M_x + M_y) \right\} dxdy - \int \left(\bar{w} K_n - \left(\frac{\partial \bar{w}}{\partial n}\right) M_n \right) ds \ . \tag{32}$$

The complementary energy Φ, per unit area, yields the inner moment-curvature relations (18) in inverted form

$$\kappa_x = - \frac{\partial \Phi}{\partial M_x} \ , \quad \kappa_y = - \frac{\partial \Phi}{\partial M_y} \ , \quad \kappa_{xy} = - \frac{1}{2} \frac{\partial \Phi}{\partial M_{xy}} \quad . \tag{33}$$

To apply the principle by varying the quantities U and V we can rely on definitions (28), but we must also express the boundary loads in terms of the stress functions. In natural boundary co-ordinates, let U_n, U_t be the components of the stress-function vector. Then

$$\Omega = \frac{1}{2}\left(\frac{\partial U_t}{\partial n} - \frac{\partial U_n}{\partial t}\right) , \tag{34}$$

and, by a set of transformation rules detailed in Appendix 2,

$$M_n = \frac{\partial U_t}{\partial t} - P_0 = \frac{\partial U_t}{\partial s} + U_n \frac{d\psi}{ds} - P_0 \quad , \tag{35}$$

$$K_n = Q_n + \frac{\partial M_{nt}}{\partial s} = \frac{\partial \Omega}{\partial t} - \frac{\partial P_0}{\partial n} - \frac{1}{2}\frac{\partial}{\partial s}\left(\frac{\partial U_t}{\partial n} + \frac{\partial U_n}{\partial t}\right) = - \frac{\partial^2 U_n}{\partial s \partial t} - \frac{\partial P_0}{\partial n}$$

or finally,

$$K_n = - \frac{\partial^2 U_n}{\partial s^2} + \frac{\partial}{\partial s}\left(U_t \frac{d\psi}{ds}\right) - \frac{\partial P_0}{\partial n} \quad . \tag{36}$$

M_n and K_n as given by equations (35) and (36) should be substituted into the boundary term of the principle. The variational derivatives for U and V are respectively

$$\left.\begin{aligned} \frac{\partial \kappa_y}{\partial x} - \frac{\partial \kappa_{xy}}{\partial y} &= - \frac{\partial \kappa_0}{\partial x} \\ - \frac{\partial \kappa_{xy}}{\partial x} + \frac{\partial \kappa_x}{\partial y} &= - \frac{\partial \kappa_0}{\partial y} \end{aligned}\right\} . \tag{37}$$

They represent necessary integrability conditions for a single-valued lateral displacement w, as defined by equation (17). When modified through equations (33) and (34), they become the pair of partial-differential equations governing the stress functions. The boundary values associated to them are (35) and (36), where K_n and M_n are specified. The complementary set, where deflections and slopes are specified, is

$$\frac{\partial^2 \bar{w}}{\partial s^2} + \left(\frac{\partial \bar{w}}{\partial n}\right) \frac{d\psi}{ds} = \kappa_t + \kappa_0 = \frac{\partial^2 w}{\partial t^2} , \tag{38}$$

stemming from the boundary variations on U_n, and

$$\frac{\partial}{\partial s} \left(\frac{\partial \bar{w}}{\partial n}\right) - \frac{\partial \bar{w}}{\partial s} \frac{d\psi}{ds} = \kappa_{nt} = \frac{\partial^2 w}{\partial n \partial t} , \tag{39}$$

stemming from the boundary variations on U_t.

This, in principle, formulates the problem to be solved for a simply connected domain. The additional statements of equilibrium required for a multi-connected domain are the continuity of K_n and M_n across the barriers. Once again, taking straight barriers for simplicity, equations (35) and (36) show that those equilibrium conditions are equivalent to

$$\left(\frac{\partial U_t}{\partial t}\right)^+ = \left(\frac{\partial U_t}{\partial t}\right)^-$$

$$\left(\frac{\partial^2 U_n}{\partial t^2}\right)^+ = \left(\frac{\partial^2 U_n}{\partial t^2}\right)^-$$

P_0 and $\partial P_0/\partial n$ are assumed to be single-valued throughout. Hence

$$\left.\begin{aligned} U_t^+ &= U_t^- + \mu_t \\ U_n^+ &= U_n^- + \mu_n - Tt \end{aligned}\right\} . \tag{40}$$

The significance of the constants μ_t, μ_n, and T appears in Appendix 1. The variational transition equations based on (40) can be written directly from equations (38) and (39). Because there are no specified displacements on the barrier, the variation on U_t^- yields

$$(\kappa_{nt})^+ = (\kappa_{nt})^-$$

and the variation on U_n^-

$$(\kappa_t + \kappa_0)^+ = (\kappa_t + \kappa_0)^- .$$

In view of definitions (17), these results are equivalent to

$$\left(\frac{\partial^2 w}{\partial n \partial t}\right)^+ = \left(\frac{\partial^2 w}{\partial n \partial t}\right)^- \quad \text{and} \quad \left(\frac{\partial^2 w}{\partial t^2}\right)^+ = \left(\frac{\partial^2 w}{\partial t^2}\right)^-$$

r, finally

$$\left.\begin{aligned} w^+ &= w^- + h_B + \omega_n t \\ \left(\frac{\partial w}{\partial n}\right)^+ &= \left(\frac{\partial w}{\partial n}\right)^- - \omega_t \end{aligned}\right\} . \tag{41}$$

As with plane elasticity, this expresses the fact that the two sides of the barrier can undergo a kinematic relative displacement: a relative vertical translation h_B and two relative rotations about the local axes n and t in B.

Again conditions (37) are not sufficient for single values of the lateral displacements and its slopes. To avoid dislocations, parameters in U and V must be adjusted to obtain

$$h_B = 0 , \quad \omega_n = 0 , \quad \omega_t = 0 , \tag{42}$$

on each barrier. Those parameters can be the arbitrary constants in the expression $(\alpha-\omega y)$ and $(\beta+\omega x)$ that can be added respectively to U and V along each internal boundary of cavity, without disturbing stress equilibrium.

The equilibrium models of finite plate-flexure elements can be generated by a parametric, single-valued, differentiable stress-function vector coupled with a parametric loading function P_0, if transverse loading modes are desired. The elements will be stress-diffusing if the stress-function vector remains single-valued at the interfaces, together with the loading function and its normal slope. This follows immediately on inspection of equations (35) and (36). In the multi-connected case U and V can be taken as single-valued in the domain cut by a set of barriers. Across those, U and V will eventually suffer a rigid-body type of discontinuity (distributed like $u_0-\omega y$, $v_0+\omega x$ along the barrier). The variational principle (32) will remain applicable to the gridwork of elements and produce upper stress-energy bounds if $\bar{w} = 0$, $(\partial\bar{w}/\partial n) = 0$, lower bounds if $\bar{q} = 0$, $\bar{K}_n = 0$, $\bar{M}_n = 0$.

Again, since in the approximation the functions U and V are not required to, and generally will not, satisfy the partial-differential equations derived from equations (37), the lateral displacement w will not be integrable. Numerical information on lateral displacements and slopes is only provided in the form of weighted averages.

The Analogies

The comparison of relations for both systems shows clearly the mathematical analogies which exist on a one-to-one basis between various quantities and equations. Table 1 shows this in detail.

Table 1 - Mathematical Analogies

(1)	⟷	(28)
$\begin{Bmatrix} \varepsilon_x \\ \varepsilon_y \\ \varepsilon_{xy} \end{Bmatrix}$	⟷	$\begin{Bmatrix} M_y \\ M_x \\ -M_{xy} \end{Bmatrix}$
αCT	⟷	P_0
$\begin{Bmatrix} u \\ v \end{Bmatrix}$	⟷	$\begin{Bmatrix} U \\ V \end{Bmatrix}$
(11)	⟷	(22)
(16)	⟷	(42)
$\begin{Bmatrix} p \\ q \\ \omega \end{Bmatrix}$	⟷	$\begin{Bmatrix} \mu_t \\ \mu_n \\ -T \end{Bmatrix}$
(7)	⟷	(17)
$\begin{Bmatrix} \sigma_x \\ \sigma_y \\ \tau_{xy} \end{Bmatrix}$	⟷	$\begin{Bmatrix} \kappa_y \\ \kappa_x \\ -\kappa_{xy} \end{Bmatrix}$
F	⟷	$-\kappa_0$
ϕ	⟷	w
(5)	⟷	(37)
(14)	⟷	(41)
$\begin{Bmatrix} N_n \\ N_t \\ M_B \end{Bmatrix}$	⟷	$\begin{Bmatrix} \omega_n \\ \omega_t \\ h_B \end{Bmatrix}$
(3)	⟷	(43)
(10)	⟷	(18)

The formal elegance of the analogies could be improved by the adoption of the following unusual notation for plate-flexure theory

M_y	and	κ_y	instead of	M_x	and	κ_x
M_x	and	κ_x	instead of	M_y	and	κ_y
$-M_{xy}$	and	$-\kappa_{xy}$	instead of	M_{xy}	and	κ_{xy} .

It will be observed that the role of stresses and strains, stress functions and displacements, equilibrium conditions and integrability conditions, become at all times reversed.

This included the analogy between physical dislocations in multi-connected domains and multiple-values of stress functions.

APPLICATIONS TO FINITE-ELEMENT TECHNIQUES

Conforming displacement models, analysed by matrix methods, have discrete elastic characteristics described in terms of a set of generalized displacements $\{q\}_e$ and a corresponding set of generalized loads $\{g\}_e$

$$\{g\}_e = [K]_e\{q\}_e \quad , \tag{43}$$

where $[K]_e$ is the stiffness matrix of the element ([1],[2],[3],[11]). Stress-diffusing equilibrium models can be described in the same way; the stiffness matrix is of the form

$$[K]_e = [C]_e[F]_e^{-1}[C]_e^T \quad , \tag{44}$$

where $[F]_e$ is the flexibility matrix of the element (the matrix of the stress energy in terms of stress parameters) and $[C]_e$ a load-connection matrix, relating generalized loads and the set $\{b\}_e$ of stress parameters

$$\{g\}_e = [C]_e\{b\}_e \quad . \tag{45}$$

This procedure can lead to elements with spurious kinematic degrees of freedom, which then require special handling in the treatment of the problem at the structural level ([1],[11],[12]).

The use of the analogies opens new possibilities in the construction of stress-diffusing equilibrium models and in their handling at the structural level.

From the analysis of the analogies it becomes clear that each conforming displacement model of a finite element produces an analogue stress-diffusing equilibrium model by identification of the parametric displacement field of the former to a parametric stress-function field of the latter. Thus a plate-flexure element generates a plane-stress (or strain) element and vice versa. A good example is the conforming plate-flexure quadrilateral which goes over into a plate-extension equilibrium element with linear-stress variations, by analogy between w and ϕ [13]. However, even the treatment of the problem at the structural level can benefit from the analogies.

The analogue matrix relation to equation (43) is

$$\{\varepsilon\}_e = [F]_e\{c\}_e \quad , \tag{46}$$

where $\{c\}_e$ is the set of local values of (ϕ and $\partial\phi/\partial n$) or (U and V) analogous to the local displacements (w and $\partial w/\partial n$) or (u and v). $[F]_e$ is now an 'extended' flexibility matrix. While the flexibility matrix in (44) is non-singular, the extended one is singular because the equation

$$[F]_e\{z\} = \{0\} \quad , \tag{47}$$

has three independent solutions, representing rigid modes of the stress functions (the analogues of kinematic displacement modes) which generate no stresses and consequently no strains. The column matrix $\{\varepsilon\}_e$ is precisely in the nature of generalized strains as can be seen in what follows. From equation (46) and

Clapeyron's theorem, the strain energy of the element is

$$\frac{1}{2}\{c\}_e^T\{\varepsilon\}_e \quad . \tag{48}$$

Generalized loads can now be defined as in equation (45) through an 'extended' load connection matrix $[C]_e$

$$\{g\}_e = [C]_e\{c\}_e \ . \tag{49}$$

There are exactly as many generalized loads as there are local values of the stress function so that, in contrast to equation (45) the load-connection matrix is *square*.

The associated generalized displacements are such the the energy of the element is also

$$\frac{1}{2}\{g\}_e^T\{q\}_e = \frac{1}{2}\{c\}_e^T[C]_e^T\{q\}_e \quad . \tag{50}$$

Comparing equation (50) with equation (48) and noting that the equality must hold for any $\{c\}_e$ we find that

$$\{\varepsilon\}_e = [C]_e^T\{q\}_e \ . \tag{51}$$

This shows how the generalized strains are deduced from the generalized displacements.

The extended load-connection matrix is singular for it is obvious that, since stresses vanish for $\{c\}_e = \{z\}$, the generalized loads in equation (49) must also vanish.

$$[C]_e^T\{z\} = \{0\} \ . \tag{52}$$

From a classical theorem of algebra, if the homogeneous system (52) has three independent non-trivial solutions, so has the system

$$[C]_e^T\{y\} = \{0\} \ . \tag{53}$$

In view of equation (51) those solutions are to be identified with the three rigid displacement modes of the element (which generate no strains).

Since there is no other solution to equation (53) there are also no spurious kinematic freedoms in the element. Hence the fact that the stress-diffusing equilibrium model derives from an analogy with a conforming displacement model is sufficient to guarantee the absence of spurious kinematic freedoms.

In the displacement models the laws for assembling elements are those of stiffness addition or load addition ([1]-[3]) producing for the whole structure a relation between all the generalized external loads {g} and the nodal displacements {q}

$$\{g\} = [K]\{q\}$$

with a master stiffness matrix

$$[K] = \sum_e [L]_e^T [K]_e [L]_e$$

generated by the 'localizing' matrices $[L]_e$ defined by

$$\{q_e\} = [L]_e \{q\} \ .$$

In the analogous equilibrium models the laws of assembly are those of flexibility addition or strain addition, producing for the whole structure a relation

$$\{e\} = [F]\{c\}$$

between the generalized strains for the whole structure and the complete array of nodal stress-function values. The master flexibility matrix

$$[F] = \sum_e [L]_e^T [F]_e [L]_e$$

is obtained by the same localizing matrices as in the analogue structure. The procedure is particularly well adapted to equilibrium models because the number of self-stressing states, represented by each element of $\{c\}$, is markedly lower than the number of generalized displacements. The equations to be solved are both well conditioned and fewer in number.

REFERENCES

[1] FRAEIJS DE VEUBEKE, B.M., "Displacement and Equilibrium Models in the Finite Element Method", *Stress Analysis,* (editors: O.C. Zienkiewicz and G.S. Holister), 1965, Chapter 9, John Wiley and Sons, Inc., New York.

[2] CLOUGH, R.W., "The Finite Element Method in Structural Mechanics", *Stress Analysis,* (editors: O.C. Zienkiewicz and G.S. Holister), 1965, Chapter 7, John Wiley and Sons, Inc., New York.

[3] ZIENKIEWICZ, O.C. and CHEUNG, Y.K., *The Finite Element Method in Structural and Continuum Mechanics,* 1967, McGraw-Hill Book Co., Inc., New York and London.

[4] MICHELL, J.H., "On the Direction Determination of Stress in an Elastic Solid with Application to the Theory of Plates", *Proc. London Math. Soc.,* Vol. 31, 1899, p. 100.

[5] MINDLIN, R.D., "The Analogy Between Multiply Connected Slices and Slabs", *Quart. Appl. Math.,* Vol. 4, 1946, p. 279; Vol. 5, 1947, p. 238.

[6] ZIENKIEWICZ, O.C. and CRUZ, C., "The Use of the Slab Analogy in the Determination of Thermal Stresses", *Int. J. Mech. Sci.,* Vol. 4, 1962, p. 285.

[7] SOUTHWELL, R.V., "On the Analogues Relating Flexure and Extension of Flat Plates", *Q. J. Mech. Appl. Math.,* Vol. 3, 1950, p. 257.

[8] FUNG, Y.C., "Bending of Thin Elastic Plates of Variable Thickness", *J. Aeronautical Sci.,* Vol. 20, 1953, p. 455.

[9] WEINGARTEN, G., cited by LOVE, A.E.H., in *Mathematical Theory of Elasticity,* 1944, 156 p., Dover Publications, Inc., New York.

10] VOLTERRA, V., "Sur l'équilibre des corps élastiques multiplement connexes", *Annls Ec. Norm.,* Vol. 24, 1907, p. 401.

[11] SANDER, G. and FRAEIJS DE VEUBEKE, B.M., "Upper and Lower Bounds to Structural Deformations by Dual Analysis in Finite Elements", Air Force Technical Report AFFDL-TR-199, 1966.

[12] FRAEIJS DE VEUBEKE, B.M., "Upper and Lower Bounds in Matrix Structural Analysis", *AGARDograph*, Vol. 72, 1964, p. 165.

[13] FRAEIJS DE VEUBEKE, B.M., "Bending and Stretching of Plates. Special Models for Upper and Lower Bounds", *Conf. Matrix Meth. Struct. Mech.*, 1965, AFFDL, Dayton, Ohio.

[14] ZIENKIEWICZ, O.C. and GERSTNER, R.W., "A Stress Function Approach to Interface and Mixed Boundary Condition Problems", *Int. J. Mech. Sci.*, Vol. 2, 1960, p. 93.

[15] MINDLIN, R. D. and SALVADORI, M.G., *Handbook of Experimental Stress Analysis*, (editor: M. Hetenyi), 1950, Chapter 16, John Wiley and Sons, Inc., New York.

APPENDIX 1

Interpretations of the Stress Functions

The Airy Stress Function $\phi(x, y)$

Let (N_x, N_y) denote the resultants of stresses generated by ϕ along a path from a reference point A to (x, y), leaving matter on the left-hand side.

$$dN_x = -\tau_{xy}dx + \sigma_x dy = d\left(\frac{\partial\phi}{\partial y}\right),$$

$$dN_y = -\sigma_y dx + \tau_{xy}dy = -d\left(\frac{\partial\phi}{\partial x}\right).$$

Hence

$$N_x = \frac{\partial\phi}{\partial y}(x, y) - \left(\frac{\partial\phi}{\partial y}\right)_A,$$

$$N_y = -\frac{\partial\phi}{\partial x}(x, y) + \left(\frac{\partial\phi}{\partial x}\right)_A.$$

Similarly for the moment M about the point (x, y)

$$dM = N_x dy - N_y dx = d\phi - \left(\frac{\partial\phi}{\partial x}\right)_A dx - \left(\frac{\partial\phi}{\partial y}\right)_A dy \quad ,$$

and

$$M = \phi(x,\ y) - \phi_A - (x-x_A)\left(\frac{\partial\phi}{\partial x}\right)_A - (y-y_A)\left(\frac{\partial\phi}{\partial y}\right)_A \quad .$$

Since the stress field determines ϕ except for the addition of an arbitrary linear form $a_0+a_1x+a_2y$, it is always possible to make ϕ and its first derivatives vanish at the chosen reference point.

In this case we have simply

$$\phi(x,\ y) = M \ , \ -\frac{\partial\phi}{\partial x}(x,\ y) = N_y \quad , \quad \frac{\partial\phi}{\partial y}(x,\ y) = N_x \ .$$

The result is the same for all reconcilable paths from A to (x, y), because the stress field generated by ϕ does not involve body loads.

Applying this to the closed path from A to B around a cavity (Figure 2), we conclude that in equations (14), N_n, N_t, and M_B are respectively the resultants in the directions of n and t and the moment about B of the loads applied to the boundary of the cavity. Should the cavity be loaded by a system statically equivalent to zero, the stress function would remain single-valued; otherwise it would not. In particular, considering the application of a concentrated load or couple at an interior point as the limit of a loading distribution inside a circular cavity of vanishing radius, we conclude that such loads cause the stress function to have more than one value.

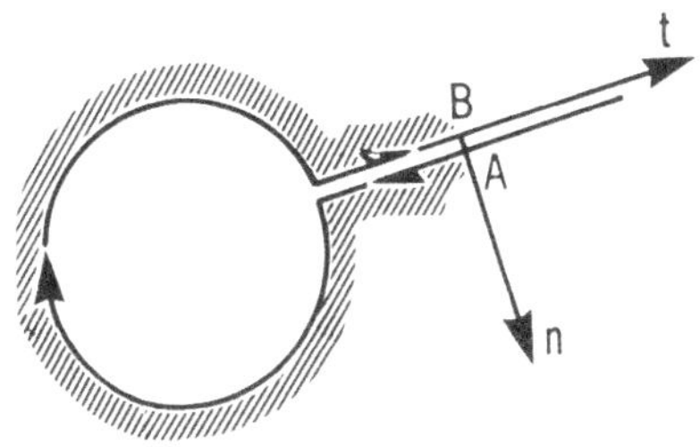

Figure 2 - Reduction of a Multi-Connected to Singly Connected Case

The Southwell Stress-Functions U and V

The total transverse load generated by the stress functions along the path from A to (x, y) (matter on the left-hand side) is denoted by T.

$$dT = -Q_y dx + Q_x dy = d\Omega$$

Hence,

$$T = \Omega(x, y) - \Omega_A .$$

The moments of the same stresses about axes parallel to x and y through the final point (x, y) are denoted by μ_x and μ_y. Then

$$d\mu_x = M_y dx - M_{xy} dy - Tdy = dU + \Omega_A dy$$

$$d\mu_y = -M_{xy} dx + M_x dy + Tdx = dV - \Omega_A dx$$

and

$$\mu_x = U(x, y) - U_A + (y-y_A)\Omega_A ,$$

$$\mu_y = V(x, y) - V_A - (x-x_A)\Omega_A .$$

Since U and V are determined except for the addition of arbitrary terms $(u_0-\omega y,\ v_0+\omega x)$, it is always possible to let U, V and Ω vanish at the reference point A. Then we have simple interpretations

$$T = \Omega(x, y) \quad , \quad \mu_x = U(x, y) \quad , \quad \mu_y = V(x, y).$$

The result is the same for all reconcilable paths because U and V without P_0 imply zero transverse pressure on the plate.

Applying it to the closed path of Figure 2, we conclude that in equations (40), T, μ_n, and μ_t are respectively the total transverse load and the bending moments about the axes n and t of the loads applied to the cavity boundary. Unless the loading system is statically equivalent to zero, the functions U and V do not have single values. In particular, considering the application of a concentrated transverse load or bending couple at an interior point of the plate as a limiting case of cavity loading, we conclude that such loads cause the stress functions to have more than one value each [14].

This raises an interesting question in the finite-element approach, when an interior point becomes a vertex common to several plate elements. Because in each element a corner load

$$Z = \Delta M_{nt}$$

will appear, owing to the jump in M_{nt} as we turn around the corner, we should infer from the foregoing considerations that if U and V have single values, the sum of all the corner loads must vanish.

The slab analogy provides an elegant proof of this. The analogue to M_{nt} being $-\varepsilon_{nt}$, the corner load is, except for a constant factor, measured by the change in wedge angle in the analogue state of plane stress described by single-valued displacement functions $u = U$ and $v = V$. Since, at an interior point, the sum of all wedge angles remains equal to 2π, the sum of all wedge-angle alterations must vanish and the corner loads consequently add up to zero.

APPENDIX 2

Boundary Transformation Rules

As the local axes (n, t) of Figure 1 are moved along the boundary, the unit vectors on them undergo the following changes

$$\frac{d\vec{t}}{ds} = -\vec{n}\frac{d\psi}{ds} \quad , \quad \frac{d\vec{n}}{ds} = \vec{t}\frac{d\psi}{ds} \quad , \tag{54}$$

which involve the radius of curvature of the boundary ($ds = R\, d\psi$). The notations $\partial/\partial s$ and $\partial/\partial t$ will indicate that we take or do not take into account the change in orientation of the local axes.

For a scalar $\phi(x, y)$ there is no difference

$$\frac{\partial \phi}{\partial t} \equiv \frac{\partial \phi}{\partial s} \quad . \tag{55}$$

For a vector $\vec{u} = u_n \vec{n} + u_t \vec{t}$, the identity

$$\frac{\partial \vec{u}}{\partial t} \equiv \frac{\partial \vec{u}}{\partial s}$$

produces in view of equation (54) the formulae for components

$$\frac{\partial u_n}{\partial t} = \frac{\partial u_n}{\partial s} - u_t \frac{d\psi}{ds} \quad , \quad \frac{\partial u_t}{\partial t} = \frac{\partial u_t}{\partial s} + u_n \frac{d\psi}{ds} \quad . \tag{56}$$

This, applied to the gradient vector

$$\text{grad } \phi = \frac{\partial \phi}{\partial n} \vec{n} + \frac{\partial \phi}{\partial t} \vec{t}$$

yields

$$\frac{\partial^2 \phi}{\partial t \partial n} = \frac{\partial^2 \phi}{\partial n \partial s} - \frac{\partial \phi}{\partial s} \frac{d\psi}{ds} \quad , \quad \frac{\partial^2 \phi}{\partial t^2} = \frac{\partial^2 \phi}{\partial s^2} + \frac{\partial \phi}{\partial n} \frac{d\psi}{ds} \quad . \tag{57}$$

For a symmetrical Cartesian tensor, we have first the rules of component transformation

$$\left.\begin{aligned} M_n &= \ell^2 M_x + m^2 M_y + 2\ell m M_{xy} \\ M_t &= m^2 M_x + \ell^2 M_y - 2\ell m M_{xy} \\ M_{nt} &= \ell m (M_y - M_x) + (\ell^2 - m^2) M_{xy} \end{aligned}\right\} \quad , \tag{58}$$

where ℓ and m are the components of the unit vector n, so that the following formulae hold

$$\frac{\partial \ell}{\partial n} = 0 \quad , \quad \frac{\partial \ell}{\partial s} = - m \frac{d\psi}{ds} \quad , \quad \frac{\partial m}{\partial n} = 0 \quad , \quad \frac{\partial m}{\partial s} = \ell \frac{d\psi}{ds} \quad . \tag{59}$$

It follows then that

$$\frac{\partial M_n}{\partial s} = \frac{\partial M_n}{\partial t} + 2\left(\ell \frac{\partial \ell}{\partial s} M_x + m \frac{\partial m}{\partial s} M_y + \left(\ell \frac{\partial m}{\partial s} + m \frac{\partial \ell}{\partial s}\right) M_{xy}\right)$$

or, using equations (59) and interpreting, we obtain

$$\frac{\partial M_n}{\partial s} = \frac{\partial M_n}{\partial t} + 2 \frac{d\psi}{ds} M_{nt} \ .$$

Similarly

$$\left.\begin{aligned} \frac{\partial M_t}{\partial s} &= \frac{\partial M_t}{\partial t} - 2 \frac{d\psi}{ds} M_{nt} \\ \frac{\partial M_{nt}}{\partial s} &= \frac{\partial M_{nt}}{\partial t} + \frac{d\psi}{ds} (M_t - M_n) \end{aligned}\right\} \ . \tag{60}$$

We note that

$$\begin{vmatrix} \frac{\partial U_n}{\partial n} & \frac{\partial U_n}{\partial t} \\ \frac{\partial U_t}{\partial n} & \frac{\partial U_t}{\partial t} \end{vmatrix}$$

is a Cartesian tensor. Hence by a similar procedure we can establish the following:

$$\frac{\partial^2 U_n}{\partial s \partial n} = \frac{\partial^2 U_n}{\partial t \partial n} + \frac{d\psi}{ds}\left(\frac{\partial U_n}{\partial t} + \frac{\partial U_t}{\partial n}\right) \ , \tag{61}$$

$$\frac{\partial^2 U_t}{\partial t^2} = \frac{\partial}{\partial s}\left(\frac{\partial U_t}{\partial t}\right) + \frac{d\psi}{ds}\left(\frac{\partial U_t}{\partial n} + \frac{\partial U_n}{\partial t}\right) \ , \tag{62}$$

$$\frac{\partial^2 U_n}{\partial t^2} = \frac{\partial}{\partial s}\left(\frac{\partial U_n}{\partial t}\right) + \frac{d\psi}{ds}\left(\frac{\partial U_n}{\partial n} - \frac{\partial U_t}{\partial t}\right) \ , \tag{63}$$

$$\frac{\partial^2 U_t}{\partial s \partial n} = \frac{\partial^2 U_t}{\partial t \partial n} + \frac{d\psi}{ds}\left(\frac{\partial U_t}{\partial t} - \frac{\partial U_n}{\partial n}\right) \ . \tag{64}$$

Further transformations of these results by equations (56) allow us to express the right-hand sides entirely in terms of s and n derivatives.

A CONFORMING FINITE ELEMENT FOR PLATE BENDING

B.M. Fraeijs de Veubeke

University of Liège, Belgium

ABSTRACT—A new, more compact, scheme is presented for building the (16×16) stiffness matrix of a finite plate element in bending. The element is of a conforming type: it satisfies continuity of deflections and slopes at the interfaces. In addition to the promise of good convergence characteristics, this property also guarantees lower bounds to influence coefficients. The arbitrary quadrilateral shape is also convenient for versatility in the applications.

1. INTRODUCTION

The difficulties involved in obtaining conforming displacement models for plate bending has been stressed in the literature [1 - 3].

Continuity of deflections at the interfaces is the prime requirement to satisfy. The next requirement is the continuity of slopes, which, if achieved, would produce a continuous, piecewise differential field of displacements in the structure. Most models, however, cannot comply rigorously with slope continuity.

Original publication, *Int. Jnl. of Solids and Structures*, Vol. 4, Pergamon Press, 1968, pp. 95-108.

Although it appears from experience that local violations in slope continuity do not necessarily prevent convergence of deflections to the correct values when mesh sizes are reduced, the convergence of the stress field is much more seriously affected. Another reason for striving to a rigorous enforcement of slope continuity is the guarantee it provides that the direct influence coefficients obtained are actually lower bounds to the true ones. This, coupled with an analysis of the same problem using equilibrium models, opens the possibility to compare those lower bounds with upper bounds and so ascertain the value of the approximations [4, 5].

A conforming element of quadrilateral shape can be obtained from a combination of cubic deflection fields in the four triangular regions delimited by the edges of the quadrilateral and its diagonals. The principle of this construction and the advantage of using the diagonals as a natural oblique co-ordinate system for assembling the fields were the subject of earlier publications [6, 7]. Further simplifications, reducing computer time, were discovered while setting up a program for the stiffness matrix of the element; they are reported here.

2. STRAIN ENERGY OF A TRIANGULAR FIELD

Consider the complete cubic deflection field

$$w = \alpha_1+\alpha_2 x+\alpha_3 y+\alpha_4 x^2+2\alpha_5 xy+\alpha_6 y^2+4(\alpha_7 x^3+\alpha_8 x^2 y+\alpha_9 xy^2+\alpha_{10} y^3) \ , \qquad (1)$$

valid in the triangular region 1 of Figure 1.

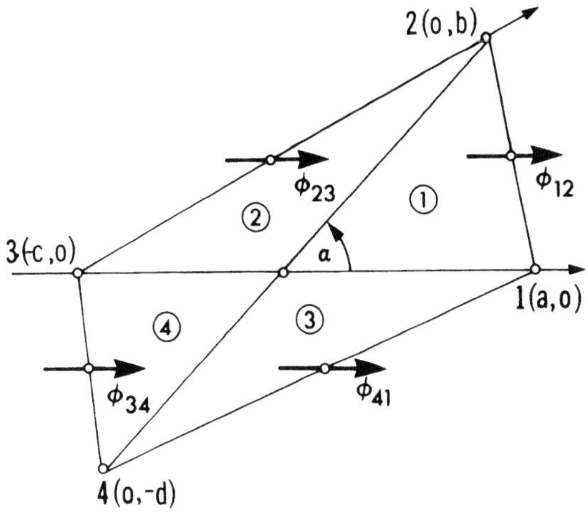

Figure 1

The ten coefficients can be determined in terms of the local deflections (w_0, w_1, w_2) at the nodes 1, 2, 0, the local slopes (ϕ_0, ϕ_1, ϕ_2), where $\phi = \partial w/\partial x$, the local slopes ($\psi_0$, ψ_1, ψ_2), where $\psi = \partial w/\partial y$, the local slope ϕ_{12} at mid distance along the edge 12.

$$\alpha_1 = w_0 \quad , \quad \alpha_2 = \phi_0 \; , \quad \alpha_3 = \psi_0 \tag{2}$$

$$\left.\begin{aligned}
a^2\alpha_4 &= -3w_0 - 2a\phi_0 + 3w_1 - a\phi_1 \; , \\
2ab\alpha_5 &= -6w_0 - 2a\phi_0 - 2b\psi_0 + 6w_1 - a\phi_1 + 2b\psi_1 + a\phi_2 - 4a\phi_{12} \; , \\
b^2\alpha_6 &= -3w_0 - 2b\psi_0 + 3w_2 - b\psi_2 \; , \\
4a^3\alpha_7 &= 2w_0 + a\phi_0 - 2w_1 + a\phi_1 \; , \\
4a^2b\alpha_8 &= 6w_0 + 2a\phi_0 + b\psi_0 - 6w_1 + a\phi_1 - b\psi_1 - a\phi_2 + 4a\phi_{12} \; , \\
4ab^2\alpha_9 &= 6w_0 + a\phi_0 + 2b\psi_0 - 6w_1 + a\phi_1 - 2b\psi_1 + 4a\phi_{12} \; , \\
4b^3\alpha_{10} &= 2w_0 + b\psi_0 - 2w_2 + b\psi_2 \; .
\end{aligned}\right\} \tag{3}$$

The bending strain energy in the oblique co-ordinates (x, y) is

$$U_1 = \frac{1}{2\sin\alpha}\iint_1 D\left[\left(\frac{w_{xx}+w_{yy}-2w_{xy}\cos\alpha}{\sin\alpha}\right)^2 - 2(1-v)(w_{xx}w_{yy}-w_{xy}^2)\right]dxdy$$

where D is the bending rigidity of the plate. To calculate it in terms of the generalized co-ordinates

$$(w_0\phi_0\psi_0w_1\phi_1\psi_1w_2\phi_2\psi_2\phi_{12}) = q_1'$$

(q_1' is the transpose of the column matrix q_1 of the generalized co-ordinates), we express the curvature as

$$\gamma = \begin{pmatrix} w_{xx} \\ w_{xy} \\ w_{yy} \end{pmatrix} = \left(\frac{1}{a^2b^2}W + \frac{2x}{a^3b^2}W_x + \frac{2y}{a^2b^3}W_y\right) q \qquad (4)$$

where the (3 x 10) matrices of constants W, W_x and W_y can be written down from

$$w_{xx} = 2\alpha_4+24x\alpha_7+8y\alpha_8$$

$$w_{xy} = 2\alpha_5+8x\alpha_8+8y\alpha_9$$

$$w_{yy} = 2\alpha_6+8x\alpha_9+24y\alpha_{10}$$

and substitution of the α_i from equations (2) and (3). Then

$$U_1 = \frac{1}{2}\iint_1 D\gamma' H\gamma dxdy \qquad (5)$$

with

$$H = \frac{1}{\sin^3\alpha}\begin{pmatrix} 1 & -2\cos\alpha & \cos^2\alpha+\nu\sin^2\alpha \\ -2\cos\alpha & 4\cos^2\alpha+2(1-\nu)\sin^2\alpha & -2\cos\alpha \\ \cos^2\alpha+\nu\sin^2\alpha & -2\cos\alpha & 1 \end{pmatrix}. \qquad (6)$$

If the bending rigidity remains constant over the area of the triangle, equation (5) can be evaluated in terms of

$$\iint_1 D\,dxdy = \frac{ab}{2}D \qquad \iint_1 Dxdxdy = \frac{a^2b}{6}D \qquad \iint_1 Dydxdy = \frac{ab^2}{6}D$$

$$\iint_1 Dx^2dxdy = \frac{a^3b}{12}D \qquad \iint_1 Dxydxdy = \frac{a^2b^2}{24}D \qquad \iint_1 Dy^2dxdy = \frac{ab^3}{12}D\ . \tag{7}$$

Then, having substituted equation (4) into (5)

$$U_1 = \frac{1}{2}\,q_1'K_1q_1 \tag{8}$$

and the stiffness matrix of the first triangle can be calculated from:

$$\frac{6a^3b^3}{D}K_1 = 3W'HW+2(W_x'HW+W'HW_x+W_y'HW+W'HW_y+W_x'HW_x+W_y'HW_y)+W_x'HW_y+W_y'HW_x\ .$$

Closer inspection of this formula reveals that, introducing the combinations

$$A_1 = W+W_x\quad ,\quad B_1 = W+W_y\quad ,\quad C_1 = W+W_x+W_y\ , \tag{9}$$

the number of products can be reduced to three

$$\frac{6a^2b^3}{D}K_1 = A_1'HA + B_1'HB + C_1'HC\quad . \tag{10}$$

Furthermore the matrices A_1, B_1 and C_1 are simpler

$$A_1 = \begin{pmatrix} 0 & -ab^2 & 0 & 0 & ab^2 & 0 & 0 & 0 & 0 & 0 \\ 0 & 0 & -ab^2 & 0 & 0 & ab^2 & 0 & 0 & 0 & 0 \\ 0 & a^3 & -2a^2b & -6a^2 & a^3 & -2a^2b & 6a^2 & 0 & -2a^2b & 4a^3 \end{pmatrix} \tag{11}$$

$$B_1 = \begin{pmatrix} 0 & -2ab^2 & b^3 & 0 & -ab^2 & -b^3 & 0 & -ab^2 & 0 & 4ab^2 \\ 0 & -a^2b & 0 & 0 & 0 & 0 & 0 & a^2b & 0 & 0 \\ 0 & 0 & -a^2b & 0 & 0 & 0 & 0 & 0 & a^2b & 0 \end{pmatrix} \tag{12}$$

$$C = \begin{pmatrix} 6b^2 & ab^2 & b^3 & -6b^2 & 2ab^2 & -b^3 & 0 & -ab^2 & 0 & 4ab^2 \\ 6ab & a^2b & ab^2 & -6ab & a^2b & -ab^2 & 0 & 0 & 0 & 4a^2b \\ 6a^2 & a^3 & a^2b & -6a^2 & a^3 & -2a^2b & 0 & 0 & a^2b & 4a^3 \end{pmatrix} . \tag{13}$$

3. ASSEMBLING THE STIFFNESS MATRICES OF THE TRIANGULAR FIELDS

In the second triangular field we can obtain continuity along the interface $x = 0$ of the deflection w (hence also continuity of $\partial w/\partial y$) and of the slope $\partial w/\partial x$ by retaining the values of the coefficients α_1, α_2, α_3, α_5, α_6, α_9 and α_{10} of the first field.

New values α_4', α_7' and α_8' can be adopted for the remaining coefficients. It can be verified that, as would be expected, the coefficients of the second field are given by the same formulas (2) and (3), provided we change a into $-c$, w_1 into w_3, ϕ_1 into ϕ_3, ψ_1 into ψ_3, ϕ_{12} into ϕ_{23}. The generalized coordinates for the second field are thus

$$(w_0\phi_0\psi_0 w_3\phi_3\psi_3 w_2\phi_2\psi_2\phi_{23}) = q_2' .$$

Proceeding in the same manner as for the first field, the strain energy

$$U_2 = \frac{1}{2}\, q_2' K_2 q_2$$

produces a stiffness matrix K_2 given by

$$\frac{6c^3b^3}{D} K_2 = A_2'HA_2 + B_2'HB_2 + C_2'HC_2$$

where A_2, B_2 and C_2 are directly obtainable from the corresponding matrices (11), (12) and (13) of the first field by changing a into $-c$.

For field 3, deflections and slopes at the interface $y = 0$ are identical with those of field 1 if the coefficients α_1,

α_2, α_3, α_4, α_5, α_7 and α_8 are retained.

The new coefficients are α'_6, α'_9 and α'_{10}. To determine the coefficients in terms of the generalized displacements

$$(w_0\phi_0\psi_0 w_1\phi_1\psi_1 w_4\phi_4\psi_4\phi_{41}) = q'_3$$

we can use formulas (2) and (3) with the modifications b into -d, w_2 into w_4, ϕ_2 into ϕ_4, ψ_2 into ψ_4 and ϕ_{12} into ϕ_{41}. The stiffness matrix K_3 is given by

$$\frac{6a^3d^3}{D} K_3 = A'_3 H A_3 + B'_3 H B_3 + C'_3 H C_3$$

where A_3, B_3, C_3 derive from A_1, B_1, C_1 by changing b into -d. Finally the fourth field conforms with fields 2 and 3 at the interfaces, if its coefficients are

$$\alpha_1,\ \alpha_2,\ \alpha_3,\ \alpha'_4,\ \alpha_5,\ \alpha'_6,\ \alpha'_7,\ \alpha'_8,\ \alpha'_9 \text{ and } \alpha'_{10}\ .$$

In terms of the generalized co-ordinates

$$(w_0\phi_0\psi_0 w_3\phi_3\psi_3 w_4\phi_4\psi_4\phi_{34}) = q'_4$$

the coefficients derive from equations (2) and (3) with the combined modifications: a into -c, b into -d

$$w_1,\ \phi_1,\ \psi_1 \quad \text{respectively into} \quad w_3,\ \phi_3,\ \psi_3$$

$$w_2,\ \phi_2,\ \psi_2 \quad \text{respectively into} \quad w_4,\ \phi_4,\ \psi_4$$

$$\phi_{12} \quad \text{into} \quad \phi_{34}\ .$$

Again, with the double change: a into -c and b into -d, the matrices (11)-(13) can be used to form

$$\frac{6c^3d^3}{D} K_4 = A'_4 H A_4 + B'_4 H B_4 + C'_4 H C_4\ .$$

The four partial stiffness matrices are combined into a (19 x 19) matrix J defined by energy addition

$$\sum_{1}^{4} q_i' K_i q_i = p' J p \qquad (14)$$

where

$$p' = (w_0 \phi_0 \psi_0 w_1 \phi_1 \psi_1 w_2 \phi_2 \psi_2 w_3 \phi_3 \psi_3 w_4 \phi_4 \psi_4 \phi_{12} \phi_{23} \phi_{34} \phi_{41}).$$

This can be done by using localizing matrices L_i

$$q_i = L_i p \qquad (15)$$

$$J = \sum_{1}^{4} L_i' K_i L_i \qquad (16)$$

or simply addressing directly the elements to their proper place in J as they are generated.

4. CONDENSATION

J is not a proper stiffness matrix since the nineteen co-ordinates in p are not independent. They depend on sixteen coefficients

$$\alpha_i \ (i = 1,2,3,\ldots,10) \quad \text{and} \quad \alpha_i' \ (i = 4,6,7,8,9,10).$$

It is interesting to observe that α_1, α_2, α_3 which represent the rigid body freedoms of the plate and α_5, which represents a torsional deformation mode, are the only coefficients valid throughout the four fields. Equating the four different field expressions obtained from equation (3) for coefficient α_5, we obtain a system of three equations that can easily be solved for the deflection and slope at the internal node:

$$w_0 = \frac{c}{a+c} w_1 + \frac{a}{a+c} w_3 - \frac{ac}{6(a+c)} \phi_1 + \frac{ac}{6(a+c)} \phi_3$$
$$+ \frac{2ac}{3(a+c)(b+d)} (-d\phi_{12}+d\phi_{23}+b\phi_{34}-b\phi_{41}) \quad ,$$

$$\phi_0 = \frac{3}{a+c} (w_1-w_3) - \frac{a}{2(a+c)} \phi_1 - \frac{c}{2(a+c)} \phi_3 + \frac{d}{2(b+d)} \phi_2 + \frac{b}{2(b+d)} \phi_4$$
$$- \frac{2}{(a+c)(b+d)} (ad\phi_{12}+cd\phi_{23}+bc\phi_{34}+ab\phi_{41}) \quad , \tag{17}$$

$$\psi_0 = \frac{c}{a+c} \psi_1 + \frac{a}{a+c} \psi_3 + \frac{2ac}{(a+c)(b+d)} (-\phi_{12}+\phi_{23}-\phi_{34}+\phi_{41}) \quad .$$

It can be verified that, with these values of w_0, ϕ_0 and ψ_0, the other equations between different expressions of the α_i and α_i' coefficients turn into identities. Hence the sixteen coefficients can be expressed in terms of the sixteen generalized co-ordinates

$$(w_1\phi_1\psi_1 w_2\phi_2\psi_2 w_3\phi_3\psi_3 w_4\phi_4\psi_4 \phi_{12}\phi_{23}\phi_{34}\phi_{41}) = q'$$

by a non-singular transformation.

If the linear relations (17) are put in matrix form

$$\begin{pmatrix} w_0 \\ \phi_0 \\ \psi_0 \end{pmatrix} = Mq \tag{18}$$

a condensation matrix N is defined by

$$N = \begin{pmatrix} M \\ \\ E \end{pmatrix} \tag{19}$$

where E is a (16 x 16) identity matrix, so that

$$p = Nq \quad . \tag{20}$$

The total energy becomes

$$\frac{1}{2}\, q'Kq \tag{21}$$

with the (proper) stiffness matrix

$$K = N'JN \,. \tag{22}$$

5. NORMALIZATION TO COMMON REFERENCE DIRECTIONS

K is now the stiffness matrix of a conforming element.

Indeed, consider the edge 12 and suppose the quantities $(w_1,\phi_1,\psi_1)(w_2,\phi_2,\psi_2)$ and ϕ_{12} to be known. From ϕ_1 and ψ_1 we can deduce the slopes s_1 and n_1 respectively in the direction 12 and normal to it in 1. Similarly from ϕ_2 and ψ_2 we can deduce s_2 and n_2. The knowledge of (w_1,w_2,s_1,s_2) is just sufficient to determine w everywhere, since it varies as a cubic along the edge. In particular the slope s_{12} at mid distance is thereby determined. From s_{12} and ϕ_{12}, n_{12} can now be determined and the knowledge of n_1, n_2 and n_{12} is just sufficient to determine the transverse slope n everywhere, since it has parabolic variation. Consequently, the knowledge of (w_1,ϕ_1,ψ_1), (w_2,ϕ_2,ψ_2) and ϕ_{12} determines completely the deflections and slopes along the edge 12. Should this edge become an interface, continuity of deflections and slopes will follow from the single-valuedness of deflections and slopes in the nodes 1 and 2 and single-valuedness of the transverse slope at mid distance.

As the natural reference directions, provided by the diagonals, can change when passing from one element to the other, it is advisable to refer the slopes to common reference directions. This facilitates pairwise identification of the co-ordinates that will in fact become the nodal displacements of the assembled structure.

For the identification of slopes at mid distance of an edge the best policy is to use as generalized co-ordinates the slopes in the directions of the outward normal (Figure 2).

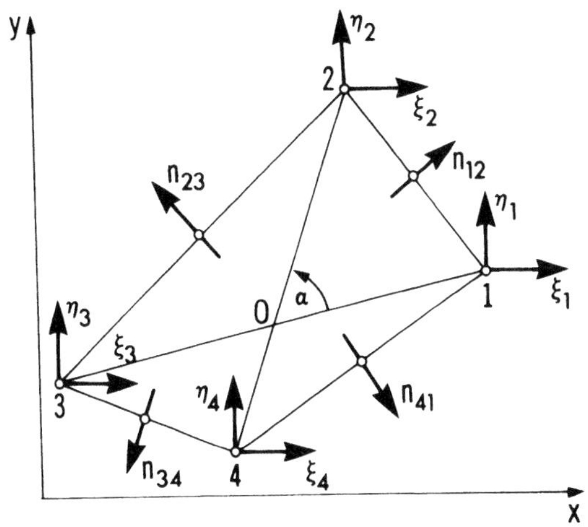

Figure 2

Then, when assembling elements, account must be taken of the reversal of sign of the reference direction. Taking the case of the edge 12, the problem consists in expressing ϕ_{12} in terms of the slope n_{12} in the direction of the outward normal and (w_1,ϕ_1,ψ_1), (w_2,ϕ_2,ψ_2).

One procedure is as follows. From ϕ_1 and ψ_1, deduce the slope s_1 in the direction 12

$$c_{12}s_1 = -a\phi_1 + b\psi_1$$

$c_{12} = (a^2+b^2-2ab\cos\alpha)^{1/2}$ the length of the edge.

Similarly in 2

$$c_{12}s_2 = -a\phi_2+b\psi_2 \ .$$

Then, considering the cubic variation of w along 12, express the slope s_{12} at mid distance

$$4c_{12}s_{12} = 6w_2-6w_1-c_{12}(s_1+s_2)$$

or, substituting s_1 and s_2,

$$4c_{12}s_{12} = 6w_2-6w_1+a\phi_1-b\psi_1+a\phi_2-b\psi_2 \ .$$

Finally express ϕ_{12} in terms of s_{12} and n_{12}

$$c_{12}\phi_{12} = b\sin\alpha\, n_{12}-(a-b\cos\alpha)s_{12} \ .$$

When s_{12} is substituted the required formula is

$$\phi_{12} = \frac{b\sin\alpha}{c_{12}} n_{12} - \frac{a-b\cos\alpha}{4c_{12}^2}(6w_2-6w_1+a\phi_1-b\psi_1+a\phi_2-b\psi_2) \ . \tag{23}$$

At the vertices, where several elements are interconnected, the slopes should be referred to a common cartesian reference frame (Figure 2). The cartesian slopes $\partial w/\partial X$ and $\partial w/\partial Y$ will be denoted by ξ and η respectively. They are related to the slopes ϕ and ψ in the reference frame of the diagonals by

$$\phi = \xi\frac{\partial X}{\partial x} + \eta\frac{\partial Y}{\partial x} \quad , \quad \psi = \xi\frac{\partial X}{\partial y} + \eta\frac{\partial Y}{\partial y} \ . \tag{24}$$

There is a choice for expressing the direction cosines, which can sometimes produce analytical simplifications:

$$\frac{\partial X}{\partial x} = \frac{x_{10}}{a} = \frac{x_{03}}{c} = \frac{x_{13}}{a+c} \quad , \quad \frac{\partial Y}{\partial x} = \frac{y_{10}}{a} = \frac{y_{03}}{c} = \frac{y_{13}}{a+c}$$

$$\frac{\partial X}{\partial y} = \frac{x_{20}}{b} = \frac{x_{04}}{d} = \frac{x_{24}}{b+d} \quad , \quad \frac{\partial Y}{\partial y} = \frac{y_{20}}{b} = \frac{y_{04}}{d} = \frac{y_{24}}{b+d} \ . \tag{25}$$

In these formulas x_{ij} stands for x_i-x_j and y_{ij} for y_i-y_j, where (x_i,y_i) are the co-ordinates of the intersection of the diagonals $(i = 0)$ and of the vertices $(i = 1,2,3,4)$ in the cartesian reference frame. The computer can generate the required goemetrical characteristics from the co-ordinates of the vertices alone by formulas such as

$$\left.\begin{aligned} x_0 &= \frac{x_1x_2y_{34}+x_2x_3y_{41}+x_3x_4y_{12}+x_4x_1y_{23}}{y_{31}x_{24}-x_{31}y_{24}} \\ y_0 &= \frac{y_1y_2x_{34}+y_2y_3x_{41}+y_3y_4x_{12}+y_4y_1x_{23}}{x_{31}y_{24}-y_{31}x_{24}} \end{aligned}\right\} \tag{26}$$

$$\left.\begin{aligned} a^2 &= x_{10}^2 + y_{10}^2 \\ (a+c)^2 &= x_{13}^2 + y_{13}^2 \\ c_{12}^2 &= x_{12}^2 + y_{12}^2 \\ ab\sin\alpha &= x_{10}y_{20} - y_{10}x_{20} \\ ab\cos\alpha &= x_{10}x_{20} + y_{10}y_{20} \\ a(a-b\cos\alpha) &= x_{10}x_{12} + y_{10}y_{12} \end{aligned}\right\} . \tag{27}$$

Our final (conforming) set of generalized co-ordinates is described by the row vector

$$r' = (w_1\eta_1\xi_1w_2\eta_2\xi_2w_3\eta_3\xi_3w_4\eta_4\xi_4n_{12}n_{23}n_{34}n_{41}) .$$

The transformation

$$q = Pr \tag{28}$$

from the old (natural) set of co-ordinates to the conforming one is made up from eight equations of type (24) applied to the slopes at the vertices, four identities $w_i = w_i$ at the vertices, and four equations of type (23) in which, however, the natural slopes at the vertices are still to be replaced by the cartesian ones. The result for ϕ_{12} turns out to be

$$\phi_{12} = \frac{x_{10}y_{20} - y_{10}x_{20}}{\sqrt{[(x_{10}^2 + y_{10}^2)(x_{12}^2 + y_{12}^2)]}} n_{12}$$

$$- \frac{x_{10}x_{12} + y_{10}y_{12}}{4(x_{12}^2+y_{12}^2)\sqrt{(x_{10}^2+y_{10}^2)}} (6w_2-6w_1+x_{12}\xi_1+x_{12}\xi_2+y_{12}\eta_1+y_{12}\eta_2) \quad (29)$$

In this form, the other ϕ_{ij} values can be obtained by a cyclic permutation of the indices (1, 2, 3, 4).

Under the linear transformation (28) the energy of the element becomes the quadratic form

$$\frac{1}{2} r'Rr$$

in the conforming co-ordinates. The final (operational) stiffness matrix R is to be calculated from

$$R = P'KP \quad \text{or} \quad R = Q'JQ \quad (30)$$

where Q = NP is the product of the condensing matrix N and the conforming matrix P. This product can be programmed numerically If an analytical form is preferred, the only algebra remaining is the substitution of equations (24) and (29) into equations (17).

6. SHAPING OR WEIGHTING FUNCTIONS

In order to translate externally applied loads into generalized loads, the transverse displacement of the plate should be expressed in terms of the generalized displacements. Suppose that in each field (i = 1,2,3,4) w is expressed in function of the co-ordinates q_i. For example, in field 1,

$$\left.\begin{aligned} w = {} & w_0W_{0,1}(x,y)+\phi_0\Phi_{0,1}(x,y)+\psi_0\Psi_{0,1}(x,y)+w_1W_{1,0}(x,y) \\ & + \phi_1\Phi_{1,1}(x,y)+\psi_1\Psi_{1,1}(x,y)+w_2W_{2,1}(x,y)+\phi_2\Phi_{2,1}(x,y) \\ & + \psi_2\Psi_{2,1}(x,y)+\phi_{12}\Phi_{12}(x,y) \ . \end{aligned}\right\} \quad (31)$$

Then, the virtual work equation for a transverse load $\pi(x,y)$ on that part of the element is

$$\sin\alpha \iint_1 \pi(x,y)\,w\,dx\,dy = q_1' f_1 \tag{32}$$

where the elements of the generalized loads matrix f_1 are in succession

$$\sin\alpha \iint_1 \pi(x,y) W_{0,1}(x,y)\,dx\,dy, \sin\alpha \iint_1 \pi(x,y)\Phi_{0,1}(x,y)\,dx\,dy, \text{ etc.}$$

The contribution of concentrated loads is simply obtained by multiplication of the load by the local values of the weighting functions

$$W_{0,1}(x,y) \text{ , etc.}$$

The total virtual work, adding the four field contributions, will be, by virtue of equation (15)

$$\sum_1^4 q_i' f_i = p' \sum_1^4 L_i' f_i = p' f \text{ .} \tag{33}$$

Hence, the generalized loads conjugate to the generalized displacements in p are obtained from

$$f = \sum_1^4 L_i' f_i \text{ .} \tag{34}$$

Again, from the invariance of the virtual work, and by virtue of relations (20), (28) and (30)

$$g = N' \left(\sum_1^4 L_i' f_i \right) \qquad \text{(virtual work } g'q\text{)} \tag{35}$$

is the loads matrix conjugate to q (defining a set of "natural" generalized displacements), while

$$h = P'g = Q'\left(\sum_{1}^{4} L_i' f_i\right) \qquad \text{(virtual work } h'r) \tag{36}$$

is the loads matrix conjugate to r (defining a set of "conforming" generalized displacements).

Hence, since the matrix Q is available, being necessary to generate the operational stiffness matrix, it is convenient to evaluate the loads by the f_i contributions.

The required weighting functions for field 1 turn out to be:

$$\left.\begin{aligned}
W_{0,1} &= \left(1 - \frac{x}{a} - \frac{y}{b}\right)^2 \left(1 + 2\frac{x}{a} + 2\frac{y}{b}\right) ; \\
\Phi_{0,1} &= x\left(1 - \frac{x}{a} - \frac{y}{b}\right)^2 ; \quad \Psi_{0,1} = y\left(1 - \frac{x}{a} - \frac{y}{b}\right)^2 ; \\
W_{1,1} &= \frac{x}{a}\left(3\frac{x}{a} + 6\frac{y}{b} - 2\frac{x^2}{a^2} - 6\frac{x}{a}\frac{y}{b} - 6\frac{y^2}{b^2}\right) ; \\
\Phi_{1,1} &= x\left(-\frac{x}{a} - \frac{y}{b} + \frac{x^2}{a^2} + \frac{x}{a}\frac{y}{b} + \frac{y^2}{b^2}\right) ; \\
\Psi_{1,1} &= y\frac{x}{a}\left(2 - \frac{x}{a} - 2\frac{y}{b}\right) ; \\
W_{2,1} &= \frac{y^2}{b^2}\left(3 - 2\frac{y}{b}\right) ; \quad \Phi_{2,1} = x\frac{y}{b}\left(1 - \frac{x}{a}\right) ; \\
\Psi_{2,1} &= y\frac{y}{b}\left(-1 + \frac{y}{b}\right) ; \\
\Phi_{12} &= 4x\frac{y}{b}\left(-1 + \frac{x}{a} + \frac{y}{b}\right) .
\end{aligned}\right\} \tag{37}$$

The modifications to introduce in order to convert those functions into the weighting functions of fields 2, 3 and 4 are exactl the same as those discussed in Section 3.

7. GENERALIZED EXTERNAL LOADS FOR UNIFORM TRANSVERSE LOADING

To evaluate the integrals involved in equation (32), which are extended over a triangular field, it is convenient to change variables as follows:

$$\frac{x}{a} = v - u \quad ; \quad \frac{y}{b} = 1 - v$$

so that equation (32) is transformed into

$$q_1' f_1 = ab \sin\alpha \int_0^1 \left\{ \int_0^v \pi(x,y) w du \right\} dv \tag{38}$$

and the weighting functions (37) become

$$\left.\begin{aligned}
W_{0,1} &= u^2(3 - 2u) \; ; \\
\Phi_{0,1} &= au^2(v - u) \; ; \\
\Psi_{0,1} &= bu^2(1 - v) \; ; \\
W_{1,1} &= v^3(3 - 2v) - u^2(3 - 2u) \; ; \\
\Phi_{1,1} &= a(v - u)(u^2 + v^2 - uv - v) \; ; \\
\Psi_{1,1} &= b(v^2 - u^2)(1 - v) \; ; \\
W_{2,1} &= (1 - v)^2(1 + 2v) \; ; \\
\Phi_{2,1} &= a(v - u)(1 - v)(1 - v + u) \; ; \\
\Psi_{2,1} &= - bv(1 - v)^2; \\
\Phi_{12} &= - 4au(v - u)(1 - v) \; .
\end{aligned}\right\} \tag{39}$$

In the case of a uniform transverse load π_1 in field one, integration produces the following generalized loads matrix

$$f_1' = \frac{\pi_1 ab \sin\alpha}{120} (18 \ 2a \ 2b \ 24 \ -5a \ 4b \ 18 \ 3a \ -4b \ -4a).$$

Similar results are immediately obtainable for the three other fields by the appropriate substitutions in the geometrical parameters a and b. The proper combinations generating the equivalent external loads for the element are then obtained through equation (36).

8. LAGRANGIAN MASS MATRICES

Let $m_1(x,y)$ be the distribution of mass over the area of the first field of the plate element. The kinetic energy, neglecting the rotary inertia, will then be

$$\sin\alpha \frac{1}{2} \iint_1 m_1(x,y)\dot{w}^2(x,y)dxdy = \frac{1}{2} ab \sin\alpha \int_0^1 \left\{ \int_0^v m_1(u,v)\dot{w}^2(u,v)du \right\} dv .$$

Substitution of the transverse velocity in terms of the shaping functions

$$\dot{w} = \dot{w}_0 W_{0,1}(u,v) + \cdots + \dot{\phi}_{12}\Phi_{12}(u,v)$$

gives the quadratic form

$$\frac{1}{2} \dot{q}_1' M_1 \dot{q}_1 .$$

With a Lagrangian mass matrix M_1 having elements defined by

$$\left.\begin{aligned} & ab \sin\alpha \int_0^1 \left\{ \int_0^v m_1(u,v) W_{0,1}^2(u,v)du \right\} dv \quad , \\ & ab \sin\alpha \int_0^1 \left\{ \int_0^v m_1(u,v) W_{0,1}\Phi_{0,1} du \right\} dv \quad \text{etc.} \end{aligned}\right\} \qquad (40)$$

Those elements have been evaluated for the case of a constant mass distribution $m_1(x,y) = m$ and the resulting Lagrangian mass matrix M_1 is given in Table 1. The corresponding matrices for the other fields follow by the appropriate modifications in the geometrical constants.

Table 1 - Lagrangian Mass Matrix for the q_1 Co-ordinates; Uniform Mass Distribution m

$$\frac{mab\sin\alpha}{10\cdot 080}\begin{pmatrix} 864 & & & & & & & & & \\ 84a & 12a^2 & & & & & \textit{Symmetrical} & & & \\ 84b & 6ab & 12b^2 & & & & & & & \\ 396 & 66a & 36b & 1224 & & & & & & \\ -90a & -15a^2 & -8ab & -240a & 50a^2 & & & & & \\ 66b & 9ab & 8b^2 & 174b & -37ab & 32b^2 & & & & \\ 252 & 18a & 48b & 396 & -90a & 96b & 864 & & & \\ 42a & 5a^2 & 6ab & 102a & -23a^2 & 22ab & 108a & 20a^2 & & \\ -66b & -5ab & -12b^2 & -102b & 23ab & -24b^2 & -168b & -24ab & 36b^2 & \\ -96a & -12a^2 & -12ab & -144a & 32a^2 & -28ab & -96a & -20a^2 & 24ab & 32a^2 \end{pmatrix}$$

The Lagrangian mass matrix of the element in conforming co-ordinates follows by addition of energies and application of the linear transformations (15), (20) and (28) as

$$M = P'N'\left(\sum_1^4 L_i'M_iL_i\right)NP = Q'\left(\sum_1^4 L_i'M_iL_i\right)Q\ . \qquad (41)$$

9. FINAL REMARKS

The element is now operational and preliminary results for the case of a rectangular plate, centrally loaded, under various boundary conditions, are extremely satisfactory from the deflections point of view. Numerical results and comparison with other reported data are given in [8].

Finally, in a private communication, Professor T.H. Pian pointed out the similarity of our technique for assembling polynomial approximations to the methods described in [9 - 11].

REFERENCES

[1] ZIENKIEWICZ, O.C. and CHEUNG, Y.K., "The Finite Element Method for Analysis of Elastic Isotropic and Orthotropic Slabs", *Proc. Inst. Civil Engrs.*, Vol. 28, 1964, pp. 471-488.

[2] BAZELEY, G.P., CHEUNG, Y.K., IRONS, B.M. and ZIENKIEWICZ, O.C., "Triangular Elements in Bending. Conforming and Nonconforming Solutions", *Proc. Conf. Matrix Methods in Structural Mechanics*, Air Force Institute of Technology, Dayton, Ohio, October, 1965.

[3] CLOUGH, R.W. and TOCHER, J.L., "Finite Element Stiffness Matrices for Analysis of Plates in Bending", *Proc. Conf. Matrix Methods in Structural Mechanics*, Air Force Institute of Technology, Dayton, Ohio, October, 1965.

[4] FRAEIJS DE VEUBEKE, B.M., "Upper and Lower Bounds in Matrix Structural Analysis", *AGARDograph 78*, Pergamon Press, 1963, pp. 165-201.

[5] FRAEIJS DE VEUBEKE, B.M., "Displacement and Equilibrium Models in the Finite Element Method", *Stress Analysis*, (editors: O.C. Zienkiewicz and G.S. Holister), Chapter 9, Wiley, 1965.

[6] SANDER, G., "Bornes supérieures et inférieures dans l'analyse matricielle des plaques en flexion-torsion", *Bull. Soc. r. Sci. Liège*, Vol. 33, 1964, pp. 456-494.

[7] FRAEIJS DE VEUBEKE, B.M., "Bending and Stretching of Plates", *Proc. Conf. Matrix Methods in Structural Mechanics*, Air Force Institute of Technology, Dayton, Ohio, October, 1965.

[8] SANDER, G. and FRAEIJS DE VEUBEKE, B.M., "Upper and Lower Bounds to Structural Deflections by Dual Analysis in Finite Elements", *Annual Summary I*, Air Force Technical Report AFFDL-TR-66-199, 1966, pp. 3-21 to 3-27.

[9] BIRKHOFF, G. and GARABEDIAN, H.L., "Smooth Surface Interpolation", *J. Math. Phys.*, Vol. 39, 1960, pp. 258-268.

[10] DE BOOR, C., "Bicubic Spline Interpolation", *J. Math. Phys.*, Vol. 41, 1962, pp. 212-218.

[11] BIRKHOFF, G. and DE BOOR, C., *Piecewise Polynomial Interpolation and Approximation*, Elsevier, 1965.

AN EQUILIBRIUM MODEL FOR PLATE BENDING

B.M. Fraeijs de Veubeke and G. Sander

University of Liège, Belgium

ABSTRACT—A conforming displacement model for plate bending was presented earlier [5]. It is of interest to have also an equilibrium model available in order to generate both upper and lower bounds to plate deflections. The theory of the triangular equilibrium model, which is presented here, is a revised version of that of [3], taking advantage of oblique co-ordinates. It is also extended to cover transverse loading modes. Because the numerical investigations required adaptability of the model to a stiffness computer-program, only the elaboration of the stiffness matrix was aimed at, following the general procedure set up in [2]. The element can nevertheless be recognized as the Southwell analogue [4] of the plane stress model with quadratic displacement field described in [2]. As such it could be handled efficiently by a force program.

Numerical results show the expected monotonic convergence of deflections from above for the equilibrium model and from below for the conforming model. The convergence rate, in terms of total number of generalized coordinates, is compared with that of other plate bending elements.

1. PLATE EQUATIONS IN OBLIQUE CO-ORDINATES

The (x,y) axes of Figure 1 are taken to lie in the middle plane of the unstrained plate; the middle surface deflections w(x,y) are taken positive upwards in the direction of the Oz axis.

Original publication, *Int. Jnl. of Solids and Structures*, Vol. 4, (in collaboration with G. Sander), Pergamon Press, 1968, pp. 447-468.

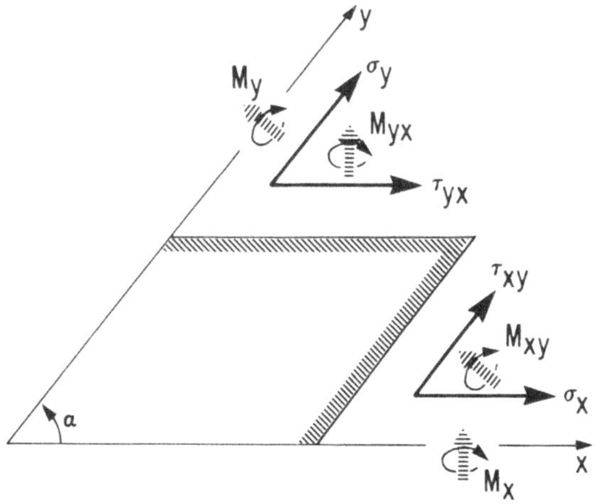

Figure 1 - Stresses and Moments in Oblique Co-ordinates

Bending and twisting moments are oriented by the right-hand screw convention and correspond to oblique stress components positive in the upper layers. Shear loads are positively oriented as the transverse shear stresses. We define covariant slopes of the middle surface by the formulas

$$\phi = - \frac{\partial w}{\partial x} = -w_{,x} \quad , \quad \psi = - \frac{\partial w}{\partial y} = -w_{,y} \tag{1}$$

and verify that they are the orthogonal projections of the local rotation vector $\underline{\omega}$, lying in the middle plane, on the directions $\underline{k} \times \underline{i}$ and $\underline{k} \times \underline{j}$ respectively conjugate to $\underline{i}$ and $\underline{j}$. $(\underline{i},\underline{j},\underline{k})$ are unit vectors along the axes 0x, 0y and 0z respectively. Hence the conjugate direction to $\underline{i}$ is obtained by turning it of 90° about 0z in the positive (right-hand screw) sense. The same for the direction conjugate to $\underline{j}$ (Figure 2).

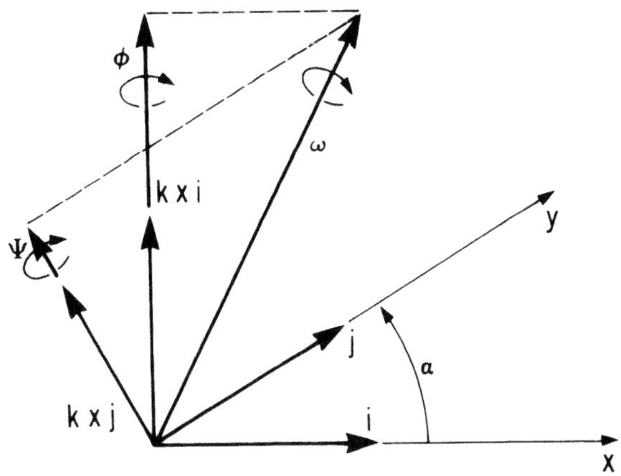

Figure 2 - Definition of Covariant Slopes in Oblique Co-ordinates

It is sufficient to carry out the proof for the $\underline{\omega}_0$ vector at the origin. In the neighbourhood of the origin the rigid body displacement field is

$$\underline{w} = \underline{\omega} \times \underline{r}$$

where

$$\underline{r} = x\underline{i} + y\underline{j} \ .$$

Hence

$$\underline{w} = (\underline{\omega}\times\underline{i})x + (\underline{\omega}\times\underline{j})y = -(\underline{i}\times\underline{\omega})x - (\underline{j}\times\underline{\omega})y \ .$$

In projection on the 0z axis

$$w = -x(\underline{i}\times\underline{\omega}).\underline{k} - y(\underline{j}\times\underline{\omega}).\underline{k}$$

and by definition (1)

$$\phi = (\underline{i}\times\underline{\omega}).\underline{k} = (\underline{k}\times\underline{i}).\underline{\omega}$$
$$\psi = (\underline{j}\times\underline{\omega}).\underline{k} = (\underline{k}\times\underline{j}).\underline{\omega} \ . \qquad (2)$$

This property of covariant slopes is convenient because the bending and twisting moments are precisely oriented along the conjugate directions so that the virtual work of moments has a natural association with the covariant components of rotation. For the elementary parallelogram of edges dx and dy the virtual work performed by moments in slope increases $\delta\phi$ and $\delta\psi$ is

$$\frac{\partial}{\partial x}(M_x\delta\phi+M_{xy}\delta\psi)dxdy + \frac{\partial}{\partial y}(M_{xy}\delta\phi+M_y\delta\psi)dxdy \ .$$

The transverse shear loads Q_x and Q_y contribute

$$\frac{\partial}{\partial x}(Q_x\delta w)\ dxdy + \frac{\partial}{\partial y}(Q_y\delta w)dy$$

and the external transverse load distribution q(x,y)

$$q\delta w \sin\alpha\ dxdy \quad .$$

The total virtual work is stored as an increase in strain energy $\delta W \sin\alpha\ dxdy$, where W denotes the strain energy per unit area. Hence, after simplification by the common factor dxdy and substitution of (1)

$$\sin\alpha\ \delta W = -\frac{\partial}{\partial x}\left(M_x\frac{\partial}{\partial x}\delta w + M_{xy}\frac{\partial}{\partial y}\delta w\right) - \frac{\partial}{\partial y}\left(M_{xy}\frac{\partial}{\partial x}\delta w + M_y\frac{\partial}{\partial y}\delta w\right)$$

$$+ \frac{\partial}{\partial x}(Q_x\delta w) + \frac{\partial}{\partial y}(Q_y\delta w) + q\sin\alpha\ \delta w \ . \qquad (3)$$

The equilibrium equations are now obtained by stating that there is no increase in strain energy when the additional deflection field is of rigid body type. Thus is δw = constant

$$\frac{\partial Q_x}{\partial x} + \frac{\partial Q_y}{\partial y} + q\sin\alpha = 0 \quad . \qquad (4)$$

If $\delta w = x \times$ constant

$$-\frac{\partial M_x}{\partial x} - \frac{\partial M_{xy}}{\partial y} + Q_x = 0 \ . \qquad (5)$$

If $\delta w = y \times$ constant

$$- \frac{\partial M_{xy}}{\partial x} - \frac{\partial M_y}{\partial y} + Q_y = 0 \ . \tag{6}$$

Those equilibrium equations are formally identical with those in cartesian coordinates, only the factor $\sin\alpha$ in (4) betrays the obliquity of the axes. The general statement (3) can now be simplified by taking the equilibrium equations into account:

$$\sin\alpha \ \delta W = -M_x \delta w_{,xx} - 2M_{xy}\delta w_{,xy} - M_y \delta w_{,yy} \ . \tag{7}$$

It shows that the strain energy per unit area is a function of the elements ($w_{,xx}$, $w_{,xy}$, $w_{,yy}$) of the curvature tensor such that

$$M_x = -\sin\alpha \frac{\partial W}{\partial w_{,xx}} \quad ; \quad 2M_{xy} = -\sin\alpha \frac{\partial W}{\partial w_{,xy}} \quad ; \quad M_y = -\sin\alpha \frac{\partial W}{\partial w_{,yy}} \ . \tag{8}$$

The complementary energy per unit area is defined, as is usual, by the Legendre transformation

$$\Phi = w_{,xx} \frac{\partial W}{\partial w_{,xx}} + w_{,xy} \frac{\partial W}{\partial w_{,xy}} + w_{,yy} \frac{\partial W}{\partial w_{,yy}} - W$$

where the right hand side is to be expressed in terms of the conjugate variables M_x, M_{xy} and M_y. Differentiating totally

$$\Phi = - \frac{1}{\sin\alpha} (M_x w_{,xx} + 2M_{xy} w_{,xy} + M_y w_{,yy}) - W \tag{9}$$

and taking (7) into account

$$\delta\Phi = - \frac{1}{\sin\alpha} (w_{,xx}\delta M_x + 2w_{,xy}\delta M_{xy} + w_{,yy}\delta M_y) .$$

So that the dual energy relations are

$$w_{,xx} = -\sin\alpha \frac{\partial \Phi}{\partial M_x} \quad ; \quad 2w_{,xy} = -\sin\alpha \frac{\partial \Phi}{\partial M_{xy}} \quad ; \quad w_{,yy} = -\sin\alpha \frac{\partial \Phi}{\partial M_y} \ . \tag{10}$$

We know [1] that, for an isotropic plate of bending rigidity D, the strain energy density is

$$W = \frac{D}{\sin^2\alpha} \quad \frac{1}{2}\,\theta_1^2 - (1-\nu)\theta_2$$

$$\theta_1 = \frac{1}{\sin\alpha}\,(w_{,xx}+w_{,yy}-2\cos\alpha\; w_{,xy}) \tag{11}$$

$$\theta_2 = w_{,xx}w_{,yy}-w^2_{,xy} \; .$$

This gives by (8) the following curvature-moment relations

$$\sin^3\alpha\; M_x = -D[w_{,xx}-2\cos\alpha\; w_{,xy}+(\cos^2\alpha+\nu\sin^2\alpha)w_{,yy}]$$

$$\sin^3\alpha\; M_{xy} = -D[-\cos\alpha(w_{,xx}+w_{,yy})+(1+\cos^2\alpha-\nu\sin^2\alpha)w_{,xy}] \tag{12}$$

$$\sin^3\alpha\; M_y = -D[w_{,yy}-2\cos\alpha\; w_{,xy}+(\cos^2\alpha+\nu\sin^2\alpha)w_{,xx}] \; .$$

From this we find that

$$J_1 = M_x + M_y + 2\cos\alpha\; M_{xy} = -D(1+\nu)\theta_1$$

$$J_2 = M_xM_y - M_{xy}^2 = \frac{D^2}{\sin^2\alpha}\,(\nu\theta_1^2+(1-\nu)^2\theta_2) \; . \tag{13}$$

Solving for θ_1 and θ_2 and replacing into (11) which then becomes the complementary energy density

$$\Phi = \frac{1}{D(1-\nu)}\left[\frac{1}{2(1+\nu)\sin^2\alpha}\,J_1^2 - J_2\right] \; . \tag{14}$$

The dual energy relations obtained from (10) are then

$$D(1-\nu^2)w_{,xx} = -\frac{1}{\sin\alpha}\,(M_x+M_y+2\cos\alpha\; M_{xy})+(1+\nu)\sin\alpha\; M_y$$

$$D(1-\nu^2)w_{,xy} = -\frac{\cos\alpha}{\sin\alpha}\,(M_x+M_y+2\cos\alpha\; M_{xy})-(1+\nu)\sin\alpha\; M_{xy} \tag{15}$$

$$D(1-\nu^2)w_{,yy} = -\frac{1}{\sin\alpha}\,(M_x+M_y+2\cos\alpha\; M_{xy})+(1+\nu)\sin\alpha\; M_x .$$

2. BENDING AND TWISTING MOMENTS AND KIRCHHOFF LOADS FOR ARBITRARY BOUNDARY ORIENTATION

The analysis of an arbitrary triangular element is conveniently carried out by making two of the edges coincident with the oblique reference axes (Figure 3). The third edge is located by its intercepts a and b; its length c is related to the lengths of the outer edges by the elementary formula

$$c^2 = a^2 + b^2 - 2ab\cos\alpha \; .$$

Other lengths which are convenient to introduce to simplify formulas are the two parts in which the third edge is cut by the perpendicular from 0:

$$u = \frac{a(a-b\cos\alpha)}{c} \quad ; \quad v = \frac{b(b-a\cos\alpha)}{c} \quad ; \quad u+v = c \; . \qquad (16)$$

In calculating the shear load Q_n, bending moment M_n and twisting moment M_{ns} corresponding to a facet of normal $\underline{n}$ in terms of Q_x, Q_y, M_x, M_{xy} and M_y at the same point, the lengths (a,b,c) are first considered as infinitesimals.

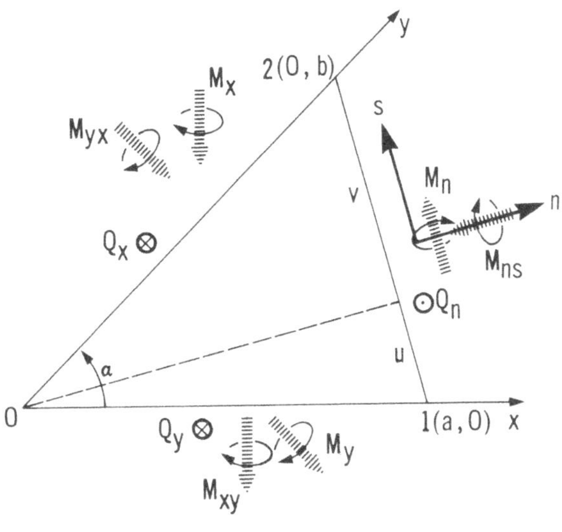

Figure 3 - Moments and Shear Load Along Slanting Edge. Shear Loads Q_x and Q_y are taken Positive Downwards, Q_n Upwards

Then, since the contribution of the transverse load q is an infinitesimal of higher order, vertical equilibrium requires

$$cQ_n = bQ_x + aQ_y \ . \tag{17}$$

Again, in rotational equilibrium, the couples due to shear loads can be neglected as infinitesimals of higher order. Then, by considering equilibrium about an axis perpendicular to Ox

$$M_n b \sin\alpha - M_{ns}(a-b\cos\alpha) = bM_x + (a+b\cos\alpha)M_{xy} + a\cos\alpha\, M_y$$

and for equilibrium about Ox

$$M_n(a-b\cos\alpha) + M_{ns} b \sin\alpha = b\sin\alpha\, M_{xy} + a\sin\alpha\, M_y \ .$$

These relations solves for M_n and M_{ns} can be written

$$M_n = \frac{\sin\alpha}{c^2}(b^2M_x+2abM_{xy}+a^2M_y) \tag{18}$$

$$M_{ns} = -\frac{u}{a}\left[\frac{b}{c}M_x + \frac{a}{c}M_{xy}\right] + \frac{v}{b}\left[\frac{a}{c}M_y + \frac{b}{c}M_{xy}\right] \ . \tag{19}$$

Formulas (17), (18) and (19) can now be applied to the third edge of the finite element, provided $(Q_x, Q_y, M_x, M_{xy}, M_y)$ are known along this edge. In particular, considering the virtual work equation

$$\int_1^2 \left(Q_n\delta w - M_{ns}\frac{\partial}{\partial s}\delta w\right)ds = -[M_{ns}\delta w]_1^2 + \int_1^2\left(Q_n + \frac{\partial}{\partial s}M_{ns}\right)\delta w ds$$

it can be concluded that the shear stresses along this edge are statically equivalent to the Kirchhoff transverse load

$$K_{12} = Q_n + \frac{\partial}{\partial s}M_{ns} \tag{20}$$

and concentrated end loads (all defined positive upwards)

$$R_{12} = M_{ns} \qquad \text{in } 1$$

$$R_{21} = -M_{ns} \qquad \text{in } 2 \ .$$

Noting that along 12

$$\frac{\partial}{\partial s} = -\frac{a}{c}\frac{\partial}{\partial x} + \frac{b}{c}\frac{\partial}{\partial y} \ .$$

Substitution of (17) and (19) into (20) produces, after some transformations by the equilibrium equations (5) and (6)

$$K_{12} = 2\frac{b}{c}Q_x + 2\frac{a}{c}Q_y - \left[\frac{v}{b}\frac{\partial}{\partial x} + \frac{u}{a}\frac{\partial}{\partial y}\right]\left[\frac{b^2}{c^2}M_x + 2\frac{a}{c}\frac{b}{c}M_{xy} + \frac{a^2}{c^2}M_y\right] \tag{21}$$

while

$$R_{12} = -\frac{u}{a}\left[\frac{b}{c}M_x(a,0) + \frac{a}{c}M_{xy}(a,0)\right] + \frac{v}{b}\left[\frac{a}{c}M_y(a,0) + \frac{b}{c}M_{xy}(a,0)\right]$$

$$R_{21} = +\frac{u}{a}\left[\frac{b}{c}M_x(0,b) + \frac{a}{c}M_{xy}(0,b)\right] - \frac{v}{b}\left[\frac{a}{c}M_y(0,b) + \frac{b}{c}M_{xy}(0,b)\right] .$$

If we let b tend to zero ($a \to c$, $u \to a$, $v/b \to -\cos\alpha$) the formulas specialize to

$$M_n = \sin\alpha \, M_y \qquad (y = 0) \tag{22}$$

$$M_{ns} = -M_{xy} - \cos\alpha \, M_y \ . \tag{23}$$

In this limiting process the material comes to lie on the wrong side of the edge $y = 0$ of the finite element. For the material on the correct side, the analytical expressions remain invariant provided the positive orientations of M_n and M_{ns} are reversed, as indicated on Figure 4. However, in the case of the transverse loads we change signs in the limiting formulas in order to keep their definition as positive upwards. We then obtain along the edge y = 0 of the finite element

$$K_{01} = -\ 2Q_y\ -\ \cos\,\alpha\,\frac{\partial}{\partial x}\,M_y\ +\ \frac{\partial}{\partial y}\,M_y$$

$$R_{10} = M_{xy}(a,0)\ +\ \cos\alpha\ M_y(a,0) \qquad (y = 0) \tag{24}$$

$$R_{01} = -\ M_{xy}(0,0)\ -\ \cos\alpha\ M_y(0,0)\quad .$$

Similarly, by letting a tend to zero ($b \to c$, $v \to b$, $u/a \to -\cos\alpha$) we find for the edge x = 0

$$M_n = \sin\alpha\ M_x \tag{25}$$

$$(x = 0)$$

$$M_{ns} = \cos\alpha\ M_x + M_{xy} \tag{26}$$

and positive upwards transverse loads

$$K_{20} = -\ 2Q_x\ -\ \cos\alpha\,\frac{\partial}{\partial y}\,M_x\ +\ \frac{\partial}{\partial x}\,M_x$$

$$R_{20} = \cos\alpha\ M_x(0,b) + M_{xy}(0,b) \qquad (x = 0) \tag{27}$$

$$R_{02} = -\ \cos\alpha\ M_x(0,0) - M_{xy}(0,0)\quad .$$

Combination of the concentrated loads at the corners produces the final corner loads

$$P_0 = R_{01}+R_{02} = -\ 2M_{xy}(0,0)\ -\ \cos\alpha(M_x(0,0)\ +\ M_y(0,0))$$

$$P_1 = R_{12}+R_{10} = 2\,\frac{v}{c}\,M_{xy}(a,0)\ -\ \frac{bu}{ac}\,M_x(a,0)\ +\ \left[\cos\alpha\ +\ \frac{av}{bc}\right]M_y(a,0)$$

$$P_2 = R_{21}+R_{20} = 2\,\frac{u}{c}\,M_{xy}(0,b)\ -\ \frac{av}{bc}\,M_y(0,b)\ +\ \left[\cos\alpha\ +\ \frac{bu}{ac}\right]M_x(0,b)\quad . \tag{28}$$

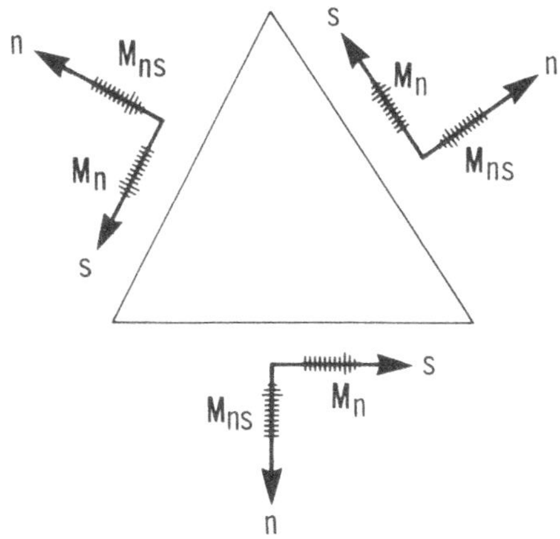

Figure 4 - Choice of Positive Orientations for Interconnection (+ element).

3. THE LINEAR FIELD EQUILIBRIUM MODEL

According to the general procedure for building up an equilibrium model [2], we start with an assumption concerning the stress field. It turns out that, if a nine parameter linear bending-twisting moments field is adopted:

$$\begin{aligned} M_x &= \beta_1+\beta_2 \frac{x}{a} + \beta_3 \frac{y}{b} \\ M_{xy} &= \beta_4+\beta_5 \frac{x}{a} + \beta_6 \frac{y}{b} \\ M_y &= \beta_7+\beta_8 \frac{x}{a} + \beta_9 \frac{y}{b} \end{aligned} \tag{29}$$

the number of generalized loads required at the boundaries is exactly twelve. These loads are connected by three overall equilibrium conditions reducing their degree of independence to nine so that the model will be free of any spurious kinematical freedoms. The same conclusion is found directly if it is recognized that this plate equilibrium model is the Southwell analogue [3] of the quadratic displacement field triangular displacement model of [2]. From (29), using equilibrium equations (5) and (6)

$$Q_x = \frac{1}{a}\beta_2 + \frac{1}{b}\beta_6 \quad , \quad Q_y = \frac{1}{a}\beta_5 + \frac{1}{b}\beta_9 \; . \tag{30}$$

Equilibrium equation (4) shows then that the element will not accept a distributed transverse load q.

Methods for dealing with distributed transverse loads will be presented in a further section of the paper. The model will presently be restricted to accept only such external loads that conform with the interface distributions of stresses. Equations (29) and (30) can now be substituted into (18), (21), (22), (24), (25), (27) and (28) to obtain the edge load distributions in terms of the stress parameters β_i.

For dimensional homogeneity all generalized loads will be defined as forces. The first three generalized loads will be the corner loads P_1, P_2 and P_3; their associated generalized displacements are obviously the local plate deflections w_1, w_2 and w_3 at the corners. The next three generalized loads will be taken to be the total transverse forces due respectively to K_{01}, K_{12} and K_{20}. Noting that those Kirchhoff-type line loads are uniform we shall have

$$P_{01} = aK_{01} \quad , \quad P_{12} = cK_{12} \quad , \quad P_{20} = bK_{20} \; , \tag{31}$$

and, by virtual work considerations, their associated generalized displacements are the simple averages of plate deflections along the edges:

$$w_{01} = \frac{1}{a}\int_0^1 w\,dx \quad , \quad w_{12} = \frac{1}{c}\int_1^2 w\,ds \quad , \quad w_{20} = \frac{1}{b}\int_0^2 w\,dy \; . \tag{32}$$

The last six generalized loads are due to the normal bending moments alond the edges. They each have linear variations and are defined by two generalized quantities.

For instance along the edge 01 $(y = 0)$

$$M_n = \sin\alpha \left[\beta_7 + \frac{x}{a}\beta_8\right] .$$

This distribution is determined by the local values

$$M_{01} = M_n(0,0) = \sin\alpha\,\beta_7$$
$$M_{10} = M_n(a,0) = \sin\alpha(\beta_7+\beta_8) \tag{33}$$

which are taken to be generalized loads. In terms of those

$$M_n = M_{01}\left[1 - \frac{x}{a}\right] + M_{10}\frac{x}{a} . \tag{34}$$

The corresponding virtual work equation

$$\int_0^a M_n\phi dx = M_{01}\phi_{01} + M_{10}\phi_{10} , \left[\phi = -\frac{\partial w}{\partial n}\right]$$

gives, after substitution of (34) and identification, the associated generalized displacements

$$\phi_{01} = a\int_0^1 \phi\left[1 - \frac{x}{a}\right] d\frac{x}{a} , \quad \phi_{10} = a\int_0^1 \phi\frac{x}{a}\, d\frac{x}{a} . \tag{35}$$

In a similar fashion we define

$$M_{02} = \beta_1\sin\alpha , \quad M_{20} = (\beta_1+\beta_3)\sin\alpha \tag{36}$$

$$\phi_{02} = b\int_0^1 \phi\left[1 - \frac{y}{b}\right] d\frac{y}{b} , \quad \phi_{20} = b\int_0^1 \phi\frac{y}{b}\, d\frac{y}{b} . \tag{37}$$

Along the edge 12 we set

$$\frac{x}{a} = \zeta , \quad \frac{y}{b} = 1 - \zeta , \quad 0 < \zeta < 1$$

so that, from (18)

$$M_{12} = \frac{\sin\alpha}{c^2}\,[b^2(\beta_1+\beta_3)+2ab(\beta_4+\beta_6)+a^2(\beta_7+\beta_9)]\ , \qquad (38)$$

$$M_{21} = \frac{\sin\alpha}{c^2}\,[b^2(\beta_1+\beta_2)+2ab(\beta_4+\beta_5)+a^2(\beta_7+\beta_8)]\ , \qquad (39)$$

$$\phi_{12} = c\int_0^1 \phi(1-\zeta)d\zeta\ , \quad \phi_{21} = c\int_0^1 \phi\zeta d\zeta\ . \qquad (40)$$

The result of expressing all generalized loads in terms of the stress parameters is summarized in the loads connection matrix C of Table 1, where

$$g = Cb\ ,$$

$$g' = (P_0P_1P_2P_{01}P_{12}P_{20}M_{01}M_{10}M_{12}M_{21}M_{20}M_{02})\ , \qquad (41)$$

$$b' = (\beta_1\beta_2\beta_3\beta_4\beta_5\beta_6\beta_7\beta_8\beta_9)\ .$$

The flexibility matrix F of the element results from

$$\iint \phi\ \sin\alpha\ dxdy = \frac{1}{2}\ \beta'F\beta$$

where the complementary energy density (14) is expressed through the invariants (13) in terms of the stress parameters.

This flexibility matrix is presented in Table 2. It is to be inverted numerically to produce [2] the stiffness matrix K of the equilibrium model

$$g = Kq\ , \quad K = CF^{-1}C'$$

$$q' = (w_1w_2w_3w_{01}w_{12}w_{20}\phi_{01}\phi_{10}\phi_{12}\phi_{21}\phi_{20}\phi_{02})\ .$$

Table 1 - The Load Connection Matrix C

	β_1	β_2	β_3	β_4	β_5	β_6	β_7	β_8	β_9
P_0	$-\cos\alpha$	0	0	-2	0	0	$-\cos\alpha$	0	0
P_1	$-\frac{bu}{ac}$	$-\frac{bu}{ac}$	0	$2\frac{v}{c}$	$2\frac{v}{c}$	0	$\cos\alpha+\frac{av}{bc}$	$\cos\alpha+\frac{av}{bc}$	0
P_2	$\cos\alpha+\frac{bu}{ac}$	0	$\cos\alpha+\frac{bu}{ac}$	$2\frac{u}{c}$	0	$2\frac{u}{c}$	$-\frac{av}{bc}$	0	$-\frac{av}{bc}$
P_{01}	0	0	0	0	-2	0	0	$-\cos\alpha$	$-\frac{a}{b}$
P_{12}	0	$\frac{b}{a}\left(2-\frac{v}{c}\right)$	$-\frac{bu}{ac}$	0	$2\frac{u}{c}$	$2\frac{v}{c}$	0	$-\frac{av}{bc}$	$\frac{a}{b}\left(2-\frac{u}{c}\right)$
P_{20}	0	$-\frac{b}{a}$	$-\cos\alpha$	0	0	-2	0	0	0
M_{01}	0	0	0	0	0	0	$\sin\alpha$	0	0
M_{10}	0	0	0	0	0	0	$\sin\alpha$	$\sin\alpha$	0
M_{12}	$\frac{b^2}{c^2}\sin\alpha$	0	$\frac{b^2}{c^2}\sin\alpha$	$\frac{2ab}{c^2}\sin\alpha$	0	$\frac{2ab}{c^2}\sin\alpha$	$\frac{a^2}{c^2}\sin\alpha$	0	$\frac{a^2}{c^2}\sin\alpha$
M_{21}	$\frac{b^2}{c^2}\sin\alpha$	$\frac{b^2}{c^2}\sin\alpha$	0	$\frac{2ab}{c^2}\sin\alpha$	$\frac{2ab}{c^2}\sin\alpha$	0	$\frac{a^2}{c^2}\sin\alpha$	$\frac{a^2}{c^2}\sin\alpha$	0
M_{20}	$\sin\alpha$	0	$\sin\alpha$	0	0	0	0	0	0
M_{02}	$\sin\alpha$	0	0	0	0	0	0	0	0

Table 2 - The Flexibility Matrix of an Isotropic Element

$$F = \frac{ab}{12(1-\nu^2)D\sin\alpha}\,G(\alpha,\nu)$$

$G(\alpha,\nu)$ is expressed below in terms of the auxiliary quantities

$$\lambda = 1+\cos^2\alpha+\nu\sin^2\alpha$$

$$\mu = \cos^2\alpha-\nu\sin^2\alpha$$

6								
2	1							
2	0.5	1			Symmetrical			
$12\cos\alpha$	$4\cos\alpha$	$4\cos\alpha$	12λ					
$4\cos\alpha$	$2\cos\alpha$	$\cos\alpha$	4λ	2λ				
$4\cos\alpha$	$\cos\alpha$	$2\cos\alpha$	4λ	λ	2λ			
6μ	2μ	2μ	$12\cos\alpha$	$4\cos\alpha$	$4\cos\alpha$	6		
2μ	μ	0.5μ	$4\cos\alpha$	$2\cos\alpha$	$\cos\alpha$	2	1	
2μ	0.5μ	μ	$4\cos\alpha$	$\cos\alpha$	$2\cos\alpha$	2	0.5	1

4. ASSEMBLED STIFFNESS MATRICES

Some minor modifications to the direct stiffness method are required for assembling the individual stiffness matrices. All transverse loads and deflections are referred to a common positive upwards direction. Hence in the localizing matrix L_k of element (k)

$$q_{(k)} = L_k q$$

the first six rows are void except for a single unit in each, identifying a vertical displacement in the element with a nodal displacement of the structure.

However, bending moments and slopes at an interface have natural reference to reciprocal directions. This follows from the natural anticlockwise sense of definition of the bending moment orientations around each element (see Figure 4), or, alternatively, from the outward normal definitions of the conjugate slopes along the edges. As a consequence, in the six last rows of a localizing matrix, the units identifying a $\phi_{ij(k)}$ slope of element (k) with a nodal slope of the structure, will have to be affected by a sign. The sign will be positive if the nodal slope has the same positive orientation, negative otherwise.

In some cases the mesh of elements and positive orientation of nodal slopes can be so devised that, for a given element, all positive slopes are codirectional with nodal slopes (+ element) or antidirectional (- element). A case in point is represented in Figure 5. The adjacent of a + element must then be a - element and this is obviously possible if the number of elements meeting at an interior point remains even.

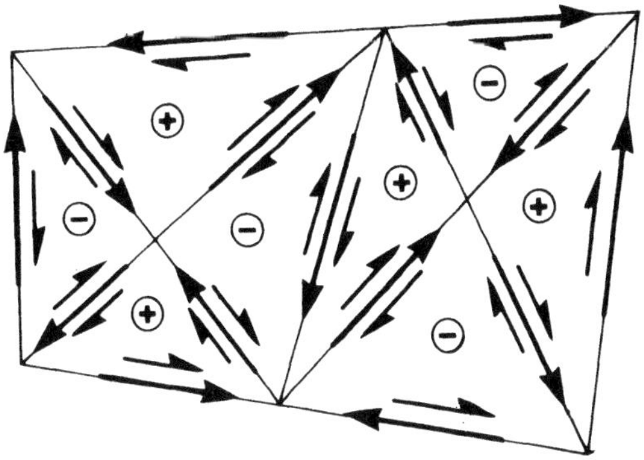

Figure 5 - A Typical Mesh of + Type and - Type Elements

The nature of the external generalized loads that the structure can accept as an equilibrium model is also clear:

- at each vertex a concentrated load can be associated with the local nodal deflection;

- along each interface segment, or boundary segment, a uniform transverse line load can be applied;

- its resultant is associated with the average deflection of the segment, or boundary segment, can be the axis of an applied couple of linearly varying intensity;

- the intensities at each end are the generalized external loads associated with the generalized slopes.

It should be noted that, just as the end intensities of the distributed couples have the physical dimensions of a force, the generalized slopes defined by (35, 36 or 40) have really the physical dimensions of a deflection.

5. TRANSVERSE SURFACE LOADS

From the viewpoint of equilibrium models the only correct way to introduce transverse surface loads is by increasing the number of generalized co-ordinates. Taking the simplest case, where the model is required to accept a uniform surface load q, let

$$p = \frac{1}{2} qab \sin\alpha \tag{42}$$

the total load, be the additional generalized co-ordinate. The associated generalized displacement, obtained from virtual work consideration, is the ordinary average of deflection over the plate area

$$\bar{w} = \frac{2}{ab} \iint w dx dy . \tag{43}$$

A particular field of bending and twisting moments and shear loads, satisfying the equilibrium equations (4), (5) and (6) under the uniform load is

$$\left.\begin{aligned} M_x &= \frac{1}{3} p \frac{a}{b} \left[1 - \frac{x}{a}\right] \frac{x}{a} \\ M_{xy} &= - \frac{1}{3} p \frac{xy}{ab} \end{aligned}\right\} \tag{44}$$

$$\left.\begin{aligned} M_y &= \frac{1}{3} p \frac{b}{a} \left[1 - \frac{y}{b}\right] \frac{y}{b} \\ Q_x &= \frac{p}{3b} \left[1 - 3 \frac{x}{a}\right] \\ Q_y &= \frac{p}{3a} \left[1 - 3 \frac{y}{b}\right] . \end{aligned}\right\} \tag{45}$$

It was derived from a general quadratic field in the bending and twisting moments by requiring

(a) that the moments vanish along the three edges of the element

$$M_x = M_{xy} = 0 \qquad \text{for } x = 0$$

$$M_{xy} = M_y = 0 \qquad \text{for } y = 0$$

$$bM_x + aM_{xy} = 0 \text{ and } aM_y + bM_{xy} = 0 \qquad \text{for } \frac{x}{a} + \frac{y}{b} = 1$$

(b) that the shear loads be constant along the edges

Q_x independent of y for $x = 0$

Q_y independent of x for $y = 0$

$cQ_n = bQ_x + aQ_y$ constant for $\frac{x}{a} + \frac{y}{b} = 1$.

The requirements are certainly sufficient to prevent the necessity of introducing new generalized interface loads. In fact the new external load p is simply reached at the interfaces by three uniform Kirchhoff shear loads

$$K_x = -\frac{p}{3b} \qquad (-Q_x \text{ for } x = 0)$$

$$K_y = -\frac{p}{3a} \qquad (-Q_y \text{ for } y = 0)$$

$$K_n = -\frac{p}{3c} \qquad \left(\frac{b}{c} Q_x + \frac{a}{c} Q_y \text{ for } \frac{x}{a} + \frac{y}{b} = 1\right) .$$

The generalized reaction loads are, consequently

$$P_{01} = P_{12} = P_{20} = -\frac{1}{3} p \tag{46}$$

$$P_0 = P_1 = P_2 = M_{01} = M_{10} = M_{12} = M_{21} = M_{20} = M_{02} = 0 . \tag{47}$$

The requirements (a) and (b) can in fact be regarded as necessary. For, if one allows the particular solution in equilibrium with a transverse pressure q(x,y) to be reached by a complete system of the (previously defined) generalized loads, superposition with a general field of type (29) can always implement the nine conditions (47) [equivalent to requirement (a)] by adjustment of the nine β_i parameters.

If p is now considered to be the tenth stress parameter

$$p = \beta_{10} \tag{48}$$

equations (46) and (47) are incorporated in an augmented load connection matrix by bordering the C matrix of Table 1 with a tenth column, hereafter written in transpose

$$\left(0 \quad 0 \quad 0 \; -\frac{1}{3} \; -\frac{1}{3} \; -\frac{1}{3} \quad 0 \quad 0 \quad 0 \quad 0 \quad 0 \quad 0\right) .$$

Since p also becomes the thirteenth generalized load, C is further extended by a thirteenth border row

$$(0 \quad 0 \quad 0 \quad 0 \quad 0 \quad 0 \quad 0 \quad 0 \quad 0 \quad 1)$$

expressing equation (48).

The flexibility matrix is also augmented and follows from computation of the strain energy as a quadratic form in the ten β_j parameters obtained by superposition of the fields (29) and (44) (where p is replaced by β_{10}). For an isotropic plate of constant rigidity D the elements of the additional column (and row) of the G matrix of Table 2 turn out to be

$$\frac{1}{3}\left[\frac{a}{b} + \mu\,\frac{b}{a} - \cos\alpha\right]$$

$$\frac{1}{30}\left[4\left(\frac{a}{b} - \cos\alpha\right) + 3\mu\,\frac{b}{a}\right]$$

$$\frac{1}{30}\left[4\left(\mu\,\frac{b}{a} - \cos\alpha\right) + 3\,\frac{a}{b}\right]$$

$$\frac{2}{3}\left[\frac{a}{b} + \frac{b}{a}\right]\cos\alpha - \frac{1}{3}\,\lambda$$

$$\frac{1}{15}\left[\left(4\,\frac{a}{b} + 3\,\frac{b}{a}\right)\cos\alpha - 2\lambda\right]$$

$$\frac{1}{15}\left[\left(3\,\frac{a}{b} + 4\,\frac{b}{a}\right)\cos\alpha - 2\lambda\right]$$

$$\frac{1}{3}\left[\frac{b}{a} + \mu\,\frac{a}{b} - \cos\alpha\right]$$

$$\frac{1}{30}\left[4\left(\mu\,\frac{a}{b} - \cos\alpha\right) + 3\,\frac{b}{a}\right]$$

$$\frac{1}{30}\left[4\left(\frac{b}{a} - \cos\alpha\right) + 3\mu\,\frac{a}{b}\right]$$

$$\frac{2}{3}\left[\frac{a^2}{b^2} + \frac{b^2}{a^2}\right] - \frac{4}{3}\left[\frac{a}{b} + \frac{b}{a}\right]\cos\alpha + \frac{1}{9}\,(10\mu+4\lambda)\ .$$

Clearly the process can be extended to cover more complicated surface load distributions. Even in the absence of surface loads, such extended models are valuable in that they yield additional information on the displacement field.

Should we wish to keep the simplicity of the original model and yet take surface loads $q(x,y)$ into account, one obvious procedure consists in replacing those surface loads by equivalent interface loads and rely on de Saint-Venant's principle to keep the effects of this substitution local. However the guarantee of upper deflection bounds is lost.

6. NUMERICAL RESULTS

In reference [7] a comprehensive numerical comparison was made between several models of plate bending elements under various loading and boundary conditions. Some of those results are again presented here in order to evaluate the performance obtained under similar conditions for the present equilibrium element and the conforming quadrilateral element of [5]. Only those elements that exhibited satisfactory convergence characteristics were retained for comparison. In the code of [7] they are:

(1) (ACM) a rectangular element developed by Adini, Clough and Melosh. It is based on a twelve parameter transverse displacement field, containing the complete cubic field (ten para-

meters) and, in addition, the x^3y and xy^3 terms. The generalized coordinates are the deflections and slopes at the four vertices. Continuity of the normal slope along the boundary is not secured, i.e., the element is not conforming.

(2) (M) a rectangular element developed by Melosh on the basis of physical analogies with beam bending. It is a hybrid element, neither of the displacement nor of the equilibrium type.

(3) (HCT) the so-called "high compatibility triangle" developed by Clough. It is a conforming model of the spline interpolation type. The major objective was to obtain a linear normal slope variation along each edge so that continuity of deflections and slopes could result from an identification of deflections and slopes at the vertices of the elements only.

(4) (CQ) is our code name for the conforming quadrangular element, described earlier [5].

(5) (EQT) is our code number for the present equilibrium triangle.

(6) (Z) stands for a non-conforming triangular displacement model developed by Bazeley et al [8]. This model was also used as the basis of conforming ones, pursuing a very similar objective to that of the HCT triangle. However this objective was reached differently. Inspection of the boundary conditions in areal co-ordinates permitted the addition of several types of correction fields, linearizing the normal slope variations. The conforming elements appear to be considerably stiffened and their convergence characteristics are difficult to assess from the numerical results reported.

The comparisons were performed for a square plate and a rectangular plate of aspect ratio 2, centrally loaded and with edges either clamped or simply supported. Symmetry allowed the treatment to be reduced to a quarter plate. The center deflections are reported in Tables 3 - 6 in the form of a dimensionless coefficient β defined by

$$w = \beta \frac{Pa^2}{D}$$

P: central load, a: size of quarter plate, D: bending rigidity. The corresponding graphs (Figures 7 - 10) are plots of

$$\mu = \frac{\beta_{computed} - \beta_{exact}}{\beta_{exact}} .$$

The exact solution was computed by a Navier type series to the required four digit accuracy. The graphs are similar to those of [7] except that the new definition of the ordinate gives a direct estimate of percentage error and the use of the number of generalized co-ordinates in place of a mesh number provides a better comparison of accuracy versus computer load.

Table 3 - Case 1: Square Plate-Simply Supported-Concentrated Load in the Center

Analysis	N = 1		N = 2		N = 4		N = 6		N = 8	
	N.G.C.	$\beta.10^3$	N.G.C.	$\beta.10$	N.G.C.	$\beta.10^3$	N.G.C.	$\beta.10^3$	N.G.C.	$\beta.10^3$
1 Exact	-	11.6008	-	11.6008	-	11.6008	-	11.6008	-	11.6008
2 ACM	12	13.75	27	12.30	75	11.81	147	11.69	273	11.64
3 M	12	12.10	27	11.85	75	11.66	147	11.61	273	11.60
4 HCT	12	8.84	27	10.48	75	11.25	147	11.44	273	11.48
5 CQ	16	10.876	39	11.439	115	11.560	231	11.588	-	-
6 EQT	19	13.770	57	12.124	193	11.674	409	11.656	-	-
7 Z	12	13.02	27	11.76	75	11.65	147	-	-	-

N.G.C. stands for Number of Generalized Co-ordinates.
N is the mesh size number.
β is defined by: $w_c = \beta(Pa^2/D)$.

Table 4 - Case 2: Rectangular Plate (b/a = 2)-Simply Supported-Concentrated Load in the Center

Analysis	N = 1		N = 2		N = 4		N = 6		N = 8	
	N.G.C.	$\beta.10^3$	N.G.C.	$\beta.10^3$	N.G.C.	$\beta.10^3$	N.G.C.	$\beta.10^3$	N.G.C.	$\beta.10^3$
1 Exact	-	16.5239	-	16.5239	-	16.5239	-	16.5239	-	16.5239
2 ACM	12	18.487	27	17.579	75	16.919	147	16.745	273	16.656
3 M	12	14.664	27	16.476	75	16.616	147	16.584	273	16.570
4 HCT	12	11.216	27	13.60	75	15.520	147	16.024	273	16.24
5 CQ	16	14.486	39	16.000	115	16.400	231	16.477	-	-
6 EQT	19	22.177	57	17.878	193	16.804	409	16.666	-	-

N.G.C. stands for Number of Generalized Co-ordinates.
N is the mesh size number.
β is defined by: $w_c = \beta(Pa^2/D)$.

Table 5 - Case 3: Square Plate-Clamped-Concentrated Load in the Center

Analysis	N = 1		N = 2		N = 4		N = 6		N = 8	
	N.G.C.	$\beta.10^3$	N.G.C.	$\beta.10^3$	N.G.C.	$\beta.10^3$	N.G.C.	$\beta.10^3$	N.G.C.	$\beta.10^3$
1 Exact	-	5.605	-	5.605	-	5.605	-	5.605	-	5.605
2 ACM	12	5.919	27	6.137	75	5.807	147	5.704	273	5.671
3 M	12	4.231	27	5.736	75	5.688	147	5.653	273	5.640
4 HCT	12	1.0	27	4.2400	75	5.192	147	5.40	273	5.496
5 CQ	16	5.208	39	5.430	115	5.5708	231	5.5966	-	-
6 EQT	19	8.2565	57	6.1939	193	5.7557	409	5.6712	-	-
7 Z	12	5.21	27	5.89	75	5.72	147	-	-	-

N.G.C. stands for Number of Generalized Co-ordinates.
N is the mesh size number.
β is defined by: $w_c = \beta(Pa^2/D)$.

Table 6 - Case 4: Rectangular Plate (b/a = 2)-Clamped-Concentrated Load in the Center

Analysis	N = 1		N = 2		N = 4		N = 6		N = 8	
	N.G.C.	$\beta.10^3$	N.G.C.	$\beta.10^3$	N.G.C.	$\beta.10^3$	N.G.C.	$\beta.10^3$	N.G.C.	$\beta.10^3$
1 Exact	-	7.215	-	7.215	-	7.215	-	7.215	-	7.215
2 ACM	12	6.3923	27	7.799	75	7.5253	147	7.3928	273	7.3342
3 M	12	2.1815	27	6.7742	75	7.2312	147	7.2538	273	7.2490
4 HCT	12	1.156	27	3.9546	75	6.1415	147	6.6943	273	6.8384
5 CQ	16	5.649	39	6.681	115	7.106	231	7.1798	-	-
6 EQT	19	13.937	57	8.7392	193	7.5892	409	7.3786	-	-

N.G.C. stands for Number of Generalized Co-ordinates.
N is the mesh size number.
β is defined by: $w_c = (Pa^2/D)$.

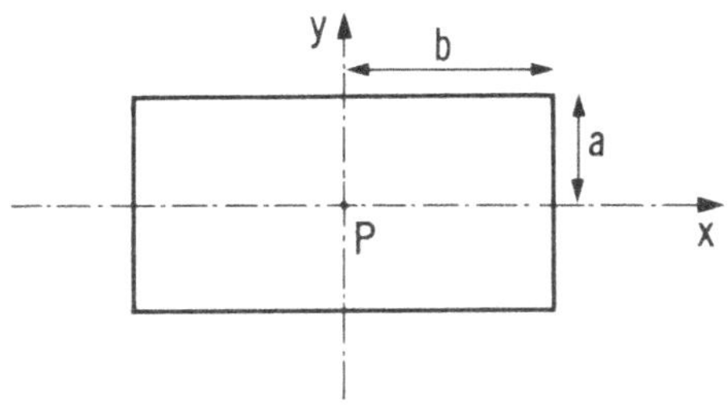

Case	b/a	Boundary Conditions
1	1	Simply supported
2	2	Simply supported
3	1	Clamped
4	2	Clamped

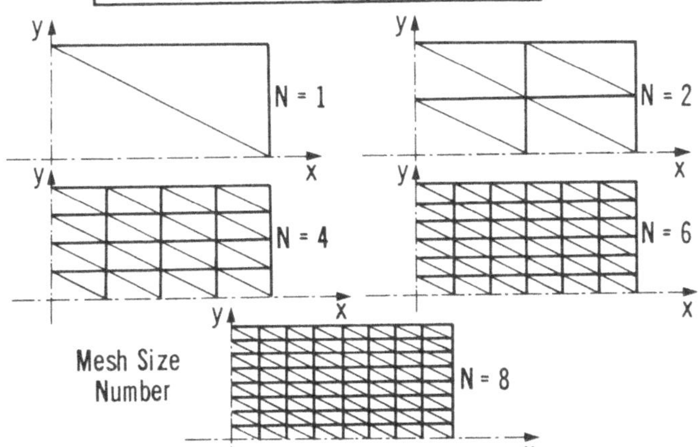

Figure 6 - Boundary Conditions and Idealization Patterns of Numerical Calculations for Centrally Loaded Plate

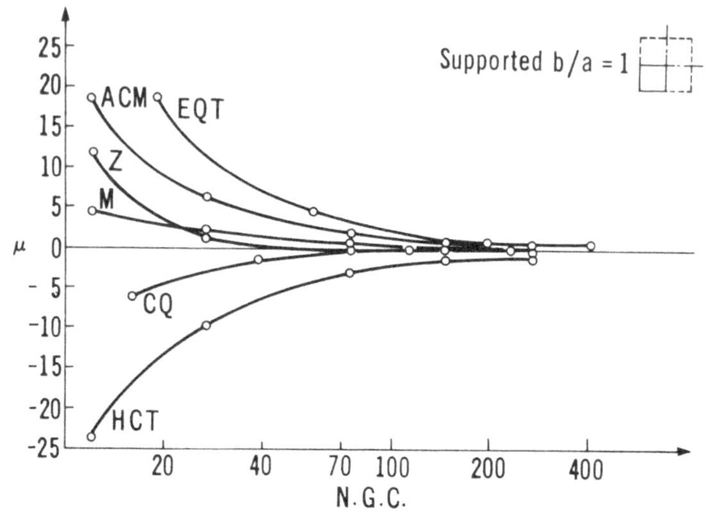

Figure 7 - Convergence Curves of Various Types of Finite Elements. Square Simply Supported Plate with Central Load

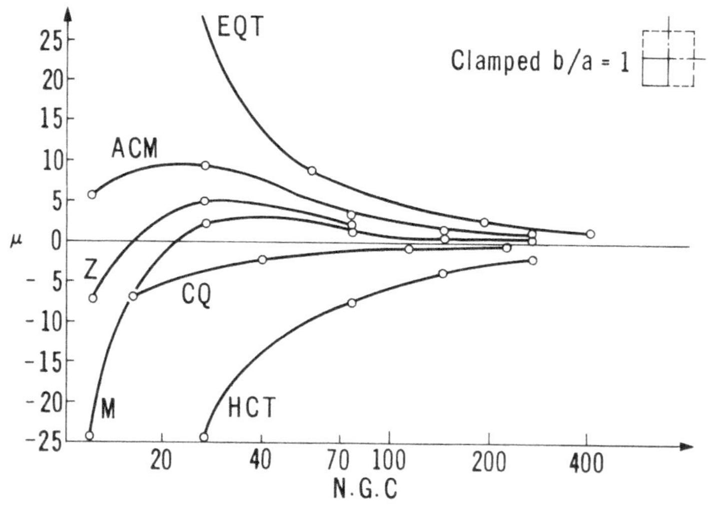

Figure 8 - Convergence Curves. Square Clamped Plate with Central Load

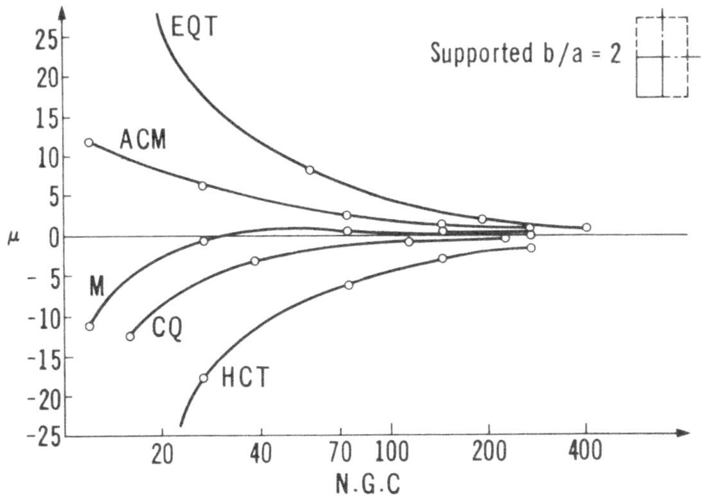

Figure 9 - Convergence Curves. Aspect Ratio 2, Simply Supported Plate with Central Load

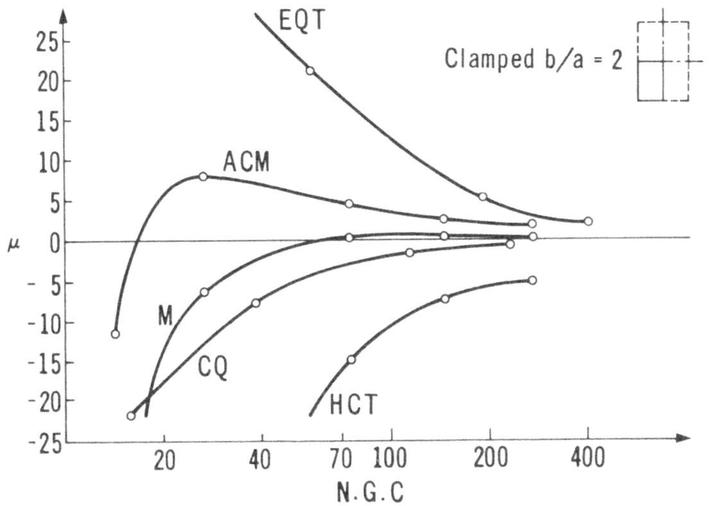

Figure 10 - Convergence Curves. Aspect Ratio 2, Clamped Plate with Central Load

Convergence of deflections to their exact value is only one aspect of the accuracy of a computation. Convergence of the stress field is in many cases still more important. By their very nature, equilibrium models should be expected to yield better stress information than displacement models. A comparison is therefore included of the computed and exact bending moments along a symmetry axis for the simply supported square plate case. The test is severe since the exact bending moment tends to infinity as the central load is approached.

Figure 11 illustrates the evolution of the bending moment distribution as the mesh size of the (CQ) element is reduced. Figure 12 illustrates the same for the (EQT) element. For clarity the exact distribution is not represented as it almost coincides with the computed solution with the finer mesh size. It is however represented in Figure 13 where the performance of both models is compared for a given mesh size. The superiority of the equilibrium model in the large stress gradient region is thereby evidenced.

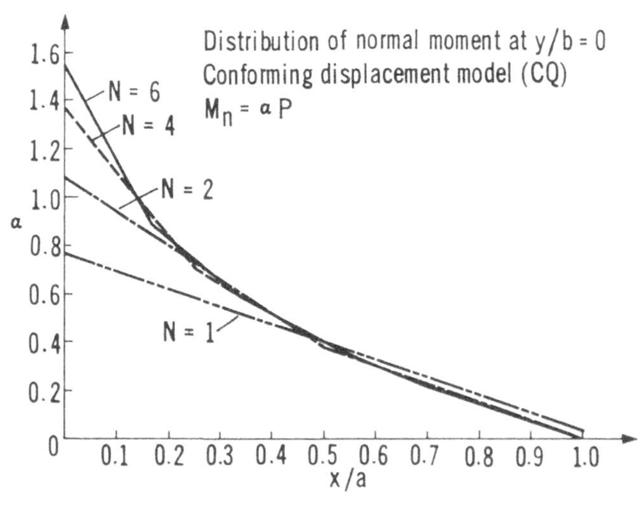

Figure 11 - Bending Moment Output of (CQ) Model Along $y = 0$. Square Simply Supported Plate

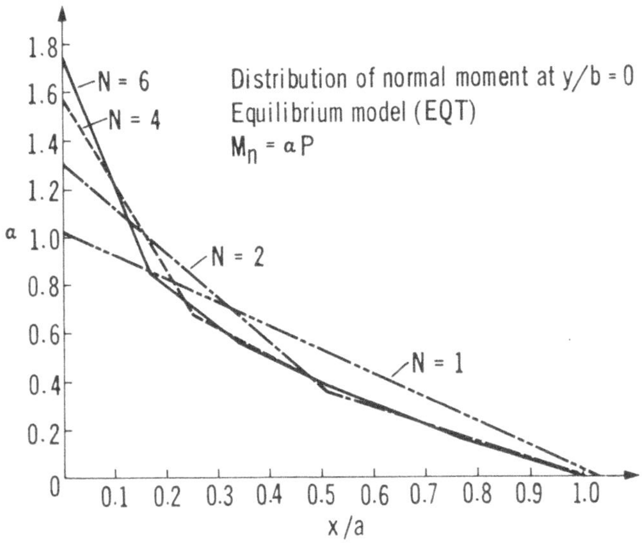

Figure 12 - Bending Moment Output of (EQT) Model Along y = 0. Square Simply Supported Plate

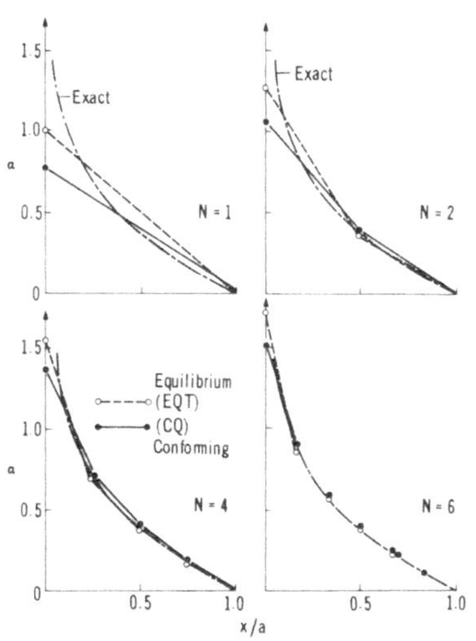

Figure 13 - Comparison of Exact Bending Moment Along y = 0 and Outputs of (CQ) and (EQT) Models for given Idealization Patterns

7. CONCLUSIONS

Numerical results confirm the expected monotonic convergence characteristics of the central deflection from below for the conforming displacement elements (HCT and CQ) and from above for the equilibrium element (EQT). In particular the gap between the deflections of (EQT) and (CQ) constitutes a convenient quantitative estimator of the state of convergence. It is a good illustration of the general principle of a dual analysis in finite elements [6].

Non-conforming and hybrid elements, such as (M) and (Z), can present excellent convergence rates, sometimes monotonic, sometimes oscillatory. However in problems where the exact solution is not accessible, the accuracy of computations based on such elements is more difficult to estimate since there is no guarantee of upper or lower bounds. The approximations based on the (CQ) element are, in most cases, the best, especially for crude idealizations; this is probably due to the additional freedom in normal slope variation along the edges. Suppressing this additional freedom by requiring a linear variation, as in the (HCT) triangle and conforming solutions based on the (Z) model, would probably make the element behave as (HCT) for a small number of generalized co-ordinates. The convergence characteristics of (EQT) are similar to those of (HCT) but from the other side. Furthermore (EQT) elements involve rapidly a high number of generalized displacements; as is generally the case with equilibrium elements it would be more efficiently handled by a Force computer program than by a Stiffness program.

Finally, both the (CQ) and the (EQT) element can provide an accurate representation of the stress field.

REFERENCES

[1] MORLEY, L.S.D., *Skew Plates and Structures*, Pergamon Press, 1963.

[2] FRAEIJS DE VEUBEKE, B.M., "Displacement and Equilibrium Models in the Finite Element Method", *Stress Analysis*, (editors: O.C. Zienkiewicz and G.S. Holister), Chapter 9, Wiley, 1965.

[3] SANDER, G., "Bornes supérieures et inférieures dans l'analyse matricielle des plaques en flexion-torsion", *Bull. Soc. r. Sci. Liège*, Vol. 33, 1964, pp. 456-494.

[4] FRAEIJS DE VEUBEKE, B.M. and ZIENKIEWICZ, O.C., "Strain Energy Bounds in Finite Element Analysis, by Slab Analogy", *J. Strain Analysis*, Vol. 2, 1967, pp. 265-271.

[5] FRAEIJS DE VEUBEKE, B.M., "A Conforming Finite Element for Plate Bending", *Int. J. Solids Struct.*, Vol. 4, 1968, pp. 95-108.

[6] SANDER, G. and FRAEIJS DE VEUBEKE, B.M., "Upper and Lower Bounds to Structural Deformations by Dual Analysis in Finite Elements", Air Force Technical Report AFFDL-TR-199, 1966.

[7] CLOUGH, R.W. and TOCHER, J.L., "Finite Element Stiffness Matrices for Analysis of Plate Bending", *Proc. Conf. Matrix Methods Structural Mechanics*, pp. 515-546, AFFDL-TR-66-80, 1966.

[8] BAZELEY, G.P., CHEUNG, Y.K., IRONS, B.M. and ZIENKIEWICZ, O.C., "Triangular Elements in Plate Bending - Conforming and Non-conforming Solutions", *Proc. Conf. Matrix Methods Structural Mechanics*, pp. 547-576, AFFDL-TR-66-80, 1966.

THE THEORETICAL DESIGN LAWS OF WARPING-FREE MULTICELLULAR BOX BEAMS

B.M. Fraeijs de Veubeke

University of Liège, Belgium

A variational method is applied to derive the theoretical laws for skin thicknesses and boom areas under which a multicellular box beam becomes free from warping effects under either torsion or bending. For total freedom under arbitrary combinations of bending and torsion the laws take a strikingly simple form. Their practical implementation is related to the geometrical convexity of a set of elementary cells, each of which contains the shear centre in its interior. In such a case the shear centre, flexure-torsion centre, and centroid are coincident, as are the principal axes of inertia and principal shear axes. The Bernoulli-Navier and de Saint-Venant theories are identical and exact solutions under arbitrary transverse loads.

1. BOX-BEAM GEOMETRY

The box beams considered consist of an arbitrarily complex arrangement of skin panels connected by stringers or spar flanges parallel to z-axis (Figure 1). In a given cross-section, the traces of skin panels are "arcs" separated by nodes of co-ordinates (x_m, y_m) which are the geometrical centres of the stringers or spar flanges. The bending and torsional rigidity of the stringers and flanges is neglected; they are uniaxially stressed (tension or compression).

Original publication, *Problems of Hydrodynamics and Continuum Mechanics*, English Edition, Contributions in Honour of the Sixtieth Birthday of Academician L.I. Sedov, Philadelphia, 1969.

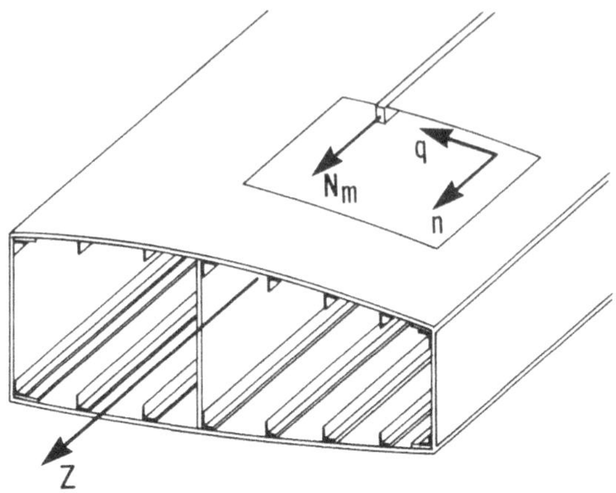

Figure 1 - Box-Beam Geometry (n - axial stress flow in skin, q - shear flow in skin, N_m - axial stringer load)

The skin thickness, effective in carrying the shear flow q, is supposed to vary according to an affinity law

$$t_q = h(z)a(s) \tag{1.1}$$

where s is measured along the arcs. In a cantilevered beam, we can take h(0) = 1 at the root section, and a(s) is the skin thickness at the root. Similarly, the skin thickness effective in carrying the normal stress flow n and the flange area carrying the normal load N_m at node (m) will be taken to vary as

$$t_n = g(z)b(s) \quad , \quad A_m = g(z)S_m \ . \tag{1.2}$$

The difference between the thicknesses allows for an orthotropic skin structure. Further, the assumption of closely spaced diaphragms, maintaining the geometrical shape of any cross-section without inducing axial restraints, is adopted (only membrane stresses in the skin are then needed to resist transverse loads). The displacement vector in a point P (x,y,z) of the skin is then resolved into three locally orthogonal components: $u_n(s,z)$ along the normal $\underline{n}$, $u_t(s,z)$ along the tangent $\underline{t}$ to the arc, $w(s,z)$ along the generator parallel to z-axis. For the second component

$$u_t = ux' + vy' + \theta r \tag{1.3}$$

where $x' = dx/ds$, $y' = dy/ds$ and

$$r = xy' - yx' \tag{1.4}$$

is the projection of OP on $\underline{n}$ (Figure 2). The diaphragm passing through P has been translated by $u(z)$ parallel to Ox, $v(z)$ parallel to Oy, and rotated through an angle $\theta(z)$ about Oz (Figure 3). The shear strain in the skin at P is

$$\gamma = w' + \dot{u}_t \tag{1.5}$$

where the "prime" denotes differentiation with respect to s, the "dot" differentiation with respect to z.

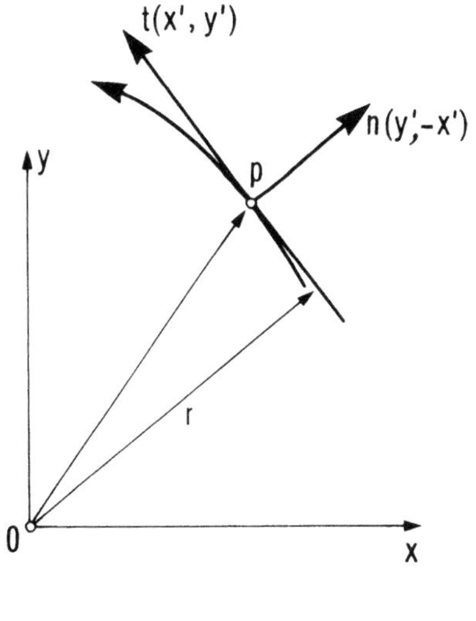

Figure 2 - Definition of r; $(\underline{n},\underline{t})$ is a rotated (Ox,Oy)

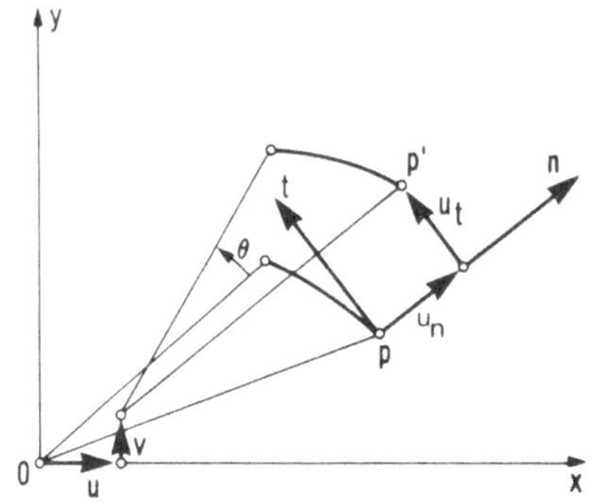

Figure 3 - Diaphragm Displacement (u,v,θ) and Definitions of u_n and u_t

2. EQUILIBRIUM EQUATIONS

In the absence of axial constraints from the diaphragms and without external axial loads on the skin, the equilibrium of a skin element in the direction of a generator is

$$q' + \dot{n} = 0 \; . \tag{2.1}$$

Similarly for a stringer element at node (m)

$$\dot{N}_m = T_m(q) \; . \tag{2.2}$$

The operator T_m is defined with respect to a conventional sense of description along each arc:

$$T_m(q) = (\Sigma q)_{conv} - (\Sigma q)_{div} \; .$$

The first sum is extended to all the values of shear flows at the node related to arcs convergent to the node, the second, to arcs divergent from the node. Note that the algebraic value of q is itself related to this conventional sense of description.

The T_m operator is extremely useful in connection with integrations extended over the totality of arcs of a cross-section.

The following formula for integration by parts is easily established

$$\int_C \alpha d\beta = \sum_m T_m(\alpha\beta) - \int_C \beta d\alpha \ . \tag{2.3}$$

Whenever one of the functions is single-valued, it can be factored out; in the case of a single-valued α, for instance,

$$\int_C \alpha d\beta = \sum_m \alpha_m T_m(\beta) - \int_C \beta d\alpha \ . \tag{2.4}$$

Transverse equilibrium of a diaphragm involves the total shear forces acting in a section:

$$S_x(z) = \int_C q dx \ , \qquad S_y(z) = \int_C q dy \ , \tag{2.5}$$

and the torsional moment about 0z:

$$M(z) = \int_C q r ds \ . \tag{2.6}$$

The bending moments are taken with respect to axes parallel to 0x and 0y through the centroid (x_G, y_G) of the section:

$$\left.\begin{aligned} M_x &= \int_C n(x-x_G)ds + \sum_m N_m(x_m-x_G) \ , \\ M_y &= \int_C n(y-y_G)ds + \sum_m N_m(y_m-y_G) \ . \end{aligned}\right\} \tag{2.7}$$

Equilibrium equations about these axes are then

$$\dot{M}_x = S_x \ , \qquad \dot{M}_y = S_y \ , \tag{2.8}$$

and can be recovered by differentiation of (2.7), using (2.1) and (2.2) and the formula for integration by parts.

Equilibrium with applied transverse loads is expressed through the resultant distributions $p_x(z)$, $p_y(z)$ and $m(z)$ as

$$\dot{S}_x = -p_x \, , \quad \dot{S}_y = -p_y \, , \quad \dot{M} = -m \, . \tag{2.9}$$

3. STRESS-STRAIN RELATIONS. BERNOULLI SHEAR CENTRE

The average shear stress $\tau = q/(ha)$ in the skin is related to the shear strain by $\tau = G\gamma$. In terms of the unknown displacements $w(s,z)$, $u(z)$, $v(z)$, and $\theta(z)$ this stress-strain relation takes the form

$$\frac{q}{Gha} = w' + \dot{u}x' + \dot{v}y' + \dot{\theta}r \tag{3.1}$$

stemming from (1.3) and (1.5). The other stress-strain relations are for the axial components in skin and stringers:

$$\frac{n}{gb} = E\dot{w} \, , \quad \frac{N_m}{gS_m} = E\dot{w}_m \, . \tag{3.2}$$

A Bernoulli-Navier type of assumption

$$w = F_0(z) + X_0(z)(x-x_G) + Y_0(z)(y-y_G) \tag{3.3}$$

that keeps the cross-section plane, suppresses in (3.1) the effect of warping and produces a shear flow

$$q_B = Gha[(X_0+\dot{u})x' + (Y_0+\dot{v})y' + \dot{\theta}r] \, . \tag{3.4}$$

The corresponding shear centre $B(x_B,y_B)$ is obtained by setting $\dot{\theta} = 0$ and expressing that the torsional moment vanishes

$$M_B = \int_C q_B r_B ds = 0$$

for arbitrary values of $(X_0+\dot{u})$ and $(Y_0+\dot{v})$. This furnishes the requirements

$$\int_C ar_B dx = 0 \quad , \quad \int_C ar_B dy = 0 \; ; \tag{3.5}$$

when we substitute here the value

$$r_B = (x-x_B)y' - (y-y_B)x' = r'-x_B y' + y_B x' \; , \tag{3.6}$$

we obtain a pair of linear equations to calculate the co-ordinates of the Bernoulli shear centre

$$\begin{aligned} -A_{xx}y_B + A_{xy}x_B &= \int_C ardx \; , \\ -A_{xy}y_B + A_{yy}x_B &= \int_C ardy \; . \end{aligned} \tag{3.7}$$

In those formulae appears the symmetrical *reduced area tensor* discovered by Drymael [2]:

$$\begin{aligned} A_{xx} &= \int_C a(x')^2 ds \; , \quad A_{xy} = \int_C ax'y'ds = A_{yx} \; , \\ A_{yy} &= \int_C a(y')^2 ds \; . \end{aligned} \tag{3.8}$$

The contracted tensor is the invariant

$$A_{xx} + A_{yy} = \int_C ads$$

proportional to the total shear resistant area. If the reference axes are translated to the Bernoulli shear centre and the complete shear flow is written as

$$q = Gha(w'+\dot{u}_B x'+\dot{v}_B y'+\dot{\theta} r_B) \; ,$$

substitution of this into equations (2.5) and (2.6) produces, in view of the properties (3.5), the formulae

$$\left.\begin{aligned} \frac{S_x}{Gh} &= \int_C aw'dx + A_{xx}\dot{u}_B + A_{xy}\dot{v}_B \,, \\ \frac{S_y}{Gh} &= \int_C aw'dy + A_{xy}\dot{u}_B + A_{yy}\dot{v}_B \,, \end{aligned}\right\} \tag{3.9}$$

$$\frac{M_B}{Gh} = \int_C ar_B w'ds + I_B\dot{\theta} \tag{3.10}$$

where

$$I_B = \int_C ar_B^2 ds \,. \tag{3.11}$$

Clearly, the quantity is directly related to the torsional stiffness of the box beam under the Bernoulli-Navier assumption. Indeed, the integral in (3.10) vanishes by virtue of (3.5) if we substitute (3.3). The Bernoulli-Navier torsional stiffness I_B is a minimum with respect to the choice of co-ordinates of the centre. For, in the case of another choice, corresponding to

$$r_C = r_B - (x_C - x_B)y' + (y_C - y_B)x' \,,$$

we obtain, again in view of (3.5),

$$I_C = \int_C ar_C^2 ds = I_B + \int_C a(y'\Delta x - x'\Delta y)^2 ds > I_B \,.$$

This minimum property was also discovered by Drymeal [2].

4. SCALAR PRODUCT. TRUE WARPING

To avoid lengthy formulae we introduce a convenient scalar product notation for two single-valued functions $\alpha(s)$ and $\beta(s)$ defined over the cross-section. By definition

$$(\alpha,\beta) = \int_C b\alpha\beta ds + \sum_m S_m\alpha_m\beta_m = (\beta,\alpha) \,. \tag{4.1}$$

With this notation (1,1) is the total root section area resisting to axial loads. The co-ordinates of the centroid are defined by the properties

$$(x-x_G,1) = 0 \ , \quad (y-y_G,1) = 0 \ , \qquad (4.2)$$

and the moments of inertia of the root section are given by

$$I_{xx} = (x-x_G,x-x_G) \ , \quad I_{xy} = (x-x_G,y-y_G) \ , \quad I_{yy} = (y-y_G,y-y_G) \ . \qquad (4.3)$$

We shall denote by (i_{xx},i_{xy},i_{yy}) the elements of the reciprocal tensor, so that

$$i_{xx}I_{xx}+i_{xy}I_{xy} = 1 \ , \quad i_{xx}I_{xy}+i_{xy}I_{yy} = i_{xy}I_{xx}+i_{yy}I_{xy} = 0$$

and

$$i_{xy}I_{xy}+i_{yy}I_{yy} = 1 \ .$$

The characteristic properties of a *true warping* can be stated in kinematical form or in terms of stresses. The function W(s) in the affinity law

$$w = f(z)W(s)$$

is a true warping if the associated axial loads from (3.2)

$$n = Eg\dot{f}bW \ , \quad N_m = Eg\dot{f}S_mW_m$$

are statically equivalent to zero. This requires the vanishing of the total axial load

$$\int_C nds + \Sigma N_m = Eg\dot{f}(W,1) = 0$$

and of the two bending moments (2.7), or

$$E g \dot{f}(W, x-x_G) = 0 \ , \quad E g \dot{f}(W, y-y_G) = 0 \ .$$

A kinematical interpretation of the same conditions

$$(W,1) = 0 \ , \quad (W, x-x_G) = 0 \ , \quad (W, y-y_G) = 0 \qquad (4.4)$$

is obtained by looking for the parameters (α,β,γ) of an average plane $w^* = \alpha(x-x_G)+\beta(y-y_G)+\gamma$ from which to measure an arbitrary distortion w of the cross-section so as to minimize the squared norm $(w-w^*, w-w^*)$. It turns out immediately that, for the best plane, $W = w-w^*$ satisfies conditions (4.4)

5. DESIGN CONDITIONS FOR ZERO WARPING

The exact solution of the integro-differential equations (2.1) - (2.2), (3.1) - (3.2) and (2.5) - (2.6), with suitable end conditions at $z = 0$ and $z = L$, is fairly complicated. With affinity laws such as (1.1) and (1.2) the problem can be solved by separation of variables as shown first by von Karman and Chien [4] for symmetrical shapes in torsion and then by Benscoter [1] for the general case. In these approaches w is expanded in a series of eigenwarpings. The first term of the solution is the Bernoulli-Navier approximation. A much faster convergence is obtained by expanding the axial stresses in eigenmodes, as shown by the author [3]. The first-order terms are then made of the de Saint-Venant solutions for bending and torsion. While the first approach can be developed as a variational principle for displacements, the second is really a variational principle for stresses and requires the equilibrium equations to be satisfied beforehand.

Our purpose here is different, we look directly for possible functions $a(s)$, $b(s)$ and S_m, such that the problem becomes freed from warping effects. The net result is that the Bernoulli-Navier and de Saint-Venant theories become identical and exact solutions for arbitrary transverse loading modes. The only loading constraints remaining are in the end sections $z = 0$ and $z = L$, if eigenwarping modes are to be completely avoided.

The simplest approach to this goal is variational in nature. Looking at equations (3.9) and (3.10) we shall try to obtain the following functional identities

$$\int_C aW'dx = 0 , \qquad \int_C aW'dy = 0 , \tag{5.1}$$

$$\int_C ar_B W'ds = \int_C ar_B dW = 0 \tag{5.2}$$

for any true warping functions $W(s)$.

6. FREEDOM FROM TORSIONAL WARPING

First take condition (5.2). We may replace in it the true warping function W by any single-valued function w, differentiable along each arc. Indeed, such a function differs from a true warping by a linear form with arbitrary coefficients (α,β,γ)

$$w = W + \alpha(x-x_G) + \beta(y-y_G) + \gamma$$

and by virtue of properties (3.5),

$$\int_C ar_B dw = \int_C ar_B dW .$$

The condition

$$\int_C ar_B dw = 0$$

just shown to be equivalent to (5.2) is now integrated by parts and gives

$$\sum_m w_m T_m(ar_B) - \int_C wd(ar_B) = 0 .$$

The necessary conditions following from the arbitrariness of w are

$$dp = 0 \quad \text{on each arc,} \tag{6.1}$$

$$T_m(p) = 0 \quad \text{at each node,} \tag{6.2}$$

where $p = ar_B$. These conditions are identical to the Kirchhoff conservation laws for currents, the arcs being thought of as the branches of an electric circuit. Their general solution is a superposition of currents circulating in a complete set of independent meshes of the circuit (mesh-currents). Let ω_i denote a mesh-current in the ith mesh with a positive sense of description fixed, for instance, by the rotation of Ox to Oy. Then the general solution of equations (6.1) and (6.2) is

$$p = \sum_{i=1}^{k} \omega_i \delta_i \ . \tag{6.3}$$

The sum is extended to a given choice of k independent meshes. The solution is expressed with reference to the original sense of description adopted for each arc so that

$$\delta_i = \begin{cases} 0 & \text{for those arcs which do not belong to mesh i;} \\ 1 & \text{for the arcs of mesh i where mesh sense and original sense coincide;} \\ -1 & \text{for the arcs of mesh i where mesh sense and original sense are opposite.} \end{cases}$$

In the design application to the torsion problem, the choice of the Bernoulli shear centre is free. Once fixed, the thickness function is defined by the law

$$a(s) = \frac{1}{r_B} p \tag{6.4}$$

where p, as given by (6.3) depends on k parameters with obvious limitations in the possible choices to maintain positive thickness everywhere.

It is then easily verified that a simple warping-free solution of the torsion problem is at hand.

For u_B, v_B, and w identically zero but an arbitrary $\theta(z)$, the shear flow

$$q = \dot{\theta}Ghp \tag{6.5}$$

and zero axial stresses (one of the semi-inverse assumptions of de Saint-Venant for torsion)

$$n = 0 , \quad N_m = 0 \tag{6.6}$$

satisfy all the equilibrium and compatibility conditions with zero shear loads:

$$\left.\begin{aligned} S_x &= \dot{\theta}Gh \int_C pdx = \dot{\theta}Gh \int_C ar_B dx = 0 , \\ S_y &= \dot{\theta}Gh \int_C pdy = \dot{\theta}Gh \int_C ar_B dy = 0 , \end{aligned}\right\} \tag{6.7}$$

but with a couple M_B related to torsion by the Bernoulli torsional rigidity (equal here to that of de Saint-Venant):

$$M_B = \dot{\theta}Gh \int_C ar_B^2 ds = \dot{\theta}GhI_B . \tag{6.8}$$

The end conditions are compatible with both the built-in case and the free-end case.

7. FREEDOM FROM WARPING IN BENDING

To facilitate subsequent derivations, we shall assume that the orientation of the reference axes is that of the principal axes of the tensor of reduced areas. Then

$$A_{xy} = \int_C ax'dy = \int_C ay'dx = 0 . \tag{7.1}$$

Let us examine the possibility of implementing the second of conditions (5.1). We can replace in it the true warping function by

$$W = w-\alpha(x-x_G) - \beta(y-y_G) - \gamma$$

and, taking (7.1) into account, obtain the condition

$$\int_C aw'dy - \beta A_{yy} = 0 \tag{7.2}$$

to be satisfied by any single-valued axial displacement function w. We also note that, by virtue of (4.2), (4.3), and (4.4),

$$(w,x-x_G) - \alpha I_{xx} - \beta I_{xy} = 0$$

$$(w,y-y_G) - \alpha I_{xy} - \beta I_{yy} = 0 \ .$$

Solving these equations for β and substituting into (7.2), the freedom condition is finally expressed entirely in terms of the arbitrary w by

$$\int_C aw'dy = \int_C ay'dw = A_{yy}[i_{xy}(w,x-x_G) + i_{yy}(w,y-y_G)] \ .$$

Integrating by parts and replacing the scalar products by their explicit definitions, we conclude from the arbitrariness of w that we should have

$$(ay')' = [i_{xy}b(x-x_G) + i_{yy}b(y-y_G)]A_{yy} \quad , \tag{7.3}$$

$$T_m(ay') = - [i_{xy}S_m(x_m-x_G) + i_{yy}S_m(y_m-y_G)]A_{yy} \ . \tag{7.4}$$

If we consider given distributions of b(s) and S_m, these equations provide design laws for a(s) in order to implement the second of conditions (5.1). They will generally conflict with the laws (6.1) and (6.2), and we shall later examine the b(s) and S_m distributions needed to avoid this conflict and obtain freedom of warping effects in both bending and torsion.

Williams and Fine [5] gave and discussed a result equivalent to (7.3) for the case of a single closed cell with a plane of symmetry (so that $i_{xy} = 0$ too).

To solve (7.3) and (7.4) for ay' we introduce the concepts of the "open" static moments

$$\left.\begin{aligned} X_0(s) &= \int_- b(x-x_G)ds + \Sigma_- S_m(x_m-x_G) \ , \\ Y_0(s) &= \int_- b(y-y_G)ds + \Sigma_- S_m(y_m-y_G) \ , \end{aligned}\right\} \quad (7.5)$$

defined as follows. The closed meshes of the set of arcs are opened by a (in principle arbitrary) choice of k cuts and a node is chosen as origin. On each arc (or segment of cut arc) there is now a unique direction to follow in order to reach the origin and this is conveniently chosen as the positive sense of shear flow. Furthermore, at each point of this tree a new cut separates the graph into two portions. They are distinguished by the convention that, in following the positive flow sense, one passes from the negative side to the positive side. The static moments (7.5) are then uniquely defined at each point by extending the integral and the sum at the portion on the negative side. By moving of ds in the positive sense, we add a small contribution to the intregrals and find

$$dX_0 = b(x-x_G)ds \ , \quad dY_0 = b(y-y_G)ds \ . \qquad (7.6)$$

By stepping over a node, coming from several convergent branches and departing along the sole divergent branch, we find that

$$-T_m(X_0) = S_m(x_m-x_G) \ , \quad -T_m(Y_0) = S_m(y_m-y_G) \ . \qquad (7.7)$$

Naturally, the open static moments are different if a different set of cuts is adopted, but they always satisfy the same equations (7.6) and (7.7). Being linear these equations show that the difference between two solutions, that is the difference between two open static moments, is a solution of the corresponding homogeneous equations (6.1) and (6.2). Hence the general solution of equations (7.6) and (7.7) in the case of X is

$$X = X_0 - \sum_{i=1}^{k} \omega_i \delta_i$$

where X corresponds to a fixed set of cuts. The arbitrariness in the solution disappears when one adds the requirements

$$\oint \left(\frac{X}{a}\right) \delta_j ds = 0 \qquad (j = 1,2,\ldots,k) \;, \qquad (7.8)$$

the integrals being taken in the positive circulation sense around each mesh that was opened. Such requirements yield in fact a system of k equations in the k unknowns ω_i:

$$\sum_{i=1}^{k} \omega_i \oint_j \frac{\delta_i \delta_j}{a} ds = \oint_j \frac{X_0}{a} ds \qquad (j = 1,2,\ldots,k) \qquad (7.9)$$

and make the solution X unique. This solution and the similar one for Y will be called the *closed static moments.*

Since a closed static moment verifies the same equations

$$dX = b(x-x_G)ds \;, \qquad T_m(X) = -S_m(x_m-x_G) \qquad (7.10)$$

as an open moment, it also satisfies the following useful relations:

$$\int_C Xdx = -I_{xx} \;, \qquad \int_C Xdy = -I_{xy} \qquad (7.11)$$

that are immediately established by integration by parts, use of (7.10) and identification with (4.3). Similarly,

$$\int_C Y dx = -I_{xy} , \quad \int_C Y dy = -I_{yy} . \tag{7.12}$$

Coming back to equations (7.3) and (7.4), one possible solution appears to be

$$ay' = (i_{xy}X + i_{yy}Y)A_{yy} .$$

It is in fact the only one which ensures that y is single-valued, since the conditions

$$\oint_j \delta_j y'ds = 0 \qquad (j = 1,2,...,k)$$

are actually implemented by (7.8) and the corresponding property for Y. Furthermore, in view of (7.11) and (7.12), we can check that

$$A_{xy} = \int_C ay'dx = -A_{yy}(i_{xy}I_{xx}+i_{yy}I_{xy}) = 0 .$$

The final form of the condition for avoiding a warping effect in bending (in one plane) is thus

$$ay' = -A_{yy}(i_{xy}X+i_{yy}Y) . \tag{7.13}$$

The right-hand side is determined by the b(s) and S_m distributions, except for the free parameter A_{yy}.

Again, if we succeed in satisfying equation (7.13) for a practical design, a simple solution of the bending problem is at hand. The axial stress distribution of this solution is that of the de Saint-Venant theory for a bending moment M_y:

$$\left.\begin{aligned} n &= M_y b[i_{xy}(x-x_G)+i_{yy}(y-y_G)] , \\ N_m &= M_y S_m[i_{xy}(x_m-x_G)+i_{yy}(y_m-y_G)] . \end{aligned}\right\} \tag{7.14}$$

The corresponding shear flow distribution in equilibrium with axial stresses is

$$q = -S_y(i_{xy}X+i_{yy}Y) \tag{7.15}$$

and has, by virtue of (7.13), the alternate expression

$$q = \frac{S_y}{A_{yy}}\, ay' \ . \tag{7.16}$$

The bending moment M_x and shear load S_x in the other plane are zero. The deformation equations are satisfied by setting

$$\theta = 0 \ , \tag{7.17}$$

$$\ddot{v} = -i_{yy}\frac{M_y}{Eg} + \frac{1}{GA_{yy}}\left(\frac{S_y}{h}\right) , \tag{7.18}$$

$$\ddot{u} = -i_{xy}\frac{M_y}{Eg} , \tag{7.19}$$

$$w = \left(\frac{S_y}{GhA_{yy}} - \dot{v}\right)(y-y_G)-\dot{u}(x-x_G) \ . \tag{7.20}$$

Hence, as expected, the cross-section remains plane, and the end conditions are again compatible with both the built-in case and the free-end case. The whole theory, repeated for implementation of the first of conditions (5.1), would yield a similar solution for bending in the other plane.

8. FREEDOM FROM WARPING IN BENDING AND TORSION

Here we examine the possibility of combining freedom from warping in both torsion and bending in one plane. As we shall see this automatically implies also freedom from warping in bending in the other plane. Suppose we have adopted a law for a(s) that suppresses the de Saint-Venant warping in torsion and let the origin of the axes be placed at the chosen Bernoulli shear centre. We then have the properties

$$d(ar) = 0 \quad \text{on each arc} \ , \tag{8.1}$$

$$T_m(ar) = 0 \quad \text{at each node} , \tag{8.2}$$

with

$$r = xy' - yx' . \tag{8.3}$$

To remove also the de Saint-Venant warping due to shear loads S_y we need, according to the results of Section 7,

$$ay' = -A_{yy}(i_{xy}X + i_{yy}Y) \quad . \tag{8.4}$$

The differential of the left-hand side along an arc is manipulated to incorporate (8.1) as follows

$$d(ay') = ard\left(\frac{y'}{r}\right) = -\frac{ay}{\rho r}\, ds$$

where

$$\frac{1}{\rho} = x'y'' - y'x'' \tag{8.5}$$

is the curvature of the arc, positive when the normal $\underline{n}$ lies on the convex side. The differential of the right-hand side follows from equations like the first of (7.10). The final result

$$\frac{ay}{\rho r} = A_{yy}[i_{xy}(x-x_G)+i_{yy}(y-y_G)]b$$

can serve the specify the b(s) distribution. To avoid infinities for $y = 0$ we are led to require furthermore that

$$i_{xy} = 0 \quad \text{or} \quad I_{xy} = 0 \quad \text{and} \quad y_G = 0 \tag{8.6}$$

which reduces the law to

$$b = \frac{I_{yy}}{A_{yy}}\frac{a}{\rho r} = -\frac{I_{yy}}{A_{yy}}\frac{(ay')'}{y} . \tag{8.7}$$

From (8.4) and equations like the second of (7.10) we obtain

$$S_m = \frac{I_{yy}}{A_{yy}} \frac{T_m(ay')}{y_m} \tag{8.8}$$

as laws governing the area of stringers.

We now develop relations of symmetry to show that the same laws suppress warping due to S_x shear loads. From the identity

$$ar = x(ay') - y(ax')$$

we obtain, in view of properties (8.1) and (8.2), that

$$xd(ay') = yd(ax') \tag{8.9}$$

and

$$x_m T_m(ay') = y_m T_m(ax') \ . \tag{8.10}$$

Hence the symmetrical result can be proved:

$$x_G = 0 \ . \tag{8.11}$$

The value of x_G is found from the property $(x-x_G,1) = 0$ or

$$x_G(1,1) = (x,1) = \int_C bxds + \sum_m S_m x_m \ .$$

But, from (8.7) and (8.8),

$$x_G(1,1) \frac{A_{yy}}{I_{yy}} = - \int_C \frac{x}{y} d(ay') + \sum_m \frac{x_m}{y_m} T_m(ay')$$

and finally, from (8.9) and (8.10),

$$x_G(1,1) \frac{A_{yy}}{I_{yy}} = - \int_C d(ax') + \sum_m T_m(ax') = 0$$

as announced. Next calculate

$$A_{xx} = \int_C ax'dx = \sum_m x_m T_m(ax') - \int_C xd(ax') \ .$$

Apply the symmetry relations (8.9) and (8.10) to find

$$A_{xx} = \sum_m \frac{x_m^2}{y_m} T_m(ay') - \int_C \frac{x^2}{y} d(ay')$$

and finally reintroduce b and S_m from (8.7) and (8.8):

$$A_{xx} = \frac{A_{yy}}{I_{yy}} \left(\sum_m S_m x_m + \int_C bx^2 ds \right) .$$

This relation, exhibited in the symmetrical form

$$\mu = \frac{A_{xx}}{I_{xx}} = \frac{A_{yy}}{I_{yy}} , \tag{8.12}$$

proves that the laws found for b and S_m can be placed in the form that would have resulted from the suppression of the warping due to S_x instead of the S_y shear loads.

To summarize: the skin thickness carrying shear flow must be distributed to satisfy equations (8.1) and (8.2), the nature of this solution was discussed in Section 6. The skin thickness carrying normal loads must be distributed according to the law

$$b = \frac{1}{\mu} \frac{a}{r\rho} \tag{8.13}$$

depending on the curvature of the skin and proportional to a desired ratio of inertia moments to reduced areas.

The area of nodal stringers is given by either of the formulae

$$S_m = \frac{1}{\mu} \frac{T_m(ax')}{x_m} = \frac{1}{\mu} \frac{T_m(ay')}{y_m} . \tag{8.14}$$

For such designs the axes Ox and Oy are both principal axes of inertia ($I_{xy} = 0$) and principal shear axes ($A_{xy} = 0$). The origin is simultaneously the shear centre, flexure-torsion centre and centroid. Under arbitrary transverse loads, and provided the end loads conform with the de Saint-Venant distribution, the section remains plane and the ordinary engineering beam theory is an exact solution.

9. THE PRACTICAL DESIGN OF WARPING-FREE SECTIONS

We still need to deal, in the simplest possible manner, with the limitations of the theory of warping-free sections represented by the requirements for positive thicknesses and areas for skin and stringers. The single closed cell section will be considered to be the simplest building block for more complex structures.

In a single cell we take

$$a = \frac{k}{r} \tag{9.1}$$

where k is a constant connected with the desired torsional rigidity

$$J = \oint ar^2 ds = k\oint r ds = 2\Omega k \tag{9.2}$$

Ω being the area of the cell. Hence k should be positive and r likewise. This limits the choice of the shear centre (origin of the co-ordinate axes) to the interior of the cell, which, will be seen next, must be convex. The skin thickness for normal loads is fixed by

$$b = \frac{a}{\mu\rho r} = \frac{k}{\mu\rho r^2} \,. \tag{9.3}$$

The parameter μ is chosen according to (8.12) to obtain a desired moment of inertia and is positive. Hence ρ must also be positive: the section must be convex. At the limit, zero curvature could be tolerated, if the property b = 0 can be represented by a corrugated sheet. At angular points a concentrated stringer area is needed and can be calculated by either one of equations (8.14). It is however more instructive to obtain it by a limiting process.

The total area of a curved skin piece is

$$\int_{M_1}^{M_2} b\,ds = \frac{k}{\mu}\int_{M_1}^{M_2}\frac{ds}{\rho r^2} = \frac{k}{\mu}\int_{\phi_1}^{\phi_2}\frac{d\phi}{r^2}$$

where $\phi_2-\phi_1$ is the angle through which the tangent to the arc has turned. In the limit of strong curvature this formula should hold for the stringer area where $\phi_2-\phi_1$ is the turning angle of the tangent at the angular point. In this case r varies according to the law $r = R\cos\phi$ (Figure 4) and we find

$$S = \frac{k}{\mu}\,\frac{\tan\phi_2-\tan\phi_1}{R^2} = \frac{k}{\mu}\,\frac{\sin(\phi_2-\phi_1)}{r_1r_2} \tag{9.4}$$

where r_1 and r_2 are the values of r when entering and leaving the angular point. This result is easily proven to be identical with (8.14). For multi-cellular sections the simplest method is to superimpose the thickness and area distributions of component single cells (Figure 5). All component cells should be convex and contain the chosen shear centre in their interior. In addition to the freedom in the choice of a shear centre and the choice of μ (which must be the same for all components), there are as many independent parameters k_i as component cells. The overall torsional stiffness of the multicellular case is readily calculated to be

$$J = 2\sum_i k_i\Omega_i\ . \tag{9.5}$$

The consideration that warping-free beams can resist rapidly varying or even discontinuous transverse loads without stress concentrations due to shear lag or torsion induced bending stresses warrants a careful design study. Admittedly this ideal can in some cases conflict with a more efficient use of the material.

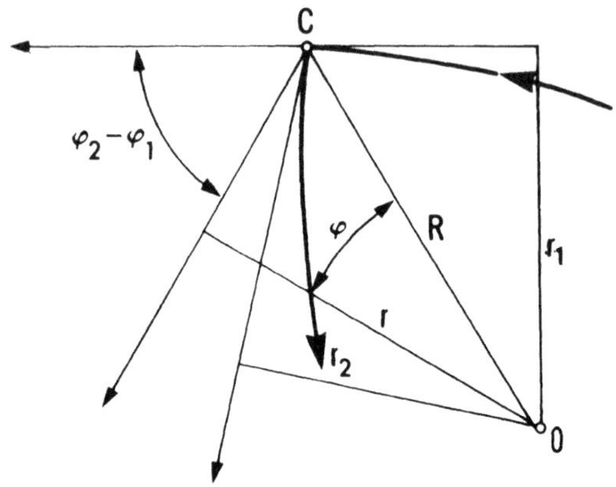

Figure 4 - Calculating the Boom Area at an Angular Point (Corner C)

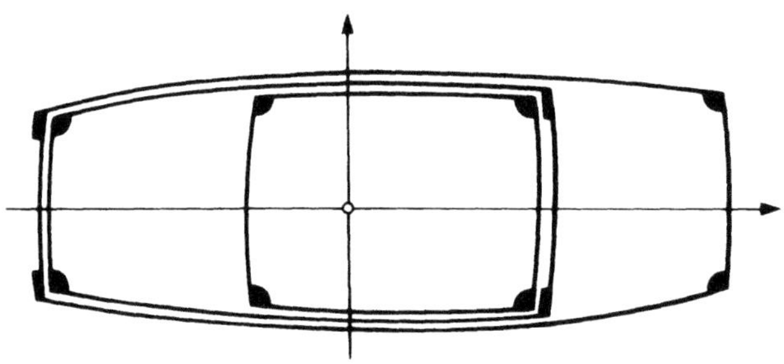

Figure 5 - Principle of Superposition of Skin Thicknesses and Boom Areas of Component Single Cells

REFERENCES

[1] BENSCOTER, S.U., "Secondary Stresses in Thin-Walled Beams with Closed Cross-Sections", NACA Techn. Notes, No. 2529, 1951.

[2] DRYMAEL, J., "Some Theoretical Considerations on Box-Beams", *J. Roy. Aeronautical Soc.*, Vol. 45:361, 1941, pp. 5-28.

[3] FRAEIJS DE VEUBEKE, B.M., "Théorie des coques prismatiques minces renforcées par lisses et diaphragmes", *Mém. Acad. Roy. Belgique, Cl. Sci.*, Vol. 32:8, No. 1729, 1961.

[4] VON KARMAN, T. and CHIEN, W.Z., "Torsion with Variable Twist", *J. Aeronautical Sci.*, Vol. 13, No. 10, 1946, pp. 503-510.

[5] WILLIAMS, D. and FINE, M., "Stress Distribution in Reinforced Flat Sheet, Cylindrical Shells and Cambered Box-Beams under Bending", ARC Reports and Memor., 1940, No. 2099.

DUAL PRINCIPLES OF ELASTODYNAMICS FINITE ELEMENT APPLICATIONS

B.M. Fraeijs de Veubeke

University of Liège, Belgium

INTRODUCTION

Up to the present time, most of the finite element applications to linear structural dynamics have been based on the displacement approach and its corresponding Hamiltonian variational principle. It is sufficient to this effect to add to the static stiffness matrix lagrangian, or so-called coherent, mass matrices.

The first really satisfactory formulation of a dual principle, in which the kinetic energy is transformed through satisfaction of the dynamic equilibrium equations into a functional expressed in terms of space derivatives of an impulse field, is due to Toupin [5]. Similar approaches were followed by Crandall [9], Yu Chen [10], and Gladwell and Zimmermann [11]. Finite element applications to the eigenvalue problems of beams and plates are due to Tabarrok, Sakaguchi and Karnopp [8, 12] and to Geradin [7].

Original publication, NATO Advanced Studies Institute Lecture Series on *Finite Element Methods in Continuum Mechanics*, Lisbon, 1971, edited by Oden and Oliveira, UAH Press, 1973.

In the following sections, there is presented a logical derivation of the dual dynamic principle through the canonical form in the spirit of Friedrichs transformations [1]. Also discussed is the general procedure for assembling equilibrium type finite elements in order to implement the dual principle. Finally, it is shown that there is no advantage in using the dual principle together with the requirement of forcing orthogonality with respect to all zero frequency modes. Instead, experience shows [7, 8] that by ignoring this unnecessary requirement eigenvalues generally converge to their exact values by lower bounds, hence giving precious accuracy estimates by comparison with the displacement approach.

HAMILTON'S PRINCIPLE

Hamilton's principle for conservative systems is a statement of stationarity of a functional of the displacement field

$$f[u] = \int_{t_1}^{t_2} (T-V)dt \tag{1}$$

where T denotes the kinetic energy

$$T = \int_R \frac{1}{2} \rho \dot{u}_j \dot{u}_j dR \tag{2}$$

and V the total potential energy of the system. V can have several distinct parts; one is the strain energy

$$V_1 = \int_R W(Du)dR \tag{3}$$

where the energy density is expressed in terms of strains, i.e.,

$$\varepsilon_{ij} = \frac{1}{2} (D_i u_j + D_j u_i) \tag{4}$$

if we restrict ourselves to the linearized case; another one can be a potential energy associated to the displacements on a part $\partial_2 R$ of the boundary

$$V_2 = -\int_{\partial_2 R} F(u,t)dS \, . \tag{5}$$

On the remaining part $\partial_1 R$ of the boundary we shall assume that the displacements are prescribed time functions

$$u_j = \bar{u}_j(t) \quad \text{on} \quad \partial_1 R \, . \tag{6}$$

For simplicity we do not take into consideration other potential energies like that associated with conservative external body forces functions of position.

In applying the principle

$$\delta f[u] = 0$$

with Hamilton's rule

$$\delta u_j \equiv 0 \quad \text{for } t = t_1 \quad \text{and} \quad t = t_2 \tag{7}$$

and $\delta u_i = 0$ at all times on $\partial_1 R$, which is a consequency of (6), the following variational derivatives are obtained

$$-\rho \ddot{u}_j + D_i \tau_{ij} = 0 \tag{8}$$

where

$$\tau_{ij} = \frac{\partial W}{\partial \varepsilon_{ij}} \tag{9}$$

are the stresses. They are the dynamic equilibrium conditions in d'Alembert's form. The natural boundary conditions supplementing (6) are found to be

$$-n_i \tau_{ij} + \frac{\partial F}{\partial u_j} = 0 \quad \text{on} \quad \partial_2 R \tag{10}$$

with the usual notation

$$t_j = n_i \tau_{ij} \tag{11}$$

for the surface tractions, they state that these surface tractions are prescribed functions of displacement and time:

$$t_j = \frac{\partial F}{\partial u_j} = \bar{t}_j(u,t) \quad \text{on} \quad \partial_2 R \; . \tag{12}$$

An important special case is that of a linear function

$$f(u,t) = u_j \bar{t}_j(t) \tag{13}$$

in which the natural boundary conditions reduce to

$$t_j = \bar{t}_j(t) \quad \text{on} \quad \partial_2 R \; . \tag{14}$$

FRIEDRICHS TRANSFORMATION. THE CANONICAL VARIATIONAL PRINCIPLE

We follow the way indicated by Friedrichs [1] to transform Hamilton's principle into canonical and, later, dual form. To this purpose we introduce a dislocation potential Δ into the functional (1) that becomes

$$\hat{f}[u,v,\varepsilon,\lambda,\alpha,\mu] = \int_{t_1}^{t_2} (T-V+\Delta)dt \; . \tag{15}$$

The dislocation potential consists of three parts:

$$\Delta_1 = \int_R \lambda_{ij}(\varepsilon_{ij} - \frac{1}{2} D_i u_j - \frac{1}{2} D_j u_i)dR \tag{16}$$

the incorporates the constraints (4) as natural variational equations for the stationarity of (15) by means of a tensor of lagrangian multipliers λ_{ij}. Correspondingly the potential energy V_1 is now expressed as

$$V_1 = \int_R W(\varepsilon)dR . \tag{3'}$$

A second part

$$\Delta_2 = \int_{\partial_1 R} \alpha_j\{u_j - \bar{u}_j(t)\}dS \tag{17}$$

incorporates the boundary constraints (6) as natural boundary conditions through the vector of lagrangian multipliers α_j. Finally

$$\Delta_3 = \int_R \mu_j(\dot{u}_j - v_j)dR \tag{18}$$

introduces the velocity vector field v_j as an independent field and the kinetic energy is now expressed as

$$T = \int_R \frac{1}{2} \rho v_j v_j dR . \tag{2'}$$

The dislocation potentials Δ_1 and Δ_2 have been used in the static problem [2, 3] to provide a logical approach to the two-field variational principles, one of which is the Reissner-Hellinger principle. The introduction of the dislocation potential Δ_3 is an essential step in the logical transformation from Hamilton's principle to the eulerian variational principles of fluid mechanics [4]. The first step in the simplification of the functional (15) consists in identifying the multipliers λ_{ij} and μ_j. This results immediately from setting the variational derivatives of ε_{ij} and v_j equal to zero:

$$\lambda_{ij} - \frac{\partial W}{\partial \varepsilon_{ij}} = 0 , \tag{19}$$

$$\rho v_j - \mu_j = 0 . \tag{20}$$

Hence the λ_{ij} tensor is identified with the stress tensor τ_{ij} related to the ε_{ij} by the elastic constitutive equations (9); the μ_j

vector is identified with the momentum per unit mass. When (19) is substituted into the functional, the expression

$$\varepsilon_{ij} \frac{\partial W}{\partial \varepsilon_{ij}} - W = \phi(\tau) \tag{21}$$

that appears on the left hand side is treated as a Legendre transformation introducing the complementary energy density ϕ, a function of the elements of the stress tensor. This transformation is possible whenever the strains can be expressed in terms of the stresses, in which case the constitutive equations (9) are equivalent to

$$\varepsilon_{ij} = \frac{\partial \phi}{\partial \tau_{ij}} \ . \tag{22}$$

Thus one obtains

$$\Delta_1 - V_1 = U_1 - \Lambda_1$$

where

$$U_1 = \int_R \phi(\tau) dR \tag{23}$$

the complementary strain energy, and

$$\Lambda_1 = \int_R \tau_{ij} D_i u_j dR \tag{24}$$

where account was taken of the symmetry of the stress tensor. Also the substitution of (20) yields

$$T + \Delta_3 = - T + \Lambda_3$$

where

$$\Lambda_3 = \int_R \rho v_j \dot{u}_j dR \ . \tag{25}$$

The last set of multipliers is identified through the natural boundary conditions stemming from variations on the displacements on $\partial_1 R$

$$\alpha_j - n_i \tau_{ij} = 0 \quad \text{on} \quad \partial_1 R \,. \tag{26}$$

This is simply used to transform (17) into

$$\Delta_2 = \int_{\partial_1 R} n_i \tau_{ij} \{u_j - \bar{u}_j(t)\} dS \,. \tag{17'}$$

The final result is a canonical variational principle in the Friedrich sense, requiring the stationarity of the functional

$$c[u,v,\tau] = \int_{t_1}^{t_2} (U_1 - T - V_2 + \Delta_2 - \Lambda_1 + \Lambda_3) dt \tag{27}$$

in which the various integrands are given respectively by (23), (2'), (5), (17'), (24) and (25). It can be used to develop approximate solutions based on independent discretizations of displacements, velocities and stresses. Its variational equations are respectively

$$D_i \tau_{ij} - \rho \dot{v}_j = 0 \tag{28}$$

in the variations on displacements (the dynamic equilibrium equations);

$$\rho(-v_j + \dot{u}_j) = 0 \tag{29}$$

in the variations on velocities, restoring the constraints between velocity and displacement fields;

$$\frac{\partial \phi}{\partial \tau_{ij}} - \frac{1}{2}(D_i u_j + D_j u_i) = 0 \tag{30}$$

in the variations on stresses, restoring the constraints between stresses and displacements. All boundary conditions are natural.

On $\partial_1 R$ they follow from the variations on the surface tractions (11) and correspond to (6); on $\partial_2 R$ they follow from the variation of displacements and correspond to (12).

For finite element applications it is important to recognize the nature of the transition conditions at interfaces. Because the only space derivatives contained in the functional are those affecting displacements, we must consider that physical continuity

$$(u_j)_+ = (u_j)_- \tag{31}$$

is an a *priori* requirement; the natural transition conditions provided by the principle and stemming from the common variations $(\delta u_j)_+ = (\delta u_j)_-$ at interfaces are then

$$(t_j)_+ + (t_j)_- = 0 \ . \tag{32}$$

From that standpoint the canonical principle (27) behaves as the classical principle of variations of displacements. Obviously, the interface constraints (31) could be incorporated into the principle though an additional dislocation potential and the corresponding multipliers identified. Then, just as was found to be the case for the boundary conditions, all transitional conditions can be cared for by the principle.

THE COMPLEMENTARY ENERGY PRINCIPLE OF ELASTODYNAMICS

It can be deduced from the canonical principle by specializations. To this effect the functional (27) is integrated by parts with respect to the displacements in both the terms Λ_1 and Λ_3 of the integrand

$$\Lambda_1 = \int_{\partial R} n_i \tau_{ij} u_j dS - \int_R u_j D_i \tau_{ij} dR$$

$$\int_{t_1}^{t_2} \Lambda_3 dt = \left[\int_R \rho v_j u_j dR\right]_{t_1}^{t_2} - \int_{t_1}^{t_2} \int_R \rho \dot{v}_j u_j dRdt \ .$$

This operation cancels part of the contribution of Δ_2. Furthermore the term at the time limits is dropped on the basis of the requirements (7) and introduction of the new requirements

$$\delta v_j \equiv 0 \qquad \text{for} \quad t = t_1 \quad \text{and} \quad t = t_2 \ . \tag{33}$$

Thus, for the following portion of the canonical functional,

$$\int_{t_1}^{t_2} (-V_2 + \Delta_2 - \Lambda_1 + \Lambda_3)dt = \int_{t_1}^{t_2} \int_R u_j (D_i \tau_{ij} - \rho \dot{v}_j) dRdt$$
$$- \int_{t_1}^{t_2} \int_{\partial_1 R} n_i \tau_{ij} \bar{u}_j (t) dSdt + \int_{t_1}^{t_2} \int_{\partial_2 R} \{F(u,t) - n_i t_{ij} u_j\} dSdt \ . \tag{34}$$

The right hand side is now simplified by making the assumption that the dynamic equilibrium equations (28) can be satisfied *a priori;* this causes the first term to vanish. In view of (11) and (12), the integrand of the last term is obviously related to another Legendre transformation

$$u_j \frac{\partial F}{\partial u_j} - F(u,t) = G(t_j,t) \tag{35}$$

provided equations (12) can be solved for the boundary displacements. This is the case in principle whenever the Hessian

$\left|\frac{\partial^2 F}{\partial u_i \partial u_j}\right|$ does not vanish. The complementary function

G is then such that

$$\frac{\partial G}{\partial t_i} = u_i \ . \tag{36}$$

With the notations

$$B_1 = \int_{\partial_1 R} t_j \bar{u}_j(t)\,dS \qquad B_2 = \int_{\partial_2 R} G(t_j, t)\,dS \tag{37}$$

the canonical functional is now reduced to one that depends only on the stress field

$$g[\tau] = \int_{t_1}^{t_2} (U_1 - T - B_1 - B_2)\,dt \tag{38}$$

provided the kinetic energy can be expressed in terms of it. The dynamical equilibrium equations (28) that must be satisfied *a priori* allow precisely to express the velocity field in terms of the stresses:

$$v_j = v_j\Big|_{t_1} + \frac{1}{\rho} D_i \int_{t_1}^{t} \tau_{ij}\,dt \,. \tag{39}$$

This suggests the introduction of an impulse field [5] T_{ij} such that

$$v_j = \frac{1}{\rho} D_i T_{ij} \tag{40}$$

hence

$$T_{ij} = T_{ij}\Big|_{t_1} + \int_{t_1}^{t} \tau_{ij}\,dt \tag{41}$$

and

$$\dot{T}_{ij} = \tau_{ij} \,. \tag{42}$$

The requirement (33) can now obviously be replaced by

$$\delta T_{ij} \equiv 0 \quad \text{for} \quad t = t_1 \quad \text{and} \quad t = t_2 \,. \tag{43}$$

A convenient formulation of the complementary energy principle of elastodynamics is thus the stationarity of the functional

$$g[T_{ij}] = \int_{t_1}^{t_2} \left\{ \int_R (\phi(\dot{T}_{ij}) - \frac{1}{2\rho} D_i T_{ij} D_m T_{mj})\,dR \right.$$

$$- \int_{\partial_1 R} \dot{T}_j \bar{u}_j(t) dS - \int_{\partial_2 R} G(\dot{T}_j, t) dS \Big\} dt \tag{44}$$

where

$$T_j = n_i T_{ij} \qquad t_j = \dot{T}_j . \tag{45}$$

The variational derivatives of this principle are

$$- \frac{d}{dt} \frac{\partial \phi}{\partial \dot{T}_{ij}} + \frac{1}{2} (D_i D_m T_{mj} + D_j D_m T_{mi}) = 0 \tag{46}$$

that, in view of (42), (22) and (40), are easily interpreted as the time derivatives of compatibility conditions:

$$- \dot{\varepsilon}_{ij} + \frac{1}{2} (D_i v_j + D_j v_i) = 0 .$$

The natural boundary conditions are seen to be

$$- \frac{1}{\rho} D_i T_{ij} + \frac{d}{dt} \bar{u}_j(t) = 0 \qquad \text{on} \qquad \partial_1 R \tag{47}$$

$$- \frac{1}{\rho} D_i T_{ij} + \frac{d}{dt} \frac{\partial G}{\partial \dot{T}_j} = 0 \qquad \text{on} \qquad \partial_2 R \tag{48}$$

Both are easily interpreted through (40) and (36).

In the special case (13), the Legendre transformation (35) yields a function G that is identically zero; the boundary conditions (14) become *a priori* boundary conditions in the complementary energy formulation, thus

$$\dot{T}_j = \bar{t}_j(t) \qquad \text{on} \qquad \partial_2 R . \tag{14'}$$

Similarly, because the displacement field vanishes completely, the transitional conditions at interfaces (32) have now to be satisfied *a priori*.

FINITE ELEMENT DISCRETIZATION USING THE COMPLEMENTARY ENERGY PRINCIPLE

The approach to discretization by finite elements can follow a very similar path to that used in statics [6]. At the level of a single finite element it is convenient to consider boundary conditions to consist only on imposed displacements.

The tensorial impulse field is discretized in the same manner as the stress field was in statics, namely

$$\tau(x,t) = R(x)c(t) + S(x)s(t) \tag{49}$$

where $\tau(x,t)$ here denotes a column matrix of the elements of the impulse tensor, ranged in the same order as the stresses in reference 6, and where the parameters ranged in the column matrices c and s are now functions of time. As was assumed in reference 6, the approximating functions in $S(x)$ are such that

$$D'S(x) = 0$$

where D' is the differential operator generating the quantities $D_i T_{ij}$. Thus the $s(t)$ coordinates do not contribute to the kinetic energy.

On the other hand each column of

$$D'R(x) = U(x)$$

furnishes the components of an inertia force distribution accounted for in the discretization. It should at least contain the inertia force distributions related to the motions of the finite element as a rigid body. The discretization (49) furnishes, when used in the evaluation of the kinetic energy, a consistent "inverse mass" or "mobility matrix" N:

$$\int_E \frac{1}{2\rho} D_i T_{ij} D_m T_{mj} dE = \frac{1}{2} \int_E \frac{1}{\rho} \{D'R(x)c\}'\{D'R(x)c\} dE = \frac{1}{2} c'Nc \tag{50}$$

$$N = \int_E \frac{1}{\rho} U'U dE \quad \text{positive definite} . \tag{51}$$

It also furnishes a stress energy

$$\int_E \Phi(\dot{T}_{ij}) dE = \frac{1}{2} \int_E \dot{\tau}' H^{-1} \dot{\tau} dE = \frac{1}{2} \dot{c}' F_{cc} \dot{c} + \dot{c}' F_{cs} \dot{s} + \frac{1}{2} \dot{s}' F_{ss} \dot{s} \tag{52}$$

with the same flexibility matrices that were already defined in reference 6.

Again, generalized boundary impulses can be defined by examination of the distributions $N'R(x)$ and $N'S(x)$ of the T_j impulses appearing along each boundary of the element; and they are finally expressible in terms of the parameters of the impulse field through the same load connection matrices G and C

$$g = Gc + Cs . \tag{53}$$

The generalized boundary displacements q^* conjugate to the generalized body loads $\dot{g}$ are defined exactly as in Section 4.2 of reference 6. Thus the discretization of (44), assuming imposed displacements along the whole boundary, leads to a requirement of stationarity of

$$\int_{t_1}^{t_2} \left\{ \frac{1}{2} \dot{c}' F_{cc} \dot{c} + \dot{c}' F_{cs} \dot{s} + \frac{1}{2} \dot{s}' F_{ss} \dot{s} - \frac{1}{2} c' N c - (\dot{c}' G' + \dot{s}' C') q^* \right\} dt$$

with δc, δs vanishing at the time limits. Hence

$$-F_{cc} \ddot{c} - F_{cs} \ddot{s} - Nc + G' \dot{q}^* = 0 \tag{54}$$

$$-F_{sc} \ddot{c} - F_{ss} \ddot{s} + C' \dot{q}^* = 0 . \tag{55}$$

When the s coordinates are taken from the second set of equations and substituted into the first, there comes

$$F\ddot{c} + Nc = (G' - F_{cs} F_{ss}^{-1} C') \dot{q}^* \tag{56}$$

with

$$F = F_{cc} - F_{cs} F_{ss}^{-1} F_{sc} . \tag{57}$$

Equations (56) can be regarded as determining the dynamics of the impulse coordinates to which an inertia is attached as excited by prescribed boundary displacements.

The finite elements are now assembled with the same procedure as in statics, allocating each generalized boundary displacement of an element E to a set w* of nodal displacements

$$q_E^* = L_E w^* , \tag{58}$$

with the help of Boolean matrices. Then, for each finite element, we write the set of equations (56) in the form:

$$F_E \ddot{c}_E + N_E c_E = J_E' \dot{w}^* \qquad (E = 1,2,\dots,N) \tag{59}$$

with

$$J_E' = (G' - F_{cs} F_{ss}^{-1} C')_E L_E . \tag{60}$$

There are as many equations in the set (59) as there are coordinates in the set of all c_E. We refer to them as the "internal coordinates" and note that their dynamics is described by (59), provided we consider them excited by known time histories of the nodal displacements. Of course we need a complementary set of dynamic equations to obtain those time histories. They are obtained by using the property of internal connexion loads $\dot{g}_E$ to be in equilibrium with the external loads applied at interfaces or at the boundaries of the assembled structure. Thus, by reference to equation (105) of reference 6,

$$\sum_E L'_E \dot{g}_E = \dot{y}_{(w)} \tag{61}$$

where $y_{(w)}$ denotes the external impulse vector conjugate to the nodal displacements. The connexion loads, or rather their time derivatives, can be obtained in terms of the internal coordinates and nodal velocities by differentiating twice with respect to time and substituting $\ddot{s}$ from (55) again. Whence, for each element

$$\ddot{g}_E = (G - CF_{ss}^{-1}F_{sc})_E\ddot{c}_E + K_EL_E\dot{w}^* \tag{62}$$

$$K_E = (CF_{ss}^{-1}C')_E \ . \tag{63}$$

We recognize in K_E the stiffness matrix of the equilibrium element as used in statics. Differentiating (61) with respect to time and substituting (62), we obtain the complementary set of dynamic equations

$$\sum_E J_E\ddot{c}_E + K\dot{w}^* = \ddot{y}_{(w)} \tag{64}$$

where in $K = \sum_E L'_EK_EL_E$ (65)

we recognize the master stiffness matrix of the assembled structure. The equations (59) and (63) constitute a complete set of dynamic equations for the unknowns. There is no difficulty in taking into account lumped inertias associated to the nodal displacements, that is applied either at interfaces or along the structural boundary, we merely have to set up the corresponding consistent mass matrix M and write

$$\dot{y}_{(w)} = -M\ddot{w}^* + f_{(w)}(t) \tag{66}$$

modifying equation (64) that becomes now

$$\sum_E J_E\ddot{c}_E + M\ddot{w}^* + K\dot{w}^* = f_{(w)}(t) \ . \tag{67}$$

The real excitation of the system takes place either from elements of $f_{(w)}(t)$ as prescribed external forces, or from elements of w^* as prescribed displacements in which case the conjugate elements in $f_{(w)}$ become unknown reactions.

FREE VIBRATIONS

Consider the problem of a modal analysis of the free vibrations. If the structure is floating, that is not subjected to homogeneous displacement boundary conditions, we can simply set $f_{(w)} = 0$ and consider all c_E and w^* elements as unknowns. Setting

$$c_E = \hat{c}_E \sin\omega t \qquad w^* = \hat{w}^* \cos\omega t$$

the eigenvalue problem is given in the form

$$(N_E - \omega^2 F_E)\hat{c}_E + \omega J'_E \hat{w}^* = 0 \qquad E = 1,2,\ldots,N \tag{68}$$

$$(K - \omega^2 M)\hat{w}^* + \omega \sum_E J_E \hat{c}_E = 0 \ , \tag{69}$$

or, with a modal amplitude vector

$$x' = (\hat{c}'_1 \hat{c}'_2 \ldots \hat{c}'_N w^{*\prime})$$

$$\begin{bmatrix} N_1-\omega^2F_1 & 0 & 0 & 0 & \omega J'_1 \\ 0 & N_2-\omega^2F_2 & 0 & 0 & \omega J'_2 \\ 0 & 0 & \ddots & 0 & \ldots \\ 0 & 0 & 0 & N_N-\omega^2F_N & \omega J'_N \\ \omega J_1 & \omega J_2 & \ldots & \omega J_N & K-\omega^2M \end{bmatrix} x = 0 \ . \tag{70}$$

In many applications there are no lumped inertias attached to the nodal points and consequently no matrix M. This makes it possible to reduce drastically the size of the eigenvalue problem by a static condensation of most or all of the nodal displacements, leaving essentially the few internal coordinates as unknowns. In the case of the floating structure this means that we should use

$$K\hat{w}^* = -\omega \sum_E J_E \hat{c}_E \tag{71}$$

to eliminate the nodal amplitudes. Because K is singular on account of the presence of structural rigid body modes or mechanisms, it will be necessary to keep enough displacement coordinates to represent those. This is in general achieved by suppressing just enough rows and corresponding columns in K to obtain a non singular submatrix. Let P and R denote column selection operators of the same size as K

$$P + R = I \;=\; \text{identity matrix}$$

and decompose $\hat{w}^*$ into

$$\hat{w}^* = Pp + Rr \,.$$

Then, with the notations

$$P'KP = K_{pp} \qquad P'KR = K_{pr} \qquad R'KP = K_{rp} \qquad R'KR = K_{rr}$$

equation (71) becomes after premultiplication by P' and R' respectively

$$K_{pp}p + K_{pr}r = -\omega P' \sum_E J_E \hat{c}_E \tag{72}$$

$$K_{rp}p + K_{rr}r = -\omega R' \sum_E J_E \hat{c}_E \,. \tag{73}$$

Let K_{pp} be non singular, r representing kinematical freedoms of the assembled structure; p can be extracted from (72) and substituted into (68) and (73) to produce an eigenvalue problem in the unknowns $\hat{c}_E$ and r only. Clearly similar procedures apply to structures that are isostatically or hyperstatically grounded by setting some of the elements of $\hat{w}$ equal to zero. However elegant, the formulation (68) does not lend itself to the classical algorithms for eigenvalue search in many degrees of freedom. The reason is of course the use of mixed displacement and stress type variables that induces a special coupling through the J_E matrices. It might prove interesting to generalize the iterative type algorithms to this type of formulation; but it is also possible, as shown by Geradin [7], to obtain a classical formulation by a change of variables.

CLASSICAL FORM OF THE EIGENVALUE PROBLEM

The change of variables consists in setting

$$-Nc = \dot{p}^* \tag{74}$$

and, in view of (54), consider p* as a displacement vector conjugate to c. Equations (54) and (55) are then brought together in the form

$$\hat{F}\begin{pmatrix}\ddot{c}\\ \ddot{s}\end{pmatrix} = \hat{G}'\begin{pmatrix}\dot{q}^*\\ \dot{p}^*\end{pmatrix} \tag{75}$$

with extended flexibility and load connexion matrices

$$\hat{F} = \begin{pmatrix}F_{cc} & F_{cs}\\ F_{sc} & F_{ss}\end{pmatrix} \qquad \hat{G} = \begin{pmatrix}G & C\\ I & 0\end{pmatrix} . \tag{76}$$

Equations (75) is solved for the $\ddot{c}$ and $\ddot{s}$ and substituted into

$$\hat{G}\begin{pmatrix}\ddot{c}\\ \ddot{s}\end{pmatrix} = \hat{G}\hat{F}^{-1}\hat{G}'\begin{pmatrix}\dot{q}^*\\ \dot{p}^*\end{pmatrix} . \tag{77}$$

However, using (53), the left hand side is also equal to

$$\begin{pmatrix}G\ddot{c} + C\ddot{s}\\ \ddot{c}\end{pmatrix} = \begin{pmatrix}\ddot{g}\\ \ddot{c}\end{pmatrix} . \tag{78}$$

Consequently, partitioning the extended stiffness matrix

$$\hat{G}\hat{F}^{-1}\hat{G}' = \begin{pmatrix}\hat{K}_{qq} & \hat{K}_{qp}\\ \hat{K}_{pq} & \hat{K}_{pp}\end{pmatrix} \tag{79}$$

we obtain

$$\ddot{g} = \hat{K}_{qq}\dot{q}^* + \hat{K}_{qp}\dot{p}^* \tag{80}$$

$$\ddot{c} = \hat{K}_{pq}\dot{q}^* + \hat{K}_{pp}\dot{p}^* \quad . \tag{81}$$

These results are applicable to each finite element. Thus, using (58) and (74) again in (81), we have a first set of dynamic equations

$$-N_E^{-1}\dddot{p}_E^* = (\hat{K}_{pq})_E L_E \dot{w}^* + (\hat{K}_{pp})_E \dot{p}_E^* \qquad (E = 1,2,\ldots,N). \tag{82}$$

Also, substituting the equations of type (80) into the time derivative of (61) and using (58) and (66)

$$\sum_E L'_E(\hat{K}_{qq})_E L_E \dot{w}^* + \sum_E L'_E(\hat{K}_{qp})_E \dot{p}_E^* = M\dddot{w} + \dot{f}_{(w)} \quad . \tag{83}$$

Equations (82) and (83) form again a complete dynamical system for the variables w^* and p_E^*.

The corresponding eigenvalue problem has now the classical form

$$(\overset{\circ}{K} - \omega^2\overset{\circ}{M})\overset{\circ}{x} = 0 \tag{84}$$

with as modal column the amplitudes of the p_E^* and w^*, as stiffness matrix

$$\overset{\circ}{K} = \begin{bmatrix} (\hat{K}_{pp})_1 & 0 & 0 & 0 & (\hat{K}_{pq})_1 L_1 \\ 0 & (\hat{K}_{pp})_2 & 0 & 0 & (\hat{K}_{pq})_2 L_2 \\ 0 & 0 & \ddots & 0 & \cdots \\ 0 & 0 & 0 & (\hat{K}_{pp})_N & (\hat{K}_{pq})_N L_N \\ L'_1(\hat{K}_{qp})_1 & L'_2(\hat{K}_{qp})_2 & \cdots & L'_N(\hat{K}_{qp})_N & \sum_E L'_E(\hat{K}_{qq})_E L_E \end{bmatrix} \tag{85}$$

and as mass matrix

$$\overset{\circ}{M} = \begin{bmatrix} N_1^{-1} & 0 & 0 & 0 & 0 \\ 0 & N_2^{-1} & 0 & 0 & 0 \\ 0 & 0 & \ddots & 0 & 0 \\ 0 & 0 & 0 & N_N^{-1} & 0 \\ 0 & 0 & 0 & 0 & M \end{bmatrix} . \tag{86}$$

Again, if M disappears, static condensation of the nodal displacements becomes possible.

THE RAYLEIGH QUOTIENT

Turning back to the dual variational principle for a continuum, the results obtained show that the eigenvalue problem can be stated in the form of the stationarity of the following Rayleigh quotient

$$\omega^2 = \frac{\int_R \frac{1}{2\rho} D_i A_{ij} D_m A_{mj} \, dR}{\int_R \phi(A_{ij}) \, dR} \tag{87}$$

where A_{mj} is either the tensor of impulse amplitudes or more simply that of stress amplitudes. Moreover

$$n_i A_{ij} = 0 \quad \text{on} \quad \partial_2 R \tag{88}$$

are stress type boundary conditions to be satisfied a *priori*, while the complementary boundary conditions, of the kinematical type,

$$D_i A_{ij} = 0 \quad \text{on} \quad \partial_1 R \tag{89}$$

are natural; they are taken care of by the principle itself.

This is the converse of the displacement formulation, where the kinematical conditions are those to be satisfied ab initio. Another contrast is the nature of the zero frequency modes; they are obviously given by stress distributions in static equilibrium with themselves or self-stressings obeying

$$D_i A_{ij} = 0 \quad \text{in} \quad R \qquad n_i A_{ij} = 0 \quad \text{on} \quad \partial_1 R \; .$$

For a continuum their number is infinite, while in the displacement formulation the zero frequency modes are related to a finite number of rigid body or kinematical degrees of freedom. Otherwise the Rayleigh quotient (87) has characteristics very similar to those of the displacement type quotient

$$\omega^2 = \frac{\int_R W(Du)dR}{\int_R \frac{1}{2} \rho u_i u_i dR} \; . \tag{90}$$

In both cases the functional in the numerator is constructed with space derivatives of the variables involved in the functional of the denominator. All minimax properties of (90), as established by Courant [1], can be translated to (87), as shown for instance by Tabarrok [8].

Thus in theory (87) will furnish an upper bound to the eigenvalue ω_1^2, provided $(A_{ij})_1$ is orthogonal to all the self-stressing modes of zero frequency. It is of interest to note that it is quite possible to construct all the stress distributions that are orthogonal to the self-stressing modes. Orthogonality means that

$$\int_R A_{ij} \hat{\varepsilon}_{ij} dR = 0 \tag{91}$$

where A_{ij} is any self-stressing and $\hat{\varepsilon}_{ij}$ is the distribution of strains associated with the orthogonal stress distribution through the constitutive equations. A sufficient condition to implement (91) is to make the strain distribution compatible:

$$\hat{\varepsilon}_{ij} = \frac{1}{2} (D_i u_j + D_j u_i) \text{ in } R \qquad u_i = 0 \quad \text{on} \quad \partial_1 R \; . \tag{92}$$

Indeed one obtains then

$$\int_R A_{ij}\hat{\varepsilon}_{ij}dR = \int_R A_{ij}D_i u_j dR = \int_{\partial R} n_i A_{ij} u_j dS - \int_R u_j D_i A_{ij} dR$$

and the right hand side vanishes in view of (88), (89) and (92).

The condition is also necessary as can be shown by satisfying (89) with the help of stress functions, integrating (91) by parts and using the arbitrariness in the stress functions to show that the strains have to satisfy the integrability conditions for the existence of displacements. Finally the most general stress distribution orthogonal to all the self-stressing states if given by

$$\hat{A}_{ij} = C^{pq}_{ij} D_p u_q \tag{93}$$

where the u_q are arbitrary save for the boundary condition

$$u_q = 0 \quad \text{on} \quad \partial_1 R . \tag{94}$$

But then, the Rayleigh quotient (87) becomes expressible entirely in terms of the auxiliary displacement field u_q

$$\omega^2 = \frac{\int_R \frac{1}{2\rho} C^{pq}_{ij} C^{rs}_{mj} D_i D_p u_q D_m D_r u_s dR}{\int_R W(Du)dR} \tag{95}$$

This new Rayleigh quotient, that leads immediately to the non zero eigenvalues of the problem, has the same kinematical boundary conditions to be satisfied *ab initio* as (90) but in addition the transformed boundary conditions (88)

$$n_i C^{pq}_{ij} D_p u_q = 0 \quad \text{on} \quad \partial_2 R . \tag{96}$$

There is also an interesting inequality relating (90) and (95).

To prepare its proof consider the following obvious inequality

$$\int_R \left(\sqrt{\rho}a_j - \frac{\lambda b_j}{\sqrt{\rho}}\right)\left(\sqrt{\rho}a_j - \frac{\lambda b_j}{\sqrt{\rho}}\right) dR \geq 0$$

valid for any value of the scalar λ; expanding it

$$\int_R \rho a_j a_j dR - 2\lambda \int_R a_j b_j dR + \lambda^2 \int_R \frac{1}{\rho} b_j b_j dR \geq 0 .$$

The minimum of the left hand side is reached for

$$\lambda = \int_R a_j b_j dR \Big/ \int_R \frac{1}{\rho} b_j b_j dR$$

hence, after substitution

$$\int_R \frac{1}{2} \rho a_j a_j dR - \left\{\int_R \frac{1}{2} a_j b_j dR\right\}^2 \Big/ \int_R \frac{1}{2\rho} b_j b_j dR \geq 0 . \qquad (97)$$

Setting

$$a_j = u_j \quad \text{and} \quad b_j = C^{pq}_{ij} D_i D_p u_q$$

we already identity two of the integrals as the denominator of (90) and the numerator of (96) respectively. For the third integral

$$\int_R \frac{1}{2} a_j b_j dR = \int_R \frac{1}{2} u_j C^{pq}_{ij} D_i D_p u_q dR$$

$$= \int_{\partial R} \frac{1}{2} u_j n_i C^{pq}_{ij} D_p u_q dS - \int_R \frac{1}{2} C^{pq}_{ij} D_i u_j D_p u_q dR$$

$$= - \int_R W(Du) dR < 0$$

since the surface integral vanishes on account of (94) and (96).

Manipulating the inequality (97), there comes

$$\omega_R^2 \leq \omega_{R'}^2 \tag{98}$$

where ω_R^2 denotes the classical Rayleigh quotient (90) and $\omega_{R'}^2$ the quotient (95).

This means that whenever a displacement field is chosen that satisfies both the kinematical and the stress boundary conditions, the classical Rayleigh quotient is always a better approximation than the one derived from the stress approach with orthogonality to all zero frequency modes. However, as experience shows, direct application of the quotient (87) is beneficial in that the frequencies are usually underestimated and converge through lower bounds when the degrees of freedom are increased [7, 8]. This is of course due to the fact that the assumed stress modes do not satisfy orthogonality with respect to all the self-stressing modes. Unfortunately there is no guarantee of this property of lower boundedness and further theoretical research is necessary to be able to incorporate this into the formalism.

REFERENCES

[1] COURANT, R. and HILBERT, D., *Methods of Mathematical Physics*, Vol. 1, Interscience, 1962.

[2] FRAEIJS DE VEUBEKE, B.M., "Diffusion des Inconnues Hyperstatiques dans les Voilures a Longerons Couplés", *Bulletin du Service Technique de l'Aéronautique*, No. 24, Bruxelles, 1951.

[3] FRAEIJS DE VEUBEKE, B.M., "Displacement and Equilibrium Models in the Finite Element Method", Chapter 9 of *Stress Analysis*, Wiley, London, 1965.

[4] FRAEIJS DE VEUBEKE, B.M., "Variational Principles in Fluid Mechanics", Symposium of Polish Academy of Sciences, Jurata, 1965. In *Fluid Dynamics Transactions*, Vol. 3, Polish Scientific Publishers, Warszawa.

[5] TOUPIN, R.A., "A Variational Principle for the Mesh-Type Analysis of a Mechanical System", *J. Appl. Mech.*, Vol. 19, June, 1952.

[6] FRAEIJS DE VEUBEKE, B.M., "Duality in Structural Analysis by Finite Elements", NATO Advanced Study Institute Lectures on Finite Element Methods in Continuum Mechanics, Lisbon, September, 1971.

[7] GERADIN, M., "Computational Efficiency of Equilibrium Models in Eigenvalue Analysis", IUTAM Symposium on High Speed Computing of Elastic Structures, Congrès et Colloques de l'Université de Liège, 1971.

[8] TABARROK, B. and SAKAGUCHI, R.L., "Calculations of Plate Frequencies from Complementary Energy Formulation", *Int. J. for Num. Meth. in Eng.*, Vol. 2, 1970, pp. 283-293.

[9] CRANDALL, S.H., "Complementary Extremum Principle for Dynamics", *Proc. 9th Int. Congress Appl. Mech.*, Vol. 5, 1957, pp. 80-87.

[10] CHEN, Y., "Remarks on Variational Principles in Elastodynamics", *J. Franklin Inst.*, Vol. 278, No. 1, 1964, pp. 1-7.

[11] GLADWELL, G.M.L. and ZIMMERMANN, G., "On Energy and Complementary Energy Formulations of Acoustic and Structural Vibration Problems", *J. of Sound and Vibration*, Vol. 3, 1966, pp. 233-241.

[12] TABARROK, B. and KARNOPP, B.H., "On Duality in the Oscillations of Framed Structures", *J. Sound and Vibration*, Vol. 8, No. 3, 1968, pp. 469-481.

DUAL ANALYSIS BY FINITE ELEMENTS LINEAR AND NON LINEAR APPLICATIONS

NON LINEAR THEORIES

NON LINEAR SHELL THEORY

B.M. Fraeijs de Veubeke
University of Liege, Belgium

NON LINEAR THEORIES

3.1. *Large displacement formulations*

All the difficulties inherent to large displacement formulations are due to the existence of finite material rotations. The strains will be kept of a very small order of magnitude (10^{-3} for example) in order to deal with purely elastic material behavior. Furthermore cartesian reference frames only will be used since this restriction does not affect any of the physical principles involved in the theory. The additional complications stemming from the use of general curvilinear coordinates, as necessary for instance in shell theory (see section 4), can be separately introduced afterwards.

It is for large displacements that the distinction between lagrangian and eulerian coordinates becomes effective. The lagrangian coordinates are here the cartesian coordinates a_i

Original publication, *Report AFFDL-TR-72-93,* Wright-Patterson A.F.B., (in collaboration with G. Sander and P. Beckers).

$(i = 1,2,3)$ of a material point in a so-called "initial" or "reference" configuration of the elastic body. The eulerian coordinates x_i $(i = 1,2,3)$ are the cartesian coordinates of particles in the "final" or "displaced" configuration.

The cartesian components of the displacement vector are then

$$u_i = x_i - a_i \ .$$

From the lagrangian point of view one writes

$$x_i = x_i\ (a_j) = a_i + u_i\ (a_j)\ . \tag{3.1}$$

From the eulerian point of view

$$a_i = a_i\ (x_j) = x_i - u_i\ (x_j)\ . \tag{3.2}$$

The reference configuration will be assumed to coincide with a stress free configuration and the stress-strain characteristics to be defined in this configuration from a knowledge of the material properties and, when the case may be, its oriented fiber type reinforcements. It must be stressed that, in case of anisotropy, the stress-strain characteristics in the final configuration do depend on the local finite rotations and cannot be deduced without solving the field equations and boundary conditions of the problem. This rules out in practice the formulation in eulerian coordinates as initiated by equations (3.2). It is only for isotropic relations, which are then independent on the rotations, that such a formulation becomes possible. Because of this limited usefulness the treatment will be here restricted to the lagrangian formulation; for considerations on the eulerian point of view reference can be made to (3.1).

3.1.1. *Lagrangian measures of strain*

Differentiating (3.1) we obtain the geometrical relationship between a neighborhood (da_i) of a point P in the reference configuration and its neighborhood $(d\,x_i)$ in the displaced configuration:

$$dx_i = D_j\, x_i\, da_j = da_i + D_j\, u_i\, da_j \tag{3.3}$$

where summation is understood for repeated subscripts and

$$D_j = \partial/\partial a_j \quad .$$

In matrix notation, the column matrices $da = \{da_i\}$ and $dx = \{dx_i\}$ are thus related by

$$dx = J\, da = (E + A)\, da \tag{3.4}$$

where E denotes the identity matrix and J and A the jacobian matrices

$$J = \{D_j\, x_i\} \qquad A = \{D_j\, u_i\} \qquad \begin{array}{l} i \text{ row subscript} \\ j \text{ column subscript} \end{array} .$$

The elementary rectangular volume element $da_1\, da_2\, da_3$ in the reference configuration becomes in the displaced configuration an oblique parallelepiped with volume equal to $|J|\, da_1\, da_2\, da_3$, where $|J|$ denotes the determinant of the jacobian matrix. The square of the distance between two neighboring points is $da'da$ in the reference and $dx'dx$ in the final configuration. According to (3.4) this alteration can be expressed as a quadratic form in da_i

$$dx'dx = da'da = da'\,(J'J - E)\, da = 2\, da'\, G_a\, da \tag{3.5}$$

introducing the cartesian representation of the Green strain tensor

$$G_a = \frac{1}{2}(J'J - E) = G_a' \tag{3.6}$$

which is the most commonly used lagrangian measure of local deformation. Using the jacobian matrix $A = J - E$ in place of J, there follows

$$G_a = \frac{1}{2}(A + A' + A'A) \tag{3.7}$$

G_a is a measure of local strain since, when the neighbourhood transformation (3.4) is of a rigid body type, the distances between neighbouring points is not altered and the components of G_a are necessarily zero. Conversely, if the components of G_a are all zero, the distances are unaltered and the neighbourhood transformation is either of rigid body type or consists of a rigid body displacement followed (or preceded) by a reflection with respect to a plane. Hence it must still be shown that reflections are ruled out by physical arguments. Polar decomposition of the transformation (3.4) is one of the methods to explain this; it has furthermore the advantage of relating the Green strain tensor to another lagrangian measure of strain, which is closer to the usual engineering definition. In the polar decomposition the transformation is split into two successive operations, first a pure strain transformation

$$dy = (E + H)\, da \qquad H = H' \tag{3.8}$$

followed by a finite rotation

$$dx = U\, dy \qquad UU' = U'U = E \qquad |U| = 1 \quad . \tag{3.9}$$

The restriction to orthogonal transformation matrices U of determinant equal to unity implies that one has a "true rotation" and not a rotation accompanied by a reflection, as would be the case for $|U| = -1$. From (3.8) and (3.9) follows, by comparison with (3.4)

$$J = U\ (E + H) \quad \text{and} \quad |J| = |E + H| \tag{3.10}$$

and it will be shown that this decomposition is always possible and unique. We first obtain from (3.10) and (3.6) that

$$G_a = \frac{1}{2}\ (E + H)\ U'U(E + H) - \frac{1}{2}\ E = \frac{1}{2}\ (E + H)^2 - \frac{1}{2}\ E$$

$$= H + \frac{1}{2}\ H^2 \quad . \tag{3.11}$$

Hence, whatever be the rotation U, the strain tensor H determines uniquely the Green strain tensor. Conversely G_a determines H uniquely as can be seen by a principle axis transformation where H reduces to diagonal form with diagonal elements h_i, whence, by virtue of (3.11), G_a is reduced to diagonal form with diagonal elements

$$\gamma_i = h_i + \frac{1}{2}\ h_i^2 \qquad (i = 1,2,3) \quad . \tag{3.12}$$

Given the γ_i, the solutions for h_i are real and unique

$$h_i = -\ 1 + \sqrt{1+2\gamma_i} \tag{3.13}$$

provided one can justify the restrictions

$$1 + 2\gamma_i > 0 \qquad 1 + h_i > 0 \qquad (i = 1,2,3)\ . \tag{3.14}$$

Indeed, from (3.6) and (3.10) we have

$$|J|^2 = |2\ G_a + E| = (1+2\gamma_1)(1+2\gamma_2)(1+2\gamma_3) \tag{3.15}$$

$$|J| = |E + H| = (1+h_1)(1+h_2)(1+h_3) \quad . \tag{3.16}$$

Physically the transformation from initial to final configuration takes place continuously. At any intermediate stage none of the factors appearing in the right hand sides of (3.15) and (3.16) may vanish without making the volume element vanish, a physical impossibility. Hence since they satisfy conditions (3.14) at the start ($\gamma_i = 0$, $h_i = 0$) they must keep their sign and still verify those conditions at the end. In particular we have

$$|2\, G_a + E| > 0 \qquad |J| = |E + H| > 0 \ . \tag{3.17}$$

The argument is valid for finite strains; in the case of very small strains equations (3.13) are immediately obvious. Given J satisfying (3.17) and once H uniquely determined as shown above by the relation

$$\frac{1}{2}\,(J'J - E) = H + \frac{1}{2}\,H^2$$

the orthogonal matrix in the polar decomposition (3.10) is uniquely determined by

$$U = J\,(E + H)^{-1}$$

and is that of a pure rotation.

In a principle axis system the principle strains of H have the usual engineering definitions

$$\gamma_i = \frac{dx_i^2 - da_i^2}{2da_i^2} = \frac{dx_i - da_i}{da_i} \cdot \frac{dx_i + da_i}{2da_i} \ . \tag{3.19}$$

The advantage of the Green strain measure is that it can be expressed immediately in terms of the jacobian matrices J or A as shown by (3.6) and (3.7). The effective computation of the engineering strain measure H requires a cumbersome principle axis transformation. Both measures become however equivalent in the case of very small strains as shown by the relation (3.11) where the components of H^2 become negligible by comparison with those of H.

If we add the assumption of very small rotations to that of small strains, the rotation matrix can be written as

$$U = E + K$$

where the components of K are of the same small order of magnitude as those of H. The orthogonality condition U'U = E becomes

$$(E + K')(E + K) = E \quad \text{or} \quad K' + K = 0$$

if we neglect the second order quantities represented by K'K. Hence to this approximation K is skew symmetrical. Then from (3.10)

$$J = E + A = (E + K)(E + H)$$

which, after neglecting again the second order quantities represented by KH, yields the usual approximations of linear elasticity theory

$$A = K + H \qquad H = \frac{1}{2}(A + A') \qquad K = \frac{1}{2}(A - A') \ . \tag{3.20}$$

Inversion of the order of polar decomposition is simple

$$J = U(E + H)U'U = (E + UHU')U = (E + \hat{H})U \ . \tag{3.21}$$

The rotation, which comes now first, is unchanged; the straining, which comes second, involves a matrix $\hat{H}$ related to H by

$$\hat{H} = U H U' = \hat{H}' \tag{3.21}$$

This suggests that H and $\hat{H}$ are in fact representations of the same tensor in two different reference frames. It is clear from the first polar decomposition that, if an observer uses the initial configuration as a frame to describe the distribution of strains incurred by the configuration change, the matrix H is the representation of the local strain tensor for the actual cartesian axes. If he uses the final configuration as frame, H is the representation of the local strain tensor for cartesian axes as turned by the material rotation; $\hat{H}$ would be the representation of the same tensor for the original orientation of the cartesian axes. Using the Green measure for strains, the same statements apply respectively to G_a and to

$$\hat{G}_a = U\,G_a U' \underset{(3.11)}{=} \hat{H} + \frac{1}{2}\hat{H}^2 \underset{(3.21)}{=} \frac{1}{2}(JJ'-E) = \frac{1}{2}(A+A' + AA') \ . \tag{3.22}$$

From (3.7) and (3.22) we deduce the following expressions of the Green tensor components

$$G_a = \{\gamma_{mn}\} \text{ and } \hat{G}_a = \{\hat{\gamma}_{mn}\}$$

$$2\gamma_{mn} = D_m x_i D_n x_i - \delta_{mn} = D_m u_n + D_n u_m + D_m u_i D_n u_i \ . \tag{3.23}$$

$$2\hat{\gamma}_{mn} = D_i x_m D_i x_n - \delta_{mn} = D_m u_n + D_n u_m + D_i u_m D_i u_n \ . \tag{3.24}$$

3.1.2. *The* a_i *as convective coordinates*

Let $\vec{e}_j$ denote the unit vectors of the cartesian reference frame. Then, for the position vector of a point P in the reference configuration

$$\vec{r} = a_j\,\vec{e}_j \tag{3.25}$$

and for the same point in the displacement configuration

$$\vec{R} = x_j \, \vec{e}_j \quad . \tag{3.26}$$

In this last configuration the surfaces a_i = constant define a set of convected curvilinear coordinates with local base vectors

$$\vec{G}_i = D_i \vec{R} = (D_i x_j) \, \vec{e}_j \ . \tag{3.27}$$

The fundamental metric tensor of this set of curvilinear coordinates is

$$G_{mn} = \vec{G}_m \cdot \vec{G}_n = (D_m x_i)\vec{e}_i \cdot (D_n x_j)\vec{e}_j = D_m x_i D_n x_i = \delta_{mn} + 2\gamma_{mn} \ . \tag{3.28}$$

This relationship with the components of the Green tensor further justifies their use as a measure of deformation. Let us now investigate the Christoffel symbols of the second kind defined by

$$D_m \vec{G}_i = \Gamma_{m\ i}^{\ n} \, \vec{G}_n \quad ,$$

or, preferably, taking the scalar product of both sides with the local base vectors, the symbols of the first kind

$$\vec{G}_p \cdot D_m \vec{G}_i = \Gamma_{m\ i}^{\ n} \, \vec{G}_n \cdot \vec{G}_p = \Gamma_{m\ i}^{\ n} G_{np} = \Gamma_{mpi} \quad . \tag{3.29}$$

Using (3.27) to evaluate the left-hand side

$$\Gamma_{mpi} = (D_p x_j)\vec{e}_j \cdot (D_m D_i x_k)\vec{e}_k = D_p x_j D_m D_i x_j \tag{3.30}$$

where the symmetry with respect to the subscripts m and i is noticeable. As one can easily verify, (3.30) can also be written as

$$\Gamma_{mpi} = D_m\gamma_{pi} + D_i\gamma_{mp} - D_p\gamma_{im} \quad . \tag{3.31}$$

3.1.3. *Gradients of the rotation and compatibility equations*

Introduce the elements α_{im} of the finite rotation operator U

$$U = \{\alpha_{im}\} \quad \begin{matrix} \text{i row index} \\ \\ \text{m column index} \end{matrix} \qquad \alpha_{im}\alpha_{jm} = \delta_{ij} \qquad \alpha_{im}\alpha_{in} = \delta_{mn} \tag{3.32}$$

and those of the engineering strain tensor $H = \{h_{jm}\}$

$$h_{jm} = h_{mj} \quad . \tag{3.33}$$

The polar decomposition (3.10) can then be expressed by

$$J_{ij} = D_j x_i = \alpha_{im} (\delta_{mj} + h_{mj}) \quad . \tag{3.34}$$

Our purpose is to analyse the gradient of the neighborhood transformation (3.4) as we proceed from a particle P, of coordinates a_q, to a nearby particle P' of coordinates $a_q + da_q$. Beginning with the rotation

$$U (a_q + da_q) = U (a_q) + D_q U da_q$$

where $D_q U$ is the gradient of the rotation operator at P. This we write in the form

$$U (a_q + da_q) = U (a_q)(E + U'(a_q)D_q U da_q) \tag{3.35}$$

so that the rotation in the nearby point P' is effected in two successive steps: first the small incremental rotation existing between P and P', then the finite rotation in P.

According to the known structure of a small rotation operator, $U'D_qU$ must be skew symmetrical and this obviously results from

$$D_q(U'U) = 0 \qquad U'D_qU + (D_qU')U = 0 \quad .$$

Introduce the following notation for the elements of $U'D_qU$, the rotation gradients

$$\alpha_{in}D_q\alpha_{im} = {}^q\omega_{nm} \qquad {}^q\omega_{mn} = - {}^q\omega_{nm} \quad , \tag{3.36}$$

take now the gradient of the complete neighborhood transformation operator (3.34)

$$D_qJ_{ij} = D_qD_jx_i = D_q\alpha_{im}(\delta_{mj} + h_{mj}) + {}_{im}D_qh_{mj}$$

and look at it, as was done in (3.35), from the viewpoint of a reference frame in P turned by the material rotation.

$$\alpha_{in}D_qJ_{ij} = \alpha_{in}D_qD_jx_i = {}^q\omega_{nm}(\delta_{mj} + h_{mj}) + D_qh_{nj} \quad . \tag{3.37}$$

It is obvious from the middle term that the right-hand side is symmetrical in the subscripts q and j; this provides a mean of eliminating x_i:

$$ {}^q\omega_{nm}(\delta_{mj} + h_{mj}) + D_qh_{nj} = {}^j\omega_{nm}(\delta_{mq} + h_{mq}) + D_jh_{nq} \quad . \tag{3.38}$$

This relation may be considered as the result of eliminating x_i by cross differentiation between equations (3.34), together with an interpretation in terms of rotational gradients.

The particular form

$$ {}^{q}\omega_{nj} + D_q h_{nj} = {}^{j}\omega_{nq} + D_j h_{nq} \tag{3.39} $$

that approximates (3.38) for very small strains, is similar to the Beltrami type of equations one obtains from the linear elasticity approximation (3.20) or

$$ D_j x_n = \omega_{nj} + h_{nj} \qquad K = \{\omega_{nj}\} \quad . $$

By cross differentiation

$$ D_q \omega_{nj} + D_q h_{nj} = D_j \omega_{nq} + D_j h_{nq} \quad . \tag{3.40} $$

Both sets (3.39) and (3.40) can be manipulated to yield the independent components of $D_q \omega_{nj}$ or ${}^{q}\omega_{nj}$ in terms of the strain gradients. Considering first (3.39), then rewriting it with the cyclic change of subscripts (qnj) to (jqn), then a third time with (njq), adding the two first and substracting the third, there comes, with due consideration to the symmetry of h_{nj} and skew symmetry of ${}^{q}\omega_{nm}$,

$$ {}^{j}\omega_{qn} = D_n h_{qj} - D_q h_{nj} \quad . \tag{3.41} $$

Similarly for the set (3.40)

$$ D_j \omega_{qn} = D_n h_{qj} - D_q h_{nj} \quad . \tag{3.42} $$

The Beltrami equations (3.42) yield the partial derivatives of the small rotations of linear elasticity theory in terms of the derivatives of strains. A second elimination process by cross differentiation of the rotations produces the linear local compatibility conditions to be satisfied by the strains:

$$D_p(D_n h_{qj} - D_q h_{nj}) = D_j (D_n h_{qp} - D_q h_{np}). \tag{3.43}$$

They are trivially satisfied for $p = j$, or $n = q$; in fact only six of them are known to be distinct.

The situation is different if one tries to eliminate the rotational gradients in the set (3.31). Returning to the definition (3.36), and applying (3.32)

$$\alpha_{jn}\alpha_{in}D_q\alpha_{im} = D_q\alpha_{jm} = \alpha_{jn}\,{}^q\omega_{nm} \quad .$$

Differentiate the last equality:

$$D_pD_q\alpha_{jm} = {}^q\omega_{nm}D_p\alpha_{jn} + \alpha_{jn}D_p\left({}^q\omega_{nm}\right),$$

observe that the left-hand side is symmetrical in p and q:

$${}^q\omega_{nm}D_p\alpha_{jn} + \alpha_{jn}D_p\left({}^q\omega_{nm}\right) = {}^p\omega_{nm}D_q\alpha_{jn} + \alpha_{jn}D_q\left({}^p\omega_{nm}\right)$$

multiply this by α_{jr} and use definition (3.36) and properties (3.32) to obtain

$${}^q\omega_{nm}\,{}^p\omega_{rn} + D_p\left({}^q\omega_{rm}\right) = {}^p\omega_{nm}\,{}^q\omega_{rn} + D_q\left({}^p\omega_{rm}\right) \quad . \tag{3.44}$$

This is the equation that, for gradients of finite rotations, replaces the simple relation $D_pD_q\omega_{rm} = D_qD_p\omega_{rm}$ in linear elasticity. A substitution of (3.41) into (3.44) yields the compatibility conditions for very small strains in the presence of finite rotations:

$$(D_nh_{qm}-D_mh_{qn})(D_rh_{pn}-D_nh_{pr}) + D_p(D_rh_{qm}-D_mh_{qr}) =$$

$$(D_nh_{pm}-D_mh_{pn})(D_rh_{qn}-D_nh_{qr}) + D_q(D_rh_{pm}-D_mh_{pr}) \quad . \tag{3.45}$$

They differ from (3.43) by the non linear terms in the first derivatives of strains; the same equations hold for the γ_{mn} measures of strain, which do not differ practically from the h_{mn}.

The general compatibility equations for finite rotations and strains will be derived here for the sake of completeness. Some preparatory work will be necessary. We introduce a convenient notation for the element of the inverse jacobian matrix

$$J^{-1} = \{\partial_j a_m\} \quad \begin{matrix} m \text{ rows} \\ j \text{ columns} \end{matrix} \qquad \partial_j = \partial/\partial x_j$$

and note the corresponding properties

$$\partial_j a_m D_n x_j = \delta_{mn} \quad \text{and} \quad \partial_j a_m D_m x_i = \delta_{ji} \; . \tag{3.46}$$

They are used to express the reciprocal of the fundamental metric tensor given by (3.28). First

$$\partial_j a_m G_{mn} = (\partial_j a_m D_m x_i) D_n x_i = D_n x_j \quad ,$$

next

$$\partial_j a_p \partial_j a_m G_{mn} = \partial_j a_p D_n x_j = \delta_{pn} \quad ,$$

from which we conclude that the reciprocal metric tensor can be written

$$G^{pn} = \partial_j a_p \partial_j a_n \; . \tag{3.47}$$

Note that since from (3.28) the elements of G_{mn} are expressible in terms of the strains γ_{mn}, the same is true of the elements of the reciprocal G^{pn}.

We now use (3.30) to obtain an expression for the second derivatives of x_r:

$$\partial_r a_p \Gamma_{mpi} = (\partial_r a_p D_p x_j) D_m D_i x_j = D_m D_i x_r \quad . \tag{3.48}$$

Differentiating (3.30)

$$D_q \Gamma_{mpi} = (D_q D_p x_j)(D_m D_i x_j) + (D_p x_j)(D_q D_m D_i x_j) \; ,$$

exchanging the subscripts q and m

$$D_m \Gamma_{qpi} = (D_m D_p x_j)(D_q D_i x_j) + (D_p x_j)(D_m D_q D_i x_j) \; ,$$

and substracting the results to eliminate the third derivatives, while substituting all second derivatives by means of (3.48)

$$D_q \Gamma_{mpi} - D_m \Gamma_{qpi} = \partial_j a_r \partial_j a_s \Gamma_{qrp} \Gamma_{msi}$$

$$- \partial_j a_r \partial_j a_s \Gamma_{mrp} \Gamma_{qsi} \qquad ,$$

we can finally introduce the reciprocal metric tensor and obtain

$$D_q \Gamma_{mpi} - D_m \Gamma_{qpi} = G^{rs} (\Gamma_{qrp} \Gamma_{msi} - \Gamma_{mrp} \Gamma_{qsi}) \tag{3.49}$$

or

$$D_q \Gamma_{mpi} - D_m \Gamma_{qpi} - \Gamma_{q\ p}^{\ s} \Gamma_{msi} + \Gamma_{m\ p}^{\ s} \Gamma_{qsi} = 0 \quad .$$

In differential geometry this result expresses that the space referred to the convected (curvilinear) coordinates is Euclidean; its curvature tensor is zero. It also constitutes the local compatibility conditions for the strains since from (3.28) and (3.31) everything in them is expressible in terms of the γ_{mn} and their derivatives.

3.1.4. *Stress tensors*

From the viewpoint of an observer using the initial configuration as reference frame, the lagrangian stresses are defined per unit original area and referred to their cartesian components. For instance a facet of direction cosines (1, 0, 0) for the outward normal gives surface traction components ($t_{11}\, da_2\, da_3$, $t_{12}\, da_2\, da_3$, $t_{13} da_2\, da_3$). Their virtual work in a material infinitesimal variation δu_j of the displacement components is thus ($t_{1j}\; \delta u_j$) $da_2\, da_3$. For all the surface tractions on the facets of an elementary parallelepiped this virtual work becomes

$$D_i(t_{ij}\delta u_j) da_1\; da_2\; da_3\; .$$

If external body forces, defined per unit volume in the initial configuration, and having cartesian components denoted by A_j, exist, their contribution to the virtual work is $A_j\; \delta u_j\; da_1\; da_2\; da_3$ and the energy balance equation reads

$$\delta W = D_i(t_{ij}\delta u_j) + A_j\; \delta u_j \tag{3.50}$$

where W is the elastic energy per unit initial volume.

In translation modes, where each δu_j is an arbitrary constant, there is no energy increase and (3.50) reduces to the translational equilibrium conditions

$$D_i t_{ij} + A_j = 0 \qquad (j = 1,2,3)\; . \tag{3.51}$$

The expanded form of (3.50)

$$\delta W = \delta u_j (D_i t_{ij} + A_j) + t_{ij}\; D_i\; \delta u_j$$

can thus be simplified; moreover, since the material variations

δ are taken under constant values of the lagrangian variables, the operators D_i and δ commute and the simplified form of the energy balance can be

$$\delta W = t_{ij}\ \delta\,(D_i u_j)\ . \tag{3.52}$$

From this it appears that the lagrangian stresses can be considered to be the partial derivatives of the energy W, considered as a function of the nine displacement gradients $D_i u_j$. This is made more precise by consideration of the rotational equilibrium conditions. Consider a displacement variation field, again of rigid body type, but representing now an infinitesimal rotation

$$\delta u_1 = x_3\ \delta w_2 - x_2\ \delta w_3$$

$$\delta u_2 = x_1\ \delta w_3 - x_3\ \delta w_1$$

$$\delta u_3 = x_2\ \delta w_1 - w_1\ \delta w_2\ .$$

Note that those additional displacements must be taken from the actual coordinates x_i of the particles and that the energy increase must again vanish. Then, substituting this field into (3.52) and equating separately to zero the coefficients of the infinitesimal rotation angles δw_i, we obtain the rotational equilibrium conditions to be satisfied by the lagrangian stresses:

$$t_{i3}\ D_i\ x_2 - t_{i2}\ D_i\ x_3 = 0$$

$$t_{i1}\ D_i\ x_3 - t_{i3}\ D_i\ x_1 = 0$$

$$t_{i2}\ D_i\ x_2 - t_{i1}\ D_i\ x_2 = 0\ . \tag{3.53}$$

They clearly express that the tensor $t_{im} D_i x_n$ is symmetrical in the subscripts m and n and not the stress tensor itself. The substitution

$$t_{im} = s_{ip} D_p x_m \tag{3.54}$$

satisfies the rotational equilibrium equations

$$s_{ip} D_p X_m D_i x_n = s_{ip} D_p x_n D_i x_m \tag{3.55}$$

provided the new stress tensor be symmetrical:

$$s_{ip} = s_{pi} \quad . \tag{3.56}$$

This is immediately seen by exchanging the dummy subscripts p and i in the right hand side of (3.55). This somewhat artificial definition receives a strong foundation through the following consequences: (3.52) becomes

$$\delta W = s_{im} D_m x_j \,\delta\, (D_i u_j) = S_{im} D_m x_j \,\delta\, (D_i x_j)$$

or, exchanging the dummy indices i and m,

$$\delta W = s_{mi} D_i x_j \,\delta\, (D_m x_j) \quad .$$

After adding the two formulas, using the symmetry property (3.56), the components of the Green strain tensor appear through (3.23):

$$2\delta W = s_{im} \,\delta\, (D_m x_j D_i x_j) \text{ or } \delta W = s_{im} \,\delta\gamma_{im} \quad . \tag{3.57}$$

This result shows that W is a function of the displacement gradients through the six independent components of the (symmetrical) Green tensor. The symmetrical "Kirchhoff-Trefftz" stress tensor is made up of the partial derivatives of the energy:

$$s_{11} = \frac{\partial W}{\partial \gamma_{11}} \qquad s_{22} = \frac{\partial W}{\partial \gamma_{22}} \qquad s_{33} = \frac{\partial W}{\partial \gamma_{33}}$$
$$s_{12} = \frac{\partial W}{\partial \eta_{12}} \qquad s_{23} = \frac{\partial W}{\partial \eta_{23}} \qquad s_{31} = \frac{\partial W}{\partial \eta_{31}} \tag{3.58}$$

where the shearing strains

$$\eta_{12} = \gamma_{12} + \gamma_{21} \qquad \eta_{23} = \gamma_{23} + \gamma_{32} \qquad \eta_{31} = \gamma_{31} + \gamma_{13} \tag{3.59}$$

were introduced for convenience. The compact formula

$$s_{ij} = \frac{\partial W}{\partial \gamma_{ij}} \tag{3.60}$$

for describing the stress-strain relations (3.58) is also acceptable, provided one distinguishes between γ_{ij} and γ_{ji} as in (3.59) and applies the chain differentiation rule.

The translational equilibrium equations satisfied by the Kirchhoff-Trefftz stresses were given by Signorini (3.2), from (3.54) and (3.51) they are

$$D_i\,(s_{im} D_m x_j) + A_j = 0 \quad . \tag{3.61}$$

A straightforward geometrical interpretation of the Kirchhoff-Trefftz stresses is obtained when transforming the surface tractions successively by (3.54) and (3.27)

$$\vec{t}_i = t_{im}\,\vec{e}_m = s_{ip}\,D_p\,x_m\,\vec{e}_m = s_{ip}\,\vec{G}_m \quad . \tag{3.62}$$

The Kirchhoff-Trefftz stresses result from the decomposition of the lagrangian surface tractions in the metric induced by the convection of the lagrangian coordinates.

We are now in a position to check on the validity of the following interpretation of equation (3.52)

$$t_{ij} = \partial W / \partial (D_i u_j) \tag{3.63}$$

where the explicit dependence of the energy on the displacement gradients is to result from its dependence on the Green strain tensor and from (3.23).

$$t_{ij} = \frac{\partial W}{\partial \gamma_{mn}} \partial \gamma_{mn} / \partial (D_i x_j) = \frac{1}{2} s_{mn} (\delta_{mi} D_n x_j + \delta_{ni} D_m x_j)$$

$$= \frac{1}{2} (s_{in} D_n x_j + s_{mi} D_m x_j) = s_{in} D_n x_j$$

which is confirmed by (3.54). In the form

$$t_{ij} = D_n x_j \frac{\partial W}{\partial \gamma_{in}} \tag{3.64}$$

it can be considered as the constitutive equations for the lagrangian stresses incorporating rotational equilibrium. Since from (3.11)

$$\gamma_{mn} = h_{mn} + \frac{1}{2} h_{mp} h_{pn} \tag{3.65}$$

the energy can also be conceived as a function of the six components h_{mn} of the engineering strain tensor H and a corresponding symmetrical stress tensor defined:

$$r_{ij} = \partial W / \partial h_{ij} \quad . \tag{3.66}$$

From the rule of chain differentiation we obtain readily

$$r_{ij} = s_m \frac{\partial \gamma_{mn}}{\partial h_{ij}} = s_{mn}(\delta_{mi}\delta_{nj} + \frac{1}{2} h_{jn}\delta_{mi} + \frac{1}{2} h_{mi}\delta_{nj})$$

$$= s_{ij} + \frac{1}{2} s_{in} h_{jn} + \frac{1}{2} s_{mj} h_{mi} \tag{3.67}$$

or, in matrix formulation,

$$R = S + \frac{1}{2} (S\ H + H\ S) = \frac{1}{2} S\ (E + H) + \frac{1}{2} (E + H)S\ . \tag{3.68}$$

It appears that the difficulty encountered in expressing H in terms of G_a has here its counterpart in the difficulty of expressing the Kirchoff-Trefftz stresses S in terms of the stress tensor R.

Looking to the relationship between R and the lagrangian stresses, we can introduce into (3.54) the polar decomposition (3.34)

$$t_{im} = s_{ip}\ \alpha_{mq}(\delta_{qp} + h_{qp})$$

and, premultiplying by α_{mr}, obtain the components of S (E + H)

$$\alpha_{mr}\ t_{im} = s_{ip}\ (\delta_{rp} + h_{rp})$$

whence follows

$$r_{ij} = \frac{1}{2} \alpha_{mj}\ t_{im} + \frac{1}{2} \alpha_{mi}\ t_{jm}\ . \tag{3.69}$$

Here again it appears difficult to solve for the lagrangian stresses. As a consequence there is no simple way to state the translational equilibrium equations satisfied by the R stress tensor. On the other hand it is obvious from the relations between the various stress tensors that they are practically equivalent for the case of very small strains.

3.1.5. *Variational principles for large elastic displacements*

The best known is that which states that the total energy, strain energy plus potential energy of dead loads, is stationary with respect to arbitrary, kinematically admissible, variations of the displacement field. It is important to note that the rotational equilibrium conditions are incorporated ab initio in the principle by considering that the energy can be expressed in terms of the Green strain tensor; otherwise the proof of the principle is merely a reformulation of our previous energy balance considerations.

$$\int_{R_a} (W - \bar{A}_j \, u_j) \, dR_a - \int_{\partial R_a} \bar{p}_j \, u_j \, d\,\partial R_a \quad \text{stationary.} \tag{3.70}$$

For simplicity the region R_a occupied by the elastic body is assumed to be simply connected. On its bounding surface ∂R_a either the surface tractions p_j are imposed, or the displacements u_j (the kinematical boundary conditions). The surface integral only includes contributions of those areas where surface traction components are specified. The integrations are performed with respect to the lagrangian coordinates and dR_a stands for $da_1 da_2 da_3$ in our cartesian reference frame. In view of the assumptions made

$$\partial W = \frac{\partial W}{\partial \gamma_{im}} \delta\gamma_{im} = s_{im} \, \delta\left(\frac{1}{2} D_i \, x_j \, D_m \, x_j - \frac{1}{2} \delta_{im}\right)$$

$$= s_{im} \frac{1}{2}\left(D_i \, x_j \, \delta D_m \, x_j + D_m \, x_j \, \delta D_i \, x_j\right).$$

Or, in view of the symmetry of the stresses

$$\delta W = s_{im} \, D_m \, x_j \, \delta D_i \, x_j = s_{im} \, D_m \, x_j \, D_i \, (\delta u_j) \quad .$$

Hence, for the variation of (3.70),

$$\int_{R_a} (s_{im} D_m x_j D_i (\delta u_j) - \bar{A}_j \delta u_j) dR_a$$

$$- \int_{\partial R_a} \bar{p}_j \delta u_j d\, \partial R_a = 0$$

with $\delta u_j = 0$ there where u_j was specified. Integration by parts yields

$$\int_{\partial R_a} (l_i s_{im} D_m x_j - \bar{p}_j) \delta u_j d\partial R_a$$

$$- \int_{R_a} \{D_i (s_{im} D_m x_j) + \bar{A}_j\} \delta u_j dR_a = 0 \quad .$$

The Signorini equations (3.61) are seen to be the Euler equations of this principle, while the natural boundary conditions

$$l_i s_{im} D_m x_j = \bar{p}_j \tag{3.71}$$

where l_i denote the direction cosines of the outward normal in the initial configuration, express surface equilibrium between the stress field and the imposed tractions.

A generalization of the principle of variation of displacements to a two-field principle is possible by the Friedrichs method (3.7).

We will consider the energy density to be a function of the components of the Green strain tensor but, instead of expressing those a priori in terms of displacement gradients, we raise them to a status of independent variables by incorporating equations (3.23), considered as differential constraints, into the principle:

$$\int_{R_a} \left\{ W(\gamma) + s_{im}\left(\frac{1}{2} D_i x_j D_m x_j - \frac{1}{2}\delta_{im} - \gamma_{im}\right) - \bar{A}_j u_j \right\} dR_a$$

$$- \int_{\partial R_a} \bar{p}_j u_j \, d\partial R_a \quad \text{stationary} \quad . \tag{3.72}$$

Here, the quantities s_{im} denote independent lagrangian multipliers, whose Euler equations restore the differential constraints. It is already clear from the euler equations (3.61) and natural boundary conditions (3.71) of this principle relative to variations on the displacements, that the set of multipliers has the nature of lagrangian stresses. This is further confirmed by the Euler equations (3.60) that are obtained from variations of the strains.

This very general three-field variational principle allows independent approximations to be made on strains, stresses and displacements. This considerable freedom makes it difficult to apply in practice. When the Euler equations (3.60) are accepted a priori, that is if the multipliers are already identified with the Kirchhoff-Trefftz stresses, the principle can be reduced to a two-field one by elimination of the strains. Assuming that the stress-strain relations (3.60) can be solved for the strains, one introduces the complementary energy density

$$\phi(s) = s_{im}\gamma_{im} - W \quad . \tag{3.73}$$

This Legendre transformation has the property

$$d\,\phi = \left(s_{im} - \frac{\partial W}{\partial \gamma_{im}}\right) d\gamma_{im} + \gamma_{im} ds_{im} = \gamma_{im} ds_{im}$$

or

$$\gamma_{im} = \frac{\partial \phi}{\partial s_{im}} \tag{3.74}$$

of inverting the stress-strain relations (3.60). It furnishes a two-field principle discovered independently by E. Reissner and B. Fraeijs de Veubeke [3.5]

$$\int_{R_a} \left\{ \frac{1}{2} s_{im}(D_i x_j D_m x_j - \delta_{im}) - \phi(s) - \overline{A}_j u_j \right\} dR_a - \int_{\partial R_a} \overline{p}_j u_j \, d\partial R_a \text{ stationary .} \tag{3.75}$$

While the Euler equations (3.61) and natural boundary conditions of this principle are unchanged the Euler equations due to variations on the stresses are now a combination of (3.74) and the original differential constraints (3.23)

$$\frac{1}{2} (D_i x_j D_m x_j - \delta_{im}) = \partial\phi/\partial s_{im} \quad . \tag{3.76}$$

While in the linear elasticity case a variational principle for stresses only, the so-called complementary energy principle, is easily derivable from (3.75), finite rotations introduce serious difficulties.

To take the case of zero body forces for simplicity, the translational equilibrium equations can be solved for the lagrangian stresses by means of stress functions A_{sj}:

$$t_{ij} = e_{irs} D_r A_{sj} \quad (e_{irs}\text{, the permutation symbol}) \tag{3.77}$$

and this indeed allows to modify principle (3.75) into one that does no more involve explicitly displacements. However, the differential constraints

$$s_{im} D_m x_j = e_{irs} D_r A_{sj} \tag{3.78}$$

derived from (3.77) are still implicitly coupling the choice of Kirchhoff-Trefftz stresses to the displacements. It is therefore generally recognized that there is no variational stress principle available in presence of finite rotations (3.6).

We shall now explore new variational principles based on the polar decomposition method (3.34). We start from the principle

$$\int_{R_a} \left\{ W(D_i u_j) - \bar{A}_j u_j \right\} dR_a - \int_{\partial R_a} \bar{p}_j u_j \, d\partial R_a \quad \text{stationary}$$

suggested by (3.52) and whose Euler equations and natural boundary conditions are

$$D_i t_{ij} + \bar{A}_j = 0 \quad \text{and} \quad l_i t_{ij} = \bar{p}_j .$$

It contains also implicitly the rotational equilibrium conditions if the energy density is a function of the nine displacement gradients through the six Green strain tensor measures.

We now remove the nine differential constraints

$$D_i x_i - \alpha_{im} (\delta_{mj} + h_{mj}) = 0$$

equivalent to the polar decomposition (3.34), by means of nine lagrangian multipliers t_{ji}

$$\int_{R_a} \left\{ W(h) + t_{ji} D_j x_i - t_{ji} \alpha_{im} (\delta_{mj} + h_{mj}) - \bar{A}_i u_i \right\} dR_a$$

$$- \int_{\partial R_a} \bar{p}_j u_j \, d\partial R_a \quad \text{stationary} \quad . \qquad (3.79)$$

The strain energy density is naturally taken to be a function of the six liberated strain measures h_{ij}, while three additional degrees of rotational freedom (Euler angles for example) are contained in the liberated α_{im} to match the original nine displacement gradients. In taking variations with respect to the strain measures, we are not allowed to distinguish between h_{mj} and h_{jm}, so the corresponding Euler equations are

$$\frac{\partial W}{\partial h_{mj}} + \frac{\partial W}{\partial h_{jm}} - t_{ji}\,\alpha_{im} - t_{mi}\,\alpha_{ij} = 0 \ , \quad \text{or}$$

$$\frac{\partial W}{\partial h_{mj}} = r_{mj} = \frac{1}{2}\,(t_{ji}\,\alpha_{im} + t_{mi}\,\alpha_{ji}) \ . \tag{3.80}$$

Comparison of this with (3.69) justifies the notation of lagrangian stresses utilized for the multipliers, and this is further confirmed by the translational equilibrium equations and natural boundary conditions they satisfy as a result of variations on the displacements. When taking variations on the rotations account must be taken of the property

$$\alpha_{im}\,\alpha_{in} = \delta_{mn}$$

from which follows

$$\alpha_{in}\,\delta\alpha_{im} + \alpha_{im}\,\delta\alpha_{in} = 0 \quad .$$

Hence we can introduce the skew symmetric variation

$$\delta\omega_{mn} = \alpha_{in}\,\delta\alpha_{im} = -\,\delta\omega_{nm}$$

or, equivalently, express the variation of the rotation matrix in terms of those three independent variations:

$$\delta\alpha_{im} = \alpha_{in}\,\delta\omega_{mn} \quad . \tag{3.81}$$

This furnishes as last Euler equations of principle (3.79)

$$\alpha_{in}\, t_{ji}\, (\delta_{mj} + h_{mj}) - \alpha_{im}\, t_{ji}\, (\delta_{nj} + h_{nj}) = 0 \quad . \qquad (3.82)$$

One brings this back to the original form of the rotational equilibrium conditions when multiplying by $\alpha_{pn}\, \alpha_{qm}$ and using (3.34):

$$t_{jp}\, D_j\, x_q - t_{jq}\, D_j\, x_p = 0 \quad .$$

It can finally be concluded that the variational principle (3.79) is very general although somewhat complicated because of the 21 independent quantities that can be varied independently. It contains the 3 translational equilibrium conditions through $\delta\alpha_{im}$, the polar decomposition constraints through variations on the 9 stresses and the 6 stress-strain relations (3.80) through variations on the strains.

Amongst the possible simplifications of the principle, consider the following group of terms in the volume integrand

$$W(h) - t_{ji}\, \alpha_{im}\, (\delta_{mj} + h_{mj}) \quad .$$

Playing on the symmetry of the strains this can be written

$$W(h) - (\delta_{mj} + h_{mj})\, \frac{1}{2}\, (t_{ji}\, \alpha_{im} + t_{mi}\, \alpha_{ij})$$

$$= W(h) - (\delta_{mj} + h_{mj})\, r_{mj}$$

and suggests that we accept (3.80) a priori, solve for the strains in terms of the stresses r_{mj} and introduce the complementary energy density

$$\psi(r) = (\delta_{mj} + h_{mj})\, r_{mj} - W \quad . \qquad (3.83)$$

This has the property

$$\frac{\partial\psi}{\partial r_{mj}} = \delta_{mj} + h_{mj} \tag{3.84}$$

We consider however the r_{mj} to act only as auxiliary quantities defined by the second equality in (3.80). The modified variational principle

$$\int_{R_a} (t_{ji} D_j x_i - \psi - \bar{A}_i u_i) dR_a - \int_{\partial R_a} \bar{p}_i u_i \; d\partial R_a \text{ stationary}$$

still contains as independent quantities subject to variations, the 3 u_i, the 9 t_{ji} and the 3 $\delta\omega_{mn}$ related to the rotations. We now substitute

$$D_j x_i = \delta_{ji} + D_j u_i$$

and integrate by parts with respect to the u_i. Then, assuming that we can manage to satisfy a priori the translational equilibrium equations and natural boundary conditions, the principle reduces to

$$\int_{R_a} (\psi - t_{jj}) \; dR_a - \int_{\partial R_a} l_j t_{ji} \bar{u}_i \; d\partial R_a \quad \text{stationary} \tag{3.85}$$

and is very similar to the complementary energy principle of linear elasticity. Taking the simple case of no body forces, the translational equilibrium equations are easily satisfied by stress functions as in (3.77). The complementary energy density is then a function of

$$r_{mj} = \frac{1}{2} (e_{jrs} \alpha_{im} D_r A_{si} + e_{mrs} \alpha_{ij} D_r A_{si}) \; , \tag{3.86}$$

while the term

$$t_{jj} = e_{jrs} D_r A_{sj}$$

goes immediately to the boundaries by integration. The Euler equations obtained from variations of the stress functions are

$$e_{mrs} D_r \{\alpha_{ij} (\delta_{jm} + h_{jm})\} = 0 \qquad (3.87)$$

and are clearly fulfilled when the polar decomposition is used:

$$e_{mrs} D_r D_m x_i = 0 \quad .$$

This shows that, since (3.87) result from elimination of the displacements between the polar decomposition equations, they can be considered as integrability conditions for such displacements.

Variations on the α_{im} produce again a form of the rotational equilibrium conditions.

In practical applications the complementary energy form of this new principle allows the translational equilibrium equations to be rigorously satisfied; the weight of the approximations falls on the integrability conditions for displacements (explaining why displacements as such have disappeared from the principle) and on the rotational equilibrium conditions.

3.2. *Elastic stability problems*

This section discusses first the classical energy criterion of elastic stability, based on the second variation test of the total potential. It leads in general to a linear eigenvalue problem. The cases of perfectly straight plates and rods under perfectly centered loads are considered next with special emphasis on the mathematical justification of the usual approximations.

3.2.1. *The second variation criterion*

Let $U = \int_{R_a} W(\gamma) dR_a$

denote the strain energy expressed in terms of the Green measures of strain as given by (3.23). In the elastic range we take explicitly

$$W = \frac{1}{2} C^{pq}_{mn} \gamma_{mn} \gamma_{pq} \tag{3.88}$$

where the elastic moduli have the following symmetry properties

$$C^{pq}_{mn} = C^{pq}_{nm} = C^{qp}_{mn} = C^{mn}_{pq} \quad .$$

Application of (3.60) gives the stress-strain relations

$$s_{ij} = \frac{\partial W}{\partial \gamma_{ij}} = C^{pq}_{ij} \gamma_{pq} \quad . \tag{3.89}$$

As we postulate that the strains are small, in fact of the order of 10^{-3} in order to remain in the elastic range of usual materials, those stress-strain relations are the usual engineering ones. However, because the rotations can be large, the stresses are the Kirchhoff-Trefftz ones defined in section 3.1.4. They are defined per unit area in the initial (unstressed) configuration but decomposed in the metric induced by the deformation. The loads are assumed to be "dead" loads, they are not affected by the change in configuration and their potential energy is a linear functional of the displacement field:

$$P(u) = - \int_{R_a} \bar{A}_m u_m dR_a - \int_{\partial R_a} \bar{p}_m u_m d\partial R_a \tag{3.90}$$

Denote by u_m° a displacement field in a configuration of equilibrium with respect to the applied loads and consider a perturbed field

$$u_m = u_m^\circ + \varepsilon \hat{u}_m \qquad (3.91)$$

where ε is a small parameter, that satisfies the kinematical boundary conditions for all values of ε. Hence $\hat{u}_m = 0$ there where the displacement u_m was prescribed on the boundary. Expanding the strains in powers of ε

$$\gamma_{mn} = \gamma_{mn}^\circ + \varepsilon \gamma_{mn}^{(1)} + \frac{1}{2} \varepsilon^2 \gamma_{mn}^{(2)} \qquad (3.92)$$

gives

$$\gamma_{mn}^\circ = \frac{1}{2} (D_m u_n^\circ + D_n u_m^\circ + D_m u_i^\circ D_n u_i^\circ)$$

$$\gamma_{mn}^{(1)} = \frac{1}{2} (D_m \hat{u}_n + D_n \hat{u}_m + D_m u_i^\circ D_n u_i + D_n u_i^\circ D_m \hat{u}_i) \qquad (3.93)$$

$$\gamma_{mn}^{(2)} = D_m \hat{u}_i D_n \hat{u}_i \ . \qquad (3.94)$$

Substitution of (3.92) into (3.88) gives

$$W = W^\circ + \varepsilon C_{mn}^{pq} \gamma_{pq}^\circ \gamma_{mn}^{(1)} + \frac{1}{2} \varepsilon^2 \left\{ C_{mn}^{pq} \gamma_{mn}^{(1)} \gamma_{pq}^{(1)} + C_{mn}^{pq} \gamma_{pq}^\circ \gamma_{mn}^{(2)} \right\}$$

where higher powers of ε were neglected and

$$W^\circ = \frac{1}{2} C_{mn}^{pq} \gamma_{mn}^\circ \gamma_{pq}^\circ$$

is the strain energy in the equilibrium configuration. Introducing the stresses at equilibrium

$$s_{mn}^\circ = C_{mn}^{pq} \gamma_{pq}^\circ \qquad (3.95)$$

and the strain energy density calculated from the strains $\gamma_{mn}^{(1)}$ only

$$W^{(1)} = \frac{1}{2} C_{mn}^{pq} \gamma_{mn}^{(1)} \gamma_{pq}^{(1)} \tag{3.96}$$

the total potential, expanded in powers of ε up to second degree, reads

$$\begin{aligned} U + P = &\int_{R_a} W^\circ \, dR_a + P\,(u^\circ) \\ &+ \varepsilon \left\{ \int_{R_a} s_{mn}^\circ \gamma_{mn}^{(1)} \, dR_a + P\,(\hat{u}) \right\} \\ &+ \varepsilon^2 \left\{ \int_{R_a} \left(W^{(1)} + \frac{1}{2} s_{mn}^\circ \gamma_{mn}^{(2)} \right) dR_a \right\} \; . \end{aligned} \tag{3.97}$$

A necessary condition for a minimum of the total potential is the vanishing of the first degree term under all kinematically admissible perturbations $\hat{u}_m$, or

$$\int_{R_a} \left(s_{mn}^\circ \gamma_{mn}^{(1)} - \bar{A}_i \hat{u}_i \right) dR_a - \int_{\partial R_a} \bar{p}_i \hat{u}_i \, d\partial R_a = 0 \; .$$

In view of the symmetry of the stress tensor, the result of a substitution of (3.93) can be written

$$\int_{R_a} \left\{ s_{mn}^\circ (\delta_{ni} + D_n u_i^\circ) D_m \hat{u}_i - \bar{A}_i \hat{u}_i \right\} dR_a - \int_{\partial R_a} \bar{p}_i \hat{u}_i \, d\partial R_a = 0 \; .$$

After an integration by parts this produces the field and boundary conditions

$$D_m \left\{ s^\circ_{mn} (\delta_{ni} + D_n u^\circ_i) \right\} + \bar{A}_i = 0$$

$$l_m \, s^\circ_{mn} (\delta_{ni} + D_n u^\circ_i) = \bar{p}_i$$

stating that the field u°_i is one that satisfies equilibrium as was assumed. It is observed that the vanishing of the first degree term in the expansion (3.97) is equivalent to the vanishing of the first variation of the total potential as discussed in section 3.1.5. on variational principles. A sufficient condition for the minimum of the total potential in that the coefficient of ε^2, the second variation of the total potential, be positive definite under all kinematically admissible perturbations

$$\int_{R_a} \left(W^{(1)} + \frac{1}{2} s^\circ_{mn} \gamma^{(2)}_{mn} \right) dR_a > 0 \quad . \tag{3.98}$$

This positive definitness criterion will be adopted for the elastic stability of the equilibrium configuration. Clearly, if satisfied, external energy is required to maintain any kinematically admissible disturbance. Conversely, should the second variation be negative for some admissible disturbance, the energy release would have to be absorbed to hold it static. Inspection of the functional (3.98) shows it to be quadratic in the displacement disturbance field, hence, coupled with a linearized kinetic energy functional, one obtains a linear small oscillations problem. It has real eigenfrequencies if (3.98) is satisfied. Stability criterion (3.98) is valid for both bifurcation and snap-through phenomena (3.8, 3.9). This generality is one of its advantages.

3.2.2. *The stability criterion as an eigenvalue problem*

The stability criterion (3.98) is an inequality for the sum of two quadratic functionals.

The first

$$f_1(\hat{u}) = \int_{R_a} \frac{1}{2} C^{pq}_{mn} \gamma^{(1)}_{mn} \gamma^{(1)}_{pq} \, dR_a \tag{3.99}$$

is by nature positive definite. It vanishes if and only if the first order strains

$$\gamma^{(1)}_{mn} = \frac{1}{2} (\delta_{ni} + D_n u^\circ_i) D_m \hat{u}_i + \frac{1}{2} (\delta_{mi} + D_m u^\circ_i) D_n \hat{u}_i \tag{3.100}$$

vanish everywhere. In that case the displacement disturbance field $\hat{u}_i$ only modifies the equilibrium strains by second order quantities; it may be suspected that this disturbance field is in the nature of a small rigid body motion. Confirmation of this is found by an interpretation of $\gamma^{(1)}_{mn} = 0$ in the metric induced by the equilibrium strains. Consider the relations

$$x^\circ_i = a_i + u^\circ_i (a_j)$$

as a change of lagrangian coordinates and introduce the notation

$$\partial_i = \partial / (\partial x^\circ_i) \quad .$$

One has then formulas such as

$$D_m \hat{u}_i = D_m x^\circ_p \, \partial_p \hat{u}_i$$

and the first order strains (3.100) can be written as

$$\gamma_{mn}^{(1)} = \frac{1}{2} (D_n x_i^\circ D_m x_p^\circ + D_m x_i^\circ D_n x_p^\circ) \partial_p \hat{u}_i \quad .$$

Multiplying both sides by $\partial_r a_n \partial_s a_m$ and using the reciprocity formulas

$$\partial_r a_n D_n x_i^\circ = \delta_{ri}$$

we find

$$\partial_r a_n \partial_s a_m \gamma_{mn}^{(1)} = \frac{1}{2} (\partial_s \hat{u}_r + \partial_r \hat{u}_s) \quad .$$

Hence the vanishing of first order strains entails that the structure of the disturbance field is that of a small rigid body motion expressed in the new lagrangian coordinates, as suits the fact that it follows the equilibrium straining.

$$u_1 = \alpha_1 + \omega_2 x_3^\circ - \omega_3 x_2^\circ$$

$$u_2 = \alpha_2 + \omega_3 x_1^\circ - \omega_1 x_3^\circ$$

$$u_3 = \alpha_3 + \omega_1 x_2^\circ - \omega_2 x_1^\circ \quad .$$

Of course the kinematical boundary conditions might be such as to prevent one or more of those six degrees of freedom.

The second functional

$$f_2(\hat{u}) = \int_{R_a} \frac{1}{2} s_{mn}^\circ \gamma_{mn}^{(2)} dR_a \qquad (3.101)$$

has in general no definite sign. Let

$$\phi(\hat{u}) = f_1(\hat{u}) + f_2(\hat{u})$$

be the value of the complete functional. To prove its positive character under all admissible disturbances it is tempting to seek the sign of its minimum value. However, since the kinematical conditions imposed are homogeneous, if $\hat{u}_i$ is admissible, so is $\alpha\, \hat{u}_i$ and $\phi\ (\alpha\, \hat{u}) = \alpha^2\ \phi\ (\hat{u})$, where α is an arbitrary real scalar. This similarity rule shows that it is meaningless to look after an absolute "critical" disturbance that would give its minimum value to ϕ, however it shows also that the stable ($\phi > 0$) or unstable ($\phi < 0$) character of a disturbance is not altered by scaling. Hence it is appropriate to look after a critical disturbance "shape" by adding a norming constraint to the disturbance scale. Because of its positive definite character

$$f_1\ (\hat{u}) = \mu^2 \quad \text{a fixed quantity} \tag{3.102}$$

is a suitable norming constraint. It insures that the disturbance will involve additional strain energy and not merely superposition of a small rigid body motion. The problem of finding the minimum of ϕ under the norming constraint is Jacobi's accessory problem (3.101). With (3.102) as norming constraint Jacobi's problem is clearly equivalent to that of finding the minimum of the ratio

$$\lambda = \frac{f_2\ (\hat{u})}{f_1\ (\hat{u})} \tag{3.103}$$

that is independent of scaling.

A necessary condition for a relative minimum is that the variation

$$\delta\ \{f_2\ (\hat{u}) - \lambda\ f_1\ (\hat{u})\} = 0 \tag{3.104}$$

for all kinematically admissible variations of the disturbance field. Expanding this condition produces

$$\int_{R_a} \left(\frac{1}{2} s^\circ_{mn} \delta\gamma^{(2)}_{mn} - \lambda s^{(1)}_{mn} \delta\gamma^{(1)}_{mn} \right) dR_a = 0$$

where

$$s^{(1)}_{mn} = C^{pq}_{mn} \gamma^{(1)}_{pq} .$$

Substitution of (3.94) and (3.100) and use of the symmetry of the stress tensors gives

$$\int_{R_a} s^\circ_{mn} D_n \hat{u}_i - \lambda s^{(1)}_{mn} (\delta_{ni} + D_n u^\circ_i) D_m \delta\hat{u}_i dR_a = 0 .$$

Hence in addition to the homogeneous kinematical boundary conditions the critical disturbance shape must satisfy the field equations

$$D_m \{ s^\circ_{mn} D_n \hat{u}_i - \lambda s^{(1)}_{mn} (\delta_{ni} + D_n u^\circ_i) \} = 0 \quad . \qquad (3.106)$$

All the equations are homogeneous of the first degree in the $\hat{u}_i$, hence non trivial solutions, defined except for scaling, will only exist for a spectrum of eigenvalues of λ. If $\hat{u}_i$ is an eigen-solution and λ its eigenvalue, we find after scaling to satisfy the norming constraint (3.102) and using (3.103)

$$\phi(\hat{u}) = f_1 (\hat{u}) + f_2 (\hat{u}) = (1 + \lambda) \mu^2 .$$

Consequently the stability criterion will be satisfied if the smallest eigenvalue is larger than -1.

3.2.3. *Stability boundaries*

The previous analysis furnishes a way to test stability under a fixed set of dead loads. In many cases one wishes to evaluate the stability margin of a structure and therefore to increase the loading up to the appearance of the critical value $\lambda = -1$, where λ is the smallest eigenvalue. Topological investigations (3.9) show that the appearance of this critical value can have several meanings: loss of stability, asymmetric bifurcation to a stable continuous post-buckling equilibrium path, stable-symmetric bifurcation possibility and unstable-symmetric bifurcation. The equilibrium paths of the structure are followed by assuming a similarity law for the loading distribution that results in a single load scaling parameter σ. The general case is fairly complicated because the equilibrium stress distribution s°_{mn} depends non linearly on the load parameter σ and incremental analysis becomes necessary.

Direct access to the critical situation becomes possible when the initial equilibrium path can be dealt with by the linear elasticity approximation, that is when both rotations and strains remain very small. Then all displacement gradients remain very small at equilibrium

$$|D_m\, u^{\circ}_i| \ll 1 \tag{3.107}$$

and (3.93) can be approximated by the formulas

$$\gamma^{(1)}_{mn} = \frac{1}{2}\,(D_m\, u_n + D_n\, \hat{u}_m) \quad . \tag{3.108}$$

Consequently the equilibrium equations for the Kirchhoff-Trefftz stresses have valid linear approximations

$$D_m\, s^{\circ}_{mn} + \bar{A}_i = 0$$

$$1_m\, s^{\circ}_{mi} = \bar{p}_i$$

and the scaling operation on the loads becomes equivalent to a similar scaling on the equilibrium stress distribution:

$$s^{\circ}_{mn} = \sigma\, S_{mn} \qquad \sigma > 0 \tag{3.109}$$

where S_{mn} is a fixed distribution.

According to the analysis of the previous section the stability condition becomes

$$\sigma \mu > -1 \tag{3.110}$$

where

$$\mu = \min \frac{\frac{1}{2}\int_{R_a} S_{mn}\, \gamma_{mn}^{(2)}\, dR_a}{\frac{1}{2}\int_{R_a} C_{mn}^{pq}\, \gamma_{mn}^{(1)}\, \gamma_{pq}^{(1)}\, dR_a} \quad . \tag{3.111}$$

We now distinguish between two cases.

CASE 1 : $\mu > 0$ There is obviously unconditional stability; σ can be as large as the limitations on the elastic behavior of the material will permit.

CASE 2 : $\mu > 0$ we then set

$$\mu = - \frac{1}{\sigma_{cr}} \qquad \sigma_{cr} > 0 \tag{3.112}$$

and the stability condition (3.110) becomes equivalent to

$$\sigma < \sigma_{cr} \tag{3.113}$$

$\sigma = \sigma_{cr}$ being the critical situation. Moreover with μ replaced by (3.112), (3.111) becomes equivalent to

$$\sigma_{cr} = \min \frac{\frac{1}{2}\int_{R_a} C_{mn}^{pq}\, \gamma_{mn}^{(1)}\, \gamma_{pq}^{(1)}\, dR_a}{-\frac{1}{2}\int_{R_a} S_{mn}\, \gamma_{mn}^{(2)}\, dR_a} \tag{3.114}$$

and the determination of the critical load parameter has been turned into an eigenvalue problem. The presence of the minus sign in the denominator accounts for the fact that positive values of σ_{cr} are normally associated with compressive stresses. If σ_{cr} as determined by (3.114) turns out to be negative, $\sigma_{cr} = -\alpha^2$, it is easily found that $\alpha^2 = \mu$ as given by (3.111). Hence, while (3.113) becomes meaningless, we are in a case of unconditional stability.

With due consideration to the simplified expressions (3.108), the Euler equations and natural boundary conditions to be satisfied by the critical disturbances of problem (3.114) are

$$D_m T_{mi} = 0 \qquad \text{in } R_a \tag{3.115}$$

$$l_m T_{mi} = 0 \qquad \text{where } \bar{p}_i \text{ specified on } \partial R_a \tag{3.116}$$

$$T_{mi} = C^{pq}_{mi} D_p \hat{u}_q + \sigma_{cr} S_{mn} D_n \hat{u}_i \quad . \tag{3.117}$$

3.2.4. *The equilibrium configuration as reference configuration*

Up to now the unstressed configuration was used as initial or reference configuration. The stability analysis can however also be carried out by choosing the actual equilibrium configuration to be tested as reference. This may present some advantages when the prevailing state of stress is not due to external loading but to internal reasons like thermal effects or self-stressing. To be clear it is necessary to distinguish between the original cartesian coordinates a_i of the particles and their coordinates at equilibrium

$$x^\circ_i = a_i + u^\circ_i (a_j) \quad . \tag{3.118}$$

Note that x_i° is not to be confused here with an eulerian coordinate but represents a new lagrangian coordinate that will be used in formulating the stability analysis. Hence (3.118) is in the nature of a change of lagrangian coordinates. Then with the notation $\partial_i = \partial/\partial x_i$, the strains incurred in a disturbance $\varepsilon\, \hat{u}_i\, (x_j^\circ)$ from the equilibrium configuration are given by

$$\beta_{mn} = \varepsilon\, \frac{1}{2}\, (\partial_m \hat{u}_n + \partial_n \hat{u}_m) + \frac{\varepsilon^2}{2}\, \partial_m \hat{u}_i\, \partial_n \hat{u}_i$$

or, to prepare further calculations,

$$\beta_{mn} = \varepsilon\, \beta^{(1)}_{mn} + \frac{\varepsilon^2}{2}\, \beta^{(2)}_{mn} \tag{3.119}$$

with

$$\beta^{(1)}_{mn} = \frac{1}{2}\, (\partial_m \hat{u}_n + \partial_n \hat{u}_m) \tag{3.120}$$

$$\beta^{(2)}_{mn} = \partial_m \hat{u}_i\, \partial_n \hat{u}_i \quad .$$

The energy per unit volume in the new reference configuration is denoted by W_x and will be a function of the strains (3.119). From a Maclaurin expansion

$$W_x = W_x^\circ + \beta_{mn} \left(\frac{\partial W_x}{\partial \beta_{mn}} \right)_\circ + \frac{1}{2}\, \beta_{mn}\, \beta_{pq} \left(\frac{\partial^2 W_x}{\partial \beta_{mn}\, \partial \beta_{pq}} \right)_\circ .$$

The stresses defined as partial derivatives of the energy will be

$$\sigma_{mn} = \left(\frac{\partial W_x}{\partial \beta_{mn}} \right)_\circ + \beta_{pq} \left(\frac{\partial^2 W_x}{\partial \beta_{mn}\, \partial \beta_{pq}} \right) .$$

$$= \sigma^\circ_{mn} + E^{pq}_{mn}\, \beta_{pq} \quad .$$

Substitution of (3.119) into

$$W_x = W_x^\circ + \sigma_{mn}^\circ \beta_{mn} + \frac{1}{2} E_{mn}^{pq} \beta_{mn} \beta_{pq}$$

and grouping of powers in ε yields

$$W_x = W_x^\circ + \varepsilon \left(\sigma_{mn}^\circ \beta_{mn}^{(1)}\right) + \varepsilon^2 \left(\frac{1}{2} E_{mn}^{pq} \beta_{mn}^{(1)} \beta_{pq}^{(1)} + \frac{1}{2} \sigma_{mn}^\circ \beta_{mn}^{(2)}\right) + \ldots$$

Since the new reference configuration is in equilibrium, the first variation of the total potential vanishes under all kinematically admissible disturbances and we easily obtain the equilibrium equations

$$\partial_m \sigma_{mn}^\circ + \overline{X}_n = 0$$

$$n_m \sigma_{mn}^\circ = \overline{q}_n$$

where $\overline{X}_n$ and $\overline{q}_n$ are the prescribed applied forces described in the new geometry and n_i the direction cosines of the outward normal to the new boundary surface ∂R_x. The stability criterion is now

$$\frac{1}{2} \int_{R_x} \left(E_{mn}^{pq} \beta_{mn}^{(1)} \beta_{pq}^{(1)} + \sigma_{mn}^\circ \beta_{mn}^{(2)}\right) dR_x > 0 \quad .$$

It bears very close resemblance to the old form and can in fact be shown to be identical if due account is taken of all effects due to the change (3.118) in lagrangian coordinates. It has apparently the advantage that the expressions (3.120) of the first order strains are simpler than their counterpart (3.93) and analogous to their simplified values (3.108). In general however this advantage is more than offset by the necessity to apply the boundary conditions to a distorted configuration.

In the case discussed under 3.2.3. where the linear elasticity approximation is adequate to follow the equilibrium path up to a critical situation, both formulations of the stability problem are identical. In the first formulation the first order strains can be approximated by (3.108), in the second one may simply ignore the change in lagrangian coordinates and perform all operations on the unstressed configuration.

3.2.5. *Stability of membrane loaded flat plates*

The assumptions are that the plate structure is perfectly symmetrical with respect to the middle plane $a_3 = 0$ and similarly for the applied set of external loads and kinematical boundary conditions. Then we are in a case where, provided we remain in the elastic range of material behavior, (3.107) and (3.108) are valid to determine the set of equilibrium stresses. The energy density $W^{(1)}$ has the classical structure of linear elasticity in the perturbation displacements $\hat{u}_i$. Moreover, because of the assumed symmetry, it must be independent of a change of orientation of the a_3 axis. Only $\gamma^{(1)}_{13}$ and $\gamma^{(1)}_{23}$ do change sign with the substitution $-a_3$ for a_3, hence in the quadratic form $W^{(1)}$ those strains are not coupled with the other ones and the quadratic form is a sum of two independant positive definite forms in each group of variables:

$$W^{(1)} = B\left(\gamma^{(1)}_{11}, \gamma^{(1)}_{12}, \gamma^{(1)}_{22}, \gamma^{(1)}_{33}\right) + A\left(\gamma^{(1)}_{13}, \gamma^{(1)}_{23}\right). \tag{3.121}$$

In order to establish with mathematical accuracy the considerable simplifications that can be introduced in the stability criterion, it will be sufficient to introduce the simplest assumptions on the structure of the perturbation displacement field:

$$\hat{u}_1 = v_1\ (a_1,\ a_2) + a_3\ \alpha_1\ (a_1,\ a_2)$$
$$\hat{u}_2 = v_2\ (a_1,\ a_2) + a_3\ \alpha_2\ (a_1,\ a_2) \qquad (3.122)$$
$$\hat{u}_3 = w\ (a_1,\ a_2)\quad .$$

They produce the following strains

$$\left.\begin{aligned} \gamma^{(1)}_{33} &= D_3\ \hat{u}_3 = 0 \\ \gamma^{(1)}_{11} &= D_1\ \hat{u}_1 = D_1\ v_1 + a_3\ D_1\ \alpha_1 \\ \gamma^{(1)}_{22} &= D_2\ \hat{u}_2 = D_2\ v_2 + a_3\ D_2\ \alpha_2 \\ 2\gamma^{(1)}_{12} &= D_1\ \hat{u}_2 + D_2\ \hat{u}_1 = (D_1\ \alpha_2 + D_2\ \alpha_1)a_3 \\ &\qquad + (D_1\ v_2 + D_2\ v_1) \end{aligned}\right\} \qquad (3.123)$$

$$\left.\begin{aligned} 2\gamma^{(1)}_{13} &= D_1\ \hat{u}_3 + D_3\ \hat{u}_1 = \alpha_1 + D_1\ w \\ 2\gamma^{(1)}_{23} &= D_2\ \hat{u}_3 + D_3\ \hat{u}_2 = \alpha_2 + D_2\ w\quad . \end{aligned}\right\} \qquad (3.124)$$

Integration across the thickness $2h(a_1,\ a_2)$ of the plate, induces uncoupling of terms due to their even or odd dependence on a_3:

$$\int_{-h}^{h} B\ da_3 = B_V\ (D_1 v_1,\ D_2 v_2,\ D_1 v_2 + D_2 v_1)$$
$$+ B_\alpha\ (D_1\alpha_1,\ D_2\alpha_2,\ D_1\alpha_2 + D_2\alpha_1) \qquad (3.125)$$

$$\int_{-h}^{h} A\ da_3 = K\ (\alpha_1 + D_1 w,\ \alpha_2 + D_2 w). \qquad (3.126)$$

Similarly we assume that the equilibrium membrane stresses have the following simple structure

$$s^{\circ}_{11}(a_1, a_2) \qquad s^{\circ}_{12}(a_1, a_2) \qquad s^{\circ}_{22}(a_1, a_2)$$

$$s^{\circ}_{13} = 0 \qquad s^{\circ}_{23} = 0 \qquad s^{\circ}_{33} = 0$$

whence that part of the stability criterion that depends on the equilibrium stresses reduces to the volume integral of

$$\frac{1}{2} s^{\circ}_{11} \gamma^{(2)}_{11} + s^{\circ}_{12} \gamma^{(2)}_{12} + \frac{1}{2} s^{\circ}_{22} \gamma^{(2)}_{22}$$

where, in accordance with the previous assumptions,

$$\gamma^{(2)}_{11} = (D_1 v_1 + a_3 D_1 \alpha_1)^2 + (D_1 v_2 + a_3 D_1 \alpha_2)^2 + (D_1 w)^2$$

$$\gamma^{(2)}_{22} = (D_2 v_1 + a_3 D_2 \alpha_1)^2 + (D_2 v_2 + a_3 D_2 \alpha_2)^2 + (D_2 w)^2$$

$$\gamma^{(2)}_{21} = (D_1 v_1 + a_3 D_1 \alpha_1)(D_2 v_1 + a_3 D_2 \alpha_1) + (D_1 v_2 + a_3 D_1 \alpha_2)$$

$$(D_2 v_2 + a_3 D_2 \alpha_2) + (D_1 w)(D_2 w) \quad .$$

Again, integration across the thickness of the plate produces uncoupling of terms

$$\frac{1}{2} \int_{-h}^{h} s^{\circ}_{11} \gamma^{(2)}_{11} + 2 s^{\circ}_{12} \gamma^{(2)}_{12} + s^{\circ}_{22} \gamma^{(2)}_{22} \quad da_3$$

$$= S_v + S_\alpha + S_w \tag{3.127}$$

where

$$S_v = h s^{\circ}_{11}\{(D_1 v_1)^2 + (D_1 v_2)^2\} + 2 s^{\circ}_{12} h \{D_1 v_1 D_2 v_1 + D_1 v_2 D_2 v_2\}$$

$$+ s^{\circ}_{22} h\{(D_2 v_1)^2 + (D_2 v_2)^2\}$$

$$\tag{3.128}$$

$$S_\alpha = \frac{1}{3} h^3 s^\circ_{11} \{(D_1\alpha_1)^2 + (D_1\alpha_2)^2\} + \frac{2}{3} h^3 s^\circ_{12}$$

$$\{D_1\alpha_1 D_2\alpha_1 + D_1\alpha_2 D_2\alpha_2\} + \frac{1}{3} h^3 s^\circ_{22} \{(D_2\alpha_1)^2 + (D_2\alpha_2)^2\}$$

$$S_w = h\, s^\circ_{11} (D_1 w)^2 + 2h\, s^\circ_{12} D_1 w\, D_2 w + h\, s^\circ_{22} (D_2 w)^2 .$$

The general stability criterion (3.98) thus assumes the form

$$\iint (B_v + S_v)\, da_1\, da_2 + \iint (B_\alpha + S_\alpha)\, da_1\, da_2$$

$$+ \iint (K + S_w)\, da_1\, da_2 > 0 \qquad (3.129)$$

where the integrals are extended over the area of the plate.

The first term depends only on the "in plane" perturbations v_1 and v_2 and is decoupled from the other two. It will now be shown that this first term can be canceled because its contribution if always positive.

We first examine the particular case of in plane perturbations only where

$$D_1 v_1 = 0 \qquad D_2 v_2 = 0 \qquad D_1 v_2 + D_2 v_1 = 0$$

and consequently $B_v = 0$. The in plane perturbation field is then of rigid body type

$$v_1 = b_1 - \omega\, a_2 \qquad v_2 = b_2 + \omega\, a_1$$

and the contribution of S_v to the criterion is, after computation, found to be

$$\iint S_v\, da_1\, da_2 = \omega^2 \iint h(s^\circ_{11} + s^\circ_{22})\, da_1\, da_2 \quad .$$

Should this be negative, the criterion would be violated at any load level and the problem would not make sense. This case must be ruled out, either because the kinematical boundary conditions do not allow the small rigid body rotation ω, or, as we shall see, by a condition imposed on the external loads. Indeed, the equilibrium equations

$$D_1 \ (2hs^\circ_{11}) + D_2 \ (2hs^\circ_{12}) + \int_{-h}^{h} \bar{A}_1 \, da_3 = 0$$

$$D_1 \ (2hs^\circ_{12}) + D_2 \ (2hs^\circ_{22}) + \int_{-h}^{h} \bar{A}_2 \, da_3 = 0$$

multiplied respectively by a_1 and a_2, added and integrated over the area of the plate, yield

$$2 \iint \{a_1 \ D_1 \ (h \ s^\circ_{11}) + a_1 \ D_2 \ (h \ s^\circ_{12}) + a_2 \ D_1 \ (h \ s^\circ_{12}) + a_2 \ D_2 \ (h \ s^\circ_{22})\} \ da_1 \ da_3 = -\iiint_{-h}^{h} (a_1\bar{A}_1 + a_2\bar{A}_2) \ da_1 \ da_2 \ da_3 \ .$$

Transforming the left-hand side by integration by parts and noting that

$$2 \ h \ (\ell_1 \ s^\circ_{11} + \ell_2 \ s^\circ_{12}) = \int_{-h}^{h} \bar{p}_1 \ da_3$$

$$2 \ h \ (\ell_1 \ s^\circ_{12} + \ell_2 \ s^\circ_{22}) = \int_{-h}^{h} \bar{p}_2 \ da_3$$

there comes

$$2\iint h(s^\circ_{11} + s^\circ_{22}) \ da_1 \ da_2 = \iiint_{-h}^{h} (a_1\bar{A}_1 + a_2\bar{A}_2) da_1 \ da_2 \ da_3 + \oint_{-h}^{h} (a_1\bar{p}_1 + a_2\bar{p}_2) da_3 \ ds \qquad (3.130)$$

and the stability condition in the absence of kinematical boundary conditions is that the imposed loads give a positive value to the right-hand side of this equation.

Returning now to the general case where all the in plane strains $D_1 v_1$, $D_2 v_2$ and $D_1 v_2 + D_2 v_1$ are not zero, the contribution from B_V is positive. It can be counterbalanced by a negative contribution from S_V only if there are equilibrium stresses of appropriate sign and of an order of magnitude equal to that of the elastic in plane moduli. Thus some of the corresponding equilibrium strains would have to be of an order of magnitude equal to unity, contradicting assumptions (3.107). The contribution of the first term in (3.129) is thus positive in any case and the criterion can be reduced to the one

$$\iint (B_\alpha + S_\alpha)\, da_1\, da_2 + \iint (K + S_W)\, da_1\, da_2 > 0 \qquad (3.131)$$

involving only the skew symmetric disturbances α_1, α_2 and w.

In a similar manner it will now be shown that the contribution from S_α may be disregarded. Again consider first the special case where

$$D_1\, \alpha_1 = 0 \qquad D_2\, \alpha_2 = 0 \qquad D_1\, \alpha_2 + D_2\, \alpha_1 = 0$$

so that $B_\alpha = 0$. The disturbance field is now of the form

$$\hat{u}_1 = a_3\, (c_1 - \phi\, a_2) \qquad \hat{u}_2 = a_3\, (c_2 + \phi a_1)$$

and represents rigid body motions of sheets parallel to the middle plane. The contribution from S_α can be worked out as due to the twist ϕ between sheets only

$$\frac{1}{3}\, \phi^2 \iint h^3\, (s^{\circ}_{11} + s^{\circ}_{22})\, da_1\, da_2 \qquad (3.132)$$

and will be compared to the contribution from K. To this purpose we take a typical form of K

$$K = \frac{1}{2} G \; 2h \; (\alpha_1^2 + \alpha_2^2)$$

where G is an average shear modulus between the sheets. Then

$$\iint K \, da_1 \, da_2 = G \iint \{ h \, (c_1 - \phi \, a_2)^2$$

$$+ h \, (c_2 + \phi \, a_1)^2 \quad da_1 \, da_2 \; .$$

The minimum of this for a twist ϕ is found by adjusting c_1 and c_2 so that the center of twist is located at the center of gravity of the middle plane. The cartesian axes have their origin at this center if

$$\iint h \, a_1 \, da_1 \, da_2 = 0 \qquad \iint h \, a_2 \, da_1 \, da_2 = 0 \; .$$

The minimal contribution is then

$$\iint K \, da_1 \, da_2 = \phi^2 \, G \iint h \, (a_1^2 + a_2^2) \, da_1 \, da_2 \; .$$

This is positive and much larger than (3.132). In fact, if s denotes the order of magnitude of an equilibrium stress, $\bar{h}$ the average half thickness of the plate and ρ the radius of gyration of the middle plane

$$\iint h \, (a_1^2 + a_2^2) \quad da_1 \, da_2 = \bar{h} \, \rho^2 \text{ Area}$$

the ratio of the two terms if

$$\frac{1}{3} \frac{s}{G} \left(\frac{\bar{h}}{\rho} \right)^2 \ll 1 \; .$$

Thus the contribution from S_α is negligible compared to that of K. In the case that the strains $D_1 \alpha_1$, $D_2 \alpha_2$ and $D_1 \alpha_2 + D_2 \alpha_1$ are not all zero, the same arguments as between S_v and B_v show that the contribution from S_α can be neglected before that of B_α.

The stability criterion thus finally reduces to

$$\iint (K + B_\alpha + S_w)\, da_1\, da_2 > 0 \qquad (3.133)$$

$K + B_\alpha$ is in fact any of the classical flexure energies per unit plate area including shearing deformations. Under the Kirchhoff hypothesis we would have furthermore

$$\alpha_1 + D_1 w = 0 \qquad \alpha_2 + D_2 w = 0$$

so that the shearing energy K would disappear also, and the flexure energy B_α would be classically expressed in terms of the second derivatives $D_1^2 w$, $D_2^2 w$, $D_1 D_2 w$ of the transverse disturbance.

3.2.6. *Other derivations of the plate stability criterion*

Another method for deriving the plate stability criterion makes use of the equilibrium configuration as reference configuration; our lagrangian coordinates are here the x_i°. The criterion states that the increase in strain energy due to a kinematically admissible disturbance must be larger than the additional work performed by the external loads. Consider then a transverse disturbance field

$$u_1 = \varepsilon\, a_3\, \alpha_1 (x_1^\circ, x_2^\circ) \qquad u_2 = \varepsilon\, a_3\, \alpha_2 (x_1^\circ, x_2^\circ)$$

$$u_3 = \varepsilon\, w\, (x_1^\circ, x_2^\circ) \qquad \text{(note that } a_3 \equiv x_3^\circ) \quad .$$

The in plane components vanish in the middle plane $a_3 = 0$ so that no additional work is performed by the external in plane loads. It will however be seen that the additional strain energy is not necessarily positive and that the stability criterion requiring it to be positive is meaningful. A first part of the additional strain energy is the bending energy (it is not coupled to the membrane energy).

Neglecting terms of order ε^4 it can be calculated from the linearized strain measures

$$\varepsilon_{11} = \partial_1 u_1 = \varepsilon\, a_3\, \partial_1 \alpha_1 \qquad \varepsilon_{22} = \partial_2 u_2 = \varepsilon\, a_3\, \partial_2 \alpha_2$$

$$2\varepsilon_{12} = \partial_1 u_2 + \partial_2 u_1 = \varepsilon\, a_3\, (\partial_1 \alpha_2 + \partial_2 \alpha_1)$$

$$2\varepsilon_{13} = \partial_1 u_3 + \partial_3 u_1 = \varepsilon\, (\alpha_1 + \partial_1 w)$$

$$2\varepsilon_{23} = \varepsilon\, (\alpha_2 + \partial_2 w)$$

and will be denoted by ε^2 B. B is in fact the equivalent in the equilibrium configuration of our $\iint (B_\alpha + K)\, da_1\, da_2$ of section 3.2.5.

Although the linearized in plane strains vanish for $a_3 = 0$, the exact membrane strains induced by the disturbance field do not. In fact the membrane strains calculated from the Green strain measures are easily found to be

$$\gamma_{11} = \frac{1}{2}\varepsilon^2 (\partial_1 w)^2 \qquad \gamma_{22} = \frac{1}{2}\varepsilon^2 (\partial_2 w)^2$$

$$\gamma_{12} = \frac{1}{2}\varepsilon^2\, \partial_1 w\, \partial_2 w \qquad (a_3 = 0)\ .$$

Those strains are easily seen to be additive to the equilibrium strains $\gamma^{\circ}_{\alpha\beta}$, because they are induced by a displacement component that is absent in the equilibrium strains. Thus the new value of the membrane energy would be (the greek summation indices run from 1 to 2)

$$\frac{1}{2}\iint E^{\lambda\mu}_{\alpha\beta}\left(\gamma^{\circ}_{\alpha\beta} + \frac{1}{2}\varepsilon^2\,\partial_\alpha w\,\partial_\beta w\right)\left(\gamma^{\circ}_{\lambda\mu} + \frac{1}{2}\varepsilon^2\partial_\lambda w\,\partial_\mu w\right)dx_1\,dx_2 \;.$$

Or, neglecting again term of order ε^4, an additional membrane strain energy (interfercnce energy) or order ε^2 appears equal to

$$\varepsilon^2\,\frac{1}{4}\iint E^{\lambda\mu}_{\alpha\beta}\left(\gamma^{\circ}_{\alpha\beta}\,\partial_\lambda w\,\partial_\mu w + \gamma^{\circ}_{\lambda\mu}\,\partial_\alpha w\,\partial_\beta w\right)dx_1\,dx_2$$

or using the symmetry properties of the elastic moduli,

$$\varepsilon^2\,\frac{1}{2}\iint E^{\lambda\mu}_{\alpha\beta}\,\gamma^{\circ}_{\lambda\mu}\,\partial_\alpha w\,\partial_\beta w\,dx_1\,dx_2$$

$$= \varepsilon^2\,\frac{1}{2}\iint \sigma^{\circ}_{\alpha\beta}\,\partial_\alpha w\,\partial_\beta w\,dx_1\,dx_2 \;.$$

Thus, if there are compressive equilibrium stresses, the transverse displacement w can actually relieve the membrane strain energy. If this energy release is higher than the bending energy, the stability criterion is violated. This means in general that the plate will "bifurcate" in a new stable equilibrium configuration involving bending. As a matter of fact the stability criterion that has been found is, after suppression of the factor

$$B + \frac{1}{2}\iint \sigma^{\circ}_{\alpha\beta}\,\partial_\alpha w\,\partial_\beta w\,dx_1\,dx_2 > 0 \qquad (3.134)$$

completely analogous to (3.133). It becomes identical to it when, as was shown to be licit, one ignores the change in metric induced by the change (3.118) of lagrangian coordinates.

This new derivation of the simplified stability criterion for plates certainly shows it to be necessary. But, inasmuch as it ignores other types of disturbances, it does not prove it to be sufficient, although engineering intuition is perhaps enough to convince that the addition of displacement disturbances of membrane type is necessarily stabilizing.

It is rather amusing to record how, by adding in fact displacement disturbances of membrane type, Timoshenko [3.10] obtains the same criterion by a devious way. Since the transverse disturbance produces membrane strains, let us cancel them by suitable membrane displacements v_1 and v_2. We should then have that

$$\partial_1 v_1 + \frac{1}{2}(\partial_1 v_1)^2 + \frac{1}{2}(\partial_1 v_2)^2 + \frac{1}{2}\varepsilon^2(\partial_1 w)^2 = 0$$

$$\partial_2 v_2 + \frac{1}{2}(\partial_2 v_1)^2 + \frac{1}{2}(\partial_2 v_2)^2 + \frac{1}{2}\varepsilon^2(\partial_2 w)^2 = 0$$

$$\partial_1 v_2 + \partial_2 v_1 + \partial_1 v_1 \partial_2 v_1 + \partial_1 v_2 \partial_2 v_2 + \varepsilon^2 \partial_1 w \partial_2 w = 0 .$$

However it is accurate enough to write

$$\partial_1 v_1 = -\frac{1}{2}\varepsilon^2(\partial_1 w)^2 \qquad \partial_2 v_2 = -\frac{1}{2}\varepsilon^2(\partial_2 w)^2$$

$$\partial_1 v_2 + \partial_2 v_1 = -\varepsilon^2 \partial_1 w \partial_2 w \tag{3.135}$$

since it appears from this that the neglected terms are of order ε^4. To be quite precise it is necessary to recall that the material rotation $\partial_1 v_2 - \partial_2 v_1$ is of the same order of magnitude as the linearized strains, so that all gradients $\partial_\alpha v_\beta$ are of the same order.

So doing, the additional membrane energy is now canceled and the additional energy in the criterion consists of $\varepsilon^2 B$ alone. However, we have now introduced an additional work performed by the external loads

$$T = \iint (\bar{X}_1 v_1 + \bar{X}_2 v_2)\, dx_1\, dx_2 + \oint (\bar{p}_1 v_1 + \bar{p}_2 v_2)\, ds$$

and the stability criterion is now

$$\varepsilon^2 B > T \quad . \tag{3.136}$$

In the expression for T we substitute from the equilibrium equations

$$\overline{X}_1 = - \partial_1 \sigma^\circ_{11} - \partial_2 \sigma^\circ_{12} \qquad \overline{X}_2 = - \partial_1 \sigma^\circ_{12} - \partial_2 \sigma^\circ_{22}$$

$$\overline{p}_1 = n_1 \sigma^\circ_{11} + n_2 \sigma^\circ_{12} \qquad \overline{p}_2 = n_1 \sigma^\circ_{12} + n_2 \sigma^\circ_{22}$$

and perform an integration by parts in the first term, then

$$T = \iint \{\sigma^\circ_{11} \partial_1 v_1 + \sigma^\circ_{12} (\partial_1 v_2 + \partial_2 v_1)$$

$$+ \sigma^\circ_{22} \partial_2 v_2\} \, dx_1 \, dx_2$$

or, using (3.135)

$$T = - \frac{1}{2} \varepsilon^2 \iint \sigma_{\alpha\beta} \partial_\alpha w \, \partial_\beta w \, dx_1 \, dx_2 \; . \tag{3.137}$$

Thus, when substituted into (3.136) we fall back again on (3.134). In this way Timoshenko is able to introduce a potential energy of the external loads in the second variation of the energy, although it is by nature absent in the general criterion (3.98).

3.2.7. *Stability of axially loaded beams*

Critical loads of space structures have been discussed in several other publications [3.12],[3.13],[3.14]. Here the conditions of stability are first derived from the general energy criterion in an attempt to bring out clearly all the assumptions underlying the usual simplified version of the stability analysis. In particular the reduction of torsional rigidity due to axial stresses is taken into account; it is significant when the members of a frame are of thin-walled open sections. This feature has been investigated both analytically and experimentally [3.15],[3.16], [3.17],[3.18],[3.19],[3.20], and can be incorporated easily in a finite element solution.

The first assumption is that a space frame member has a perfectly straight axis of centroids and is perfectly axially loaded in the equilibrium state. This restriction gives direct access to the critical point and avoids the iterative or incremental procedures that become necessary otherwise [3.20]. Instead of the notations of the general theory we shall use here (x,y,z) for the lagrangian coordinates in the unstressed configuration. The x axis is the line of centroids, the y and z axes are principals of inertia in each cross section; thus

$$\iint y \, dy \, dz = 0 \qquad \iint z \, dy \, dz = 0 \qquad \iint yz \, dy \, da = 0 \quad . \tag{3.138}$$

The disturbance field is resolved in terms of rigid body motions of each cross section (two translations $\varepsilon v(x)$ and $\varepsilon w(x)$ and a rotation $\varepsilon\phi(x)$ about the (convected) x axis plus an axial disturbance field $\varepsilon u(x,y,z)$ that is able to take into account axial translations, bending rotations and warping. From the total displacement field

$$u_t = u^\circ (x) + \varepsilon u (x, y, z)$$

$$v_t = \varepsilon v (x) + y (\cos \varepsilon\phi - 1) - z \sin \varepsilon\phi$$

$$w_t = \varepsilon w (x) + y \sin \varepsilon\phi + z (\cos \varepsilon\phi - 1)$$

follows after expansion in powers of ε

$$\hat{u} = u (x, y, z) \qquad \hat{v} = v(x) - z\phi (x) \qquad \hat{w} = w(x) + y \phi (x) \tag{3.139}$$

In contrast to the assumptions of the general theory, equation

(3.91), there are also ε^2 and higher order terms in the displacement disturbances for v_t and w_t; they have however no effect on the second variation.

We assume the following strain energy density

$$W = \frac{E}{2}\varepsilon_x^2 + \frac{G}{2}\left(\gamma_{xy}^2 + \gamma_{xz}^2 + \gamma_{yz}^2\right)$$

so that the stress-strain relations are

$$\sigma_x = E\,\varepsilon_x \qquad \sigma_y = 0 \iint \sigma_x = 0$$

$$\tau_{xy} = G\,\gamma_{xy} \qquad \tau_{xz} = G\,\gamma_{xz} \iint \tau_{yz} = G\,\gamma_{yz}$$

and are related to an isotropic material together with the Saint Venant's semi-inverse assumptions $\sigma_y = 0 \quad \sigma_z = 0$.

Calculation of the first order strains

$$\varepsilon_x^{(1)} = \partial\hat{u}/\partial x = \partial u/\partial x$$

$$\gamma_{xy}^{(1)} = \partial\hat{u}/\partial y + \partial\hat{v}/\partial x = \partial u/\partial y + \dot{v} - z\,\dot{\phi}$$

$$\gamma_{xz}^{(1)} = \partial\hat{u}/\partial z + \partial\hat{w}/\partial x = \partial u/\partial z + \dot{w} + y\,\dot{\phi}$$

$$\gamma_{yz}^{(1)} = \partial\hat{v}/\partial z + \partial\hat{w}/\partial y = 0$$

(a dot denotes the derivative of functions of x alone), gives us first term of the stability criterion

$$f_1 = \frac{1}{2}\iint\left\{E\left(\frac{\partial u}{\partial x}\right)^2 + G\left(\frac{\partial u}{\partial y} + \dot{v} - z\,\dot{\phi}\right)^2 + G\left(\frac{\partial u}{\partial z} + \dot{w} + y\,\dot{\phi}\right)^2\right\} dx\,dy\,dz \quad . \qquad (3.139')$$

The second term requires the determination of the coefficient of

$\frac{1}{2}\,\varepsilon^2$ in the axial Green strain measure

$$\frac{\partial u_t}{\partial x} + \frac{1}{2}\left(\frac{\partial u_t}{\partial x}\right)^2 + \frac{1}{2}\left(\frac{\partial v_t}{\partial x}\right)^2 + \frac{1}{2}\left(\frac{\partial w_t}{\partial x}\right)^2$$

this turns out to be

$$\varepsilon_x^{(2)} = \left(\frac{\partial u}{\partial x}\right)^2 + (\dot{v} - z\,\dot{\phi})^2 + (\dot{w} + y\,\dot{\phi})^2 \quad .$$

The second term of the stability criterion is then

$$f_2 = \frac{1}{2}\int \{s(x)\ (\dot{v} - z\,\dot{\phi})^2 + s(x)(\dot{w} + y\,\dot{\phi})^2\}\ dx\ dy\ dz \tag{3.140}$$

where s(x) stands for the axial stress $\sigma_x^{\circ}(x)$ at equilibrium and where the term involving $(\partial u/\partial x)^2$ has been neglected by comparison with the similar one in f_1 because $|s| \ll E$.

3.2.8. *The contribution of warping to the critical disturbance*

We wish to solve the problem up to its formulation in terms of functions of x alone. To this effect the nature of the u(x,y,z) disturbance must be elucidated. Since it concerns only the functional f_1, we have by taking variation of f_1 with respect to this unknown function

$$\int \Big\{ E\,\frac{\partial u}{\partial x}\,\frac{\partial}{\partial x}\,\delta u + G\left(\frac{\partial u}{\partial y} + \dot{v} - z\,\dot{\phi}\right)\frac{\partial}{\partial y}\,\delta u$$
$$+ G\left(\frac{\partial u}{\partial z} + \dot{w} + y\,\dot{\phi}\right)\frac{\partial}{\partial z}\,\delta u \Big\}\ dx\ dy\ dz = 0 \quad .$$

After integration by parts it turns out that u should satisfy the following partial differential equation

$$E \frac{\partial^2 u}{\partial x^2} + G \quad \frac{\partial^2 u}{\partial y^2} + \frac{\partial^2 u}{\partial z^2} \quad = 0 \tag{3.141}$$

the lateral boundary condition

$$\frac{\partial u}{\partial n} + \frac{\partial y}{\partial n} (\dot{v} - z\, \dot{\phi}) + \frac{\partial z}{\partial n} (\dot{w} + y\, \dot{\phi}) = 0 \tag{3.142}$$

where $(\partial y/\partial n,\ \partial z/\partial n)$ are the direction cosines of the outward normal to the cross-sectional boundary, and the end conditions

$$\frac{\partial u}{\partial x}\, \delta u = 0 \qquad \text{in } x = 0 \quad \text{and} \quad x = 1 \quad . \tag{3.143}$$

The lateral boundary condition, which expresses tangency of the shear stresses, suggests that we set

$$u = -\, y\, \dot{v} - z\, \dot{w} + \dot{\phi}\, X\, (y,\ z) + \omega\, (x,\ y,\ z) \ . \tag{3.144}$$

The warping function X would then satisfy

$$\frac{\partial^2 X}{\partial y^2} + \frac{\partial^2 X}{\partial z^2} = 0 \tag{3.145}$$

with

$$\frac{\partial X}{\partial n} = z\, \frac{\partial y}{\partial n} - y\, \frac{\partial z}{\partial n} \tag{3.146}$$

while the "residual" warping function ω would satisfy

$$E \frac{\partial^2 \omega}{\partial x^2} + G \left(\frac{\partial^2 \omega}{\partial y^2} + \frac{\partial^2 \omega}{\partial z^2} \right) = E\, y\, \dddot{v} + E\, z\, \dddot{w} - X\, E\, \dddot{\phi} \tag{3.147}$$

with

$$\partial \omega / \partial n = 0 \quad . \tag{3.148}$$

In (3.144) the first two terms represent rotation of the cross-sections that keep them orthogonal to the neutral fiber, X (y, z) will be later recognized as the free Saint-Venant warping under twist. From (3.147) it appears that the residual warping is induced by the rates of change of curvature and by the third derivative of the torsion.

At this stage it might be possible to further amplify the definition of warping by introducing transverse shear warping functions B (y, z) and C (y, z) and a secondary warping function under twist A (y, z) satisfying differential equations and boundary conditions like

$$G\left(\frac{\partial^2 B}{\partial y^2} + \frac{\partial^2 B}{\partial z^2}\right) = E\,y \qquad G\left(\frac{\partial^2 C}{\partial y^2} + \frac{\partial^2 C}{\partial z^2}\right) = E\,z$$

$$G\left(\frac{\partial^2 A}{\partial y^2} + \frac{\partial^2 A}{\partial z^2}\right) = E\,X$$

$$\partial B/\partial n = 0 \qquad \partial C/\partial n = 0$$

$$\partial A/\partial n = 0 \quad .$$

Then setting $\omega = \dddot{v}\,B + \dddot{w}\,C - \dddot{\phi}\,A$.

One would satisfy (3.147) and (3.148) except for terms involving the fifth derivative of v, w and ϕ.

In what follows we will be satisfied with the approximation that neglects the residual warping ω. Then introducing the moments of inertia

$$I_y = \iint y^2\,dy\,dz \qquad I_z = \iint z^2\,dy\,dz \quad ,$$

the polar moment of inertia

$$I_p = I_y + I_z \quad ,$$

the cross-sectional area

$$A = \iint dy\, dz$$

the integrals over the torsional warping function

$$P = \iint X^2\, dy\, dz \qquad Q = \iint Xy\, dy\, dz \qquad R = \iint Xz\, dy\, dz\ , \tag{3.149}$$

and the free torsional rigidity

$$J = \iint \left\{ \left(\frac{\partial X}{\partial y} - z \right)^2 + \left(\frac{\partial X}{\partial z} + y \right)^2 \right\} dy\, dz\ , \tag{3.150}$$

the stability criterion appears in a form involving only functionals over functions of x alone

$$\int_0^L \frac{E}{2} \left(I_y \ddot{v}^2 + I_z \ddot{w}^2 + P \ddot{\phi}^2 - 2 Q \ddot{v} \ddot{\phi} - 2 R \ddot{w} \ddot{\phi} \right) dx + \int_0^L \frac{G}{2} J \dot{\phi}^2\, dx + \int_0^L \frac{s}{2} \left(A \dot{v}^2 + A \dot{w}^2 + I_p \dot{\phi}^2 \right) dx > 0 \tag{3.151}$$

For $\phi \equiv 0$, $w \equiv 0$, $s(x) = s$ constant, $E I_y = E I$ constant and A constant, this turns into the classical stability condition for the Euler strut

$$\frac{EI}{2} \int_0^L \ddot{v}\ dx + \frac{sA}{2} \int_0^L \dot{v}^2\, dx > 0.$$

For $v \equiv 0$, $w \equiv 0$, P, J and I_p constant, one has the purely torsional buckling problem

$$\frac{EP}{2} \int_0^L \ddot{\phi}^2\, dx + \frac{GJ}{2} \int_0^L \dot{\phi}^2\, dx + \frac{sI_p}{2} \int_0^L \dot{\phi}^2\, dx > 0\ .$$

If one deals with a massive cross section or a box beam, J is of the same order of magnitude as I_p so that in the elastic range where $|s| << G$, the criterion for pure torsion is always satisfied. For the same reason the last term of the last integral in (3.151) can be dropped in the general problem. It is only for thin-walled open sections that the order of magnitude of J becomes so small that purely torsional buckling is a possibility.

The P term takes approximately into account the restraints against torsional warping due to end conditions and changes in the rate of twist presented by the critical disturbance. For instance if the section x = 0 is not allowed to warp, the kinematical boundary condition $\dot{\phi}(0) = 0$ must be applied. This condition is $\ddot{\phi}(0) = 0$ if it is free to warp. We shall now investigate the significance of the Q and R terms that produce coupling between bending and torsion apart from the coupling that could accrue from the end conditions.

It will be necessary therefore to recapitulate briefly the properties of the torsional warping function X (y, z) by reference to the de Saint-Venant torsion problem [3.21].

3.2.9. *The torsional warping function and the flexure-torsion center*

In the Barre de Saint-Venant torsion problem all quantities are independent of x, all stresses are zero except for τ_{xy} and τ_{xz}. The twist rate θ being constant, one usually introduces the Prandtl torsion function $\Theta(y,z)$ in order to free the equilibrium equations in volume

$$\tau_{xz} = - G\theta \frac{\partial \Theta}{\partial y} \qquad \tau_{xy} = + G\theta \frac{\partial \Theta}{\partial z} \quad . \tag{3.152}$$

The complementary energy is then

$$\frac{1}{2G}\iint\left(\tau_{xz}^2 + \tau_{xy}^2\right) dy\ dz = \frac{LG}{2}\theta^2 \iint\left[\left(\frac{\partial\Theta}{\partial y}\right)^2 + \left(\frac{\partial\Theta}{\partial z}\right)^2\right] dy\ dz\ .$$

Setting the displacements $v = 0$ and $w = 0$ in section $x = 0$ and $v = -L\,\theta\, z \quad w = +L\,\theta\, y$ in the end section $x = L$, the complementary potential energy is

$$-\iint\left(-L\theta z\tau_{xy} + L\theta y\tau_{xz}\right) dy\ dz = LG\theta^2 \iint\left(y\frac{\partial\Theta}{\partial y} + z\frac{\partial\Theta}{\partial z}\right) dy\ dz.$$

The total complementary energy (divided by the constant $LG\theta^2$)

$$\iint\left[\frac{1}{2}\left(\frac{\partial\Theta}{\partial y}\right)^2 + \frac{1}{2}\left(\frac{\partial\Theta}{\partial z}\right)^2 + y\frac{\partial\Theta}{\partial y} + z\frac{\partial\Theta}{\partial z}\right] dy\ dz \quad \min \tag{3.153}$$

should be minimum under the additional assumption that the surface equilibrium condition

$$\ell\,\tau_{xy} + m\tau_{xz} = \frac{\partial y}{\partial n}\tau_{xy} + \frac{\partial z}{\partial n}\tau_{xz} = 0$$

be satisfied along the outward limiting contour c_0 and along the inner contours c_i of inside cavities, if the section is multiply-connected.

$$\text{Since} \quad \frac{\partial y}{\partial n} = +\frac{\partial z}{\partial s} \qquad \frac{\partial z}{\partial n} = -\frac{\partial y}{\partial s} \tag{3.154}$$

and in view of (3.152), this is equivalent to requiring

$$\frac{\partial\Theta}{\partial s} = 0 \qquad \text{along } c_0 \text{ and } c_i\ .$$

Without loss of generality one can take

$$\begin{aligned} \Theta &= 0 && \text{along } c_0 \\ \Theta &= \alpha_i && \text{unknown constant along } c_i\ . \end{aligned} \tag{3.155}$$

The minimization of (3.153) produces the Euler equation

$$\frac{\partial^2\Theta}{\partial y^2} + \frac{\partial^2\Theta}{\partial z^2} + 2 = 0 \qquad (3.156)$$

and the natural conditions stemming from the variation of the unknown constants α_i

$$\int_{C_i} \frac{\partial}{\partial n}\left(\Theta + \frac{y^2 + z^2}{2}\right) ds = 0 \; . \qquad (3.157)$$

Altogether (3.156), (3.155) and (3.157) determine uniquely the Prandtl torsion function. Its relation with the warping function $X(y,z)$ under torsion about the x axis is obtained by considering the displacements

$$u = \theta X \qquad v = - xz\theta \qquad w = + zy\theta \qquad (3.158)$$

and computing the stresses

$$\tau_{xy} = G\left(\frac{\partial u}{\partial y} + \frac{\partial v}{\partial x}\right) \qquad \tau_{xz} = G\left(\frac{\partial u}{\partial z} + \frac{\partial w}{\partial x}\right) \; .$$

Comparison with (3.152) yields

$$-\frac{\partial\Theta}{\partial y} = \frac{\partial X}{\partial z} + y \qquad \frac{\partial\Theta}{\partial z} = \frac{\partial X}{\partial y} - z \; . \qquad (3.159)$$

Those equations are easily put in the form of Cauchy-Riemann

$$\frac{\partial X}{\partial y} = \frac{\partial}{\partial z}\left(\Theta + \frac{y^2 + z^2}{2}\right) \qquad \frac{\partial X}{\partial z} = -\frac{\partial}{\partial y}\left(\Theta + \frac{y^2 + z^2}{2}\right) \qquad (3.159')$$

and (3.156) follows by elimination of X, while we obtain (3.145) by elimination of Θ. Also, multiplying (3.159') respectively by $\partial y/\partial n$

and $\partial z/\partial n$, adding and using (3.154)

$$\frac{\partial X}{\partial n} = \frac{\partial \Theta}{\partial s} + z\frac{\partial y}{\partial n} - y\frac{\partial z}{\partial n} \qquad \text{along } c_0 \text{ and } c_i \ .$$

This reduces to (3.146) since Θ is constant along the contours. Finally multiplying the same equations by $\frac{\partial z}{\partial n}$ and $-\frac{\partial y}{\partial n}$ and adding

$$\frac{\partial}{\partial n}\left(\Theta + \frac{y^2 + z^2}{2}\right) = -\frac{\partial X}{\partial s} \quad .$$

Hence conditions (3.157) are transformed into the univaluedness conditions for the axial displacement

$$\int_{c_i} \frac{\partial X}{\partial s}\, ds = 0 \quad . \tag{3.160}$$

If the torsion takes place about the axis $y = y_0$, $z = z_0$, the warping of the cross-section is modified by addition of a rigid body motion

$$X^* = X - z_0\, y + y_0\, z$$

this is easily verified by replacing (3.158) by

$$u = \theta X^* \qquad v = -\,x(z - z_0)\theta \qquad w = x\,(y - y_0)\theta .$$

The actual position (y_0, z_0) of the torsion center under a built in end condition $u \equiv 0$ was defined by Weinstein [3.23] as furnishing a least square property

$$\iint (X - z_0\, y + y_0\, z)^2\, dy\, dz \quad \text{minimum} \quad . \tag{3.161}$$

With due consideration to the assumptions (3.138) and definitions (3.149) it gives

$$Q = z_0 \; I_y \qquad R = - y_0 \; I_z \tag{3.162}$$

Which relate Q and R to the "built in" center of torsion, and also determine implicitly the arbitrary axial translation contained in X. The Weinstein definition may look somewhat arbitrary, it is however backed up by the fact that the same center is obtained as flexure center, that is as the locus where shear forces are to be applied to obtain flexure without torsion [3.21]. Our definition of flexure without torsion is not the unhappy one of Timoshenko [3.22], that gave rise to so many misunderstandings, but that of Trefftz [3.24] whose object was to make flexure and torsion energies additive. Thus if $\hat{\tau}_{xy}$ and $\hat{\tau}_{xz}$ denote the shear stresses due to flexure and (3.152) those due to torsion, the condition of additivity is the vanishing of the interference energy

$$\iint \left(\hat{\tau}_{xy} \frac{\partial \Theta}{\partial z} - \hat{\tau}_{xz} \frac{\partial \Theta}{\partial y} \right) dy \; dz = 0 \quad . \tag{3.163}$$

Integration by parts

$$\int_{c_0, c_i} \left(\hat{\tau}_{xy} \frac{\partial z}{\partial n} - \hat{\tau}_{xz} \frac{\partial y}{\partial n} \right) \Theta \; ds + \iint \left(\frac{\partial \hat{\tau}_{xy}}{\partial y} - \frac{\partial \hat{\tau}_{xz}}{\partial z} \right) dydz = 0.$$

Consideration of (3.154) and of

$$\frac{\partial \hat{\tau}_{xz}}{\partial y} - \frac{\partial \hat{\tau}_{xy}}{\partial z} = G \frac{\partial}{\partial x} \left(\frac{\partial \hat{w}}{\partial y} - \frac{\partial \hat{v}}{\partial z} \right) = 2 \; G \frac{\partial}{\partial x} \; \hat{\omega}_x$$

modify this to read

$$- \int_{c_0, c_i} \Theta(\hat{\tau}_{xy} \; dy + \hat{\tau}_{xz} \; dz) + 2G \frac{d}{dx} \iint \Theta \; \hat{w}_x \; dy \; dz = 0 \quad .$$

Now, considering (3.155), the first term generates the quantities

$$\int_{c_i} \hat{\tau}_{xy}\, dy + \hat{\tau}_{xz}\, dz = \iint_{\Omega_i} \left(\frac{\partial \hat{\tau}_{xz}}{\partial y} - \frac{\partial \hat{\tau}_{xy}}{\partial z} \right) dydz$$

$$= 2\, G \frac{d}{dx} \iint_{\Omega_i} \hat{\omega}_x\, dy\, dz$$

where due account was taken of the negative sense of description of the contours c_i around the inner cavities Ω_i, within which the displacement field $\hat{v}$ and $\hat{w}$ is analytically defined. The condition of additivity of the flexure and torsion energies can thus be placed in a kinematical form discovered by Fraeijs de Veubeke [3.21]

$$\iint_{\Omega} \hat{\Theta}\, \hat{\omega}_x\, dydz = 0 \qquad (3.164)$$

the integral being extended to the whole domain Ω limited by c_0 and where

$\hat{\Theta} = \alpha_i$ inside the cavity Ω_i

$\hat{\Theta} = \Theta$ elsewhere.

The Timoshenko condition corresponds to

$\hat{\Theta} = 0$ inside the cavities

$\hat{\Theta} = 1$ elsewhere.

Using (3.159), condition (3.163) is also

$$\iint \left(y\, \hat{\tau}_{xz} - z\, \hat{\tau}_{zy} \right) dy\, dz + \iint \left(\hat{\tau}_{xy} \frac{\partial x}{\partial y} + \hat{\tau}_{xz} \frac{\partial X}{\partial z} \right) dydz = 0\ .$$

The first term is the moment M shear stresses with respect to the origin; the second will be modified by integration by parts

$$M = - \int_{c_0, c_i} \left(\hat{\tau}_{xy} \frac{\partial y}{\partial n} + \hat{\tau}_{xz} \frac{\partial z}{\partial n} \right) X \, ds$$

$$+ \iint X \left(\frac{\partial \hat{\tau}_{xy}}{\partial y} + \frac{\partial \hat{\tau}_{xz}}{\partial z} \right) dy \, dz \quad .$$

The contour integrals vanish by viture of (3.153), the surface integral can be calculated by reference to the axial equilibrium equation

$$\frac{\partial \hat{\tau}_{xy}}{\partial y} + \frac{\partial \hat{\tau}_{xz}}{\partial z} = - \frac{\partial \hat{\sigma}_x}{\partial x} = - \frac{T_y}{I_y} y - \frac{T_z}{I_z} z$$

where the variation of the axial stress distribution is expressed in terms of the shear loads T_y and T_z. Thus

$$M = - \frac{T_y}{I_y} \iint X \, y \, dy \, dz - \frac{T_z}{I_z} \iint X \, z \, dy \, dz$$

or introducing the coordinates of Weinstein's center of torsion (3.162),

$$M = - z_0 T_y + y_0 T_z$$

showing that this point is also the center of flexure.

REFERENCES

[3.1] FRAEIJS de VEUBEKE, B.,"Large displacement formulations for elastic bodies," *AFFDL-TR-70-34*, March, 1970.

[3.2] MURNAGHAN, F.D., *Finite Deformation of an Elastic Solid*, J. Wiley & Sons, 1951.

[3.3] PRAGER, W., *Introduction to Mechanics of Continua*, Boston: Ginn and Co., 1961.

[3.4] FUNG, Y.C., *Foundations of Solid Mechanics*, Prentice-Hall, 1965.

[3.5] REISSNER, E., "On a variational theorem for finite elastic deformations," *Journal of Mathematics and Physics*, 32, 2-3, 1953, pp. 129-135.

[3.6] WASHIZU, K., *Variational Methods in Elasticity and Plasticity*, Pergamon Press, 1968, Chap. 3.

[3.7] COURANT, R. and HILBERT, D., *Methods of Mathematical Physics*, Vol. I, Chap. IV, Interscience, 1962.

[3.8] MARGUERRE, K., "On the application of the energy method to stability problems," *NACA Technical Memorandum 1138*, October, 1947.

[3.9] THOMPSON, J.M., "A general theory for the equilibrium and stability of discrete conservative systems," *ZAMP*, 20-6, 1969, pp. 797-846.

[3.10] TIMOSHENKO, S. and GERE, J., *Theory of Elastic Stability*, McGraw-Hill, 1961.

[3.11] FRAEIJS de VEUBEKE, B., *The Second Variation Test with Algebraic and Differential Constraints. Advanced Problems and Methods for Space Flight Optimization.* Pergamon Press, 1969, p. 192.

[3.12] GOODIER, J.N., "Torsional and flexural buckling of bars of thin-walled open section under compressive and bending loads," *Transactions ASME*, 64, A-103, 1942.

[3.13] HOFF, N.J., "A strain energy derivation of the torsional-flexural buckling loads of straight columns of thin-walled sections," *Quarterly of Appl. Math.*, I-4, January 1944, pp. 341-345.

[3.14] WAGNER, H., *Verdrehung und Knickung von offenen Profilen*, 25th Anniversary Volume of the Technische Hochschule Danzig, 1929, p. 329 (Verlag A.W. Kaffeman, Danzig 1929).

[3.15] ENGEL, H.L. and GOODIER, J.N., "Measurements of torsional stiffness changes and instability due to tension, compression and bending," *J. of Appl. Mech.*, December 1953, pp. 553-560.

[3.16] MASSONNET, C., "Flambement spatial des barres droites," *CECM Notes Techniques*, B-10.55, Bruxelles, September 1958.

[3.17] MASSONNET, C., "Théorie du flambement par torsion," *I.A. for Bridge Struct. Eng. Publications*, Vol. 6, pp. 211-231.

[3.18] ZARGMAMEE, M.S. and SHAH, J.M., "Stability of space frames," *Jl. of the Eng. Mech. Div., ASCE*, Vol. 94, E12, Proc. Paper 5878, April 1968.

[3.19] PEKOZ, T.B. and WINTER, G., "Torsional-flexural buckling of thin-walled sections under eccentric load," *Jl. ASCE*, 95-ST5, Proc. Paper 6571, 1969.

[3.20] TECZAN, S.S. and OVUNC, B., "An iteration method for the non-linear buckling of framed structures," *Proc. Intern. Conf. on Space Structures*, Univ. Surrey, 1966-F9.

[3.21] FRAEIJS de VEUBEKE, B., "Aspects cinématique et énergétique de la flexion sans torsion," *Acad. Roy. de Belgique, Mémoire* 1657, 1955.

[3.22] TIMOSHENKO, S. and GOODIER, J.N., *Theory of Elasticity*, McGraw Hill, 1951.

[3.23] WEINSTEIN, A., *Quart. Appl. Math.*, V, 1947, p. 97.

[3.24] TREFFTZ, E., *ZAMM*, 15-, 1935, pp. 220-225.

NON LINEAR SHELL THEORY

1. *Coordinate system*

We begin by considering a shell to derive from a flat plate through a finite geometrical transformation. Let x^i $(i = 1,2,3)$ denote cartesian coordinates of a point in the flat plate configuration, $x^3 = 0$ being the middle plane. In the shell configuration the same point has cartesian coordinates y^j (x^1, x^2, x^3). We can conceive the x^i as lagrangian or convective curvilinear coordinates for the shell. From the position vector of a point of the shell

$$\vec{R} = y^i \vec{e}_i$$

where $\vec{e}_i$ are the cartesian base vectors, follows the definition of the base vectors associated with the curvilinear coordinates

$$\vec{g}_j = D_j \vec{R} = (D_j y^i)\, \vec{e}_i \qquad D_j = \frac{\partial}{\partial x^j}$$

and the corresponding fundamental metric tensor

$$g_{jm} = D_j y^i \, D_m y^n \, \vec{e}_i \cdot \vec{e}_n = D_j y^i \, D_m y^i \, .$$

A considerable simplification of the analysis will be obtained by requiring that the base vector $\vec{g}_3$ remains everywhere orthogonal to $\vec{g}_1$ and $\vec{g}_2$. This condition, expressed by

$$g_{3\alpha} = D_3 y^i \ D_\alpha y^i = 0 \qquad (\alpha = 1,2) \tag{1.1}$$

can be satisfied by adopting the following structure for the geometrical transformation of coordinates

$$y^i = r^i \ (x^1, \ x^2) + x^3 \ b^i(x^1, \ x^2) \tag{1.2}$$

and setting

$$b^i \ (D_\alpha r^i + x^3 \ D_\alpha b^i) = 0 \qquad (\alpha = 1,2) \ . \tag{1.3}$$

Since this should hold for all x^3 values within the thickness boundaries of the shell, we must have separately

$$b^i \ D_\alpha r^i = 0 \qquad (\alpha = 1,2) \tag{1.4}$$

$$b^i \ D_\alpha b^i = 0 \qquad (\alpha = 1,2) \ . \tag{1.5}$$

The general solution of (1.4), considered as a linear system in the b^i, is

$$b^i = e_{ijk} \ D_1 r^j \ D_2 r^k \ \ c \ (x^1, \ x^2) \tag{1.6}$$

where e_{ijk} is the permutation symbol and $c \ (x^1 \ , \ x^2)$ an arbitrary function. Simultaneously (1.5) requires

$$g_{33} = b^i \ b^i = \text{constant} \ . \tag{1.7}$$

Without sacrificing generality, the constant can be set equal to unity. Then, substitution of (1.6) into (1.7) determines the function c through the equation

$$1 = c^2 \, e_{ijk} \, e_{ipq} \, D_1 r^j \, D_2 r^k \, D_1 r^p \, D_2 r^q$$

$$= c^2 \, (D_1 r^p \, D_1 r^p \, D_2 r^q \, D_2 r^q - D_1 r^p \, D_1 r^q \, D_2 r^p \, D_2 r^q).$$

In conclusion, the functions b^i are completely determined by the geometrical transformation $y^i = r^i \, (x^1, x^2)$ of the middle plane of the plate into the middle surface of the shell. This derivation justifies the classical choice of curvilinear coordinates for shells:

$$\vec{R} = \vec{r} \, (x^1, x^2) + x^3 \, \vec{a}_3 \, (x^1, x^2) \tag{1.8}$$

where $\vec{r}$ is the position vector of a middle surface point, and where $\vec{a}_3$ can be chosen such that

$$\vec{a}_3 \cdot D_\alpha \vec{r} = 0 \qquad (\alpha = 1,2) \quad . \tag{1.9}$$

$$\vec{a}_3 \cdot \vec{a}_3 = 1 \quad . \tag{1.10}$$

2. *Base vectors for surface and space tensors*

In what follows the greek letter indices will always refer to the surface curvilinear coordinates x^1 and x^2, the latin indices to the complete set of space curvilinear coordinates x^1, x^2 and x^3.

The base vectors, tangent to the middle plane $x^3 = 0$, are given by

$$\vec{a}_\alpha = D_\alpha \vec{r} \quad . \tag{2.1}$$

They generate the surface metric tensor

$$a_{\alpha\beta} = \vec{a}_\alpha \cdot \vec{a}_\beta = a_{\beta\alpha} \qquad (2.2)$$

and are the coefficients of the first fundamental form

$$d\vec{r} \cdot d\vec{r} = a_{\alpha\beta} dx^\alpha \, dx^\beta \quad . \qquad (2.3)$$

The conjugate base vectors $\vec{a}^\alpha$ are related by

$$\vec{a}_\beta = a_{\alpha\beta}\vec{a}^\alpha \quad \text{or} \quad \vec{a}^\alpha \cdot \vec{a}_\gamma = \delta^\alpha_\gamma \quad \text{(Kronecker symbol)} \qquad (2.4)$$

and generate the conjugate surface metric tensor

$$a^{\alpha\beta} = \vec{a}^\alpha \cdot \vec{a}^\beta \quad . \qquad (2.5)$$

The surface metric tensors are used to raise or lower indices of surface tensors.

The determinant of the surface metric will be denoted

$$a = \left| a_{\alpha\beta} \right| \qquad \frac{1}{a} = \left| a^{\alpha\beta} \right| \quad . \qquad (2.6)$$

The first fundamental form being positive definite a is a positive quantity. Let $\varepsilon_{\alpha\beta}$ denote a twice covariant antisymmetric surface tensor, it has only one strict component $\varepsilon_{12} = \varepsilon$ ($\varepsilon_{21} = -\varepsilon$). Hence if $e_{\alpha\beta} = e^{\alpha\beta}$ denotes the corresponding permutation symbol ($e_{12} = 1$, $e_{21} = -1$, $e_{11} = e_{22} = 0$) we have

$$\varepsilon_{\alpha\beta} = \varepsilon \; e_{\alpha\beta}.$$

The twice contravariant components of this tensor are related by

$$\varepsilon_{\alpha\beta} = a_{\alpha\gamma} \, a_{\beta\mu} \, \varepsilon^{\gamma\mu} = (a_{\alpha\gamma} \, a_{\beta\mu} \, e^{\gamma\mu})\sigma$$

where σ is the strict component of $\varepsilon^{\gamma\mu}$ ($\sigma = \varepsilon^{12}$). The quantity between brackets is $e_{\alpha\beta}a$. Hence, comparing the two expressions of $\varepsilon_{\alpha\beta}$

$$\varepsilon = \sigma a \quad . \tag{2.7}$$

Further we have

$$\varepsilon^{\alpha\beta} \varepsilon_{\alpha\beta} = \sigma\varepsilon \, e^{\alpha\beta} e_{\alpha\beta} = 2\sigma\varepsilon$$

and since this is obviously invariant, it can be equated to a fixed quantity. From the invariant $\sigma\varepsilon = \kappa$ and (2.7) we conclude that all antisymmetric second order surface tensors are of the form

$$\varepsilon_{\alpha\beta} = \sqrt{\kappa a}\, e_{\alpha\beta} \qquad \varepsilon^{\alpha\beta} = \sqrt{\frac{\kappa}{a}}\, e^{\alpha\beta}$$

where a is the determinant of the surface metric tensor. The usual ε-tensor for the surface corresponds to the choice $\kappa = 1$:

$$\varepsilon_{\alpha\beta} = -\varepsilon_{\beta\alpha} \qquad \varepsilon_{12} = \sqrt{a} \tag{2.8}$$

$$\varepsilon^{\alpha\beta} = -\varepsilon^{\beta\alpha} \qquad \varepsilon^{12} = 1/\sqrt{a} \quad .$$

Like the space metric tensor it enjoys the property that its twice covariant and twice contravariant representations are reciprocal: $\varepsilon^{\alpha\beta}\varepsilon_{\beta\gamma} = \delta^{\alpha}_{\gamma}$. The unit vector $\vec{a}_3 = \vec{a}^3$ being normal to the surface

$$\vec{a}_3 \cdot \vec{a}^{\alpha} = 0 \qquad \vec{a}_3 \cdot \vec{a}_{\alpha} = 0 \qquad \vec{a}_3 \cdot \vec{a}_3 = 1 \quad . \tag{2.9}$$

It can be defined by

$$\vec{a}_3 = \varepsilon^{\alpha\beta}\, \vec{a}_{\alpha} \times \vec{a}_{\beta} \qquad \vec{a}_{\alpha} \times \vec{a}_{\beta} = \varepsilon_{\alpha\beta}\, \vec{a}_3 \tag{2.10}$$

as is easily verified by scalar multiplication by $\vec{a}_{\gamma}$ and $\vec{a}_3$.

Turning now to the base vectors for space tensors, they are defined by

$$\vec{g}_i = D_i\vec{R} \quad . \tag{2.11}$$

In view of (1.8) they are related to the surface base vectors through the following equations

$$\vec{g}_\alpha = \vec{a}_\alpha + x^3 D_\alpha \vec{a}_3 \tag{2.12}$$

$$\vec{g}_3 = \vec{a}_3$$

and generate the fundamental metric tensor

$$\vec{g}_\alpha \cdot \vec{g}_\beta = g_{\alpha\beta} = a_{\alpha\beta} - 2x^3 b_{\alpha\beta} + (x^3)^2 c_{\alpha\beta} \tag{2.13}$$

$$\vec{g}_\alpha \cdot \vec{g}_3 = g_{\alpha 3} = 0 \tag{2.14}$$

$$\vec{g}_3 \cdot \vec{g}_3 = g_{33} = 1 \tag{2.15}$$

(2.14) and (2.15) follow directly from (2.9). (2.13) contains the definitions of the following, obviously symmetric, surface tensors

$$-2b_{\alpha\beta} = \vec{a}_\alpha \cdot D_\beta \vec{a}_3 + \vec{a}_\beta \cdot D_\alpha \vec{a}_3 \tag{2.16}$$

$$c_{\alpha\beta} = D_\alpha \vec{a}_3 \cdot D_\beta \vec{a}_3 \quad . \tag{2.17}$$

Simpler formulas for the first are obtained by transformations such as

$$\vec{a}_\alpha \cdot D_\beta \vec{a}_3 = D_\beta (\vec{a}_\alpha \cdot \vec{a}_3) - \vec{a}_3 \cdot D_\beta \vec{a}_\alpha = - \vec{a}_3 \cdot D_\beta \vec{a}_\alpha$$

justified by (2.9) and also

$$D_\beta \vec{a}_\alpha = D_\beta D_\alpha \vec{r} = D_\alpha \vec{a}_\beta \tag{2.18}$$

there comes

$$b_{\alpha\beta} = \vec{a}_3 \cdot D_\beta \vec{a}_\alpha = - \vec{a}_\alpha \cdot D_\beta \vec{a}_3 \tag{2.19}$$

The elements of this tensor are the coefficients of the second fundamental form

$$- d\vec{r} \cdot d\vec{a}_3 = b_{\alpha\beta}\, dx^\alpha\, dx^\beta .$$

They appear also when expressing the surface derivative of the unit normal vector to the surface

$$D_\beta \vec{a}_3 = - b_{\alpha\beta}\, \vec{a}^\alpha . \tag{2.20}$$

This formula is immediately justified by (2.4), (2.9) and (2.19) when taking scalar multiplication with the surface base vectors. Substituting (2.20) into (2.17) it is found that the coefficients of the third fundamental form

$$d\vec{a}_3 \cdot d\vec{a}_3 = c_{\alpha\beta}\, dx^\alpha\, dx^\beta \tag{2.21}$$

are expressible in terms of those of the second

$$c_{\alpha\beta} = b_{\gamma\alpha}\, b_{\varepsilon\beta}\, \vec{a}^\gamma \cdot \vec{a}^\varepsilon = b_{\gamma\alpha}\, b_{\varepsilon\beta}\, a^{\gamma\varepsilon} = b^\varepsilon_\alpha\, b_{\varepsilon\beta} . \tag{2.22}$$

It is obvious from the geometrical significance of (2.20) or from its equivalent formulation in mixed components, known as the differential equation of Weingarten,

$$D_\beta \vec{a}_3 = - b^\alpha_\beta\, \vec{a}_\alpha \tag{2.23}$$

that the surface tensor generated by the second fundamental form characterizes completely the curvature of the middle surface. With respect to a change of curvilinear surface coordinates, the fundamental invariants of this tensor are

$$H = \frac{1}{2} b^{\alpha}_{\alpha} \tag{2.24}$$

the so-called mean curvature, and

$$K = \frac{1}{2} \varepsilon^{\alpha\beta} \varepsilon_{\lambda\mu} b^{\lambda}_{\alpha} b^{\mu}_{\beta} = b^{1}_{1} b^{2}_{2} - b^{2}_{1} b^{1}_{2} \tag{2.25}$$

the Gaussian curvature.

3. *Covariant surface derivatives*

The surface Christoffel symbols are defined by the following decomposition of the derivatives with respect to the surface curvilinear coordinates of the surface base vectors themselves:

$$D_{\beta}\vec{a}_{\alpha} = \overset{\circ}{\Gamma}{}^{\gamma}_{\beta\alpha} \vec{a}_{\gamma} + \overset{\circ}{\Gamma}{}^{3}_{\beta\alpha} \vec{a}_{3} \quad . \tag{3.1}$$

In view of (2.18) the Christoffel symbols are symmetrical with respect to the lower subscripts. Scalar multiplication of (3.1) by $\vec{a}_3$ and consideration of the orthogonality of $\vec{a}_3$ with respect to the $\vec{a}_{\gamma}$ and of (1.10) yields

$$\overset{\circ}{\Gamma}{}_{\beta}{}^{3}{}_{\alpha} = \vec{a}_3 \cdot D_{\beta}\vec{a}_{\alpha} = b_{\alpha\beta} \tag{3.2}$$

as shown by (2.19). Hence (3.1) will be replaced by

$$D_{\beta}\vec{a}_{\alpha} = \overset{\circ}{\Gamma}{}_{\beta}{}^{\gamma}{}_{\alpha} \vec{a}_{\gamma} + b_{\alpha\beta} \vec{a}_{3} \quad . \tag{3.3}$$

In this form it is known as the differential equation of Gauss.

The remaining Christoffel symbols can be calculated from the derivations of the fundamental metric surface tensor. Indeed, scalar multiplication of (3.1) by $\vec{a}_\varepsilon$ produces

$$\vec{a}_\varepsilon \cdot D_\beta \vec{a}_\alpha = \overset{\circ}{\Gamma}{}_\beta{}^\gamma{}_\alpha \, a_{\gamma\varepsilon} = \overset{\circ}{\Gamma}{}_{\beta\varepsilon\alpha} \, .$$

However, the left hand side is also

$$D_\beta(\vec{a}_\varepsilon \cdot \vec{a}_\alpha) - \vec{a}_\alpha \cdot D_\beta \vec{a}_\varepsilon = D_\beta a_{\varepsilon\alpha} - \overset{\circ}{\Gamma}{}_{\beta\alpha\varepsilon} \, .$$

Hence by comparison,

$$D_\beta a_{\varepsilon\alpha} = \overset{\circ}{\Gamma}{}_{\beta\alpha\varepsilon} + \overset{\circ}{\Gamma}{}_{\beta\varepsilon\alpha} \, . \tag{3.4}$$

Similarly, by changing the order of the subscripts,

$$D_\varepsilon a_{\alpha\beta} = \overset{\circ}{\Gamma}{}_{\varepsilon\beta\alpha} + \overset{\circ}{\Gamma}{}_{\varepsilon\alpha\beta}$$

$$D_\alpha a_{\beta\varepsilon} = \overset{\circ}{\Gamma}{}_{\alpha\varepsilon\beta} + \overset{\circ}{\Gamma}{}_{\alpha\beta\varepsilon} \, .$$

Combining those relations and using the symmetry properties of the Christoffel symbols

$$D_\beta a_{\varepsilon\alpha} - D_\varepsilon a_{\alpha\beta} + D_\alpha a_{\beta\varepsilon} = 2\, \overset{\circ}{\Gamma}{}_{\beta\varepsilon\alpha} \, . \tag{3.5}$$

In this formula, the central subscript can further be raised by multiplication by the reciprocal surface metric tensor $a^{\varepsilon\gamma}$, producing

$$2\, \overset{\circ}{\Gamma}{}_\beta{}^\gamma{}_\alpha = a^{\varepsilon\gamma} \, (D_\beta a_{\varepsilon\alpha} - D_\varepsilon a_{\alpha\beta} + D_\alpha a_{\beta\varepsilon}) \, . \tag{3.6}$$

Consider now a vector defined on the middle surface

$$\vec{u} = u^\alpha \, (x^1, x^2) \, \vec{a}_\alpha + u^3(x^1, x^2) \, \vec{a}_3$$

and calculate its surface derivative

$$D_\beta \vec{u} = (D_\beta u^\alpha)\, \vec{a}_\alpha + (D_\beta u^3)\, \vec{a}_3 + u^\alpha D_\beta \vec{a}_\alpha + u^3 D_\beta \vec{a}_3 \quad .$$

Substituting (2.23) and (3.3) and reorganizing the terms

$$D_\beta \vec{u} = (u^\alpha || - b^\alpha_\beta u^3)\, \vec{a}_\alpha + (D_\beta u^3 + b_{\alpha\beta} u^\alpha)\, \vec{a}_3 \tag{3.7}$$

where

$$u^\alpha||_\beta = D_\beta u^\alpha + u^\gamma \overset{\circ}{\Gamma}{}_\beta{}^\alpha{}_\gamma \tag{3.8}$$

is, by definition, the covariant surface derivative of u^α.

In a change of curvilinear coordinates on the middle surface, $\vec{a}_3$ and u^3 are invariant, $D_\beta \vec{u}$, $b^\alpha_\beta \vec{a}_\alpha$, $D_\beta u^3$ and $b_{\alpha\beta} u^\alpha$ are obviously once covariant and the same must then be true for $u^\alpha ||_\beta \vec{a}_\alpha$. Consequently $u^\alpha||_\beta$ is a mixed surface tensor.

Let us now look after a formula analogous to (3.1) for the surface derivatives of the dual surface base vectors. Scalar multiplication of (3.1) by $\vec{a}^\varepsilon$ yields in view of (2.4) and (2.9)

$$\overset{\circ}{\Gamma}{}_\beta{}^\varepsilon{}_\alpha = \vec{a}^\varepsilon \cdot D_\beta \vec{a}_\alpha = D_\beta \delta^\varepsilon_\alpha - \vec{a}_\alpha \cdot D_\beta \vec{a}^\varepsilon = \vec{a}_\alpha \cdot D_\beta \vec{a}^\varepsilon \tag{3.9}$$

while $\quad \vec{a}_3 \cdot D_\beta \vec{a}^\varepsilon = - \vec{a}^\varepsilon \cdot D_\beta \vec{a}_3 \quad ,$

or, considering (2.23)

$$\vec{a}_3 \cdot D_\beta \vec{a}^\varepsilon = b^\alpha_\beta\, \vec{a}_\alpha \cdot \vec{a}^\varepsilon = b^\alpha_\beta \delta^\varepsilon_\alpha = b^\varepsilon_\beta \quad . \tag{3.10}$$

Finally (3.9) and (3.10) are equivalent to the required formula

$$D_\beta \vec{a}^\varepsilon = - \overset{\circ}{\Gamma}{}_\beta{}^\varepsilon{}_\alpha\, \vec{a}^\alpha + b^\varepsilon_\beta\, \vec{a}^3 \quad . \tag{3.11}$$

If we now express the surface vector u by its covariant components

$$\vec{u} = u_\varepsilon \vec{a}^3 + u_3 \vec{a}^3$$

and use the results (3.11) and (2.20) ($\vec{a}^3 = \vec{a}_3$) there comes

$$D_\beta \vec{u} = \left(u_{\alpha||\beta} - u_3 b_{\alpha\beta}\right) \vec{a}^\alpha + \left(D_\beta u_3 + b^\alpha_\beta u_\alpha\right) \vec{a}^3 \tag{3.12}$$

where

$$u_{\alpha||\beta} = D_\beta u_\alpha - \overset{\circ}{\Gamma}{}^\gamma_{\beta\alpha} u_\gamma \tag{3.13}$$

is a twice covariant surface tensor: the covariant surface derivative of u_α. Equations (3.8) and (3.13) can be generalized to express the covariant surface derivative of a surface tensor of any order. Take the particular case of the twice covariant fundamental metric tensor $a_{\varepsilon\alpha}$:

$$a_{\varepsilon\alpha||\beta} = D_\beta a_{\varepsilon\alpha} - \overset{\circ}{\Gamma}{}_\beta{}^\gamma{}_\varepsilon a_{\gamma\alpha} - \overset{\circ}{\Gamma}{}_\beta{}^\gamma{}_\alpha a_{\varepsilon\gamma} .$$

This is equivalent to

$$a_{\varepsilon\alpha||\beta} = D_\beta a_{\varepsilon\alpha} - \overset{\circ}{\Gamma}_{\beta\alpha\varepsilon} - \overset{\circ}{\Gamma}_{\beta\alpha\varepsilon} = 0 \tag{3.14}$$

by virtue of (3.4). This result is known as Ricci's Lemma. The same result is easily established by direct calculation for $a^{\alpha\beta}$, $\varepsilon^{\alpha\beta}$ and $\varepsilon_{\alpha\beta}$. It shows that those tensors behave as constants when calculating surface covariant derivatives.

Another important result concerns the covariant surface differentiation of the $b_{\alpha\beta}$ tensor. From direct application of (3.13)

$$b_{\alpha\beta||\gamma} = D_\gamma b_{\alpha\beta} - \overset{\circ}{\Gamma}{}_\gamma{}^\varepsilon{}_\alpha b_{\varepsilon\beta} - \overset{\circ}{\Gamma}{}_\gamma{}^\varepsilon{}_\beta b_{\alpha\varepsilon} \quad . \tag{3.15}$$

If we now differentiate (2.20) and use (3.11)

$$D_\gamma D_\beta \vec{a}_3 = - D_\gamma b_{\alpha\beta} \vec{a}^\alpha - b_{\alpha\beta} \left(- \overset{\circ}{\Gamma}{}^\alpha_{\gamma\varepsilon} \vec{a}^\varepsilon + b^\alpha_\gamma \vec{a}^3 \right)$$

the result can be put in the form

$$D_\gamma D_\beta \vec{a}_3 = - b_{\alpha\beta||\gamma} \vec{a}^\alpha - \overset{\circ}{\Gamma}{}_{\gamma\ \beta}^{\ \varepsilon} b_{\alpha\varepsilon} \vec{a}^\alpha - c_{\gamma\beta} \vec{a}^3 \quad .$$

The left-hand side and the two last terms of the right-hand side are symmetrical with respect to the subscripts β and γ. The same must then be true of the remaining term. This establishes the Mainardi-Codazzi equation

$$b_{\alpha\beta||\gamma} = b_{\alpha\gamma||\beta} \quad . \tag{3.16}$$

Should we have started from (2.23) instead of (2.20) we would have obtained

$$b^\varepsilon_{\beta||\gamma} = b^\varepsilon_{\gamma||\beta} \quad . \tag{3.17}$$

That this is an equivalent formulation of the Mainardi-Codazzi equation is obvious when (3.16) is multiplied by $a^{\alpha\varepsilon}$ and use made of Ricci's Lemma. The Mainardi-Codazzi equation shows that the curvature tensor $b_{\alpha\beta}$ of a surface imbedded in euclidean space is not independent of the surface metric. From this viewpoint we should consider the system of differential equations of Weingarten (2.23) and Gauss (3.3) and investigate all the integrability conditions to be satisfied for the existence of a field of surface base vectors. In this light the Mainardi-Codazzi equation appears as the integrability condition for the unit normal vector.

The integrability conditions for the other base vectors is obtained by a similar statement of symmetry of second differentials.

Differentiating (3.3) and repeating the use of (2.23) and (3.3)

$$D_\gamma D_\beta \vec{a}_\alpha = \left\{ D_\gamma \overset{\circ}{\Gamma}{}_{\beta\ \alpha}^{\ \phi} + \overset{\circ}{\Gamma}{}_{\beta\ \alpha}^{\ \varepsilon} \overset{\circ}{\Gamma}{}_{\gamma\ \varepsilon}^{\ \phi} - b_{\alpha\beta} b_\gamma^{\ \phi} \right\} \vec{a}_\phi + \left\{ b_{\gamma\varepsilon} \overset{\circ}{\Gamma}{}_{\beta\ \alpha}^{\ \varepsilon} + D_\gamma b_{\alpha\beta} \right\} \vec{a}_3 \quad .$$

Considering the symmetry of the left-hand side, the brackets in the right-hand side should also be symmetrical with respect to the interchange of β and α. That this is true for the bracket of $\vec{a}_3$ is immediately seen by substituting its second term from (3.15), producing

$$b_{\alpha\beta||\gamma} + \overset{\circ}{\Gamma}_{\beta\ \alpha}^{\ \varepsilon} b_{\gamma\varepsilon} + \overset{\circ}{\Gamma}_{\gamma\ \alpha}^{\ \varepsilon} b_{\beta\varepsilon} + \overset{\circ}{\Gamma}_{\gamma\ \beta}^{\ \varepsilon} b_{\alpha\varepsilon} \quad .$$

While the group of the 3 last terms is clearly symmetrical with respect to the interchange of β and γ, the same is true of the first term by virtue of the Mainardi-Codazzi condition. The symmetry condition applied to the first bracket yields after re-arranging terms,

$$b_{\alpha\beta}b_{\gamma}^{\phi} - b_{\alpha\gamma}b_{\beta}^{\phi} = R^{\phi}_{\cdot\alpha\gamma\beta}$$

where

$$R^{\phi}_{\cdot\alpha\gamma\beta} = D_{\gamma}\ \overset{\circ}{\Gamma}_{\beta\ \alpha}^{\ \phi} - D_{\beta}\ \overset{\circ}{\Gamma}_{\gamma\ \alpha}^{\ \phi} + \overset{\circ}{\Gamma}_{\beta\ \alpha}^{\ \varepsilon}\ \overset{\circ}{\Gamma}_{\gamma\ \varepsilon}^{\ \phi} - \overset{\circ}{\Gamma}_{\gamma\ \alpha}^{\ \varepsilon}\ \overset{\circ}{\Gamma}_{\beta\ \varepsilon}^{\ \phi} \tag{3.18}$$

depends only on the surface metric. Premultiplication by $a^{\alpha\psi}$, producing

$$b_{\beta}^{\psi}\ b_{\gamma}^{\phi} - b_{\gamma}^{\psi}\ b_{\beta}^{\phi} = R^{\phi\psi}_{\cdot\cdot\gamma\beta} \tag{3.19}$$

brings out clearly the antisymmetry with respect to the pair of upper and to the pair of lower indices. As a consequence there is only one scalar condition

$$K = b_1^1\ b_2^2 - b_2^1\ b_1^2 = R^{21}_{\cdot\cdot 21} \quad . \tag{3.20}$$

This is a celebrated equation discovered by Gauss from which it appears that the Gaussian curvature depends on the surface metric only and is consequently accessible to measurements made on the surface only. It appears here as the second integrability condition placed on the curvature tensor $b_{\alpha\beta}$.

4. *Covariant space derivatives*

The space Christoffel symbols appear as the covariant or contravariant components of derivatives of the space base vectors

$$D_j \vec{g}_i = \Gamma_{j\ i}^{\ h} \vec{g}_h = \Gamma_{jhi} \vec{g}^h \quad . \tag{4.1}$$

Since $D_j \vec{g}_i = D_j D_i \vec{R} = D_i \vec{g}_j$ they are also symmetrical with respect to their first and last subscripts. Furthermore, from $D_j(\vec{g}_i \cdot \vec{g}^k) = D_j \delta_i^k = 0$, we obtain

$$\vec{g}_i \cdot D_j \vec{g}^k = - \vec{g}^k \cdot D_j \vec{g}_i = - \Gamma_{jhi} \vec{g}^k \cdot \vec{g}^h = - \Gamma_{jhi} g^{kh} = - \Gamma_{j\ i}^{\ k}$$

or equivalently,

$$D_j \vec{g}^k = - \Gamma_{j\ i}^{\ k} \vec{g}^i \quad . \tag{4.2}$$

Then

$$D_j g_{ih} = D_j(\vec{g}_i \cdot \vec{g}_h) = \vec{g}_i \cdot D_j \vec{g}_h + \vec{g}_h \cdot D_j \vec{g}_i = \Gamma_{jih} + \Gamma_{jhi} \tag{4.3}$$

from which follows by the same type of linear combinations as used in the surface case:

$$2\Gamma_{jih} = D_j g_{ih} - D_i g_{jh} + D_h g_{ji} \tag{4.4}$$

$$\Gamma_{j\ h}^{\ k} = g^{ik} \Gamma_{jih} \quad .$$

Consider now a space vector defined by its contravariant components

$$\vec{u} = u^i \vec{g}_i \quad .$$

Its derivative formulated by means of (4.1) is

$$D_j \vec{u} = u^i|_j \, \vec{g}_i \tag{4.5}$$

where

$$u^i|_j = D_j u^i + \Gamma_j{}^i{}_k \, u^k \tag{4.6}$$

is defined as the covariant space derivative of u^i, a mixed space tensor. The same vector defined by its covariant components $u = u_k \vec{g}^k$ yields for the same derivative, using (4.2)

$$D_j \vec{u} = u_{k|j} \, \vec{g}^k \tag{4.7}$$

with

$$u_{k|j} = D_j u_k - \Gamma_j{}^m{}_k \, u_m \quad . \tag{4.8}$$

The extension of formulas (4.6) and (4.8) to form covariant derivatives of tensors of higher order is immediate.

Clearly, since (4.5) and (4.7) represent the same quantity, expanded first in the system of covariant base vectors and later in the dual system, we must have

$$u_{k|j} = g_{ki} \, u^i|_j \quad .$$

However the left-hand side is also

$$(g_{ki} \, u^i)_{|j} = g_{ki|j} \, u^i + g_{ki} u^i{}_{|j} \quad .$$

From the comparison and the arbitrariness of the vector u it can be concluded that

$$g_{ki|j} = 0 \quad . \tag{4.9}$$

This is Ricci's Lemma for the space case. Its direct computational proof relies on (4.3)

$$g_{ih|j} = D_j g_{ih} - \Gamma_{j\ i}^{\ m} g_{mh} - \Gamma_{j\ h}^{\ m} g_{im}$$

$$= D_j g_{ih} - \Gamma_{jhi} - \Gamma_{jih} = 0 \quad .$$

Let $e_{ijk} = e^{ijk}$ denote the permutation symbol, equal to unity if ijk is some even permutation of the sequence 123, to minus unity if the permutation is odd, to zero for all other cases. Any completely antisymmetric tensor is of the form

$$\varepsilon_{ijk} = \varepsilon \ e_{ijk}$$

where $\varepsilon = \varepsilon_{123}$ is the only strict component of the tensor. Similarly the contravariant form of this tensor will have a single strict component

$$\varepsilon^{mnp} = \sigma \ e^{mnp} \qquad \sigma = \varepsilon^{123} \quad .$$

The relation between the covariant and contravariant forms

$$\varepsilon_{ijk} = \varepsilon^{mnp} g_{mi} g_{nj} g_{pk} = \sigma \left(e^{mnp} g_{mi} g_{nj} g_{pk} \right)$$

becomes equivalent to $\varepsilon = \sigma g$, when the quantity between brackets is recognized as the expansion of the determinant g of the fundamental metric tensor

$$e^{mnp} g_{mi} g_{nj} g_{pk} = e_{ijk} g \quad .$$

Furthermore, the product of the strict components is easily recognized to be an invariant

$$\sigma\varepsilon = \frac{1}{6}\,\varepsilon^{ijk}\,\varepsilon_{ijk} \quad .$$

All completely antisymmetric tensors of this type differ only by the value of this invariant. For the choice $\sigma\varepsilon = 1$, corresponding to

$$\varepsilon = \sqrt{g} \qquad\qquad \sigma = 1/\sqrt{g}$$

we have the usual definition of the space ε- tensor:

$$\varepsilon_{ijk} = \sqrt{g}\, e_{ijk} \qquad \varepsilon^{mnp} = \frac{1}{\sqrt{g}}\, e^{mnp} \quad . \tag{4.10}$$

An equivalent definition is

$$\varepsilon_{ijk} = (\vec{g}_i \times \vec{g}_j) \cdot \vec{g}_k \qquad \varepsilon^{mnp} = (\vec{g}^m \times \vec{g}^n) \cdot \vec{g}^p \tag{4.11}$$

justified by the complete antisymmetry of the mixed products and the fact that when the base vectors are cartesian, $g = 1$ and (4.11) reduces effectively to the permutation symbol.

In conjunction with (4.1) or (4.2) the expressions (4.11) are convenient to compute derivatives of the ε tensor

$$\begin{aligned} D_h\, \varepsilon_{ijk} &= (D_h\vec{g}_i \times \vec{g}_j) \cdot \vec{g}_k + (\vec{g}_i \times D_h\vec{g}_j) \cdot \vec{g}_k + (\vec{g}_i \times \vec{g}_j) \cdot D_h\vec{g}_k \\ &= (\vec{g}_j \times \vec{g}_k) \cdot D_h\vec{g}_i + (\vec{g}_k \times \vec{g}_i) \cdot D_h\vec{g}_j + (\vec{g}_i \times \vec{g}_j) \cdot D_h\vec{g}_k \\ &= \varepsilon_{jkm}\, \Gamma_{h\ i}^{\ m} + \varepsilon_{kim}\, \Gamma_{h\ j}^{\ m} + \varepsilon_{ijm}\, \Gamma_{h\ k}^{\ m} \quad . \end{aligned} \tag{4.12}$$

This result is equivalent to stating that the ε-tensor also obeys Ricci's Lemma

$$\varepsilon_{ijk|h} = 0 \quad . \tag{4.13}$$

In fact we need only consider the case i = 1, j = 2, k = 3, whence, retaining only the non zero terms on the right hand side

$$D_h \varepsilon_{123} = \varepsilon_{231} \Gamma_{h\ 1}^{\ 1} + \varepsilon_{321} \Gamma_{h\ 2}^{\ 2} + \varepsilon_{123} \Gamma_{h\ 3}^{\ 3} = \varepsilon_{123} \Gamma_{h\ m}^{\ m} \tag{4.14}$$

a result that will be useful later.

5. *The structure of space metric in terms of surface metric for a shell*

Between space and surface base vectors exist the following relations deduced from (2.12) and (2.23)

$$\vec{g}_\alpha = \mu_\alpha^\beta \vec{a}_\beta \tag{5.1}$$

$$\vec{g}_3 = \vec{a}_3 \tag{5.2}$$

where although dependent on x^3,

$$\mu_\alpha^\beta = \delta_\alpha^\beta - x^3 b_\alpha^\beta \tag{5.3}$$

is a mixed surface tensor, i.e. exhibiting the covariant and contravariant behavior associated with a change of curvilinear coordinates on the surface. The correspondant relations between metric tensors are

$$g_{\alpha\beta} = \mu_\alpha^\beta \mu_\beta^\lambda a_{\gamma\lambda} \tag{5.4}$$

equivalent to (2.13).

Taking the determinant on both sides and noting that

$g = |g_{ij}| = |g_{\alpha\beta}|$ in view of (2.14) and (2.15)

$$g = \mu^2 a \tag{5.5}$$

where

$$\mu = |\mu^\gamma_\alpha| = 1 - 2\,H\,x^3 + (x^3)^2\,K \tag{5.6}$$

as is easily verified by direct expansion of this determinant from (5.3) and consideration of the definitions (2.24) and (2.25). We also need the tensor reciprocal to μ^γ_α and denote it, following Naghdi [1], by

$$(\mu^{-1})^\beta_\alpha = \frac{1}{\mu}\, e_{\alpha\gamma} e^{\beta\varepsilon}\, \mu^\gamma_\varepsilon = \frac{1}{\mu}\, \{\delta^\beta_\alpha + x^3\,(b^\beta_\alpha - 2\,H\,\delta^\beta_\alpha)\} \tag{5.7}$$

Then from (5.1) in succession

$$(\mu^{-1})^\alpha_\gamma\, \vec{g}_\alpha = \delta^\beta_\gamma\, \vec{a}_\beta = \vec{a}_\gamma$$

$$(\mu^{-1})^\alpha_\gamma\, \vec{g}_\alpha \cdot \vec{g}^\varepsilon = (\mu^{-1})^\alpha_\gamma\, \delta^\varepsilon_\alpha = (\mu^{-1})^\varepsilon_\gamma = \vec{a}_\gamma \cdot \vec{g}^\varepsilon$$

and finally

$$\vec{g}^\varepsilon = (\mu^{-1})^\varepsilon_\gamma\, \vec{a}^\gamma \tag{5.8}$$

while

$$\vec{g}^3 = \vec{a}^3\ (= \vec{a}_3 = \vec{g}_3)\ . \tag{5.9}$$

The reciprocal space metric tensor becomes

$$g^{\varepsilon\phi} = (\mu^{-1})^\varepsilon_\gamma\, (\mu^{-1})^\phi_\lambda\, a^{\lambda\gamma} \quad . \tag{5.10}$$

$$g^3 = 0 \qquad g^{33} = 1 \quad . \tag{5.11}$$

We now proceed to calculate the space Christoffel symbols:

$$\Gamma_\alpha{}^\gamma{}_\beta = \vec{g}^\gamma \cdot D_\alpha \vec{g}_\beta = (\mu^{-1})^\gamma_\lambda\, \vec{a}^\lambda \cdot D_\alpha(\mu^\varepsilon_\beta\, \vec{a}_\varepsilon)$$

$$= (\mu^{-1})^\gamma_\lambda\, \{D_\alpha\, \mu^\lambda_\beta + \mu^\varepsilon_\beta\, \overset{\circ}{\Gamma}_\alpha{}^\lambda{}_\varepsilon\} \quad .$$

Use was here made in succession of (5.1), (5.8) and (3.1) and (2.9). The expression in brackets is modified to introduce the surface covariant derivative

$$\mu^{\lambda}_{\beta||\alpha} = D_{\alpha}\mu^{\lambda}_{\beta} + \mu^{\varepsilon}_{\beta}\,\overset{\circ}{\Gamma}{}_{\alpha}{}^{\lambda}{}_{\varepsilon} - \mu^{\lambda}_{\varepsilon}\,\overset{\circ}{\Gamma}{}_{\alpha}{}^{\varepsilon}{}_{\beta}$$

This produces

$$\Gamma_{\alpha}{}^{\gamma}{}_{\beta} = (\mu^{-1})^{\gamma}_{\lambda}\left(\mu^{\lambda}_{\beta||\alpha} + \mu^{\lambda}_{\varepsilon}\,\overset{\circ}{\Gamma}{}_{\alpha}{}^{\varepsilon}{}_{\beta}\right) = \overset{\circ}{\Gamma}{}_{\alpha}{}^{\gamma}{}_{\beta} + (\mu^{-1})^{\gamma}_{\lambda}\,\mu^{\lambda}_{\beta||\alpha}. \tag{5.12}$$

Finally, in view of (5.7) and (5.3) this can be placed in the form

$$\Gamma_{\alpha}{}^{\gamma}{}_{\beta} = \overset{\circ}{\Gamma}{}_{\alpha}{}^{\gamma}{}_{\beta} - \frac{x^3}{\mu}(1 - 2\,H\,x^3)\,b^{\gamma}_{\beta||\alpha} - \frac{(x^3)^2}{\mu}\,b^{\gamma}_{\lambda}\,b^{\lambda}_{\beta||\alpha} \tag{5.13}$$

where the dependence on x^3 is more apparent. In particular, as the notation suggested

$$\overset{\circ}{\Gamma}{}_{\alpha}{}^{\gamma}{}_{\beta} = (\Gamma_{\alpha}{}^{\gamma}{}_{\beta})_{x^3 = 0} \quad .$$

Moreover the symmetry in the subscripts α and β is guaranteed by the Mainardi-Codazzi condition (3.17).

Next calculate

$$\Gamma_{\alpha}{}^{3}{}_{\beta} = \vec{g}^{\,3}\cdot D_{\alpha}\vec{g}_{\beta} = \vec{a}^{\,3}\cdot D_{\alpha}\;(\mu^{\varepsilon}_{\beta}\,\vec{a}_{\varepsilon}) \quad .$$

Expanding the derivative and using the orthogonality between $\vec{a}_{\varepsilon}$ and $\vec{a}^{\,3}$,

$$\Gamma_{\alpha}{}^{3}{}_{\beta} = \mu_{\beta}{}^{\varepsilon}\,\vec{a}^{\,3}\cdot(\overset{\circ}{\Gamma}{}_{\alpha}{}^{\lambda}{}_{\varepsilon}\,\vec{a}_{\lambda} + b_{\alpha\varepsilon}\,\vec{a}_{3}) = \mu^{\varepsilon}_{\beta}\;b_{\alpha\varepsilon} \tag{5.14}$$

or, as explicit function of x^3,

$$\Gamma_{\alpha\ \beta}^{\ 3} = b_{\alpha\varepsilon}\ (\delta^{\varepsilon}_{\beta} - x^3\ b^{\varepsilon}_{\beta}) = b_{\alpha\beta} - x^3\ \ c_{\alpha\beta}\ . \tag{5.15}$$

For $\Gamma_{\beta\ 3}^{\ \alpha} = \Gamma_{3\ \beta}^{\ \alpha}$ we have

$$\Gamma_{\beta\ 3}^{\ \alpha} = \vec{g}^{\alpha} \cdot D_{\beta}\vec{a}_{3} = (\mu^{-1})^{\alpha}_{\lambda}\ \vec{a}^{\lambda} \cdot (-b^{\gamma}_{\beta}\ \vec{a}_{\gamma}) = -\ b^{\lambda}_{\beta}\ \ (\mu^{-1})^{\alpha}_{\lambda} \tag{5.16}$$

or again as an explicit function of x^3

$$\Gamma_{\beta\ 3}^{\ \alpha} = -\ \frac{1 - 2\ H\ x^3}{\mu}\ b^{\alpha}_{\beta} - \frac{x^3}{\mu}\ b^{\lambda}_{\beta}\ \ b^{\alpha}_{\lambda}\ . \tag{5.17}$$

Finally it is easily verified that the remaining space Christoffel symbols vanish

$$\Gamma_{3\ 3}^{\ \alpha} = 0 \qquad \Gamma_{3\ \beta}^{\ 3} = 0 \qquad \Gamma_{3\ 3}^{\ 3} = 0\ . \tag{5.18}$$

6. *The divergence operator and the divergence theorems*

The divergence theorem

$$\int_S\ (\vec{n}\cdot\vec{a})\ ds = \int_V\ \mathrm{div}\ \vec{a}\ dV \tag{6.1}$$

applied to a vector field $\vec{a}$ within a volume bounded by a simple closed surface S, provides an intrinsic definition of the divergence operator

$$\mathrm{div}\ \vec{a} = \lim_{dV \to 0}\ \frac{\int_S(\vec{n}\cdot\vec{a})dS}{dV} \tag{6.2}$$

that can be used to obtain its formulation in general curvilinear coordinates. As elementary volume we take the parallelepiped built on three infinitesimal vectors $\vec{g}_1\ dx^1$, $\vec{g}_2\ dx^2$, $\vec{g}_3\ dx^3$, issued from the point P where the divergence of the field is to be calculated.

The surface vector $(\vec{g}_1 \times \vec{g}_2)\, dx^1\, dx^2$ proportional to the area of one of the facets of the parallelepiped through P, is oriented as the inward normal to this facet. Hence the entering flux of $\vec{a}$ through this facet is

$$(\vec{g}_1 \times \vec{g}_2)\, dx^1\, dx^2 \cdot (a^i\, \vec{g}_i) = a^3\, dx^1\, dx^2\, (\vec{g}_1 \times \vec{g}_2) \cdot \vec{g}_3$$

$$= dx^1\, dx^2\, a^3\, \varepsilon_{123} \quad .$$

The outgoing flux through the opposite facet is

$$dx^1\, dx^2\, a^3\, \varepsilon_{123} + dx^1\, dx^2\, dx^3\, D_3\, (a^3\, \varepsilon_{123})$$

and the net contribution to the outcoming flux is

$$dx^1\, dx^2\, dx^3\, D_3\, (a^3\, \varepsilon_{123}) \quad .$$

Taking account of the analogous contributions from the two other pairs of facets and dividing by the volume of the parallelopiped

$$dV = dx^1\, dx^2\, dx^3\, (\vec{g}_1 \times \vec{g}_2) \cdot \vec{g}_3 = dx^1\, dx^2\, dx^3\, \varepsilon_{123}$$

the application of definition (6.2) yields

$$\operatorname{div} \vec{a} = \frac{1}{\varepsilon_{123}}\, D_i\, (a^i\, \varepsilon_{123}) \quad .$$

Expanding the derivative and using our previous result (4.14)

$$\operatorname{div} \vec{a} = D_i\, a^i + a^i\, \Gamma_{i\ m}^{\cdot\ m} = a^i|_i \quad . \tag{6.3}$$

This allows to write the divergence theorem in the form

$$\int_S (n_i a^i)\, dS = \int_V a^i|_i\, dV \tag{6.4}$$

where n_i are the covariant components of the outward unit normal to the surface. A similar result will prove to be useful for the flux of surface vectors. Let an elementary parallelogram be drawn from the vectors $\vec{a}_1\, dx^1$ and $\vec{a}_2\, dx^2$ in a tangent plane to the middle surface of the shell, and let $\vec{h} = h^\alpha \vec{a}_\alpha$ be a vector field defined on this middle surface. We calculate the outgoing flux $\Sigma\, \vec{\nu}\cdot\vec{h}\, ds$, where $\vec{\nu}$ is the outward unit normal to the contour of the parallelogram. Along the side $\vec{a}_1\, dx^1$ the contribution is $(\vec{a}_1 \times \vec{a}_2)\, dx^1 \cdot h^\alpha \vec{a}_\alpha = dx^1\, h^2\, (\vec{a}_1 \times \vec{a}_3)\cdot\vec{a}_2 = -\, dx^1\, h^2\, \varepsilon_{12}$ where use was made of

$$\varepsilon_{12} = (\vec{a}_1 \times \vec{a}_2)\cdot\vec{a}_3 \tag{6.5}$$

which is a direct consequence of (2.10). Adding the contribution from the opposite side, the net result is $dx^1\, dx^2\, D_2\, (h^2\, \varepsilon_{12})$.

Finally, adding the analogous contribution from the other pair of sides:

$$dx^1\, dx^2\, D_\alpha\, (h^\alpha\, \varepsilon_{12}) = \Sigma\, \vec{\nu}\cdot\vec{h}\, ds \quad . \tag{6.6}$$

The next result is a direct consequence of (2.23), (3.3) and (2.9):

$$D_\alpha\, \varepsilon_{12} = D_\alpha\, \{(\vec{a}_1 \times \vec{a}_2)\cdot\vec{a}_3\} = \varepsilon_{12}\, \mathring{\Gamma}_{\alpha}{}^{\beta}{}_{\beta} \quad . \tag{6.7}$$

It constitutes also the verification that $\varepsilon_{\alpha\beta}$ obeys Ricci's Lemma for the surface derivative. Substituted into (6.6) it gives

$$\begin{aligned}\Sigma\, \vec{\nu}\cdot\vec{h}\, ds &= (D_\alpha\, h^\alpha + \mathring{\Gamma}_{\alpha}{}^{\beta}{}_{\beta}\, h^\alpha)\, \varepsilon_{12}\, dx^1\, dx^2 \\ &= h^\alpha||_\alpha\, \varepsilon_{12}\, dx^1\, dx^2 \quad .\end{aligned} \tag{6.8}$$

Then, by the usual agrument of decomposition into elementary cells, we obtain the surface divergence theorem

$$\oint_c \nu_\alpha h^\alpha\, ds = \int_A h^\alpha||_\alpha\, dA \tag{6.9}$$

relating a contour integral to an integral extended over an area A of the middle surface. The surface element is indeed easily observed to be

$$dA = \varepsilon_{12}\, dx^1\, dx^2 \tag{6.10}$$

7. *Three-dimensional elasticity in curvilinear coordinates*

The approach towards a shell theory will consist in reducing the general three-dimensional elasticity equations to two-dimensional ones by suitable assumptions concerning the structure of the problem with respect to the x^3 coordinate, normal to the curved middle surface. Consequently the three-dimensional elasticity equations in curvilinear coordinates will play an important role and will be briefly recapitulated hereunder.

The basic stress definition will be lagrangian in nature. This implies that infinitesimal forces, that have to be defined as acting on infinitesimal areas of the strained elastic body, are kept invariant but, for the purpose of defining stresses, are manipulated in the metric of a "reference configuration". In the reference configuration, also called initial configuration, the body is by definition unstrained. The geometrical considerations to follow will then apply essentially to the reference configuration.

An oriented surface vector is defined in general as the cross product

$$d\vec{r}_{(1)} \times d\vec{r}_{(2)} = (\vec{g}_i\, dx^i_{(1)}) \times (\vec{g}_j\, dx^j_{(2)})$$

between two infinitesimal vectors. In modulus it is proportional to the area of the parallelogram constructed on these vectors; in orientation it is normal to the plane of the parallelogram.

Consider now an elementary tetrahedron, three lateral facets of which are generated by infinitesimal vectors taken along the local base vectors and issued from a point P.

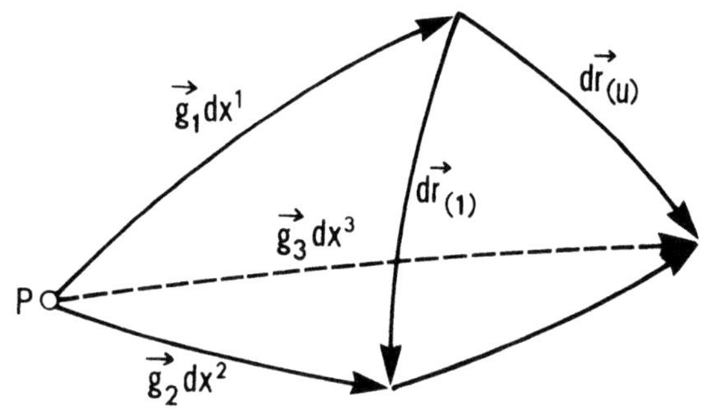

The fourth facet, called the "inclined" facet, is half the parallelogram constructed on the vectors

$$d\vec{r}_{(1)} = \vec{g}_2\, dx^2 - \vec{g}_1\, dx^1 \qquad d\vec{r}_{(2)} = \vec{g}_3\, dx^3 - \vec{g}_1\, dx^1$$

and has an oriented surface vector that, considering the antisymmetric properties of the cross product, can be written in cyclic form as

$$d\vec{r}_{(1)} \times d\vec{r}_{(2)} = (\vec{g}_1 \times \vec{g}_2)\, dx^1\, dx^2 + (\vec{g}_2 \times \vec{g}_3)\, dx^2\, dx^3 + (\vec{g}_3 \times \vec{g}_1)\, dx^3\, dx^1 \quad . \tag{7.1}$$

Each term on the right represents the surface vector of one of the lateral facets; moreover, if the orientation of the surface vector of the inclined facet is outwards, those of the lateral facets are inwards. Hence (7.1) expresses that the sum of all outward surface vectors of the tetrahedron vanishes.

In (7.1) the left hand side can be replaced by $2\,\vec{n}\,dS$, where $\vec{n}$ is the unit outward normal and dS the area of the inclined facet, while the first term in the right-hand side can be replaced by

$$\vec{g}_1 \times \vec{g}_2 \;\; dx^1\, dx^2 = \varepsilon_{123} \;\; dx^1\, dx^2\, \vec{g}^3 = 2\,\frac{\vec{g}^3}{\sqrt{g^{33}}}\, dS_3$$

since $\vec{g}^3/\sqrt{g^{33}}$ is obviously the (inward) unit normal, dS_3 denoting the area, of the lateral facet generated by $\vec{g}_1\, dx^1$ and $\vec{g}_2\, dx^2$. Thus (7.1) becomes

$$\vec{n}\, dS = \frac{\vec{g}^1}{\sqrt{g^{11}}}\, dS_1 + \frac{\vec{g}^2}{\sqrt{g^{22}}}\, dS_2 + \frac{\vec{g}^3}{\sqrt{g^{33}}}\, dS_3$$

or, introducing the covariant components n_i of the unit normal $\vec{n}$

$$\vec{n} = n_i\, \vec{g}^i \tag{7.2}$$

and identifying

$$dS_i = \sqrt{g^{ii}}\; n_i\; dS \qquad \text{(without summation)} \quad . \tag{7.3}$$

Now let $\vec{t}$ and $\vec{t}^i$ denote respectively the stress vectors acting on the inclined facet and on the lateral facets. Imagining the tetrahedron to be convected in some strained state of the body, those stresses are defined as the ratio of the force acting in the strained state on each facet to the area of the unstrained facet. Hence the translational equilibrium condition of the tetrahedron is, the resultant of mass forces being negligibly small,

$$\vec{t}dS = \vec{t}^i\, dS_i \quad . \tag{7.4}$$

The substitution of (7.3) into (7.4) suggests that a lagrangian stress tensor t^{ij} be defined according to equation

$$\sqrt{g^{ii}}\,\vec{t}^{\,i} = t^{ij}\,\vec{g}_j \tag{7.5}$$

so that the equilibrium condition is now expressed by

$$\vec{t} = n_i t^{ij}\,\vec{g}_j \quad . \tag{7.6}$$

It follows that the t^{ij} are not physical stress components. The physical components derived from definition (7.5) are given by

$$\sqrt{\frac{g_{jj}}{g^{ii}}}\,t^{ij} \qquad \text{(no summation)} \ . \tag{7.7}$$

If the points of the strained body are submitted to an additional small displacement field $\delta\vec{u}\ (x^1, x^2, x^3)$, the statement of energy conservation is

$$\int_S \vec{t}\cdot\delta\vec{u}\ dS + \int_V \vec{X}\cdot\delta\vec{u}\ dV = \int_V \delta W\ dV \tag{7.8}$$

where surface and volume integrals are taken in the initial configuration. W is the elastic strain energy and $\vec{X}$ the body force field per unit volume in this initial configuration. In curvilinear coordinates this becomes

$$\int_S (n_i t^{ij}\ \delta u_j)\ dS + \int_V X^j\ \delta u_j\ dV = \int_V \delta W\ dV \quad . \tag{7.9}$$

When the surface integral is transformed in a volume integral by the divergence theorem (6.4), the local statement of conservation of energy (per unit volume) is

$$(t^{ij}\ \delta u_j)\big|_i + X^j\ \delta u_j = \delta W \qquad . \tag{7.10}$$

If the additional displacement field is of a rigid body type, there is no energy increase; this can be used to extract from (7.10) the translational and rotational equilibrium equations. In the translational case

$$D_i\,(\delta u_j\,\vec{g}^j) = \delta u_{j|i}\,\vec{g}^j = 0 \qquad \text{or} \qquad \delta u_{j|i} = 0 \qquad (7.11)$$

so that (7.10) with $\delta W = 0$ reduces to

$$(t^{ij}{}_{|i} + X^j)\,\delta u_j = 0$$

and, on account of the arbitrariness of the translation vector

$$t^{ij}{}_{|i} + X^j = 0 \qquad (j = 1,2,3)\ . \qquad (7.12)$$

Those equilibrium equations satisfied by the lagrangian stresses are linear; they allow a first simplification of the energy equation (7.10)

$$t^{ij}\,\delta u_{j|i} = \delta W\ . \qquad (7.13)$$

For the rotational case we can write the additional displacement field as

$$\delta\vec{u} = d\vec{\omega} \times (\vec{R} + \vec{u}) \qquad (7.14)$$

where $d\vec{\omega}$ is a constant infinitesimal rotation vector and $\vec{u}$ the already prevailing displacement field from the initial configuration. Then

$$D_m\,\delta\vec{u} = d\vec{\omega} \times (\vec{g}_m + D_m\vec{u}) \qquad \text{or}$$

$$\delta u_{i|m}\,\vec{g}^i = \left(\delta^p_m + u^p{}_{|m}\right)\,d\vec{\omega} \times \vec{g}_p \qquad .$$

Scalar multiplication by g_j and transformation of

$$(d\vec{\omega} \times \vec{g}_p)\cdot\vec{g}_j = d\omega^r \, (\vec{g}_r \times \vec{g}_p)\cdot\vec{g}_j = \varepsilon_{rpj} \, d\omega^r$$

furnishes

$$\delta u_{j|m} = \varepsilon_{rpj} \left(\delta^p_m + u^p{}_{|m} \right) d\omega^r \; . \tag{7.15}$$

For this value of the covariant derivative of δu_j, the energy increase vanishes again in (7.13) and, since the $d\omega^r$ are arbitrary

$$\varepsilon_{rpj} \left(t^{pj} + t^{ij} \, u^p{}_{|i} \right) = 0 \quad \text{all } r \text{ and } j \; .$$

Considering the complete antisymmetry of ε_{rpj} this rotational equilibrium condition is satisfied if and only if the tensor between brackets is symmetrical in the superscripts p and j. This produces the final rotational equilibrium equations connecting the lagrangian stress tensor components

$$t^{pj} + t^{ij} \, u^p{}_{|i} = t^{ip} \, u^j{}_{|i} \;R + t^{jp} \quad . \tag{7.16}$$

There are only three distinct ones, since they are identically satisfied for $p = j$. They reduce to a statement of symmetry of the lagrangian stress tensor only in the case where the covariant derivatives of the contravariant displacement tensor are negligibly small compared to unity (geometrical linearity). A method for automatically satisfying (7.16) consists in writing

$$t^{ij} = \left(\delta^j_q + u^j{}_{|q} \right) s^{qi} \quad \text{with} \quad s^{qi} = s^{iq} \quad . \tag{7.17}$$

The new symmetrical stress tensor s^{qi} is that of Kirchhoff and Trefftz, it transforms the energy equation (7.13) into

$$s^{qi}\left(\delta u_{q|i} + u^j|_q\, \delta u_{j|i}\right) = \delta W$$

or, making use of the symmetry of the new stress tensor

$$\frac{1}{2}\, s^{qi}\left(\delta u_{q|i} + \delta u_{i|q} + u^j|_q\, \delta u_{j|i} + u^j|_i\, \delta u_{j|q}\right) = \delta W\ . \tag{7.18}$$

Because the formulation is lagrangian, the order sequence of the δ operator and the derivatives with respect to the coordinates x^i may be interchanged. It is then recognized that the tensor between brackets is a perfect differential, provided one also recognizes, e.g., by applying Ricci's Lemma, the equivalence

$$u^j|_i\, \delta u_{j|q} = u_{j|i}\, \delta u^j|_q \quad .$$

In fact, introducing the symmetrical tensor

$$\gamma_{qi} = \frac{1}{2}\left(u_{q|i} + u_{i|q} + u^j|_q\, u_{j|i}\right) \tag{7.19}$$

the energy statement (7.18) reduces to

$$\delta W = s^{qi}\, \delta\gamma_{qi} \quad . \tag{7.20}$$

It shows that the definition (7.17) is not altogether artificial since it now appears that the strain energy density is a function of the displacement gradients through six distinct combinations of them only, the components of a symmetrical tensor γ_{qi}, and that the Kirchhoff-Trefftz stresses are the corresponding partial derivatives. Obviously the γ_{qi} must represent strains, they are in fact the most commonly used lagrangian measures of strain and are usually derived from a statement of conservation of distance between neighboring points in a rigid body displacement. It is easily verified that

$$d(\vec{R} + \vec{u}) \cdot d(\vec{R} + \vec{u}) - d\vec{R} \cdot d\vec{R} = 2\gamma_{qi}\, dx^q\, dx^i \quad . \tag{7.21}$$

While the lagrangian viewpoint offers many advantages, the hybrid nature of the stress tensor definitions it requires is well illustrated by the Kirchhoff-Trefftz tensor. They can be interpreted as the same lagrangian stress vectors but decomposed in the metric of the strained body. The base vectors of this metric are naturally defined by

$$\vec{G}_p = D_q(\vec{R} + \vec{u}) = \vec{g}_q + u^j|_q\, \vec{g}_j = \left(\delta_q^{\ j} + u^j|_q\right)\vec{g}_j \ . \tag{7.22}$$

Then, starting from definition (7.6) of the lagrangian stress tensor and substituting (7.17)

$$\vec{t} = n_i\, s^{qi}\left(\delta_q^j + u^j|_q\right)\vec{g}_j = n_i\, s^{qi}\, \vec{G}_q \tag{7.23}$$

which proves the assertion. This geometrically hybrid character is reflected in the nature of the translational equilibrium equations they satisfy, the Signorini equations, which depend on the displacement field:

$$s^{ij}|_i + \left(s^{qi}\, u^j|_q\right)\Big|_i + X^j = 0 \quad . \tag{7.24}$$

Due to the symmetry of the Kirchhoff-Trefftz stresses, (7.20) shows that the energy density is in fact a function of the six strain measures

$$\begin{aligned} \gamma_{11},\ \gamma_{22},\ \gamma_{33},\ \text{and}\ \ \eta_{12} &= \gamma_{12} + \gamma_{21} \\ \eta_{23} &= \gamma_{23} + \gamma_{32} \\ \eta_{31} &= \gamma_{31} + \gamma_{13} \end{aligned} \tag{7.25}$$

with the following general form of the stress-strain relations

$$\sigma^{ii} = \frac{\partial W}{\partial \gamma_{ii}} \qquad \sigma^{ij} = \frac{\partial W}{\partial \eta_{ij}} \quad (i \neq j) \quad . \tag{7.26}$$

If, however, it is agreed to distinguish γ_{ij} from γ_{ji} in (7.25), the general formula

$$\sigma^{ij} = \frac{\partial W}{\partial \gamma_{ij}} \tag{7.27}$$

holds true by the chain rule of differentiation.

In writing down an explicit form of the energy density we shall limit ourselves to the generalized Hooke situation (physical linearity).

The energy is then a homogeneous quadratic form in the components of the strain tensor:

$$W = \frac{1}{2} E^{ijkl} \gamma_{ij} \gamma_{kl} \quad . \tag{7.28}$$

Because the tensor of elastic moduli is symmetrical with respect to the pair of indices ij and kl and also with respect to the interchange of those pairs, its total number of independant components is not larger than 21.

In the isotropic case, the energy is explicitly

$$W = G \frac{1 - \nu}{1 - 2\nu} I_1^2 - 2GI_2 \tag{7.29}$$

with G the shear modulus and ν Poisson's ratio. The invariants I_1 and I_2 are obtained as coefficients of the powers of γ in the expansion of the determinant

$$\begin{vmatrix} \gamma_1^1 - \gamma & \gamma_1^2 & \gamma_1^3 \\ \gamma_2^1 & \gamma_2^2 - \gamma & \gamma_2^3 \\ \gamma_3^1 & \gamma_3^2 & \gamma_3^3 - \gamma \end{vmatrix} = -\gamma^3 + I_1 \gamma^2 - I_2 \gamma + I_3$$

$$I_1 = \gamma_1^1 + \gamma_2^2 + \gamma_3^3 = \gamma_i^i$$

$$I_2 = \gamma_1^1\gamma_2^2 + \gamma_2^2\gamma_3^3 + \gamma_3^3\gamma_1^1 - \gamma_1^2\gamma_2^1 - \gamma_2^3\gamma_3^2 - \gamma_3^1\gamma_1^3 = \frac{1}{2}\left(\gamma_i^i\gamma_j^j - \gamma_i^j\gamma_j^i\right).$$

Going back to the doubly covariant components of strain

$$I_1 = g^{ij}\,\gamma_{ij} \qquad I_2 = \frac{1}{2}\,g^{mi}\,g^{nj}\,(\gamma_{mi}\gamma_{nj} - \gamma_{mj}\gamma_{ni}) \qquad (7.30)$$

and applying formally (7.27)

$$\sigma^{pq} = E^{pqij}\,\gamma_{ij}$$

with

$$E^{pqij} = G\left(g^{pi}\,g^{qj} + g^{pj}\,g^{qi} + \frac{2\nu}{1 - 2\nu}\,g^{pq}\,g^{ij}\right) . \qquad (7.31)$$

8. *Stress assumption for shell theory*

The basic stress assumption will be that the direct Kirchhoff-Trefftz σ^{33} normal to the middle surface is zero; it generalizes the assumption common to plate theories. Then, in cartesian local axes, the third axis being oriented like $\vec{a}_3$, the isotropic stress-strain relations reduce to

$$\sigma_1^1 = \frac{E}{1 - \nu^2}\left(\gamma_1^1 + \nu\gamma_2^2\right) , \quad \sigma_2^2 = \frac{E}{1 - \nu^2}\left(\gamma_2^2 + \nu\gamma_1^1\right) , \quad \sigma_3^3 = 0$$

$$\sigma_i^j = \frac{E}{1 - \nu^2}\,(1 - \nu)\gamma_i^j \qquad (i \neq j)$$

where E is Young's modulus. Building up the corresponding energy density by means of Clapeyron's theorem

$$W = \frac{1}{2}\,\sigma_i^j\,\gamma_j^i$$

the resulting expression is easily put in the form

$$W = \frac{E}{2(1-\nu^2)}\left\{\left(\gamma_\alpha^\alpha\right)^2 + 2(1-\nu)\gamma_\alpha^3\gamma_3^\alpha - (1-\nu)\left(\gamma_\alpha^\alpha\gamma_\beta^\beta - \gamma_\alpha^\beta\gamma_\beta^\alpha\right)\right\}.$$

It is clearly valid also in the curvilinear system of coordinates of the shell in view of the invariance of the different terms with respect to any change in metric not involving the third axis.

Turning then to the covariant strains measures

$$W = \frac{E}{2(1-\nu^2)}\left\{\left(g^{\alpha\beta}\gamma_{\alpha\beta}\right)^2 + 2(1-\nu)\,g^{\alpha\beta}\gamma_{\alpha 3}\gamma_{3\beta} - \right.$$
$$\left. - (1-\nu)\,g^{\alpha\lambda}g^{\beta\mu}\left(\gamma_{\alpha\lambda}\gamma_{\beta\mu} - \gamma_{\alpha\mu}\gamma_{\beta\lambda}\right)\right\}. \tag{8.1}$$

By taking formal derivatives it is found that

$$\left.\begin{aligned}
E^{\lambda\mu\alpha\beta} &= \frac{E}{2(1-\nu^2)}\left\{2\nu g^{\lambda\mu}g^{\alpha\beta} + (1-\nu)\left(g^{\lambda\alpha}g^{\mu\beta} + g^{\lambda\beta}g^{\mu\alpha}\right)\right\}\\
E^{\lambda 3\alpha 3} &= 2\,G\,g^{\lambda\alpha}\\
E^{\lambda\mu\alpha 3} &= 0 \qquad E^{\lambda\mu 33} = 0 \qquad E^{\lambda 333} = 0 \qquad E^{3333} = 0.
\end{aligned}\right] \tag{8.2}$$

9. *Displacement assumptions for a shell theory*

Following the basic stress assumption $\sigma^{33} = 0$, there are, like in plate theory, different ways open for reaching the goal of reducing a shell theory to essentially two dimensions. Further assumptions can be made on stress distributions, generalizing E. Reissner's approach to the plate bending formulation. Displacement assumptions can be used instead, as in Hencky's plate bending theory, or both systems can be mixed to take advantage of two-field variational principles.

We shall here discuss a displacement formulation, first because it is more widely in use, secondly because it can be specialized to include a discussion of the Kirchhoff-Love hypothesis. The simplest assumption is that of linearity across the shell thickness of the displacement vector:

$$\vec{u} = \vec{u}_{(0)} + x^3 \, \vec{u}_{(1)} \tag{9.1}$$

where the two component vectors depend only on the middle surface coordinates. Covariant components of surface displacement tensors are consequently introduced by the definitions

$$\vec{u}_{(0)} = v_\alpha \, \vec{a}^\alpha + w \, \vec{a}^3 \tag{9.2}$$

$$\vec{u}_{(1)} = w_\alpha \, \vec{a}^\alpha + p \, \vec{a}^3 \quad . \tag{9.3}$$

The v_α are middle surface displacements, w is a transversal displacement normal to the middle surface, the w_α are sectional rotations and finally p is a transverse "pinch". The pinch component is usually omitted; it is however essential to preserve the ability of (9.1) to represent a finite rigid body rotation and moreover it lends more symmetry to the subsequent derivations. We next introduce the surface tensors generated by taking surface derivatives of the component vectors

$$D_\beta \vec{u}_{(0)} = \lambda_{\gamma\beta} \vec{a}^\gamma + \phi_\beta \vec{a}^3 = \lambda^{\gamma}_{\cdot\beta} \vec{a}_\gamma + \phi_\beta \vec{a}_3 \tag{9.4}$$

$$D_\beta \vec{u}_{(1)} = \rho_{\alpha\beta} \vec{a}^\gamma + \psi_\beta \vec{a}^3 = \rho^{\gamma}_{\cdot\beta} \vec{a}_\gamma + \psi_\beta \vec{a}^3 \tag{9.5}$$

where, from application of formula (3.12), we find

$$\lambda_{\gamma\beta} = v_{\gamma||\beta} - b_{\gamma\beta}\, w \tag{9.6}$$

$$\rho_{\gamma\beta} = w_{\gamma||\beta} - b_{\gamma\beta}\, p \tag{9.7}$$

$$\phi_\beta = D_\beta w + b^\varepsilon_\beta\, v_\varepsilon \tag{9.8}$$

$$\psi_\beta = D_\beta p + b^\varepsilon_\beta\, w_\varepsilon \quad . \tag{9.9}$$

To prepare the computation of the Green strain tensor, we calculate the various derivatives $u_{i|q}$

$$\left.\begin{aligned}
u_{\alpha|\beta} &= \vec{g}_\alpha \cdot D_\beta \vec{u} = \mu^\varepsilon_\alpha\, \vec{a}_\varepsilon \cdot \{(\lambda_{\gamma\beta} + x^3\rho_{\gamma\beta})\vec{a}^\gamma + (\phi_\beta + x^3\psi_\beta)\vec{a}^3\} \\
&= \mu^\gamma_\alpha\, (\lambda_{\gamma\beta} + x^3\rho_{\gamma\beta}) \\
u_{\alpha|3} &= \vec{g}_\alpha \cdot D_3 \vec{u} = \mu^\varepsilon_\alpha\, \vec{a}_\varepsilon \cdot \vec{u}_{(1)} = \mu^\varepsilon_\alpha\, w_\varepsilon \\
u_{3|\beta} &= \vec{g}_3 \cdot D_\beta \vec{u} = \vec{a}_3 \cdot D_\beta \vec{u} = \phi_\beta + x^3\psi_\beta \\
u_{3|3} &= \vec{g}_3 \cdot D_3 \vec{u} = \vec{a}_3 \cdot \vec{u}_{(1)} = p \quad .
\end{aligned}\right] \tag{9.10}$$

Similar calculations, using (5.8), produce

$$\left.\begin{aligned}
u^\alpha{}_{|\beta} &= (u^{-1})^\alpha_\gamma\, (\lambda^{\gamma}_{\cdot\beta} + x^3\rho^{\gamma}_{\cdot\beta}) \\
u^\alpha{}_{|3} &= (u^{-1})^\alpha_\varepsilon\, w^\varepsilon \\
u^3{}_{|\beta} &= u_{3|\beta} \\
u^3{}_{|3} &= u_{3|3}
\end{aligned}\right] \tag{9.11}$$

Substitution into the definitions (7.19) gives then

$$\left.\begin{aligned}
2\gamma_{\alpha\beta} &= \lambda_{\alpha\beta} + \lambda_{\beta\alpha} + \lambda_{\varepsilon\alpha}\lambda^{\varepsilon}_{\cdot\beta} + \phi_{\alpha}\phi_{\beta} \\
&\quad + x^3(\rho_{\alpha\beta} + \rho_{\beta\alpha} - b^{\varepsilon}_{\alpha}\lambda_{\varepsilon\beta} - b^{\varepsilon}_{\beta}\lambda_{\varepsilon\alpha} + \rho_{\varepsilon\alpha}\lambda^{\varepsilon}_{\cdot\beta} + \rho_{\varepsilon\beta}\lambda^{\varepsilon}_{\cdot\alpha} + \\
&\qquad + \phi_{\alpha}\psi_{\beta} + \phi_{\beta}\psi_{\alpha}) \\
&\quad + (x^3)^2\,(-b^{\varepsilon}_{\alpha}\rho_{\varepsilon\beta} - b^{\varepsilon}_{\beta}\rho_{\varepsilon\alpha} + \rho_{\varepsilon\alpha}\rho^{\varepsilon}_{\cdot\beta} + \psi_{\alpha}\psi_{\beta}) \\
2\gamma_{\alpha 3} &= w_{\alpha} + \phi_{\alpha} + w_{\varepsilon}\lambda^{\varepsilon}_{\cdot\alpha} + p\phi_{\alpha} + x^3\,(\psi_{\alpha} - b^{\varepsilon}_{\alpha}w_{\varepsilon} + w_{\varepsilon}\rho^{\varepsilon}_{\cdot\alpha} + p\psi_{\alpha}) \\
2\gamma_{33} &= 2p + p^2 + w^{\varepsilon}w_{\varepsilon}\ .
\end{aligned}\right\} \quad (9.12)$$

Some additional physical interpretation of the surface tensors in equations (9.4) and (9.5) can be derived in the case of small material rotations. The incremental displacement field in some infinitesimal neighborhood of a point can then be analyzed, following Helmholtz, as a superposition of a pure strain and a small rotation:

$$d\vec{u} = \vec{\omega} \times d\vec{R} + \gamma_{iq}\ dx^q\ \vec{g}^i \quad \text{with} \quad \gamma_{qi} = \gamma_{iq}\ . \qquad (9.13)$$

The rotational component can be explicited, either by introducing an antisymmetrical tensor ω_{iq}

$$\vec{\omega} \times d\vec{R} = \omega_{iq}\ dx^q\ \vec{g}^i \qquad (9.14)$$

$$\omega_{qi} = -\ \omega_{iq} \qquad (9.15)$$

or a pseudo-tensor

$$\vec{\omega} = \omega^m\ \vec{g}_m \quad . \qquad (9.16)$$

Then

$$\vec{\omega} \times d\vec{R} = \omega^m \, dx^q \, \vec{g}_m \times \vec{g}_q = \varepsilon_{mqr} \, \omega^m \, dx^q \, \vec{g}^r \quad .$$

Hence the relationship

$$\omega_{rq} = \varepsilon_{mqr} \, \omega^m \quad . \tag{9.17}$$

The antisymmetrical tensor is more convenient because its dependance on x^3 is simpler. Inserting (9.15) into (9.14), replacing

$$d\vec{u} = D_q \vec{u} \, dx^q$$

and comparing coefficients of dx^q, there comes

$$D_\beta \vec{u} = (\omega_{i\beta} + \gamma_{i\beta}) \, \vec{g}^i$$

$$D_3 \vec{u} = (\omega_{i3} + \gamma_{i3}) \, \vec{g}^i \quad .$$

Scalar multiplication of those equations by $\vec{g}_j$ and evaluation of the lefthand sides as in the previous computation of the strain components, yields

$$\omega_{\alpha\beta} + \gamma_{\alpha\beta} = \mu_\alpha^\varepsilon \, (\lambda_{\varepsilon\beta} + x^3 \rho_{\varepsilon\beta})$$

$$\omega_{3\beta} + \gamma_{3\beta} = \phi_\beta + x^3 \psi_\beta$$

$$\omega_{\beta 3} + \gamma_{\beta 3} = \mu_\beta^\varepsilon w_\varepsilon$$

$$\gamma_{33} = p \quad .$$

From here, playing on the symmetry characteristics,

$$
\begin{aligned}
2\gamma_{\alpha\beta} &= \mu_\alpha^\varepsilon\,(\lambda_{\varepsilon\beta} + x^3\rho_{\varepsilon\beta}) + \mu_\beta^\varepsilon\,(\lambda_{\varepsilon\alpha} + x^3\rho_{\varepsilon\alpha}) \\
2\gamma_{\beta 3} &= \phi_\beta + x^3\psi_\beta + \mu_\beta^\varepsilon w_\varepsilon \\
2\gamma_{33} &= 2p
\end{aligned}
\qquad (9.18)
$$

and

$$
\begin{aligned}
2\omega_{\alpha\beta} &= \mu_\alpha^\varepsilon\,(\lambda_{\varepsilon\beta} + x^3\rho_{\varepsilon\beta}) - \mu_\beta^\varepsilon\,(\lambda_{\varepsilon\alpha} + x^3\rho_{\varepsilon\alpha}) \\
2\omega_3 &= \phi_\beta + x^3\psi_\beta - \mu_\beta^\varepsilon w_\varepsilon \quad .
\end{aligned}
\qquad (9.19)
$$

Expressions (9.18) can be recognized as the linearized versions of (9.11), (9.12) and (9.13); they are formed from all the linear terms of the general strain components. The surface tensor of the "rotation of the normal" is, by definition

$$(\omega_{3\beta})_{x^3=0} = \frac{1}{2}\,(\phi_\beta - w_\beta) \qquad (9.20)$$

while the "rotation about the normal" is

$$(\omega_{12})_{x^3=0} = \frac{1}{2}\,(\lambda_{12} - \lambda_{21}) = \frac{1}{2}\,(v_{1||2} - v_{2||1}) . \qquad (9.21)$$

The rate of change in the direction of the normal of the rotation about the normal, or "normal twist", plays an important role in plate theory where it can be shown to be responsible for boundary layer effects, while it vanishes under a Kirchhoff-Love hypothesis. Its expression in shell theory can be defined from

$$D_3\,(\omega^3 \vec{g}_3) = (D_3\,\omega^3)\,\vec{g}_3$$

as measured by $D_3\,\omega^3$. Since according to (9.17)

$$\omega_{21} = \varepsilon_{312}\,\omega^3 = \sqrt{g}\,\omega^3 \qquad (9.22)$$

and in view of (5.5), (5.6) and (2.24) and (9.19)

$$2\sqrt{a}(1-2Hx^3 + (x^3)^2K)\omega^3 = \lambda_{21} - \lambda_{12} + x^3(\rho_{21} - \rho_{12} - b^{\varepsilon}_{2}\lambda_{\varepsilon 1} + b^{\varepsilon}_{1}\lambda_{\varepsilon 2}) - (x^3)^2 (b^{\varepsilon}_{2}\rho_{\varepsilon 1} - b^{\varepsilon}_{1}\rho_{\varepsilon 2})$$

$$\omega^3 = \omega^3_{(0)} + x^3\omega^3_{(1)} + (x^3)^2\omega^3_{(2)} + \dots$$

where

$$2\sqrt{a}\,\omega^3_{(0)} = \lambda_{21} - \lambda_{12}$$

$$2\sqrt{a}\,(-2H\omega^3_{(0)} + \omega^3_{(1)}) = \rho_{21} - \rho_{12} - b^{\varepsilon}_{2}\lambda_{\varepsilon 1} + b^{\varepsilon}_{1}\lambda_{\varepsilon 2}$$

$$\dots .$$

Hence for the normal twist on the middle surface

$$\omega^3_{(1)} = \frac{1}{2\sqrt{a}}\left\{b^{\varepsilon}_{\varepsilon}(\lambda_{21} - \lambda_{12}) + \rho_{21} - \rho_{12} - b^{\varepsilon}_{2}\lambda_{\varepsilon 1} + b^{\varepsilon}_{1}\lambda_{\varepsilon 2}\right\}$$

$$= \frac{1}{2\sqrt{a}}\left\{\rho_{21} - \rho_{12} + b^{\varepsilon}_{1}\lambda_{2\varepsilon} - b^{\varepsilon}_{2}\lambda_{1\varepsilon}\right\} = \frac{1}{2}\,\varepsilon^{\alpha\beta}(\rho_{\beta\alpha} + b^{\varepsilon}_{\alpha}\lambda_{\beta\varepsilon}). \tag{9.23}$$

The Kirchhoff-Love hypothesis requires $\gamma_{\beta 3} = 0$. It can be satisfied in the linearized case by setting

$$w_{\beta} = -\phi_{\beta} = -D_{\beta}w - b^{\varepsilon}_{\beta}v_{\varepsilon} \tag{9.24}$$

thus relating the sectional rotations to the middle surface and transverse displacement. Also we should have

$$\psi_{\beta} - b^{\varepsilon}_{\beta}w_{\varepsilon} = D_{\beta}p = 0 \tag{9.25}$$

a condition identically satisfied by suppressing the pinch component in the original displacement assumptions. Under those conditions

$$\rho_{21} = w_{2||1} = - D_1 D_2 w - (b_2^{\varepsilon} v_{\varepsilon})_{||1}$$

and

$$\rho_{21} - \rho_{12} = b_1^{\varepsilon} v_{\varepsilon||2} - b_2^{\varepsilon} v_{\varepsilon||2} + v_{\varepsilon} (b_{1||2}^{\varepsilon} - b_{2||1}^{\varepsilon}) \quad .$$

The last term vanishes by virtue of the Mainardi-Codazzi condition, hence, using (9.6)

$$\rho_{21} - \rho_{12} = b_1^{\varepsilon} \lambda_{\varepsilon 2} - b_2^{\varepsilon} \lambda_{\varepsilon 1}$$

$$\omega^3_{(1)} = \frac{1}{2\sqrt{a}} \left\{ b_1^{\varepsilon} (\lambda_{\varepsilon 2} + \lambda_{2\varepsilon}) - b_2^{\varepsilon} (\lambda_{\varepsilon 1} + \lambda_{1\varepsilon}) \right\}$$

$$= \frac{1}{2} \varepsilon^{\alpha\beta} b_{\alpha}^{\varepsilon} (\lambda_{\varepsilon\beta} + \lambda_{\beta\varepsilon}) \quad . \qquad (9.26)$$

Hence, under a Kirchhoff-Love hypothesis, there remains a normal twist due to the curvature of the middle surface. Subtracted from (9.23) it furnishes the "additional normal twist"

$$\Delta\omega^3_{(1)} = \frac{1}{2} \varepsilon^{\alpha\beta} (\rho_{\beta\alpha} - b_{\alpha}^{\varepsilon} \lambda_{\varepsilon\beta}) \quad . \qquad (9.27)$$

Again, in view of the antisymmetry of $\varepsilon^{\alpha\beta}$ and the Mainardi-Codazzi condition, this can be put in the form

$$\Delta\omega^3_{(1)} = \frac{1}{2} \varepsilon^{\alpha\beta} (w_{\beta} - b_{\beta}^{\varepsilon} v_{\varepsilon})_{||\alpha} \quad . \qquad (9.28)$$

10. *The Lagrangian Stress Resultants*

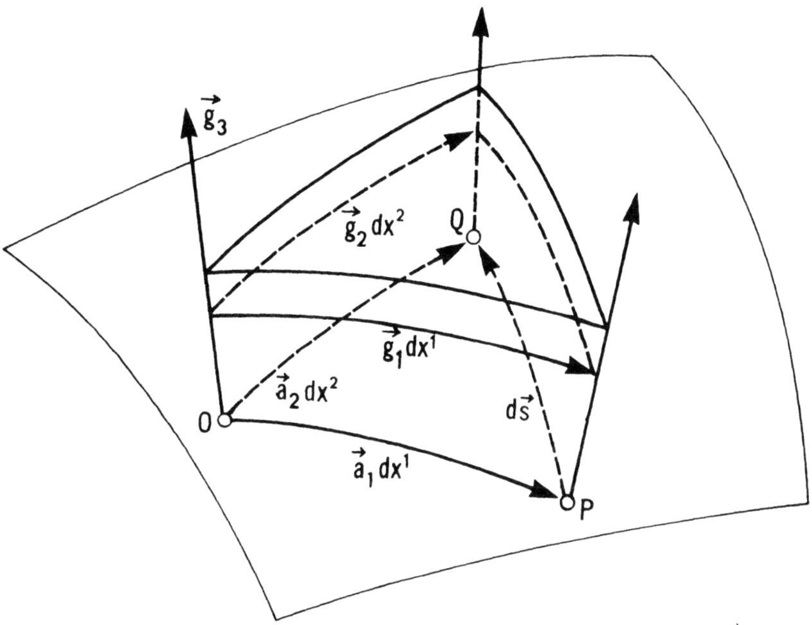

Let O, P and Q be three points in the middle surface, $\vec{OP}$ is taken along the coordinate curve x^2 = constant, $\vec{OQ}$ along the coordinate curve x^1 = constant. The "inclined" side of the triangle, $\vec{PQ}$ is an element $d\vec{s}$ of a contour drawn on the middle surface. Erecting normals to the middle surface in those three points, and measuring equal distances x^3 and dx^3 on each, we define a triangular slab of which the translational equilibrium will be expressed. Obviously at the limit when both P and Q tend to coincide with O, the only forces to take into account will be due to the stresses acting on the lateral sides of the slab. The first lateral side has a surface vector $\vec{g}_1 \times \vec{g}_3\, dx^1\, dx^3 = - \sqrt{g}\, \vec{g}^2\, dx^1\, dx^2$ oriented outwards. The stress vector on this is $\vec{t}^2$ and the resultant force applied to it

$$- \sqrt{g}\sqrt{g^{22}}\; \vec{t}^2\, dx^1\, dx^3 = - \sqrt{g}\; t^{2j}\, \vec{g}_j\, dx^1\, dx^3 \quad .$$

Similarly for the second lateral side with outwardly oriented surface vector

$$- \vec{g}_2 \times \vec{g}_3 \, dx^2 \, dx^3 = - \sqrt{g} \, \vec{g}^1 \, dx^2 \, dx^3$$

on which the resultant force is

$$- \sqrt{g} \sqrt{g^{11}} \, \vec{t}^1 \, dx^2 \, dx^3 = - \sqrt{g} \, t^{1j} \, \vec{g}_j \, dx^2 \, dx^3 \, .$$

Then, if $d\vec{f}$ denotes the force acting on the third (inclined) lateral side, equilibrium demands

$$d\vec{f} = \sqrt{g} \, (t^{2j} \, \vec{g}_j \, dx^1 + t^{1j} \, \vec{g}_j \, dx^2) \, dx^3 \quad . \tag{10.1}$$

Let, as in section 5, $\vec{\nu}$ denote the outward normal to a closed contour described in the positive sense on the middle surface, then

$$\vec{\nu} \, ds = d\vec{s} \times \vec{a}_3 = (- \vec{a}_1 \, dx^1 + \vec{a}_2 \, dx^2) \times \vec{a}_3$$

$$= \sqrt{a} \, (\vec{a}^2 \, dx^1 + \vec{a}^1 \, dx^2)$$

or, introducing the covariant components of $\vec{\nu}$ by

$$\vec{\nu} = \nu_\alpha \, \vec{a}^\alpha$$

and comparing

$$\nu_1 \, ds = \sqrt{a} \, dx^2 \qquad \nu_2 \, ds = \sqrt{a} \, dx^1 \quad . \tag{10.2}$$

This result can be substituted into (10.1) after using (5.5)

$$d\vec{f} = \mu(\nu_2 t^{2j} + \nu_1 t^{1j}) \, \vec{g}_j \, ds \, dx^3 = \mu \nu_\alpha t^{\alpha j} \, \vec{g}_j \, ds \, dx^3$$

or, in view of (5.1) and (5.2),

$$d\vec{f} = \mu(\nu_\alpha t^{\alpha\beta} \, \mu_\beta^\gamma \vec{a}_\gamma + \nu_\alpha t^{\alpha 3} \, \vec{a}_3) \, ds \, dx^3 \quad . \tag{10.3}$$

In order to see what are the stress resultants to introduce, we form the scalar product of this elementary force with the incremental displacement vector corresponding to the assumptions (9.1), (9.2) and (9.3)

$$\delta\vec{u} = (\delta v_\varepsilon + x^3 \delta w_\varepsilon)\,\vec{a}^\varepsilon + (\delta w + x^3 \delta p)\,\vec{a}^3$$

then integrate over the thickness of the shell:

$$\int_{-h}^{h} d\vec{f} \cdot \delta\vec{u}\, dx^3 = (\nu_\alpha N^{\alpha\varepsilon} \delta v_\varepsilon + \nu_\alpha M^{\alpha\varepsilon} \delta w_\varepsilon + \nu_\alpha Q^\alpha dw + \nu_\alpha P^\alpha \delta p)\, ds \tag{10.4}$$

with the membrane resultants

$$N^{\alpha\varepsilon} = \int_{-h}^{h} \mu\mu_\beta^\varepsilon t^{\alpha\beta} dx^3 \tag{10.5}$$

the bending and twisting moments

$$M^{\alpha\varepsilon} = \int_{-h}^{h} \mu\mu_\beta^\varepsilon t^{\alpha\beta} x^3 dx^3 \tag{10.6}$$

the shear loads

$$Q^\alpha = \int_{-h}^{h} \mu t^{\alpha 3}\, dx^3 \tag{10.7}$$

and pinching loads

$$p^\alpha = \int_{-h}^{h} \mu t^{\alpha 3}\, x^3\, dx^3 \tag{10.8}$$

If (10.4) is integrated over a closed contour (c) on the middle surface and transformed into an integral over the enclosed area A by means of theorem (5.9) the contribution of the stresses to the energy increase per unit initial area of the middle surface is obtained in the form

$$\delta F_1 = (N^{\alpha\varepsilon}\delta v_\varepsilon + M^{\alpha\varepsilon}\delta w_\varepsilon + Q^\alpha\delta w + P^\alpha\delta p)\,||\alpha \ . \qquad (10.9)$$

There is also a contribution from the externally applied loads, either through body forces, or through surface tractions on the faces $x^3 = \pm h$, of the form

$$\delta F_2 = N^\varepsilon\delta v_\varepsilon + M^\varepsilon\delta w_\varepsilon + Q\delta w + P\delta p \ . \qquad (10.10)$$

From the energy equation $\delta F = \delta F_1 + \delta F_2$ the equilibrium equations and the constitutive equations can be deduced by a procedure analogous to that of the general three dimensional equations in section 7.

a. *Translational equilibrium equations for a piece of shell.*

An additional displacement field is of translational type if $D_3\ \delta\vec{u} = 0$, which implies $\delta w_\beta = 0$ and $\delta p = 0$ (10.11)

$D_\beta\ \delta\vec{u} = 0$, which implies $\delta\lambda_{\alpha\beta} = \delta v_{\alpha||\beta} - b_{\alpha\beta}\delta w = 0$ (10.12)

and $\delta\phi_\beta = D_\beta\delta w + b_\beta^\varepsilon\delta v_\varepsilon = 0$. (10.13)

The energy equation can be reduced to the form

$$\delta F = N^{\alpha\varepsilon}{}_{||\alpha}\delta v_\varepsilon + b_{\varepsilon\alpha}N^{\alpha\varepsilon}\delta w + Q^\alpha{}_{||\alpha}\delta w - Q^\alpha b_\alpha^\varepsilon\delta v_\varepsilon + N^\varepsilon\delta v_\varepsilon + Q\delta w$$

and since $\delta F = 0$, equating separately to zero the coefficients of δv_ε and δw

$$N^{\alpha\varepsilon}{}_{||\alpha} - Q^\alpha b_\alpha^\varepsilon + N^\varepsilon = 0 \quad (\varepsilon = 1,2) \ . \qquad (10.14)$$

$$Q^\alpha{}_{||\alpha} + b_{\varepsilon\alpha}N^{\alpha\varepsilon} + Q = 0 \ . \qquad (10.15)$$

b. *Rotational equilibrium equations*

In an expanded form of (10.9), the covariant derivatives of the membrane forces and transverse shear forces can now be substituted from the equilibrium equations (10.14) and (10.15). If, likewise, the covariant derivatives of the incremental displacements are taken from the incremental form of definitions (9.6), (9.7), (9.8) and (9.9)

$$\delta F = N^{\alpha\varepsilon}\delta\lambda_{\varepsilon\alpha} + M^{\alpha\varepsilon}\delta\rho_{\varepsilon\alpha} + (M^{\alpha\varepsilon}{}_{||\alpha} - b^{\varepsilon}_{\alpha}P^{\alpha} + M^{\varepsilon})\delta w_{\varepsilon}$$

$$+ Q^{\alpha}\delta\phi_{\alpha} + P^{\alpha}\delta\psi_{\alpha} + (P^{\alpha}{}_{||\alpha} + b_{\varepsilon\alpha}M^{\alpha\varepsilon} + P)\delta p \quad . \qquad (10.16)$$

It is now necessary to derive the particular values taken by the incremental displacement parameters in a small rigid body rotation superimposed on the strained state of the elastic body. Starting from (7.14) and separating the contributions with respect to the x^3 distribution

$$\delta\vec{u}_{(0)} = d\vec{w} \times (\vec{r} + \vec{u}_{(0)}) \qquad (10.17)$$

$$\delta\vec{u}_{(1)} = d\vec{w} \times (\vec{a}_3 + \vec{u}_{(1)}) \quad . \qquad (10.18)$$

In these equations we substitute (9.2) and (9.3), together with

$$\vec{r} = r^{\alpha}\vec{a}_{\alpha} + r^3\,\vec{a}_3$$

$$d\vec{w} = dw^{\lambda}\,\vec{a}_{\lambda} + dw^3\,\vec{a}_3$$

and make use of

$$\vec{a}_{\lambda} \times \vec{a}_{\alpha} = \varepsilon_{\lambda\alpha}\,\vec{a}^3 \qquad \vec{a}_{\lambda} \times \vec{a}_3 = \varepsilon_{\alpha\lambda}\,\vec{a}^{\alpha} \quad .$$

Then, comparing covariant components

$$\delta v_{\beta} = \varepsilon_{\beta\lambda}\,(r^3 + w)\,dw^{\lambda} + \varepsilon_{\alpha\beta}\,(r^{\alpha} + v^{\alpha})\,d\omega^3$$

$$\delta w = \varepsilon_{\lambda\alpha}\,(r^{\alpha} + v^{\alpha})\,d\omega^{\lambda}$$

$$\left.\begin{aligned} \delta w_\beta &= \varepsilon_{\beta\lambda}\,(1+p)\,d\omega^\lambda + \varepsilon_{\alpha\beta} w^\alpha d\omega^3 \\ \delta p &= \varepsilon_{\lambda\alpha}\, w^\alpha d\omega^\lambda \quad . \end{aligned}\right\} \tag{10.19}$$

Next, taking partial derivatives with respect to x^β on both sides of (10.17) and (10.18) and remembering (9.4) and (9.5)

$$\delta\lambda_{\gamma\beta}\vec{a}^\gamma + \delta\phi_\beta\vec{a}^3 = \Lambda^\alpha_{\bullet\beta}\, d\vec{\omega} \times \vec{a}_\alpha + \phi_\beta d\vec{\omega} \times \vec{a}_3$$

$$\delta\rho_{\gamma\beta}\vec{a}^\gamma + \delta\psi_\beta\vec{a}^3 = R^\alpha_{\bullet\beta}\, d\vec{\omega} \times \vec{a}_\alpha + \psi_\beta d\vec{\omega} \times \vec{a}_3$$

where, for concision, the following notations were introduced

$$\Lambda^\alpha_{\bullet\beta} = \delta^\alpha_\beta + \lambda^\alpha_{\bullet\beta} \qquad R^\alpha_{\bullet\beta} = \rho^\alpha_{\bullet\beta} - b^\alpha_\beta \quad . \tag{10.20}$$

From those equations we deduce

$$\left.\begin{aligned} \delta\lambda_{\gamma\beta} &= \varepsilon_{\alpha\gamma}\Lambda^\alpha_{\bullet\beta}\, d\omega^3 + \varepsilon_{\gamma\lambda}\phi_\beta\, d\omega^\lambda \\ \delta\phi_\beta &= \varepsilon_{\lambda\alpha}\Lambda^\alpha_{\bullet\beta} d\omega^\lambda \\ \delta\rho_{\gamma\beta} &= \varepsilon_{\alpha\gamma}R^\alpha_{\bullet\beta} d\omega^3 + \varepsilon_{\gamma\lambda}\psi_\beta\, d\omega^\lambda \\ \delta\psi_\beta &= \varepsilon_{\lambda\alpha}R^\alpha_{\bullet\beta}\, d\omega^\lambda \quad . \end{aligned}\right\} \tag{10.21}$$

The results (10.19) and (10.21) are to be inserted into (10.16) and since $\delta F = 0$ again, while $d\vec{w}$ is an arbitrary constant vector, the following general conditions for rotational equilibrium are found:

b.1. *equilibrium about the normal to the middle surface, obtained by equating to zero the coefficient of* $d\omega^3$

$$\varepsilon_{\gamma\beta}[N^{\alpha\beta}\Lambda^\gamma_{\bullet\alpha} + M^{\alpha\beta}R^\gamma_{\bullet\alpha} + S^\beta w^\gamma] = 0 \tag{10.22}$$

with the definition

$$S^{\beta} = M^{\alpha\beta}{}_{||\alpha} - b^{\beta}_{\alpha}P^{\alpha} + M^{\beta} \quad . \tag{10.23}$$

Considering the antisymmetry of $\varepsilon_{\gamma\beta}$, this is equivalent to require the symmetry of the bracket in (10.22) with respect to β and γ.

b.2. *equilibrium about axes tangent to the middle surface, expressed by equating to zero the coefficient of* $\varepsilon_{\gamma\lambda}\, d\omega^{\lambda}$

$$N^{\alpha\gamma}\phi_{\alpha} + M^{\alpha\gamma}\psi_{\alpha} + S^{\gamma}\,(1 + p) - Q^{\alpha}\Lambda^{\gamma}_{\bullet\alpha} - P^{\alpha}R^{\gamma}_{\bullet\alpha} - Tw^{\gamma} = 0 \qquad (\gamma = 1,2) \tag{10.24}$$

with the definition

$$T = P^{\alpha}{}_{||\alpha} + b_{\varepsilon\alpha}M^{\alpha\varepsilon} + P \quad .$$

As in the case of the general three-dimensional equilibrium equations satisfied by lagrangian stresses, the lagrangian resultants obey linear translational equilibrium conditions (10.14) and (10.15), but their rotational equilibrium conditions (10.22) and (10.24) also involve deformations.

An attempt to calculate δF directly from the general result (7.13) and the relation

$$\delta W dV = \delta W\sqrt{g}\, dx^1\, dx^2\, dx^3 = \delta W\mu\sqrt{a}\, dx^1\, dx^2\, dx^3 = \delta W\mu dx^3 dS$$

or

$$\delta F = \int_{-h}^{h} \mu\delta W\, dx^3$$

gives, in view of definitions (10.5) to (10.8) and results (9.10)

$$\delta F = N^{\beta\gamma}\delta\lambda_{\gamma\beta} + M^{\beta\gamma}\delta\rho_{\gamma\beta} + Q^{\beta}\delta\phi_{\beta} + P^{\beta}\delta\psi_{\beta} + S^{\beta}\delta w_{\beta} + T\delta p \tag{10.26}$$

which is similar to (10.16) and shows that the tensors defined by (10.23) and (10.25) are also resultants

$$S^{\beta} = \int_{-h}^{h} \mu\mu^{\beta}_{\alpha} \; t^{3\alpha} \, dx^3 \qquad T = \int_{-h}^{h} \mu t^{33} \, dx^3 \tag{10.27}$$

They do not, however, constitute internal loads, since they involve lagrangian stresses which are defined at interfaces between shell pieces.

As a matter of fact, the rotational equilibrium condition about the displaced normal to the middle surface does not contain S^{β} nor T.

To show it we calculate the base vectors in the deformed metric from (7.22) and (9.11) obtaining

$$\vec{G}_{\gamma} = (\Lambda^{\varepsilon}_{\bullet\gamma} + x^3 R^{\varepsilon}_{\bullet\gamma}) \, \vec{a}_{\varepsilon} + (\phi_{\gamma} + x^3 \, \psi_{\gamma}) \, \vec{a}_3 \tag{10.28}$$

$$\vec{G}_3 = w^{\varepsilon} \vec{a}_{\varepsilon} + (1 + p) \, \vec{a}_3 \quad . \tag{10.29}$$

The displaced base vectors on the middle surface are then given by

$$\vec{A}_{\gamma} = \Lambda^{\varepsilon}_{\bullet\gamma} \vec{a}_{\varepsilon} + \phi_{\gamma} \vec{a}_3 \tag{10.30}$$

$$\vec{A}_3 = w^{\lambda} \vec{a}_{\lambda} + (1 + p) \, \vec{a}_3 = \vec{G}_3 \quad . \tag{10.31}$$

Thus, if we denote by $R^{\gamma\beta}$ the bracket in (10.22) and by R^{γ} the left-hand side of (10.24)

$$\varepsilon_{\gamma\beta} R^{\gamma\beta} \, d\omega^3 + \varepsilon_{\gamma\lambda} R^{\gamma} \, d\omega^{\lambda} = 0$$

is the general rotational equilibrium condition, while for a rotation $d\Omega^3$ about the displaced normal $\vec{A}^3$

$$d\omega^{\lambda} = w^{\lambda} d\Omega^3 \qquad\qquad d\omega^3 = (1 + p) \, d\Omega^3$$

and the corresponding condition is

$$(1 + p)\varepsilon_{\gamma\beta} R^{\gamma\beta} + w^{\lambda}\varepsilon_{\gamma\lambda}R^{\gamma} = 0$$

or symmetry of

$$(1 + p) R^{\gamma\beta} + w^{\beta}R^{\gamma} \quad .$$

In this linear combination S^{β} and T do precisely appear in symmetrical groupings

$$(1 + p) (w^{\gamma}S^{\beta} + w^{\beta}S^{\gamma}) - Tw^{\gamma}w^{\beta}$$

and may consequently be dropped leaving the condition

$$\begin{aligned}(1 + p) N^{\alpha\beta}\Lambda^{\gamma}_{\bullet\alpha} + w^{\beta}\phi_{\alpha}N^{\alpha\gamma} + (1 + p) M^{\alpha\beta}R^{\gamma}_{\bullet\alpha} + w^{\beta}\psi_{\alpha}M^{\alpha\gamma}& \\ - Q^{\alpha}w^{\beta}\Lambda^{\gamma}_{\bullet\alpha} - P^{\alpha}w^{\beta}R^{\gamma}_{\bullet\alpha} \quad \text{symmetrical in } (\beta,\gamma)&\end{aligned} \tag{10.32}$$

The equilibrium conditions (10.24) can now be incorporated into the energy equation (10.26) by using them to eliminate S^{β}. This gives

$$\begin{aligned}\delta F = {}& N^{\beta\gamma}\left(\delta\lambda_{\gamma\beta} - \frac{1}{1 + p}\phi_{\beta}\delta w_{\gamma}\right) + M^{\beta\gamma}\left(\delta\rho_{\gamma\beta} - \frac{1}{1 + p}\psi_{\beta}\,\delta w_{\gamma}\right) \\ & + Q^{\beta}\left(\delta\phi_{\beta} + \frac{1}{1 + p}\Lambda^{\gamma}_{\bullet\beta}\delta w_{\gamma}\right) + P^{\beta}\left(\delta\psi_{\beta} + \frac{1}{1 + p}R^{\gamma}_{\bullet\beta}\delta w_{\gamma}\right) \\ & + T\left(\delta p + \frac{1}{1 + p}w^{\gamma}\delta w_{\gamma}\right) \quad .\end{aligned} \tag{10.33}$$

This new form of the energy equation cannot be expected to turn into a canonical form similar to (7.20), from which constitutive equations can be rationally derived, until the last rotational equilibrium equation (10.32) can be solved and the constraint it constitutes between the lagrangian stress resultants can be liberated.

11. *Kirchhoff-Trefftz stress parameters*

Since the Kirchhoff-Trefftz stress tensor contains implicitly the rotational equilibrium conditions in its property of symmetry, an immediate solution to our problem consists in transforming the definitions (10.5) to (10.8) by means of the relations (7.17). In our case, considering equations (9.11), we can substitute

$$
\begin{aligned}
t^{\alpha\beta} &= s^{\alpha\beta} + u^{\beta}|_{\varepsilon} s^{\varepsilon\alpha} + u^{\beta}|_{3} s^{3\alpha} \\
&= s^{\alpha\beta} + (\mu^{-1})^{\beta}_{\gamma} (\lambda^{\gamma}_{\bullet\varepsilon} + x^{3}\rho^{\gamma}_{\bullet\varepsilon})\, s^{\varepsilon\alpha} + (\mu^{-1})^{\beta}_{\gamma} w^{\gamma} s^{3\alpha} \\
t^{\alpha 3} &= s^{\alpha 3} + u^{3}|_{\beta} s^{\beta\alpha} + u^{3}|_{3} s^{3\alpha} \\
&= s^{\alpha 3} + (\phi_{\beta} + x^{3}\psi_{\beta}) s^{\beta\alpha} + p s^{3\alpha} \\
t^{3\alpha} &= s^{3\alpha} + u^{\alpha}|_{\beta} s^{\beta 3} + u^{\alpha}|_{3} s^{33} \\
&= s^{3\alpha} + (\mu^{-1})^{\alpha}_{\gamma} (\lambda^{\gamma}_{\bullet\beta} + x^{3}\rho^{\gamma}_{\bullet\beta}) s^{\beta 3} + (\mu^{-1})^{\alpha}_{\gamma} w^{\gamma} s^{33} \\
t^{33} &= s^{33} + u^{3}|_{\beta} s^{\beta 3} + u^{3}|_{3} s^{33} \\
&= s^{33} + (\phi_{\beta} + x^{3}\psi_{\beta})\, s^{\beta 3} + p\, s^{33}
\end{aligned}
\tag{11.1}
$$

The lagrangian stress resultants are then expressed in terms of the following Kirchhoff-Trefftz stress parameters

$$n^{\alpha\beta} = \int_{-h}^{h} \mu s^{\alpha\beta} dx^3 = n^{\beta\alpha}$$

$$m^{\alpha\beta} = \int_{-h}^{h} \mu s^{\alpha\beta} x^3 dx^3 = m^{\beta\alpha}$$

$$l^{\alpha\beta} = \int_{-h}^{h} \mu s^{\alpha\beta} (x^3)^2 \, dx^3 = l^{\beta\alpha}$$

$$q^{\beta} = \int_{-h}^{h} \mu s^{\beta 3} \, dx^3 \tag{11.2}$$

$$p^{\beta} = \int_{-h}^{h} \mu s^{\beta 3} \, x^3 \, dx^3$$

$$n = \int_{-h}^{h} \mu s^{33} \, dx^3 \quad .$$

There comes

$$N^{\alpha\varepsilon} = \Lambda^{\varepsilon}_{\bullet\gamma} \, n^{\gamma\alpha} + R^{\varepsilon}_{\bullet\gamma} \, m^{\gamma\alpha} + w^{\varepsilon} q^{\alpha}$$

$$M^{\alpha\varepsilon} = \Lambda^{\varepsilon}_{\bullet\gamma} \, m^{\gamma\alpha} + R^{\varepsilon}_{\bullet\gamma} \, l^{\gamma\alpha} + w^{\varepsilon} p^{\alpha}$$

$$Q^{\alpha} = (1 + p) \, q^{\alpha} + \phi_{\gamma} \, n^{\gamma\alpha} + \psi_{\gamma} \, m^{\gamma\alpha} \tag{11.3}$$

$$P^{\alpha} = (1 + p) \, p^{\alpha} + \phi_{\gamma} \, m^{\gamma\alpha} + \psi_{\gamma} \, l^{\gamma\alpha}$$

and also

$$S^{\beta} = \Lambda^{\beta}_{\bullet\alpha} q^{\alpha} + R^{\beta}_{\bullet\alpha} p^{\alpha} + w^{\beta} n$$

$$T = (1 + p) \, n + \phi_{\alpha} q^{\alpha} + \psi_{\alpha} p^{\alpha} \quad . \tag{11.4}$$

It is easily verified that all the rotational equilibrium conditions (10.22) and (10.24) or (10.32) are thereby satisfied. Furthermore the energy equation (10.26) takes the canonical form

$$\delta F = n^{\gamma\alpha}\delta\overset{\circ}{\gamma}_{\gamma\alpha} + m^{\gamma\alpha}\delta\gamma^{1}_{\gamma\alpha} + l^{\gamma\alpha}\delta\gamma^{2}_{\gamma\alpha} + 2q^{\alpha}\delta\overset{\circ}{\gamma}_{\alpha 3} + 2p^{\alpha}\delta\gamma^{1}_{\alpha 3} + n\delta\gamma_{33} \tag{11.5}$$

where the strains are the coefficients of the powers in x^3 of the Green strain tensor (9.12). As a matter of fact, this result is equivalent to a straight forward application of a Rayleigh-Ritz process, based on the displacement assumptions (9.1) to (9.3) to the expression

$$\delta F = \int_{-h}^{h} \mu\delta W \, dx^3$$

where δW is given by (7.20). Applied to the form (8.1) of the energy density it introduces the following rigidity coefficients for the shell:

$$D^{\alpha\beta\ \gamma\delta}_{(m)} = \frac{E}{1-\nu^2}\int_{-h}^{h} \mu g^{\alpha\beta} g^{\gamma\delta} (x^3)^m \, dx^3 \qquad (m = 0,1,\ldots 4) \tag{11.6}$$

$$C^{\alpha\beta}_{(m)} = \frac{E}{1-\nu^2}\int_{-h}^{h} \mu g^{\alpha\beta} (x^3)^m \, dx^3 \qquad (m = 0,1,2) \tag{11.7}$$

Taking due account of the symmetries, the total number of rigidity coefficients of type (11.6) is 6 × 5 = 30, of type (11.7) 3 × 3 = 9.

The energy per unit area is then expressible as

$$
\begin{aligned}
2\,F = {} & D^{\alpha\lambda\ \beta\mu}_{(0)}\ \{\nu\overset{\circ}{\gamma}_{\alpha\lambda}\overset{\circ}{\gamma}_{\beta\mu} + (1-\nu)\overset{\circ}{\gamma}_{\alpha\mu}\overset{\circ}{\gamma}_{\beta\lambda}\} \\
& + 2\,D^{\alpha\lambda\ \beta\mu}_{(1)}\ \{\nu\overset{\circ}{\gamma}_{\alpha\lambda}\gamma^{1}_{\beta\mu} + (1-\nu)\overset{\circ}{\gamma}_{\alpha\mu}\gamma^{1}_{\beta\lambda}\} \\
& + D^{\alpha\lambda\ \beta\mu}_{(2)}\ \{2\nu\overset{\circ}{\gamma}_{\alpha\lambda}\gamma^{2}_{\beta\mu} + \nu\gamma^{1}_{\alpha\lambda}\gamma^{1}_{\beta\mu} + 2(1-\nu)\overset{\circ}{\gamma}_{\alpha\mu}\gamma^{2}_{\beta\lambda} + (1-\nu)\gamma^{1}_{\alpha\mu}\gamma^{1}_{\beta\lambda}\} \\
& + 2\,D^{\alpha\lambda\ \beta\mu}_{(3)}\ \{\nu\gamma^{1}_{\alpha\lambda}\gamma^{2}_{\beta\mu} + (1-\nu)\gamma^{1}_{\alpha\mu}\gamma^{2}_{\beta\lambda}\} \\
& + D^{\alpha\lambda\ \beta\mu}_{(4)}\ \{\nu\gamma^{2}_{\alpha\lambda}\gamma^{2}_{\beta\mu} + (1-\nu)\gamma^{2}_{\alpha\mu}\gamma^{2}_{\beta\lambda}\} \\
& + 2(1-\nu)\ \{C^{\alpha\beta}_{(0)}\ \overset{\circ}{\gamma}_{\alpha 3}\overset{\circ}{\gamma}_{\beta 3} + 2\,C^{\alpha\beta}_{(1)}\overset{\circ}{\gamma}_{\alpha 3}\gamma^{1}_{\beta 3} + C^{\alpha\beta}_{(2)}\gamma^{1}_{\alpha 3}\gamma^{1}_{\beta 3}\}
\end{aligned}
\tag{11.8}
$$

The constitutive equations of the shell follow by taking the partial derivatives of (11.8) as indicated in the total differential (11.5).

In particular, since (11.8) is independant of γ_{33}, we find $n = 0$ in accordance with the assumption $\sigma^{33} = 0$.

The integrals in (11.6) and (11.7) are not easily carried out in closed form because of the complicated dependence of the reciprocal metric tensor (5.10) on the coordinate x^3. They can be evaluated by numerical quadrature methods or expanded in powers of a (small) parameter.

It follows from (11.5) that the Kirchhoff-Trefftz stress parameters are internal conjugate variables to the arguments of the strain energy per unit surface:

$$
\begin{aligned}
n^{\gamma\alpha} &= \frac{\partial F}{\partial \overset{\circ}{\gamma}_{\gamma\alpha}} \qquad & m^{\gamma\alpha} &= \frac{\partial F}{\partial \gamma^{1}_{\gamma\alpha}} \qquad & l^{\gamma\alpha} &= \frac{\partial F}{\partial \gamma^{2}_{\gamma\alpha}} \\
2q^{\alpha} &= \frac{\partial F}{\partial \overset{\circ}{\gamma}_{\alpha 3}} \qquad & 2p^{\alpha} &= \frac{\partial F}{\partial \gamma^{1}_{\alpha 3}} \qquad & n &= \frac{\partial F}{\partial \gamma_{33}}\ .
\end{aligned}
\tag{11.9}
$$

If those arguments as given by (9.12) are expressed through definitions (9.6) to (9.9) in terms of v_α, w_α, w, p and their derivatives, it is easily found by the chain rule of differentiation and by the results (11.3) and (11.4) that

$$\frac{\partial F}{\partial v_{\alpha||\beta}} = N^{\beta\alpha} \qquad \frac{\partial F}{\partial w_{\alpha||\beta}} = M^{\beta\alpha}$$

$$\frac{\partial F}{\partial w_{||\alpha}} = Q^{\alpha} \qquad \frac{\partial F}{\partial p_{||\alpha}} = P^{\alpha} \tag{11.10}$$

$$\frac{\partial F}{\partial p} = T - b_{\alpha\beta} M^{\alpha\beta} \qquad \frac{\partial F}{\partial w_\alpha} = b^{\alpha}_{\varepsilon} P^{\varepsilon} + S^{\alpha}$$

$$\frac{\partial F}{\partial w} = - b_{\alpha\beta} N^{\alpha\beta} \qquad \frac{\partial F}{\partial v_\alpha} = b^{\alpha}_{\varepsilon} Q^{\varepsilon} \quad . \tag{11.11}$$

Equations (11.10) can then be considered as the constitutive equations of the shell in terms of Lagrangian resultants. Since they involve (11.3) and (11.4) they automatically satisfy the rotational equilibrium equations. They also allow to apply in a simple fashion the variational principle

$$\delta\{ \int_S F dS - \int_S (N^{\alpha} v_\alpha + M^{\alpha} w_\alpha + Qw + Pp) dS + F_3 \} = 0 \tag{11.12}$$

where F_3 denotes the potential energy of external loads applied at the boundary. From (11.10) and (11.11) there comes

$$\int_S \{ N^{\beta\alpha} \delta v_{\alpha||\beta} + M^{\beta\alpha} \delta w_{\alpha||\beta} + Q^{\alpha} \delta w_{||\alpha} + P^{\alpha} \delta p_{||\alpha}$$

$$+ \, (b^{\alpha}_{\varepsilon} \, Q^{\varepsilon} - N^{\alpha}) \, \delta v_{\alpha} + (S^{\alpha} + b^{\alpha}_{\varepsilon} \, P^{\varepsilon} - M^{\alpha}) \, \delta w_{\alpha}$$

$$- \, (Q + b_{\alpha\beta} N^{\alpha\beta}) \, \delta w + (T - b_{\alpha\beta} \, M^{\alpha\beta} - P) \, \delta p\} \, dS + \delta F_{3} = 0 \, .$$

After integration by parts, the following Euler-Lagrange equilibrium equations are obtained

$$\text{for } \delta v_{\alpha} \qquad - N^{\beta\alpha}{}_{||\beta} + b^{\alpha}_{\varepsilon} \, Q^{\varepsilon} - N^{\alpha} = 0 \tag{11.13}$$

$$\delta w_{\alpha} \qquad - M^{\beta\alpha}{}_{||\beta} + S^{\alpha} + b^{\alpha}_{\varepsilon} \, P^{\varepsilon} - M^{\alpha} = 0 \tag{11.14}$$

$$\delta w \qquad - Q^{\alpha}{}_{||\alpha} - Q - b_{\alpha\beta} \, N^{\alpha\beta} = 0 \tag{11.15}$$

$$\delta p \qquad - P^{\alpha}{}_{||\alpha} + T - b_{\alpha\beta} \, M^{\alpha\beta} - P = 0 \, . \tag{11.16}$$

Associated with boundary conditions resulting from the variational equation

$$\delta F_3 + \oint \nu_{\beta} (N^{\beta\alpha} \delta v_{\alpha} + M^{\beta\alpha} \delta w_{\alpha} + Q^{\alpha} \delta w + P^{\alpha} \delta p) \, ds = 0. \tag{11.17}$$

In equations (11.13) and (11.15) we recognize the translational equilibrium equations already obtained in (10.14) and (10.15), while (11.14) and (11.16) are respectively equivalent to the definitions (10.23) and (10.25) of S^{α} and T.

The field equations of the two-dimensional non linear shell problem consist of the 6 linear equilibrium equations (11.13) to (11.16) together with the 15 resultants-displacement derivatives equations (11.10). Those are to be worked out by using in succession (11.8), (9.12) and (9.6) to (9.9). The total number of unknowns is also 21: the 6 displacements (v_{α}, w_{α}, w, p) and the 15 components of resultants in

$N^{\alpha\beta}$, $M^{\alpha\beta}$, Q^{α}, P^{α}, S^{α}, T because of the identities

$$\frac{\partial F}{\partial v_\alpha} = b^\alpha_\varepsilon \frac{\partial F}{\partial \phi_\varepsilon} = b^\alpha_\varepsilon \frac{\partial F}{\partial w_{||\varepsilon}}$$

and

$$\frac{\partial F}{\partial w} = - b_{\alpha\beta} \frac{\partial F}{\partial \lambda_{\alpha\beta}} = - b_{\alpha\beta} \frac{\partial F}{\partial v_{\alpha||\beta}}$$

stemming from (9.9) and (9.6), equations (11.11) must not be counted in the constitutive equations.

REFERENCES

[1] NAGHDI, P.M., *Progress in Solid Mechanics*, Vol. IV, North-Holland, 1963, 1-90 pp.

[2] KOITER, W.T., "Theory of thin elastic shells", *Proceedings of IUTAM Symposium on the theory of thin elastic shells*, Delft, August 1959, North-Holland, 1961.

[3] BUDIANKI, B. and SANDERS, J.L., "On the best first order linear shell theory," *Prager Ann. Volume*, Mcmillan, 1963, 129-140 pp.

DUAL ANALYSIS FOR HEAT CONDUCTION PROBLEMS BY FINITE ELEMENTS

B.M. Fraeijs de Veubeke and M.A. Hogge

University of Liège, Belgium

SUMMARY—An alternative approach to the usual finite element treatment of steady-state temperature problems is presented, using approximations for the field of the dual variables.

The appropriate extremum principle is established and its minimization is discussed in connection with a plane triangular finite element process. Original heat flow elements are derived: in conjunction with temperature elements, they enable dual analysis of a given structure and an important estimate of the convergence to the true solution by upper and lower bounds to the dissipation function, as illustrated by means of several examples.

INTRODUCTION

Steady-state heat conduction is one of the many field problems of engineering that can receive a variational formulation. The basic formulation [1] is in terms of the temperature field alone and the discretization in finite elements makes use of parametric approximations of the temperature within each element with continuity secured across the interfaces. In the present paper an alternative formulation is presented in terms of the dual (vector) variable: the heat flow. In the corresponding finite element models the heat flow satisfies thermal equilibrium within each

Original publication, *International Journal of Numerical Methods in Engineering*, Vol. 5, (in collaboration with M.A. Hogge), John Wiley & Sons Limited, Baffins Lane, Chichester, Sussex, U.K., 1972, pp. 65-82.

element and flux continuity is maintained at the interfaces. In this manner a dual analysis of a steady-state heat conduction problem extends to the benefit of the dissipation function a numerical estimate of the convergence by upper bounds (temperature models) and lower bounds (heat flow models) when no heat sources are prescribed. When the body is in contact with a uniform temperature bath taken as zero level, the role of bounds is reversed. The general case where both sources and non-uniform outside temperatures prevail can be treated by superposition if convergence estimates are required.

BASIC TEMPERATURE FORMULATION

The temperature functional in a simply connected domain D

$$I(T) = \int_D (\frac{1}{2} k_{ij}\partial_i T \partial_j T - \bar{Q}T)dD + \int_{\partial_2 D} \bar{q}T dS + \int_{\partial_3 D} \frac{1}{2} h(T-\bar{T}_e)^2 dS, \quad (1)$$

where $S = \partial_1 D + \partial_2 D + \partial_3 D$, denotes its boundary, $\partial_i T$ the temperature gradient vector, k_{ij} a symmetrical heat conductivity tensor, $\bar{Q}$ a prescribed volume heat source distribution, $\bar{q}$ a prescribed surface heat sink, h a convection coefficient and $\bar{T}_e$ the local prescribed external temperature, is stationary under small perturbations δT of the temperature field subject to the boundary conditions

$$T = \bar{T}_e \ , \quad \delta T = 0 \quad \text{on} \quad \partial_1 D \quad (2)$$

when the Euler equation

$$\partial_i (k_{ij}\partial_j T) + \bar{Q} = 0 \quad (3)$$

and the natural boundary conditions (n_i direction cosines of the outward normal)

$$n_i k_{ij}\partial_j T + \bar{q} = 0 \quad \text{on} \quad \partial_2 D \quad (4)$$

$$n_i k_{ij} \partial_j T + h(T-\overline{T}_e) = 0 \quad \text{on} \quad \partial_3 D \tag{5}$$

are satisfied. The Euler equation is a statement of thermal equilibrium for the unit volume; under Fourier's law for the heat flow vector

$$q_i = -k_{ij} \partial_j T \tag{6}$$

it becomes

$$-\partial_i q_i + \overline{Q} = 0 \ . \tag{7}$$

Three types of boundary conditions are distinguished. The surface heat sink condition (4) on part $\partial_2 D$ of the boundary, that can also be written

$$-n_i q_i + \bar{q} = 0 \quad \text{on} \quad \partial_2 D \quad , \tag{8}$$

the convection loss condition (5) or

$$-n_i q_i + h(T-\overline{T}_e) = 0 \quad \text{on} \quad \partial_3 D \tag{9}$$

and the prescribed surface temperature condition (2) that can be regarded as a limiting case $h \to \infty$ of equation (9).

The functional of this 'principle of variation of temperatures' is the sum of a 'dissipation functional'

$$F(T) = \int_D \frac{1}{2} k_{ij} \partial_i T \partial_j T dD + \int_{\partial_2 D} \frac{1}{2} h(T-\overline{T}_e)^2 dS \tag{10}$$

of the whole system body plus surrounding, [2], i.e., including the boundary heat convection effect, and a 'potential energy of prescribed thermal loads'

$$P(T) = - \int_D \overline{Q} T dD + \int_{\partial_2 D} \bar{q} T dS \ . \tag{11}$$

EXTENSION BY RECIPROCAL TRANSFORMATION

As was done in the case of elasticity theory [3] the basic formulation will not be extended through a reciprocal transformation [4]. Fourier's law is used to eliminate the temperature gradients in the dissipation function that becomes

$$F(\vec{q},T) = \int_D \frac{1}{2} r_{im} q_i q_m dD + \int_{\partial_2 D} \frac{1}{2} h(T-\overline{T}_e)^2 dS \ . \tag{12}$$

The heat resistivity tensor r_{im} is reciprocal to the heat conductivity one

$$r_{im} k_{ij} = \delta_{mj} \tag{13}$$

so that Fourier's law, solved for the temperature gradients, reads

$$r_{im} q_i = -r_{im} k_{ij} \partial_j T = -\partial_m T \ .$$

Considered as differential constraints

$$\partial_m T + r_{im} q_i = 0 \tag{14}$$

between the temperature and heat flow fields those relations are incorporated into a 'dislocation functional' involving Lagrange multipliers λ_m

$$D(\vec{q},T,\vec{\lambda},\mu) = -\int_D \lambda_m (\partial_m T + r_{im} q_i) dD + \int_{\partial_1 D} \mu (T-\overline{T}_e) dS \tag{15}$$

the second part of which also removes the necessity of satisfying *a priori* the boundary constraint (2).

The 'three-field' variational principle

$$F(\vec{q},T) + D(\vec{q},T,\vec{\lambda},\mu) + P(T) \qquad \text{stationary} \tag{16}$$

has the following variational equations and natural boundary conditions:

$$r_{im}(q_m - \lambda_m) = 0 \quad \text{or} \quad q_m = \lambda_m \tag{17}$$

as the variational equation for the heat flow field,

$$\partial_m \lambda_m - \bar{Q} = 0 \tag{18}$$

$$-n_m \lambda_m + \mu = 0 \quad \text{on} \quad \partial_1 D \tag{19}$$

$$-n_m \lambda_m + \bar{q} = 0 \quad \text{on} \quad \partial_2 D \tag{20}$$

$$-n_m \lambda_m + h(T - \bar{T}_e) = 0 \quad \text{on} \quad \partial_3 D \tag{21}$$

as variational equation and natural boundary conditions for the temperature field, and finally (14) and (2) for the variations on the Lagrange multipliers. Equations (17) and (19) identify those multipliers with the heat flow field and its normal component on $\partial_1 D$. When incorporated into the other equations they express all the thermal equilibrium conditions and boundary conditions of the problem. When incorporated *ab initio* into (16) they reduce this principle to the thermal analogue of the Hellinger-Reissner principle of elasticity:

$$- \int_D (q_m \partial_m T + \frac{1}{2} r_{im} q_i q_m + \bar{Q}T) dD + \int_{\partial_1 D} n_m q_m (T - \bar{T}_e) dS + \int_{\partial_2 D} \bar{q} T dS$$

$$+ \int_{\partial_3 D} \frac{1}{2} h (T - \bar{T}_e)^2 dS \qquad \text{stationary} \tag{22}$$

It contains only two independent fields, the temperature and the heat flow. Its corresponding variational equations are respectively (7) and (14), while (8), (9) and (2) all appear as natural boundary conditions. An integration by parts on the first term of (22) puts it into the equivalent form

$$\int_D T(\partial_m q_m - \bar{Q})dD - \int_D \frac{1}{2} r_{im} q_i q_m dD - \int_{\partial_1 D} n_m q_m \bar{T}_e dS + \int_{\partial_2 D} (\bar{q} - n_m q_m) T dS$$

$$+ \int_{\partial_3 D} \{\frac{1}{2} h(T-\bar{T}_e)^2 - n_m q_m T\} dS \quad \text{stationary.} \tag{23}$$

THE DUAL OR HEAT FLOW FUNCTIONAL

The presence of the temperature field can be completely eliminated from the functional (23) by assuming that we satisfy *a priori* the volume heat equilibrium equation (7) and the corresponding surface equation (8), whereby the first and fourth terms disappear.

However, since the last term still contains T on $\partial_3 D$, equation (9) is solved for T and back substituted. This produces, after a change of sign the 'principle of variation of heat flow':

$$J(\vec{q}) = \int_D \frac{1}{2} r_{im} q_i q_m dD + \int_{\partial_1 D} n_m q_m \bar{T}_e dS + \int_{\partial_3 D} \{(n_m q_m)^2/2h + n_m q_m \bar{T}_e\} dS$$
$$\text{stationary.} \tag{24}$$

It is the thermal analogue of the complementary energy principle of elasticity. It is again the sum of the dissipation functional (10) expressed however in terms of the heat flow through Fourier's law and (9)

$$F(\vec{q}) = \int_D \frac{1}{2} r_{im} q_i q_m dD + \int_{\partial_3 D} \frac{1}{2h} (n_m q_m)^2 dS \tag{25}$$

and a 'dual or complementary potential energy'

$$Q(\vec{q}) = \int_{\partial_1 D + \partial_3 D} \bar{T}_e n_m q_m dS \, . \tag{26}$$

The upper and lower bound character to the dissipation functional of approximations based on either

$$I(T) = F(T) + P(T) \quad \text{minimum} \tag{27}$$

or

$$J(\vec{q}) = F(\vec{q}) + Q(\vec{q}) \quad \text{minimum} \tag{28}$$

is briefly discussed in Appendix 1.

FINITE ELEMENT ANALYSIS

The temperature models of finite elements have been fully described elsewhere [1] as based on a continuous, piecewise differentiable temperature field with a finite number of degrees of freedom. Note, however, an essential difference in the treatment of the heat convection boundary condition (5). This case is no longer treated by additional generalized fluxes but results in essentially different elements because of the addition of the second term in the dissipation functional (10). This modification is necessary to obtain bounds to this functional. With temperature models, temperature boundary specifications $\overline{T}_e$ can be accounted for exactly if expressible in terms of the approximating shaping functions of the models. Heat flux boundary specifications $\bar{q}$ are translated into equivalent generalized fluxes, and boundary conditions such as (4) are only satisfied on the average. The same is true for thermal equilibrium in volume, equation (3) and there are discontinuities in heat flux transmission between adjacent elements. This behaviour is analogous to the behaviour of approximate stresses in conforming displacement models for structural analysis.

The heat flow models based on the dual principle (28) must, by contrast, satisfy in detail the thermal equilibrium conditions (7) and (8). In general, the finite degree of freedom approximations on the heat flow do not satisfy the integrability conditions for the temperature field required by equations (14). Similarly, temperature boundary conditions such as (2) and (9) are only averaged. This averaging process results automatically from the definitions of generalized temperature co-ordinates as weighted averages along the boundaries of the element or in its interior, so that no restrictive conditions exist on the prescribed temperatures input. This situation is again completely analogous to the one governing the theory of equilibrium models in elasticity [3]. Particular emphasis will be given here on the heat flow models and, for brevity, only internally unloaded ($\overline{Q} = 0$) elements will be considered.

HEAT FLOW ELEMENTS GENERATION

Let E denote the simply connected domain occupied by the element and ∂E its boundary. The boundary consists of several parts $\partial_\sigma E$ ($\sigma = 1,2,\ldots,s$) corresponding to various adjacent elements or to pieces of the boundary of the complete structure.

The heat flow vector in E is taken as a linear superposition of 'modes' $M_{ij}(x)$, each in thermal equilibrium without internal sources:

$$q_i = \alpha_j M_{ij}(x) \tag{29}$$

$$\partial_i M_{ij} = 0 \quad (j = 1,2,\ldots,n) \ . \tag{30}$$

Consider now the normal heat flux generated along part $\partial_\sigma E$ of the boundary by each of the heat flow modes

$$f_j(x) = n_i(x) M_{ij}(x) \ , \qquad x \in \partial_\sigma E \ . \tag{31}$$

Determine a maximal set of independent vectors γ_{jm} ($m = 1,2,\ldots,M_\sigma$) such that

$$\gamma_{jm} f_j(x) = 0 \quad x \in \partial_\sigma E \ . \tag{32}$$

Then, any complementary set of vectors η_{jp} $(p = 1,2,\ldots,P_\sigma;$ $M_\sigma + P_\sigma = n)$ such that the n × n matrix, j indexing the columns,

$$\underline{G} = \begin{pmatrix} \gamma_{jm} \\ \\ \eta_{jp} \end{pmatrix}$$

is non-singular, will generate a set of non-zero and independent boundary flux modes along $\partial_\sigma E$

$$\eta_{jp} f_j(x) = b_p(x) \ , \ x \in \partial_\sigma E \quad (p = 1,2,\ldots,P_\sigma) \ . \tag{33}$$

Introducing the matrix notations

$$\underline{f}_\sigma^T(x) = (f_1(x)\ldots f_n(x)) \ , \quad x \in \partial_\sigma E$$

$$\underline{b}_\sigma^T(x) = (b_1(x)\ldots b_{P_\sigma}(x)) \ , \ x \in \partial_\sigma E$$

(32) and (33) are equivalent to the matrix equation

$$\underline{G}_\sigma \underline{f}_\sigma(x) = \begin{bmatrix} 0 \\ \\ \underline{b}_\sigma(x) \end{bmatrix} , \quad x \in \partial_\sigma E \quad . \tag{34}$$

Inversion of this produces

$$\underline{f}_\sigma(x) = \underline{B}_\sigma^T \underline{b}_\sigma(x) \ , \qquad x \in \partial_\sigma E \tag{35}$$

where $\underline{B}_\sigma^T$, which is a part of $\underline{G}_\sigma^{-1}$, is a $n \times P_\sigma$ matrix of rank P_σ. Thus, along $\partial_\sigma E$, the general parametric heat flow will generate a boundary flux of type

$$\underline{a}^T \underline{f}_\sigma(x) = (\underline{B}_\sigma \underline{a})^T \underline{b}_\sigma(x) \ , \quad x \in \partial_\sigma E \quad . \tag{36}$$

The set of 'generalized flux co-ordinates' for $\partial_\sigma E$ is defined by

$$\underline{g}_\sigma = \underline{B}_\sigma \underline{a} \ , \tag{37}$$

so that (36) becomes equivalent to

$$\alpha_j f_j(x) = g_p b_p(x) \ , \quad x \in \partial_\sigma E \ . \tag{38}$$

The generalized flux co-ordinates are thus intensities of the boundary heat flux modes; their knowledge determines the complete heat flux distribution along $\partial_\sigma E$ so that the continuity of heat fluxes at interfaces can be secured simply by reciprocity of corresponding generalized flux co-ordinates. Any predetermined set of boundary heat flux modes may be replaced by a new set of P_σ linearly independent combinations of the elements of the old set. Advantage can be taken of this freedom to give simple physical significance to the generalized flux co-ordinates, either as local values or moments of various order.

The set of relations (37) for all parts of the boundary ∂E results in a global relation

$$\underline{g} = \underline{B}\ \underline{a} \tag{39}$$

where the boundary flux co-ordinates are listed in $\underline{g}$ in some conventional order and $\underline{B}$ correspondingly partitioned. As $\underline{a}$ determines $\underline{g}$, we should have conversely that a compatible set $\underline{g}$ determines $\underline{a}$ uniquely. The condition therefore is that $\underline{B}\ \underline{a} = 0$ possesses only the trivial solution $\underline{a} = \underline{0}$. If this is not true there are 'bubble heat flow modes' that were not taken into consideration. Bubble heat flow modes are linear combinations of the modes $M_{ij}(x)$ that produce no heat flux along any part of the boundary ∂E and are therefore not accounted for by any of the generalized boundary heat flux co-ordinates. This happens only for models involving a large number of parameters; if the $M_{ij}(x)$ form a complete set of polynomials of degree $\leq N$ satisfying (30), the bubble modes begin to appear for N = 2. In the examples only N = 0 and N = 1 have been considered.

It is sufficient to state here that an internal generalized heat flow co-ordinate can be assigned to each bubble mode and relation (39) completed so that $\underline{B}$ becomes really of rank n and $\underline{B}\ \underline{a} = \underline{0}$ has no other that the trivial solution. The internal co-ordinates can be expressed later in terms of the boundary co-ordinates by minimizing the functional (28) at the element level with respect to the internal co-ordinates.

The boundary co-ordinates themselves are not independent: the homogeneous adjoint relation to (39)

$$\underline{B}^T \underline{z} = \underline{0} \tag{40}$$

has at least one non-trivial solution $\underline{z} = \underline{v}$. The corresponding necessary condition for (39) to be solvable for $\underline{a}$

$$\underline{v}^T \underline{g} = 0 \tag{41}$$

expresses the vanishing of the total outgoing heat flux; it always exists because of the detailed equilibrium conditions (30) and is the analogue of the overall equilibrium conditions between generalized forces in elasticity. If equation (40) possesses no other non-trivial solution, equation (41) is also sufficient for the existence of $\underline{a}$, the uniqueness of which was already discussed. This is the case for the heat flow models presented as examples in this paper.

If equation (4) has other non-trivial solutions we have a situation analogous to that of 'spurious kinematic freedoms'[3]; techniques to deal with such situations are similar to those of the equilibrium models in elasticity but will not be presented here.

Consider now the discretization of the functional (28) resulting from (29) or its equivalent matrix form

$$\underline{q} = \{q_i\} = \underline{M}\ \underline{a} \tag{42}$$

where $\underline{M}$ is the matrix of the modal functions $M_{ij}(x)$. The boundary fluxes needed to evaluate the second part of the dissipation functional are given by combination of equations (29), (32) and (36) as

$$n_m q_m = \underline{a}^T \underline{B}_\sigma^T \underline{b}_\sigma(x) \ , \qquad x \varepsilon \partial_\sigma E \ . \tag{43}$$

The dissipation functional (25) becomes a positive definite quadratic form in the parameters α_j

$$F(\vec{q}) = \frac{1}{2} \underline{a}^T (\underline{R}+\underline{R}_c)\underline{a} \tag{44}$$

with generalized resistivity matrices

$$\underline{R} = \int_E \underline{M}^T \underline{C}^{-1} \underline{M} dE \ , \qquad \underline{C} = (k_{ij}) \tag{45}$$

$$\underline{R}_c = \sum_\sigma \underline{B}_\sigma^T \left(\int_{\partial\sigma E} \frac{1}{h} \underline{b}_\sigma \underline{b}_\sigma^T dS \right) \underline{B}_\sigma \ . \tag{46}$$

Advantage is taken of the subdivision of the structure in elements to incorporate eventual heat convection losses at interfaces where bodies of different physical nature are in contact and not only along the external boundary $\partial_3 D$ of the complete domain D. Each element E is therefore provided with a (variable if the case may be) transfer coefficient h along its boundaries ∂E.

At an interface without external heat supply it is readily seen that the effective transfer coefficient is given by

$$\frac{1}{h} = \frac{1}{h_+} + \frac{1}{h_-}$$

where h_+ and h_- have been allocated at each face. Moreover an interface can be kept at a prescribed temperature $\overline{T}_e$ by supplying or removing heat at a surface rate that will result from the computation, or this supply rate will be prescribed and the temperature

T_e will result from the computation. In both cases it is necessary to distinguish h_+ and h_-.

For faces without convection losses h can be set equal to infinity. The complementary potential energy (26) is then calculated by assuming the external temperature T_e to be defined along the whole boundary ∂E; it then becomes formally

$$Q(\vec{q}) = \underline{a}^T \sum_\sigma \left(\underline{B}_\sigma^T \int_{\partial\sigma E} T_e \underline{b}_\sigma dS \right) . \tag{47}$$

We now define generalized temperature co-ordinates along each boundary by

$$t_p = \int_{\partial\sigma E} T_e b_p(x) dS \quad , \quad x \varepsilon \partial_\sigma E \quad (p = 1,2,\ldots,P_\sigma) . \tag{48}$$

Collected in a column matrix in the same order as for $\underline{g}_\sigma$ they yield

$$\underline{t}_\sigma = \int_{\partial\sigma E} T_e \underline{b}_\sigma dS \tag{49}$$

so that we can write

$$Q(\vec{q}) = \underline{a}^T \sum_\sigma \underline{B}_\sigma^T \underline{t}_\sigma . \tag{50}$$

Again if the co-ordinates of each boundary are collected in a single matrix $\underline{t}$, in the same order as the $\underline{g}_\sigma$ into $\underline{g}$,

$$\sum_\sigma \underline{B}_\sigma^T \underline{t}_\sigma = \underline{B}^T \underline{t} \tag{51}$$

and finally

$$Q(\vec{q}) = \underline{a}^T \underline{B}^T \underline{t} = \underline{g}^T \underline{t} . \tag{52}$$

This shows that the generalized temperature co-ordinates are conjugate variables to the generalized heat fluxes. Definition (48) shows that they are weighted averages of the external temperature distribution, the weighting functions being the boundary heat flux modes.

The minimization of the discretized functional

$$\frac{1}{2}\,\underline{a}^T(\underline{R}+\underline{R}_c)\underline{a} + \underline{a}^T\underline{B}^T\underline{t}$$

with respect to the parameters furnishes

$$(\underline{R}+\underline{R}_c)\underline{a} = -\underline{B}^T\underline{t} \tag{53a}$$

or

$$\underline{a} = -(\underline{R}+\underline{R}_c)^{-1}\underline{B}^T\underline{t}\ . \tag{53b}$$

Combined with (39) this result allows to relate the generalized boundary heat fluxes to the generalized temperatures through a generalized conductivity matrix $\underline{K}$

$$\underline{g} = \underline{Kt} \tag{54}$$

with

$$\underline{K} = -\underline{B}(\underline{R}+\underline{R}_c)^{-1}\underline{B}^T\ . \tag{55}$$

The complete structural conductivity matrix is built up by properly addressing the elemental matrices $\underline{K}$ just as for stiffness matrices in structural analysis. The principle used here is the identification of the interface external temperature co-ordinates, so that corresponding heat flux co-ordinates become either reciprocal or balance external interface heat inputs. It should be observed that the structure (54) of the generalized conductivity matrix is such that the compatibility condition (41) is automatically satisfied by $\underline{v}^T\underline{B} = \underline{0}$.

Appendices II and III present the detailed matrices for constant and linearly varying two-dimensional heat flow elements, with orthotropic conductivity characteristics.

EXAMPLES

The following examples illustrate the main features described above in the case of constant and linear heat flow elements.

Example I

Let us consider a large composite wall with an isothermal surface at T_0, the other suffering convection losses (h,T_f). The uni-dimensional heat flow is analysed through a mesh of twelve linear heat flow elements. In this case the temperature gradient field is itself integrable so that the generalized temperatures obtained yield the local ones (Figure 1).

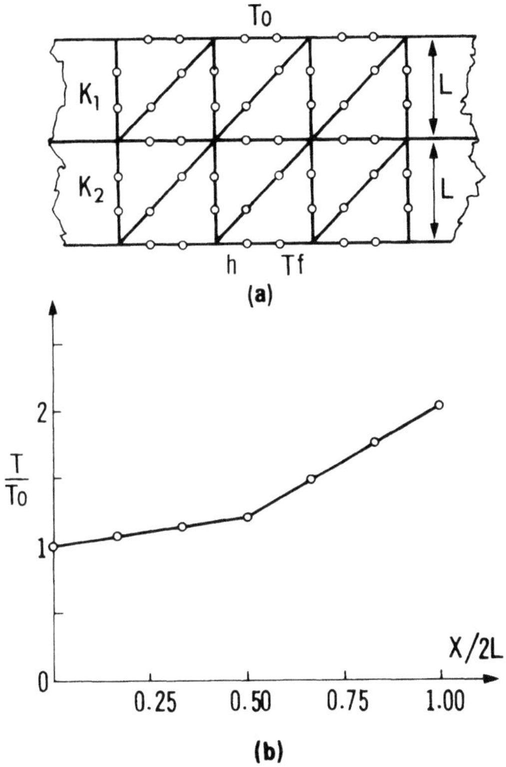

Figure 1 - An Infinite Composite Wall. (a) Mesh Idealization with Boundary Conditions; forty-six degrees of freedom. (b) Unidimensional Temperature plot showing perfect accuracy towards the exact solution (analytical).

Example 2

This attempts to show that the generalized temperatures give a good idea of local values even in case of non-integrable gradient fields. Figure 2 illustrates a conduction-cooled turbine blade having its root x = 0 maintained at a uniform temperature T_0 and operating in an effective gas with temperature T_f (exact solution under unidimensional heat flow assumption, Reference 5).

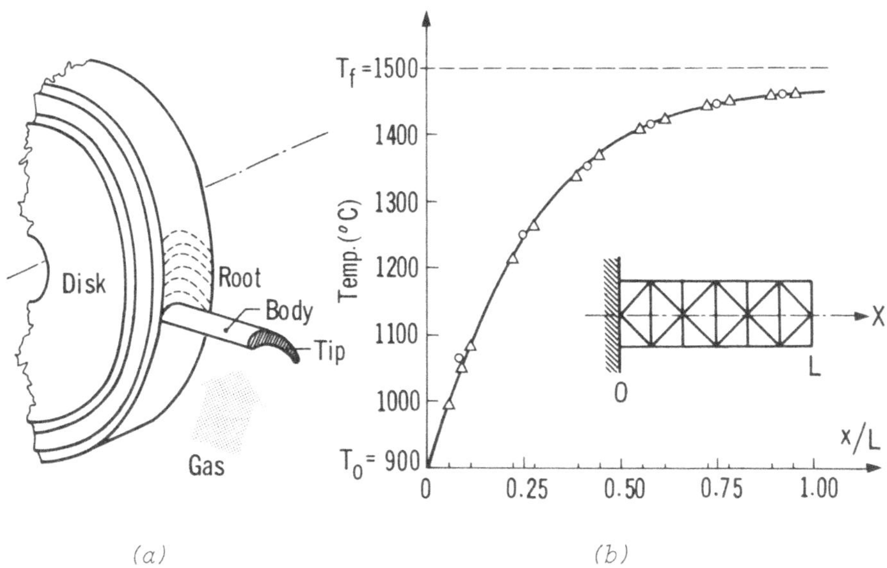

Figure 2 - Turbine Blade. (a) Blade on cooled turbine disk, with insulated tip, body in ambient gas at T_f and isothermal root disk at T_0. (b) Finite element idealization and temperature plots for blade core: -, exact; 0, constant heat flow model; Δ, linear heat flow model.

A dual analysis of the structure yields upper and lower bounds (Table 1). The larger dissipation of the exact treatment is due to the assumption of unidimensionality.

Table 1 - Dissipation Bounds for Several Models of the Blade

Finite Element Model	Degrees of Freedom	10^{-4} E
Linear Temperature Field	23	20.91
Parabolic Temperature Field	67	20.53
Linear Heat Flow Field	88	20.47
Constant Heat Flow Field	44	19.55
Exact Dissipation Under Unidimensional Heat Flow Assumption		20.76

Example 3

In this example, a dual analysis of a laminated slab is presented (Figure 3).

No exact solution is known, so that temperature mapping is compared to a cubic temperature model treatment, which is nearly exact with respect to energy bounds. Heat flow models tend to converge faster than temperature ones; however, they need more degrees of freedom for a given grid and equal degrees in the polynomial discretization. Table 3 shows remarkable accuracy for heat flux equilibrium at the boundary.

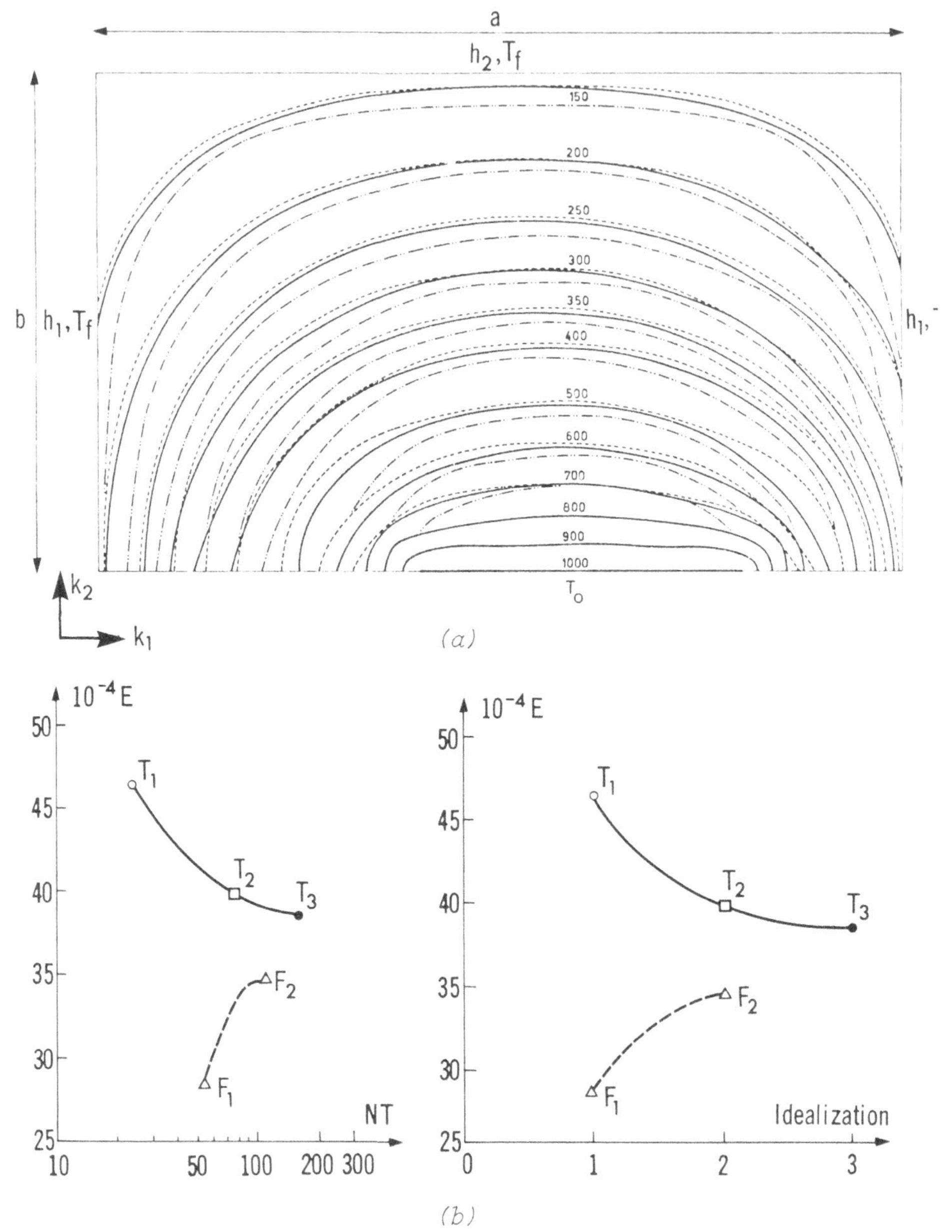

Figure 3 - Laminated Slab. (a) Slab with boundary conditions ($k_1 = 0.8$ - $k_2 = 0.4$ W/cm°C; $h_1 = 0.9$ - $h_2 = 0.5$ W/cm^2°C; $T_o = 1{,}000$ - $T_f = 100$°C; $a = 10$ - $b = 6$ cm) and temperature maps: -, cubic temperature model; --, parabolic temperature model; -··-, linear heat flow model. (b) Dissipation bounds showing convergence versus degrees of freedom or degrees of polynomial idealization.

Table 2 - Notation and Number of Degrees of Freedom for Figure 3

Elements	Symbol	Degrees of Freedom
Linear Temperature Field	O	27
Parabolic Temperature Field	□	80
Cubic Temperature Field	⊙	163
Constant Heat Flow Field	Δ	53
Linear Heat Flow Field	∇	106

Table 3 - Computed Heat Flux Equilibrium for the Slab (watt/cm)

Idealization	Total Outgoing Heat Flux Through the Convection Boundary	Isothermal Boundary (part of lower edge)
T_1	1031·607	-1031·607
T_2	887·318	-887·318
T_3	858·147	-858·147
F_2	772·272	-772·272
F_1	636·289	-636·288

Example 4

This final example considers a heat-exchanger tube with a non-uniform surface conductance on its outer surface (Figure 4).

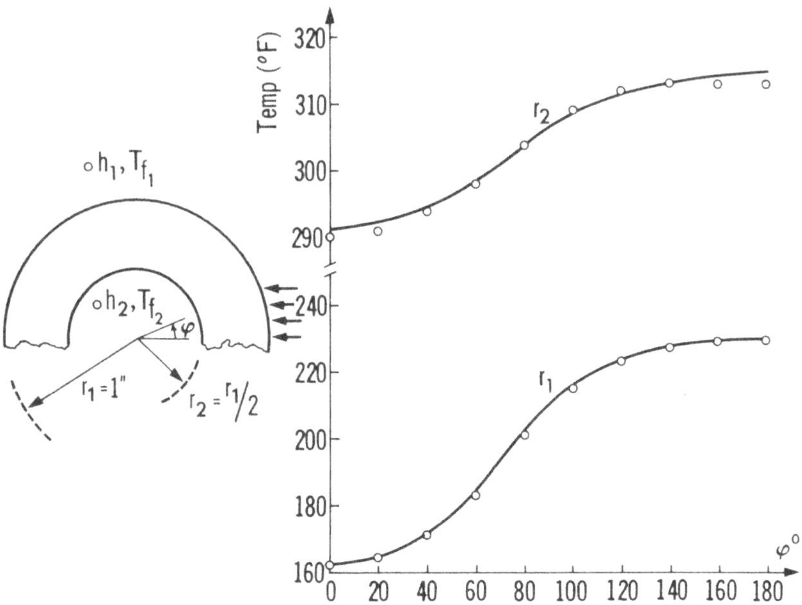

Figure 4 - Temperature Distribution in a Cross-Flow Heat-Exchanger Tube of Unit Thermal Conductivity, with High Temperature Gas at T_{f2} = 370°F and Cooling Gas at T_{f1} = 70°F, Flowing Normal to the Tube (⊖, Relaxation; -, Finite Elements)

Upper and lower bounds (Table 5) allow to consider the T_2-solution as exact. The corresponding solution is plotted versus a finite difference relaxation solution (Reference 5).

Table 4 - Distribution of Surface Conductances

ϕ^0	h_1 (Btu/h ft^2.°F)	h_2
0-10	25	55
10-30	24	55
30-50	22	55
50-70	18	55
70-90	12	55
90-110	10	55
110-180	9	55

Table 5 - Dissipation Bounds for the Exchanger Tube

Finite Element Model	Degrees of Freedom	$2 \times 10^{-4}E$
Linear Temperature Field	35	13•970
Parabolic Temperature Field	107	13•875
Linear Heat Flow Field	144	13•871
Constant Heat Flow Field	72	13•677

REFERENCES

[1] ZIENKIEWICZ, O.C. and CHEUNG, Y.K., *The Finite Element Method in Structural and Continuum Mechanics*, McGraw-Hill, London, 1967, Chap. 10.

[2] BIOT, M.A., "New Methods in Heat Flow Analysis with Application to Flight Structures", *J. Aeronaut. Sci.*, Vol. 24, 1957, pp. 857-873.

[3] FRAEIJS DE VEUBEKE, B.M., "Displacement and Equilibrium Models", in *Stress Analysis*, Wiley, London, 1965, Chap. 9.

[4] COURANT, R. and HILBERT, D., *Methods of Mathematical Physics*, Vol. 1, Wiley-Interscience, New York, 1953.

[5] SCHNEIDER, P.J., *Conduction Heat Transfer*, Addison-Wesley, Reading, 1957.

APPENDIX I

Upper and Lower Bounds

We now sketch briefly the theory of functional bounding by the dual approach. Let q denote a vector of a linear space F, representing a field of heat flow in each element of the structure. Define a scalar product in F by

$$(q,\hat{q}) = \int_D r_{ij} q_i \hat{q}_j dD + \int_{\Sigma\partial E} \frac{1}{h} (n_i q_i)(n_m \hat{q}_m) dS = (\hat{q},q) \ . \tag{56}$$

The squared norm (q,q) is equal to twice the dissipation functional for the whole structure and its surrounding.

$$(q,q) = 0 \quad \text{implies} \quad q = 0$$

i.e., the complete vanishing of all heat flows.

A field will be called integrable if a temperature field exists in each element such that

$$\partial_m T = -r_{im} q_i \tag{14'}$$

and such that the external face temperature

$$T_e = T - (n_i q_i / h) \tag{57}$$

be continuous at interfaces. If i denotes an integrable field, we have from equation (56)

$$(i,\hat{q}) = - \int_D \hat{q}_j \partial_j T dD + \int_{\Sigma\partial E} (T - T_e) n_m \hat{q}_m dS$$

or, after integration by parts,

$$(i,\hat{q}) = \int_D T \partial_j \hat{q}_j dD - \int_{\Sigma\partial E} T_e n_m \hat{q}_m dS \ . \tag{58}$$

A temperature compatible field is an integrable field with the additional conditions

$$T_e = \overline{T}_e \quad \text{along} \quad \Sigma_1 \partial E \tag{59}$$

where $\Sigma_1 \partial E$ comprises the set of interfaces and boundary faces of ∂D where the external temperature is prescribed. Any particular temperature compatible field is denoted by c_0. The set of all temperature compatible fields satisfying the homogeneous conditions

$$T_e = 0 \qquad \text{along } \Sigma_1 \partial E \tag{60}$$

where $T_e = 0$ is some uniform reference temperature, is a subspace of F denoted by C. An arbitrary member of C is denoted by c. Thus c_0+c is an arbitrary temperature compatible field for equation (59) and belongs to the linear manifold c_0+C.

A balanced heat flow field is one for which

$$\partial_j \hat{q}_j = \overline{Q} \quad \text{within each element} \tag{61}$$

and

$$n_m \hat{q}_m = \bar{q} \quad \text{along } \Sigma_2 \partial E \tag{62}$$

where $\Sigma_2 \partial E$ comprises the boundary faces of ∂D where surface sources $\bar{q}$ are prescribed and the set of interfaces where surface sources $(\bar{q}_+ + \bar{q}_-)$ are prescribed. Note that a positive $\bar{q}$ is really a heat sink. The set $\Sigma_2 \partial E$ is complementary to $\Sigma_1 \partial E$. Any particular balanced heat flow field is denoted by h_0. The set of all balanced heat flow fields satisfying the corresponding homogeneous conditions

$$\partial_j \hat{q}_j = 0 \quad \text{in} \quad D \,, \quad n_m \hat{q}_m = 0 \quad \text{along } \Sigma_2 \partial E \tag{63}$$

is a subspace of F denoted by H. An arbitrary member of H is denoted by h. Thus h_0+h is an arbitrary heat flow field balanced with respect to conditions (61) and (62) and belongs to the linear manifold h_0+H.

From equations (58) - (63) we readily find

$$(c_0,h) = -\int_{\Sigma_1 \partial E} \overline{T}_e n_m \hat{q}_m dS \tag{64}$$

$$(c,h_0) = \int_D T\overline{Q}dD - \int_{\Sigma_2 \partial E} T_e \bar{q} dS \tag{65}$$

$$(c_0,h_0) = (c_0,h) + (c,h_0) \tag{66}$$

$$(c,h) = 0 \ . \tag{67}$$

This last result shows that subspaces C and H are orthogonal.

The exact solution s must satisfy simultaneously equations (14), (57), (59), (61) and (62); it is the intersection of the linear manifolds c_0+C and h_0+H. In approaching the solution from two sides either by temperature compatible fields or by heat balanced flows, it is thus natural to require that the squared distance between each approximation

$$\rho^2 = (c_0 + c - h_0 - h,\ c_0 + c - h_0 - h)$$

be minimum. Using property (67) it is easily verified that

$$\rho^2 = (c_0+c-h_0,\ c_0+c-h_0) + (h_0+h-c_0,\ h_0+h-c_0) - (h_0-c_0,\ h_0-c_0) \ .$$

The last term is constant; the first term is positive and depends only on the choice of the arbitrary element c; the second is also positive and depends only on the choice of the arbitrary element h. Thus we can minimize separately the first and second term. In the first we drop the constant (h_0, h_0) and, dividing by 2, obtain a statement equivalent to the minimization of the temperature functional:

$$\frac{1}{2}(c_0+c,\ c_0+c) - (c_0+c,\ h_0) \quad \text{minimum} \ . \tag{68}$$

In the second we drop the constant (c_0, c_0), divide by 2, and obtain a statement equivalent to the minimization of the dual function:

$$\frac{1}{2}(h_0+h,\ h_0+h) - (h_0+h,\ c_0) \qquad \text{minimum}\ . \tag{69}$$

(Complete analogous statements to equations (27) and (28) are in fact:

(1) Principle of variation of temperatures:

$$\frac{1}{2}(c_0+c-h_0,\ c_0+c-h_0) - \frac{1}{2}(h_0-c_0,\ h_0-c_0)\ ,\ \text{minimum}$$

or

$$\frac{1}{2}(c_0+c,\ c_0+c) - (c,\ h_0),\ \text{if we drop the constant}\ (c_0,\ c_0).$$

(2) Principle of variation of heat flow:

$$\frac{1}{2}(h_0+h-c_0,\ h_0+h-c_0) - \frac{1}{2}(h_0-c_0,\ h_0-c_0)\ ,\ \text{minimum}$$

or

$$\frac{1}{2}(h_0+h,\ h_0+h) - (h,\ c_0),\ \text{if we drop the constant}\ (h_0,\ h_0).)$$

Any approximate solution to (68) by discretization in n degrees of freedom consists in taking $c = \alpha_i c_i$, where the c_i are chosen members of C, the summation index running from 1 to n, and determining the best coefficients α_i^*. Then (68) becomes

$$\frac{1}{2}\alpha_i\alpha_j(c_i,c_j) + \alpha_i(c_i,c_0-h_0) + \frac{1}{2}(c_0,\ c_0) - (h_0,\ c_0)\ ,\ \text{minimum}$$

and furnishes the set of linear equations in the unknowns α_i^*

$$\alpha_j^*(c_i,c_j) + (c_i,c_0-h_0) = 0 \quad (i = 1,2,\ldots,n)\ . \tag{70}$$

With $c_0^* = c_0+\alpha_i^* c_i$ denoting the best approximation, equation (70) can also be placed in the form

$$(c_i, c_0^* - h_0) = 0 \quad (i = 1,2,\ldots,n) .$$

Then, multiplying each equation by α_i^* and adding

$$(c_0^* - c_0,\ c_0^* - h_0) = 0 . \tag{71}$$

A similar treatment of problem (69) yields

$$(h_0^* - c_0,\ h_0^* - h_0) = 0 . \tag{72}$$

Obviously the exact solution obeys the same relation

$$(s - c_0,\ s - h_0) = 0 . \tag{73}$$

This can in fact be considered as an application of (67) since $s - c_0$ belongs to C and $s - h_0$ belongs to H.

Consider now the inequalities

$$\frac{1}{2}(c_0^*,\ c_0^*) - (c_0^*,\ h_0) \geq \frac{1}{2}(s,\ s) - (s,\ h_0) , \tag{74}$$

$$\frac{1}{2}(h_0^*,\ h_0^*) - (h_0^*,\ c_0) \geq \frac{1}{2}(s,\ s) - (s,\ c_0) , \tag{75}$$

resulting from the fact that the minima of (68) and (69) are reached by the exact solution. Eliminating in them (s, s) through (73) and combining with (71) and (72) there comes

$$(c_0^*,\ h_0 - c_0) \leq (s,\ h_0 - c_0) \leq (h_0^*,\ h - c_0) . \tag{76}$$

This upper and lower bound characteristic for the linear functional

$$(s, h_0 - c_0) = (s, h_0) - (s, c_0) = (c + c_0, h_0) - (h + h_0, c_0)$$

$$= (c, h_0) - (h, c_0)$$

$$= \int_D T\bar{Q}\,dD + \int_{\Sigma_1 \partial E} \bar{T}_e n_m q_m\,dS - \int_{\Sigma_2 \partial E} T_e \bar{q}\,dS$$

can be used directly in solving a mixed boundary value problem but does not guarantee convergence in norm. To obtain a measure of convergence in norm it is necessary to solve the problem in two steps by superposition.

First step

$c_0 = 0$, the prescribed temperatures are set equal to zero, only the prescribed heat sources are taken into account. Then $c_0^* = c^*$ and the previous result becomes

$$(c^*,h_0) \leq (s,h_0) \leq (h_0^*,h_0) \; .$$

However from expressions (71)-(73) also

$$(c^*,c^*) = (c^*,h_0) \; , \; (s,s) = (s,h_0) \; , \; (h_0^*,h_0^*) = (h_0^*,h_0^*)$$

so that we obtain upper and lower bounds for the quadratic dissipation functional

$$(c^*,c\;) \leq (s,s) \leq (h_0^*,h_0^*) \quad .$$

Second step

$h_0 = 0$, the flow rate of the prescribed heat sources is set equal to zero but account is taken of the prescribed temperatures. Then $h_0^* = h^*$ and (76) becomes

$$(c_0^*,\; c_0) \geq (s,\; c_0) \geq (h^*,\; c_0) \; .$$

Furthermore from expressions (71) - (73)

$$(c_0^*,c_0) = (c_0^*,c_0^*), (s,c_0) = (s,s), (h^*,c_0) = (h^*,h^*)$$

and the dissipation functional of that case is now bracketed in the reverse way

$$(c_0^*,c_0^*) \geq (s,s) \geq (h^*,h^*) \; .$$

APPENDIX II

Constant Heat Flow Element

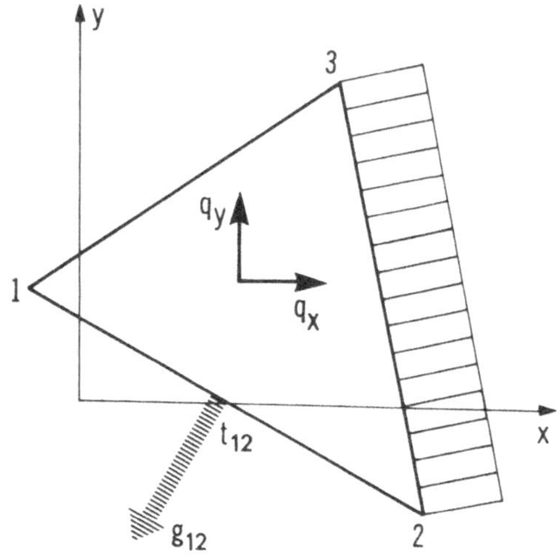

Figure 5 - Constant Heat Flow Element

$$q_x = \alpha_1, \quad q_y = \alpha_2, \qquad n = 2, \quad \underline{M} = \begin{pmatrix} 1 & 0 \\ 0 & 1 \end{pmatrix} = \underline{E}, \quad \underline{a}^T = (\alpha_1 \alpha_2)$$

$$M_\sigma = 1, \quad \underline{f}_\sigma^T = \frac{1}{L_{ij}}(y_{ij} x_{ij}), \qquad \underline{G}_\sigma = \begin{pmatrix} x_{ij} & -y_{ji} \\ y_{ji}/L_{ij}^2 & x_{ij}/L_{ij}^2 \end{pmatrix}$$

$$P_\sigma = 1, \quad \underline{b}_\sigma^T = \frac{1}{L_{ij}}$$

$$\underline{B}_\sigma = (y_{ji} x_{ij}), \quad \underline{g}_\sigma^T = \alpha_1 y_{ji} + \alpha_2 x_{ij},$$

$$\underline{t}_\sigma^T = \frac{1}{L_{ij}} \int_{\partial_\sigma E = ij} T_e dS,$$

where

L_{ij} = Length of edge i-j, $x_{ij} = x_i - x_j$, $y_{ij} = y_i - y_j$

$\underline{g}^T = (g_{12} g_{23} g_{31})$, $\underline{t}^T = (t_{12} t_{23} t_{31})$

$$\underline{B} = \begin{pmatrix} y_{21} & x_{12} \\ y_{32} & x_{23} \\ y_{13} & x_{31} \end{pmatrix}, \quad R = \Delta \begin{pmatrix} 1/k_x & 0 \\ 0 & 1/k_y \end{pmatrix},$$

$$\underset{\sim}{R}_c = \begin{pmatrix} \sum_\sigma \frac{y_{ii}^2}{h_{ij}L_{ij}} & \text{SYM} \\ \sum_\sigma \frac{x_{ij}y_{ji}}{h_{ij}L_{ij}} & \sum_\sigma \frac{x_{ij}^2}{h_{ij}L_{ij}} \end{pmatrix}$$

where

Δ = Area of the element, h_{ij} = Uniform transfer coefficient for edge i-j (= ∞ if $(T)_{ij} = (T_e)_{ij}$).

APPENDIX III

Linear Heat Flow Element

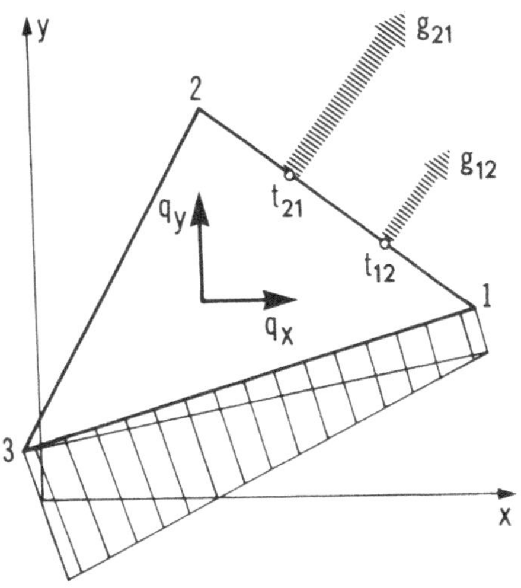

Figure 6 - Linear Heat Flow Element

$$q_x = \alpha_1 + \alpha_5 x + \alpha_4 y\,, \qquad n = 5\,, \qquad \underline{M} = \begin{pmatrix} 1 & 0 & 0 & y & x \\ 0 & 1 & x & 0 & -y \end{pmatrix},$$

$$q_y = \alpha_2 + \alpha_3 x - \alpha_5 y\,,$$

$$\underline{a}^T = (\alpha_1 \alpha_2 \alpha_3 \alpha_4 \alpha_5)$$

$$M_\sigma = 3 , \quad \underline{f}_\sigma^T = \frac{1}{L_{ij}} (y_{ji} \;\; x_{ij} \;\; xx_{ij} \;\; yy_{ji} \;\; xy_{ji}-yx_{ij})$$

$$P_\sigma = 2 , \quad \underline{b}_\sigma^T = \frac{2}{x_{ij}L_{ij}} (x-x_j \;\; x_i-x)$$

$$= \frac{2}{y_{ji}L_{ij}} (y_j-y \;\; y-y_i)$$

$$= \frac{2}{L_{ij}^2} (L_{ij}-s \;\; s)$$

$$\underline{B}_\sigma = \frac{1}{2}\begin{pmatrix} y_{ji} & x_{ij} & x_i x_{ij} & y_i y_{ji} & x_i y_{ji}-y_i x_{ij} \\ y_{ji} & x_{ij} & x_j x_{ij} & y_j y_{ji} & x_j y_{ji}-y_j x_{ij} \end{pmatrix}$$

$$\underline{g}_\sigma = \frac{1}{2}\begin{pmatrix} \alpha_1 y_{ji}+\alpha_2 x_{ij}+\alpha_3 x_i x_{ij}+\alpha_4 y_i y_{ji}+\alpha_5(x_i y_{ji}-y_i x_{ij}) \\ \alpha_1 y_{ji}+\alpha_2 x_{ij}+\alpha_3 x_j x_{ij}+\alpha_4 y_j y_{ji}+\alpha_5(x_j y_{ji}-y_j x_{ij}) \end{pmatrix}$$

$$\underline{t}_\sigma^T = \frac{2}{L_{ij}^2}\left(\int_{\partial_\sigma E=ij} T_e(L_{ij}-s)ds \quad \int_{\partial_\sigma E=ij} T_e s ds \right)$$

$$\underline{g}^T = (g_{12} \;\; g_{21} \;\; g_{23} \;\; g_{32} \;\; g_{31} \;\; g_{13})$$

$$\underline{t}^T = (t_{12} \;\; t_{21} \;\; t_{23} \;\; t_{32} \;\; t_{31} \;\; t_{13})$$

$$\underline{B} = \frac{1}{2}\begin{pmatrix} y_{21} & x_{12} & x_1 x_{12} & y_1 y_{21} & x_1 y_{21}-y_1 x_{12} \\ y_{21} & x_{12} & x_2 x_{12} & y_2 y_{21} & x_2 y_{21}-y_2 x_{12} \\ y_{32} & x_{23} & x_2 x_{23} & y_2 y_{32} & x_2 y_{32}-y_2 x_{23} \\ y_{32} & x_{23} & x_3 x_{23} & y_3 y_{32} & x_3 y_{32}-y_3 x_{23} \\ y_{13} & x_{31} & x_3 x_{31} & y_3 y_{13} & x_3 y_{13}-y_3 x_{31} \\ y_{13} & x_{31} & x_1 x_{31} & y_1 y_{13} & x_1 y_{13}-y_1 x_{31} \end{pmatrix}$$

$$\underline{R} = \begin{pmatrix} \Delta/k_x & & & & \\ 0 & \Delta/k_y & & \text{SYM} & \\ 0 & S_x/k_y & & & \\ S_y/k_x & 0 & 0 & I_y/k_x & \\ S_x/k_x & -S_y/k_y & -J_{xy}/k_y & J_{xy}/k_x & I_x/k_x+I_y/k_y \end{pmatrix}$$

where

$$\Delta = \int_E dxdy\ , \quad S_x = \int_E xdxdy\ , \quad I_x = \int_E x^2dxdy\ , \quad J_{xy} = \int_E xydxdy\ , \text{ etc.}$$

$$\underline{R}_c = \begin{bmatrix} R_{11} & & & & \\ R_{21} & R_{22} & & \text{SYM} & \\ R_{31} & R_{32} & R_{33} & & \\ R_{41} & R_{42} & R_{43} & R_{44} & \\ R_{51} & R_{52} & R_{53} & R_{54} & R_{55} \end{bmatrix}$$

where

$$R_{11} = \sum_\sigma \frac{y_{ji}^2}{h_{ij}L_{ij}}\ , \quad R_{22} = \sum_\sigma \frac{x_{ij}^2}{h_{ij}L_{ij}}\ , \quad R_{21} = \sum_\sigma \frac{x_{ij}y_{ji}}{h_{ij}L_{ij}} = R_{12}$$

$$R_{31} = \sum_\sigma \frac{(x_i+x_j)x_{ij}y_{ji}}{2h_{ij}L_{ij}} = R_{13}\ , \quad R_{32} = \sum_\sigma \frac{(x_i+x_j)x_{ij}^2}{2h_{ij}L_{ij}} = R_{23}$$

$$R_{33} = \sum_\sigma \frac{(x_i^2+x_ix_j+x_j^2)x_{ij}^2}{3h_{ij}L_{ij}}\ , \quad R_{44} = \sum_\sigma \frac{(y_i^2+y_iy_j+y_j^2)y_{ji}^2}{3h_{ij}L_{ij}}$$

$$R_{41} = \sum_\sigma \frac{(y_i+y_j)y_{ji}^2}{2h_{ij}L_{ij}} = R_{14}\ , \quad R_{42} = \sum_\sigma \frac{(y_i+y_j)x_{ij}y_{ji}}{2h_{ij}L_{ij}} = R_{24}$$

$$R_{43} = \sum_\sigma \frac{(2x_iy_i+x_iy_j+x_jy_i+2x_jy_j)x_{ij}y_{ji}}{6h_{ij}L_{ij}} = R_{34}$$

$$R_{51} = \sum_\sigma \frac{(x_i+x_j)y_{ji}^2-(y_i+y_j)x_{ij}y_{ji}}{2h_{ij}L_{ij}} = R_{15}$$

$$R_{52} = \sum_{\sigma} \frac{(x_i+x_j)x_{ij}y_{ji}-(y_i+y_j)x_{ij}^2}{2h_{ij}L_{ij}} = R_{25}$$

$$R_{53} = \sum_{\sigma} \frac{2(x_i^2+x_ix_j+x_j^2)x_{ij}y_{ji} - (2x_iy_i+x_iy_j+x_jy_i+2x_jy_j)x_{ij}^2}{6h_{ij}L_{ij}} = R_{35}$$

$$R_{54} = \sum_{\sigma} \frac{(2x_iy_i+x_iy_j+x_jy_i+2x_jy_j)y_{ji}^2 - (2y_i^2+y_iy_j+y_j^2)x_{ij}y_{ji}}{6h_{ij}L_{ij}} = R_{45}$$

$$R_{55} = \sum_{\sigma} \frac{(y_i^2+y_iy_j+y_j^2)x_{ij}^2 + (x_i^2+x_ix_j+x_j^2)y_{ji}^2 - (2x_iy_i+x_iy_j+x_jy_i+2x_jy_j)x_{ij}y_{j}}{3h_{ij}L_{ij}}$$

A NEW VARIATIONAL PRINCIPLE FOR FINITE ELASTIC DISPLACEMENTS

B.M. Fraeijs de Veubeke

Aerospace Laboratory
University of Liège, Belgium

ABSTRACT—Variational principles for finite elastic displacements have been formulated in terms of Green strain and Kirchhoff-Trefftz stress tensors. The first is a functional of the displacement field only and implies stationarity of the total potential. The second is a canonical principal, in the sense of Friedrichs [1], involving both stresses and displacements and generalizing Reissner's principle [5]. In contrast with the geometrically linearized elasticity theory, it cannot be reduced to a complementary energy principle involving equilibrium stresses only [9-10]. The paper discusses the Levinson [9] and Zubov [11] formulation in terms of displacement gradients and the Piola stress tensor, which, however interesting from a theoretical viewpoint, does not appear suitable for practical applications. A new set of variational principles, of displacement, canonical or complementary energy types, is found to derive from the use of the polar decomposition theorem of the jacobian. It involves the engineering strain tensor and the material rotation. The complementary energy formulation is discussed in terms of first order stress functions. The presence of the rotational degrees of freedom opens the possibility of discretizing the rotational equilibrium equations in approximate solutions.

1. EULERIAN AND LAGRANGIAN STRESS TENSORS

Assume, for simplicity, the initial or reference configuration of an elastic body and its strained or final configuration to be referred to the same cartesian frame. Let x_i (i = 1,2,3) denote the material coordinates, or coordinates of a material point in

Original publication, *Int. Jnl. Engng. Sci.*, Vol. 10, Pergamon Press, 1972, pp. 745-763.

the final configuration. The components of the displacement vector

$$\xi_i - x_i = u_i(x_j) \tag{1}$$

are considered to be functions of the material coordinates. In this Lagrangian point of view, the local transformation from initial to final configurations is governed by the Jacobian matrix

$$\partial\xi_j/\partial x_i = D_i\xi_j \quad . \tag{2}$$

The determinant of this matrix is denoted by J.

An oriented volume element dV in the initial configuration is constructed on three infinitesimal vectors $d_{(1)}\vec{x}, d_{(2)}\vec{x}$ and $d_{(3)}\vec{x}$ as

$$dV = e_{mnp}d_{(1)}x_m d_{(2)}x_n d_{(3)}x_p$$

where e_{mnp} is the alternating tensor. In the final configuration they become respectively $d_{(1)}\vec{\xi}, d_{(2)}\vec{\xi}$ and $d_{(3)}\vec{\xi}$ and generate the the volume element

$$\begin{aligned} d\Omega &= e_{ijk}d_{(1)}\xi_i d_{(2)}\xi_j d_{(3)}\xi_k \\ &= e_{ijk}D_m\xi_i D_n\xi_j D_p\xi_k d_{(1)}x_m d_{(2)}x_n d_{(3)}x_p \\ &= J\ e_{mnp}d_{(1)}x_m d_{(2)}x_n d_{(3)}x_p = J\ dV \quad . \end{aligned} \tag{3}$$

Hence J is a measure of the change in volume. Physically the change in configuration takes place continuously and the volume element cannot collapse to zero nor change its orientation, so that

$$J > 0 \quad \text{everywhere.} \tag{4}$$

Thus the Jacobian matrix is invertible; the elements of its inverse are

$$\partial x_i/\partial \xi_m = \partial_m x_i \tag{5}$$

and yield the following useful formulas

$$\partial_m x_i D_i \xi_j = \delta_{mj} \qquad D_i \xi_j \partial_j x_m = \delta_{im} \; . \tag{6}$$

The oriented surface elements in the initial and final configuration are respectively

$$e_{mnp} d_{(1)} x_n d_{(2)} x_p = n_m dS \tag{7}$$

$$e_{ijk} d_{(1)} \xi_j d_{(2)} \xi_k = \nu_i d\Sigma \tag{8}$$

where n_m and ν_i denote the direction cosines of the normal, taken positive outwards when the surface belongs to the boundary of the elastic body under consideration. Since

$$(D_m \xi_i)\nu_i d\Sigma = e_{ijk} D_m \xi_i D_n \xi_j D_p \xi_k d_{(1)} x_n d_{(2)} x_p$$

$$= J \; e_{mnp} d_{(1)} x_n d_{(2)} x_p = J \; n_m dS$$

the surface elements are related by

$$n_m dS = \frac{1}{J} (D_m \xi_i)\nu_i d\Sigma \quad , \quad \nu_i d\Sigma = J(\partial_i x_m) n_m \; dS \; . \tag{9}$$

In the strained configuration, the 'true' or 'Eulerian' stress tensor τ_{ij} is defined by the force element

$$dF_j = \tau_{ij} \nu_i d\Sigma \tag{10}$$

acting on the oriented surface element. This equation contains the statement of equilibrium of an infinitesimal tetrahedron, three faces of which are parallel to the coordinate planes. It can also be written in the equivalent 'surface equilibrium' form

$$\tau_j = \tau_{ij}\nu_i \tag{11}$$

where the left-hand side represents the 'surface traction' components.

In some infinitesimal change of configuration δu_i, the virtual work performed by the surface tractions, acting on a closed surface Σ bounding the volume Ω, is measured by

$$\int_\Sigma dF_j \delta u_j = \int_\Sigma \nu_i \tau_{ij} \delta u_j d\Sigma = \int_\Omega \partial_i(\tau_{ij}\delta u_j) d\Omega \ , \tag{12}$$

the last form deriving from an application of the divergence theorem. The statement of conservation of energy for the hyperelastic body can thus be written in the form

$$\delta \int_\Omega \rho U \, d\Omega = \int_\Omega \{\rho g_i \delta u_j + \partial_i(\tau_{ij}\delta u_j)\} d\Omega \tag{13}$$

where U is the specific strain energy (per unit mass), ρ the mass per unit volume in the strained configuration and g_i the acceleration components of body forces. Noting that $\rho \, d \, \Omega$, which is the mass of a particle, is an invariant,

$$\delta \int_\Omega \rho U \, d\Omega = \int_\Omega \rho \delta U \, d\Omega$$

and (13) implies

$$\rho \delta U = \rho g_j \delta u_j + \partial_i(\tau_{ij}\delta u_j) \quad . \tag{14}$$

In particular, when δu_j is a constant vector, the relative translation of the volume element produces no increase in strain energy and we get the translational equilibrium equations

$$\rho g_j + \partial_i \tau_{ij} = 0 \quad , \tag{15}$$

that reduce (14) to the simpler form

$$\rho \delta U = \tau_{ij} \partial_i \delta u_j \quad . \tag{16}$$

There is likewise no strain energy increase if

$$\delta u_j = e_{jmn} (d\omega_m) \xi_n \quad , \tag{17}$$

representing an infinitesimal rigid body rotation about the origin, superimposed on the strained configuration and characterized by the constant but arbitrary rotation vector $d\omega_m$. In this case (16) reduces to the rotational equilibrium conditions:

$$\tau_{ij} e_{jmn} \partial_i \xi_n = \tau_{ij} e_{jmi} = 0 \quad (m = 1,2,3) \quad , \tag{18}$$

which are equivalent to the symmetry statements

$$\tau_{ij} = \tau_{ji} \qquad i \neq j \quad . \tag{19}$$

Because the equilibrium conditions for the Eulerian stress tensor reduce to simple symmetry (19) or to differential equations (15) that are linear in space (final) coordinates, the Eulerian formulation, so commonly used in Fluid Mechanics, is also tempting at first sight to deal with Elasticity problems involving finite displacements. It is however, like Fluid Mechanics, restricted to isotropic constitutive equations. For fibrous materials, for instance, the constitutive equations can only be set up in a configuration where the local orientation of fibers is known; this is usually the initial configuration. In the strained configuration the fibers are reoriented by the local material rotations, whose amplitudes are not restricted in the non linear theory. As a consequence the constitutive equations depend not only on the Eulerian or Almansi strain measure but also on the local rotations, which complicates matters considerably. Furthermore the boundary conditions are also modified by the motion of the bounding surfaces.

These considerations motivate a return to material coordinates. This is achieved by expressing the force element in (10) in terms of the initial oriented surface element. The combination of (9) and (10)

$$dF_j = \tau_{ij} J(\partial_i x_m) n_m dS$$

introduces the Lagrangian or Piola stress tensor t_{mj}

$$t_{mj} = \tau_{ij} J \partial_i \, x_m \tag{20}$$

so that

$$dF_j = t_{mj} n_m dS \quad . \tag{21}$$

In this operation the force element has been simply translated from the final to the initial surface element and the virtual work has been kept invariant. Application of the divergence theorem on the initial configuration yields now

$$\int_S dF_j \delta u_j = \int_S n_m t_{mj} \delta u_j dS = \int_V D_m(t_{mj} \delta u_j) dV \ . \tag{22}$$

With ρ_0 denoting the mass per unit volume in the initial configuration, conservation of energy can be stated in the form

$$\delta \int_V \rho_0 U \ dV = \int_V \{\rho_0 g_j \delta u_j + D_m(t_{mj} \delta u_j)\} \ dV$$

and implies

$$\delta W = \rho_0 g_j \delta u_j + D_m(t_{mj}\delta u_j) \tag{23}$$

where, for concision of notation, the strain energy per unit initial volume $W = \rho_0 U$ was introduced.

The translational equilibrium equations are again obtained as the special case $\delta W = 0$, when the δu_j are arbitrary constants, yielding

$$\rho_0 g_j + D_m t_{mj} = 0 \quad (j = 1,2,3) \quad . \tag{24}$$

Hence the Piola tensor satisfies linear differential equations in material coordinates; they reduce (23) to

$$\delta W = t_{qj} D_q \delta u_j \quad . \tag{25}$$

The rotational equilibrium equations are however non linear; they are obtained by substitution of (17) with again $\delta W = 0$, whence

$$e_{jmn} t_{qj} D_q \xi_n = 0 \quad (m = 1,2,3) \quad . \tag{26}$$

The last factor is no more a simple Kronecker delta but, in view of (1),

$$D_q \xi_m = D_q x_m + D_q u_m = \delta_{qm} + D_q u_m \quad . \tag{27}$$

Geometrical linearization of elasticity theory implies that we consider the displacement gradients as negligible before unity

$$|D_q u_m| \ll 1 \ , \tag{28}$$

in which case the second term in the right-hand side of (27) can be neglected when substituted into (26) and we would again obtain a statement of symmetry for the Piola stress tensor. As a matter of fact the linearizing assumption is easily seen to make the Piola tensor and the Eulerian stress tensor undistinguishable. Even in the case of infinitesimal strains, finite rotations causes assumptions (28) not to be satisfied and the general form (26) of rotational equilibrium, showing the Piola tensor not to be symmetrical, must be retained. It can be stated in the equivalent form

$$t_{qj} D_q \xi_m = t_{qm} D_q \xi_j \quad . \tag{29}$$

2. KIRCHHOFF-TREFFTZ TENSOR

Material coordinates allow to perform all integrations and to state boundary conditions on the known and fixed initial configuration; in addition they allow to commute partial derivatives and variations

$$D_q \delta u_j = D_q \{\hat{u}_j(x_m) - u_j(x_m)\} = D_q \hat{u}_j - D_q u_j = \delta D_q u_j$$

so that equation (25) of conservation of energy becomes that of a perfect differential

$$\delta W = t_{qj} \delta e_{qj} \tag{30}$$

$$e_{qj} = D_q u_j \quad . \tag{31}$$

The energy per unit volume is thus a function of the nine displacement gradients e_{qj} and

$$t_{qj} = \partial W / \partial e_{qj} \quad . \tag{32}$$

Additional information as to the structure of the function W is provided by the rotational equilibrium equations. One way to obtain this information is to rewrite (29), using (27),(31) and (32), as

$$\frac{\partial W}{\partial e_{mj}} + e_{qm}\frac{\partial W}{\partial e_{qj}} = \frac{\partial W}{\partial e_{jm}} + e_{qj}\frac{\partial W}{\partial e_{qm}} \quad . \tag{33}$$

For m = j those equations are trivially satisfied, for m ≠ j they constitute a set of three linear partial differential equations for W, that are easily solved by the method of characteristics.

For instance, for m = 1 and = 2, the differential equations of the characteristics are

$$\frac{de_{12}}{1+e_{11}} = \frac{-de_{21}}{1+e_{22}} = \frac{de_{22}}{e_{21}} = \frac{de_{32}}{e_{31}} = \frac{-de_{11}}{e_{12}} = \frac{-de_{31}}{e_{32}}$$

$$= \frac{de_{13}}{0} = \frac{de_{23}}{0} = \frac{de_{33}}{0} = \frac{dW}{0} \quad . \tag{34}$$

Simple first integrals of those equations, that is functions of the e_{mn} that are arbitrary constants under those differential relations, are readily found to be

$$2e_{11} + e_{11}^2 + e_{12}^2 \ , \quad 2e_{22} + e_{21}^2 + e_{22}^2 \ , \quad e_{12} + e_{21} + e_{11}e_{21} + e_{12}e_{22}$$

$$e_{31} + e_{11}e_{31} + e_{12}e_{32} \ , \quad e_{32} + e_{21}e_{31} + e_{22}e_{32} \ , \ e_{31}^2 + e_{32}^2$$

and

$$e_{13} \ , \quad e_{23} \ , \quad e_{33} \quad .$$

The symmetrical Green or Lagrangian strain tensor g_{mn}, issued from the quadratic form

$$d\xi_j d\xi_j - dx_n dx_n = 2g_{mn} dx_m dx_n \quad , \tag{35}$$

has six elements

$$g_{nm} = g_{mn} = \frac{1}{2}(D_m\xi_i D_n\xi_i - \delta_{mn}) = \frac{1}{2}(e_{mn}+e_{nm}+e_{mi}e_{ni}) \tag{36}$$

that are algebraic combinations of the preceding first integrals and are, consequently, first integrals themselves. They are the only combinations that remain first integrals of the two other sets ($m=2$, $j=3$ and $m=3$, $j=1$).

Moreover W is seen to be constant itself. The conclusion is that the strain energy per unit initial volume is a function of the nine displacements gradients e_{qj} through the six independent coordinates of the Green strain tensor g_{mn}.

A different proof of this result introduces in a natural way the symmetrical Kirchhoff-Trefftz stress tensor s_{qp}. The rotational equilibrium equations (29) are satisfied by setting

$$t_{qj} = s_{qp} D_p \xi_j \tag{37}$$

provided

$$s_{pq} = s_{qp} \ . \tag{38}$$

Equation (30) becomes then

$$\delta W = s_{qp} D_p \xi_j \delta e_{qj} \tag{39}$$

but also, exchanging the dummy subscripts q and p and using the symmetry of s_{qp},

$$\delta W = s_{qp} D_q \xi_j \delta e_{pj}$$

or adding both results,

$$2\delta W = s_{qp}(D_p\xi_j\delta e_{qj} + D_q\xi_j\delta e_{pj}) \ .$$

The bracket, however, is, in view of (27),

$$\delta e_{qp} + e_{pj}\delta e_{qj} + \delta e_{pq} + e_{qj}\delta e_{pj}$$

$$= \delta(e_{qp}+e_{pq}+e_{pj}+e_{qj}) = 2\delta g_{qp}$$

and we obtain finally

$$\delta W = s_{qp}\delta g_{qp} \quad . \tag{40}$$

Provided we convene to distinguish the order of the subscripts in g_{qp} and to express W symmetrically in terms of the six quantities $\frac{1}{2}(g_{qp}+g_{pq})$, the constitutive equations or stress-strain relations can be summarized by

$$s_{qp} = \partial W/\partial g_{qp}. \tag{41}$$

They contain the symmetry statements $s_{qp}=s_{pq}$, equivalent to the rotational equilibrium conditions. By the chain rule of differentiation and use of this symmetry

$$t_{qj} = \frac{\partial W}{\partial g_{mn}}\frac{\partial g_{mn}}{\partial e_{qj}} = s_{mn}\frac{1}{2}(\delta_{mq}\delta_{nj} + \delta_{nq}\delta_{mj} + \delta_{mq}e_{nj} + \delta_{nq}e_{mj})$$

$$= s_{mn}\delta_{mq}(\delta_{nj}+e_{nj}) = s_{qn}D_n\xi_j$$

we retrieve definition (37).

Trefftz [3] gave the geometrical interpretation of the stress tensor s_{qp}. In the strained configuration the convected material coordinates constitute a set of curvilinear coordinates with local base vectors

$$\vec{G}_i = D_i\vec{\xi} = D_i\xi_j\vec{e}_j \quad . \tag{42}$$

The corresponding fundamental metric tensor is

$$G_{im} = \vec{G}_i \cdot \vec{G}_m = D_i\xi_j D_m\xi_k \vec{e}_j \cdot \vec{e}_k$$

and, since the base vectors $\vec{e}_j$ of the cartesian frame are orthonormal,

$$G_{im} = D_i\xi_j D_m\xi_j = 2g_{im} + \delta_{im} \; . \tag{43}$$

This fundamental metric tensor is in fact Cauchy's definition of the strain measure and is thus not fundamentally different from Green's strain tensor. Using in succession equations (21),(37) and (42), the force element expression that is obtained

$$d\vec{F} = dF_j\vec{e}_j = t_{qj}n_q dS\vec{e}_k = s_{qp}n_q dSD_p\xi_j\vec{e}_j = s_{qp}n_q dS\vec{G}_p \tag{44}$$

still associates the Kirchhoff-Trefftz tensor with a definition per unit initial area, but this time in the metric induced by the deformation.

3. THE PRINCIPLE OF VARIATION OF DISPLACEMENTS

With kinematical boundary conditions imposed on displacements

$$u_j = \bar{u}_j \rightarrow \delta u_j = 0 \quad \text{on} \quad S_1 \tag{45}$$

and Lagrangian surface tractions t_j imposed on the complementary part S_2 of the surface S bounding the simply connected domain V, the sum of strain energy and potential energy of loads

$$P = \int_V W\, dV - \int_V \rho_0 g_j u_j dV - \int_{S_2} t_j u_j dS \tag{46}$$

is a functional of the displacement field that is stationary under arbitrary variations δu_j respecting (45). As proof we can use (25) and obtain, after integration by parts, the Euler equations

$$D_q t_{qj} + \rho_0 g_j = 0 \quad (j = 1,2,3)$$

and natural boundary conditions

$$n_q t_{qj} = t_j \quad (j = 1,2,3) \quad \text{on} \quad S_2 \;, \tag{47}$$

expressing all equilibrium conditions, except rotational equilibrium. As observed earlier, rotational equilibrium becomes integrated in the principle provided W be considered to depend on the displacement gradients through the Green strain tensor. This is equivalent to replace (25) by (39) and (31), or

$$\delta W = s_{qp} D_p \xi_j D_q \delta u_j \tag{48}$$

and produces the other equilibrium equations in the form given by Signorini:

$$D_q(s_{qp} D_p \xi_j) + \rho_0 g_j = 0 \tag{49}$$

$$n_q s_{qp} D_p \xi_j = t_j \quad \text{on} \quad S_2 \quad . \tag{50}$$

4. THE CANONICAL PRINCIPLE BASED ON KIRCHHOFF-TREFFTZ STRESSES AND GREEN STRAINS

Starting from this last formulation, a 'canonical' variational principle in the sense of Friedrichs [1] (see also Courant and Hilbert [2]) is obtained by addition of a 'dislocation potential'

$$D = \int_V \frac{1}{2} s_{qp}(D_q u_p + D_p u_q + D_q u_j D_p u_j - 2g_{qp})dV$$

in which the s_{qp} are momentarily considered to be a set of Lagrangian multipliers. Their object is to transfer the differential constraints between Green strains and displacements obtained by substitution of (31) into (36), from the status of *a priori* equations necessary to obtain (48) to a status of Euler equations. They are indeed the variational equations associated to variation of the multipliers in the extended principle $\delta(P+D) = 0$.

This principle is a three-field principle, since, in addition to the field of Lagrangian multipliers, the field of the Green tensor g_{mn} has now become independent from the displacement field. Consequently W is now considered to depend on the g_{mn}. The variational equations associated to variations on the g_{mn} turn out to be identical to equations (41), thus identifying the Lagrangian multipliers with the Kirchhoff-Trefftz stresses related to the strains in the constitutive equations.

This allows the three-field principle to be simplified by accepting the constitutive equations to be satisfied *a priori*. Indeed, grouping all terms containing the Green strain tensor, the following contact transformation is suggested

$$s_{qp}g_{qp} - W = S(s_{mn}) \tag{51}$$

which would remove the presence of the g_{mn} and lead to the 'canonical form'

$$\int_V \left\{\frac{1}{2} s_{qp}(D_q u_p + D_p u_q + D_q u_j D_p u_j) - S\right\} dV$$
$$- \int_V \rho_0 g_j u_j dV - \int_{S_2} t_j u_j dS \quad \text{stationary.} \tag{52}$$

The function S of the Kirchhoff-Trefftz tensor is a 'complementary energy density'. Its existence and uniqueness imply that the constitutive equations (41) establish a one to one correspondance between stresses and strains, they must have a unique inverse. This inverse can then be written as

$$g_{qp} = \partial S/\partial s_{qp} \tag{53}$$

provided S be expressed in a symmetrical form.

The variational equations of the canonical two-field principle (52) are the Signorini equilibrium equations (49) and (50) and, associated to the variations of the Kirchhoff-Trefftz stresses, the compatibility equations

$$\frac{1}{2}\left(D_q u_p + D_p u_q + D_q u_j D_p u_j\right) = \partial S/\partial s_{pq} \quad . \tag{54}$$

This canonical principle obviously constitutes a generalization to finite displacements of the Reissner principle of linear elasticity theory. It was discovered independently by Reissner [5] and the author.

In the linearized case, assumptions (28) allow to discard the term $D_q u_j D_p u_j$ in (52) and to replace $D_p \xi_j$ by the Kronecker δ_{pj} in the Signorini equations. Then, if the functional (52) is integrated by parts and if the stresses are assumed to satisfy the Signorini equations *a priori*, one obtains the complementary energy principle

$$-\int_V S\, dV + \int_{S_1} n_q s_{qp} \bar{u}_p dS \quad \text{stationary} \; .$$

It is a single-field principle involving only stresses in equilibrium. Its generalization, when assumptions (28) are discarded, is known to fail when attempted on the present formulation [9,10].

5. VARIATIONAL FORMULATIONS BASED ON THE PIOLA AND DISPLACEMENT GRADIENT TENSORS

Levinson [9] pointed out that a complementary energy principle for finite displacements could be obtained on the condition that the contact transformation

$$t_{qj} e_{qj} - W = T(t_{mn}) \tag{55}$$

would exist, in which case

$$e_{qj} = \partial T / \partial t_{qj} \ . \tag{56}$$

This would furnish a canonical principle, akin to (52)

$$\int_V (t_{qj} D_q u_j - T) dV - \int_V \rho_0 g_j u_j dV - \int_{S_2} t_j u_j dS \quad \text{stationary.} \tag{57}$$

Then, integration by parts and *a priori* satisfaction of the equilibrium equations (24) and (47) would give the complementary energy principle

$$-\int_V T \, dV + \int_{S_1} n_q t_{qj} \bar{u}_j dS \quad \text{stationary .} \tag{58}$$

Zubov [11] proposed a method for computing T in the case of an isotropic medium and performed the computation for a semi-linear material. However, in the general case, it is not known whether a one to one correspondance between g_{mn} and s_{mn} will be sufficient to guarantee the existence of contact tranformation (55). In this connection the following observations can be made. Equations (30) or (32) hold true, whether or not the rotational equilibrium equations are satisfied. If they are, we know that W has a special structure, which must be reflected in a corresponding structure for T. Indeed, if T exists, we can use (56) to place the rotational equilibrium equations (29), or, equivalently,

$$t_{qj}(\delta_{qm} + e_{qm}) = t_{qm}(\delta_{qj} + e_{qj})$$

in the form

$$t_{mj} + t_{qj}\frac{\partial T}{\partial t_{qm}} = t_{jm} + t_{qm}\frac{\partial T}{\partial t_{qj}} \quad j \neq m \tag{59}$$

of a set of 3 partial differential equations to be satisfied by T. Integration by the method of characteristics yields the result

$$T(t_{mn}) = -(t_{11} + t_{22} + t_{33}) + F(t_{mi}t_{ni}) . \tag{60}$$

Thus the structure of T comprises a function to be determined in the elements of the symmetrical tensor

$$T_{mn} = t_{mi}t_{ni} . \tag{61}$$

In view of (37) and (53) it can be expressed in terms of the Kirchhoff-Trefftz tensor

$$T_{mn} = s_{mq}s_{np}D_q\xi_i D_p\xi_i = s_{mq}s_{np}(2g_{qp}+\delta_{qp}), \quad \text{or,}$$

$$T_{mn} = 2s_{mq}s_{np}\frac{\partial S}{\partial s_{qp}} + s_{mq}s_{nq} . \tag{62}$$

Furthermore, Levinson's contact transformation can be rewritten in the form

$$t_{qj}D_q\xi_j - W = t_{qj}(e_{qj} + \delta_{qj}) - W = T + t_{qq} = F$$

and using (37) once again,

$$s_{qp}D_p\xi_j D_q\xi_j - W = s_{qp}(2g_{qp} + \delta_{qp}) - W = F$$

or

$$F = 2s_{qp}g_{qp} + s_{qq} - W \ . \tag{63}$$

If we accept the existence of the contact transformation (51), the function F, is expressible in terms of the Kirchhoff-Trefftz stresses. We have thus obtained the following result: the existence of contact transformation (51), which is in the nature of a physical assumption, guarantees the existence of contact transformation (55), provided equations (62) are invertible. The existence and unicity problem for the inversion of nonlinear equations (62) can be handled theoretically by the implicit functions theorem. The result would however be useless for practical applications and this approach will not be pursued.

6. A CANONICAL VARIATIONAL PRINCIPLE BASED ON THE POLAR DECOMPOSITION

The Jacobian matrix (2) has a unique decomposition

$$D_i\xi_j = \alpha_{jm}(\delta_{mi}+h_{mi}) \tag{64}$$

in terms of a Lagrangian measure of strain

$$h_{mi} = h_{im} \tag{65}$$

and a rigid body rotation operator α_{jm}:

$$\alpha_{jm}\alpha_{jn} = \delta_{mn} \ , \quad \alpha_{jm}\alpha_{pm} = \delta_{jp} \ , \quad |\alpha_{jm}| = 1 \ . \tag{66}$$

Through this decomposition a neighborhood of the medium is first strained and afterwards rotated. The strain tensor h_{mj} is related to the Green tensor through

$$D_i\xi_j D_p\xi_j = \alpha_{jm}\alpha_{jn}(\delta_{mi}+h_{mi})(\delta_{np}+h_{np})$$

$$= \delta_{mn}(\delta_{mi}+h_{mi})(\delta_{np}+h_{np})$$

$$= (\delta_{ni}+h_{ni})(\delta_{np}+h_{np})$$

or, finally, using the symmetry properties (65),

$$2g_{ip} = 2h_{ip} + h_{in}h_{np} \ . \tag{67}$$

Because of (4) the relation is known to be one-one. The tensors are co-axial and, in a set of principal axes, their diagonal elements are related as follows:

$$2g_i = 2h_i + h_i^2 \quad (i = 1,2,3) \ . \tag{68}$$

Then, because (4) implies

$$1 + 2g_i > 0 \qquad \text{and} \qquad 1+h_i > 0 \ , \tag{69}$$

the unique inverse to (68) is

$$h_i = -1 + \sqrt{1 + 2g_i} \ . \tag{70}$$

For commodity the tensor h_{mi} will be called the 'engineering strain tensor'; this is justified by the fact that the principal strains can be defined, as in the linearized case, by the usual engineering formula

$$h_i = \frac{d\sigma_i - ds_i}{ds_i} \ ,$$

ds_i and $d\sigma_i$ being the elementary distances measured before and after straining along the principal directions. In fact, for infinitesimal strains, (67) shows that the Green and engineering strain measures become identical.

The one to one correspondance that exists, even for finite strains, establishes that W can also be considered to be a function of the engineering strains and that a new symmetrical stress tensor can be introduced by

$$r_{mi} = \partial W / \partial h_{mi} \ . \tag{71}$$

From (67)

$$\delta g_{ip} = \delta h_{ip} + \frac{1}{2} h_{in} \delta h_{np} + \frac{1}{2} h_{np} \delta h_{in}$$

and (40) can be manipulated as follows

$$\delta W = s_{ip} \delta g_{ip} = s_{ip} \delta h_{ip} + \frac{1}{2} s_{ip} h_{in} \delta h_{np} + \frac{1}{2} s_{ip} h_{np} \delta h_{in} \ .$$

In the second right hand term exchange the dummy subscripts n and i, and p and n in the third, then

$$\delta W = (s_{ip} + \frac{1}{2} s_{np} h_{ni} + \frac{1}{2} s_{in} h_{pn}) \delta h_{ip} \ .$$

The bracket is clearly symmetrical in the subscripts i and p, so that by identification with

$$\delta W = r_{ip} \delta h_{ip} \tag{72}$$

the following relationship is established between the new stress tensor and the Kirchhoff-Trefftz one:

$$r_{ip} = s_{ip} + \frac{1}{2} h_{in} s_{np} + \frac{1}{2} s_{in} h_{np} \ . \tag{73}$$

In the case of infinitesimal strains, but even for finite rotations, both tensors are clearly coalescent. Then, invertible stress-strain relations of type (41) are equivalent to invertible relations of type (71) and the following diagram of one to one relationship

$$s_{ip} \leftrightarrow g_{ip} \leftrightarrow h_{ip} \leftrightarrow r_{ip} \tag{74}$$

shows (73) to be invertible.

Introducing the polar decomposition (64) into definition (37)

$$t_{qj} = s_{qp}\alpha_{jm}(\delta_{mp}+h_{mp}) \tag{75}$$

showing that, in view of (74), the nine components of the Piola tensor are determined by the six components of the Kirchhoff-Trefftz tensor and the three effective parameters of the material rotation.

From (75) follows also

$$t_{qj}\alpha_{jn} = s_{qp}(\delta_{np}+h_{np}) = s_{qn} + s_{qp}h_{pn}$$

and, in view of (73),

$$r_{qn} = \frac{1}{2}(t_{qj}\alpha_{jn}+t_{nj}\alpha_{jq}) \quad . \tag{76}$$

To the functional P, given by (46), we add a dislocation potential that removes the differential constraints (64) of the polar decomposition and consider W to be a function of the engineering strain tensor. Since the use of (71) implies that W be expressed symmetrically with respect to h_{mi} and h_{im}, we take care to do the same for the dislocation potential:

$$\Delta = \int_V t_{ij}\left\{D_i\xi_j-\alpha_{jm}\left(\delta_{mi}+\frac{h_{mi}+h_{im}}{2}\right)\right\} dV \quad . \tag{77}$$

The variational equation corresponding to δh_{qn} in $\delta(P+\Delta) = 0$ is

$$\partial W/\partial h_{qn} = \frac{1}{2}(t_{qj}\alpha_{jn}+t_{nj}\alpha_{jq})$$

which corresponds to (76), r_{qn} being regarded as given by the constitutive equations in the form of (71). This can serve to identify the set of multipliers of the dislocation potential as the Piola stress tensor. The variational equations corresponding to the rotational degrees of freedom are obtained as follows:

$$\alpha_{jm}\alpha_{jn} = \delta_{mn} \rightarrow \alpha_{jn}\delta\alpha_{jm} = -\alpha_{jm}\delta\alpha_{jn} = \delta\omega_{mn} \quad \text{skew-symmetric}$$

and consequently,

$$\delta\alpha_{jm} = \alpha_{jn}\delta\omega_{mn} \quad . \tag{78}$$

Substituting this into $\delta(P+\Delta) = 0$, there comes

$$t_{ij}\alpha_{jn}(\delta_{mi}+h_{mi})\delta\omega_{mn} = 0$$

or, on account of the skew-symmetry of $\delta\omega_{mn}$,

$$t_{ij}\alpha_{jn}(\delta_{mi}+{}_{mi}) = t_{ij}\alpha_{jm}(\delta_{ni}+h_{ni}) \quad .$$

Multiplying by $\alpha_{pn}\alpha_{qm}$, it becomes

$$t_{ip}\alpha_{qm}(\delta_{mi}+h_{mi}) = t_{iq}\alpha_{pn}(\delta_{ni}+h_{ni})$$

and, in view of (64), reduces to the known rotational equilibrium form

$$t_{ip}D_i\xi_q = t_{iq}D_i\xi_p \quad .$$

Thus, as could be suspected, the rotational degrees of freedom in the principle have the rotational equilibrium conditions as variational equations.

We now simplify the principle by grouping the terms containing the h_{mi}

$$t_{ij}\alpha_{jm}\frac{h_{mi}+h_{im}}{2} - W \quad .$$

Exchange the dummy subscripts i and m in the second term, to obtain in view of (76)

$$h_{mi}\frac{1}{2}(t_{ij}\alpha_{jm} + t_{mj}\alpha_{ji}) - W = h_{mi}r_{mi} - W$$

which suggests again a contact transformation:

$$h_{mi}r_{mi} - W = R(r_{pq}) \ , \ h_{mi} = \partial R/\partial r_{mi} \quad . \tag{79}$$

For infinitesimal strains this transformation is identical to (51) and will be assumed to exist; in other words (71) is assumed to have a unique inverse. Note also that for $q = n = m$, (76) gives

$$r_{11} + r_{22} + r_{33} = r_{mm} = t_{mj}\alpha_{jm} \ ,$$

and that

$$t_{ij}D_i\xi_j = t_{ij}(\delta_{ij}+D_iu_j) = t_{jj} + t_{ij}D_iu_j \ .$$

The canonical form of the variational principle can now be given as

$$\int_V (t_{ij}D_iu_j+t_{jj}-r_{mm}-R)dV - \int_V \rho_0 g_j u_j dV - \int_{S_2} t_j u_j dS \quad \text{stationary.} \tag{80}$$

It contains as independent variables, the displacements u_j, the Piola stress tensor t_{ij} and the 3 rotational degrees of freedom that appear in the 'auxiliary' variables (76). Its variational equations are (24) and (47), together with

$$D_p u_q + \delta_{pq} - \alpha_{qp} - h_{ij} \frac{\partial r_{ij}}{\partial t_{pq}} = 0 \quad \text{for} \quad \delta t_{pq} ,$$

that is easily recognized to be a disguised form of the polar decomposition

$$D_p \xi_q = \alpha_{qn}(\delta_{np} + h_{nq}) \quad .$$

Also for

$$\delta\alpha_{uv} = \alpha_{up}\delta\omega_{vp}$$

$$\left(\frac{\partial r_{mm}}{\partial \alpha_{uv}} + \frac{\partial R}{\partial r_{qn}} \frac{\partial r_{qn}}{\partial \alpha_{uv}}\right)\alpha_{up}\delta\omega_{vp} = 0$$

or

$$[t_{vu} + \frac{1}{2}(h_{qv}t_{qu} + h_{vn}t_{nu})]\alpha_{up}\delta\omega_{vp} = 0$$

or

$$t_{nu}(\delta_{nv} + h_{nv})\alpha_{up} = t_{nu}(\delta_{np} + h_{np})\alpha_{uv} ,$$

a disguised form of the rotational equilibrium conditions.

7. THE COMPLEMENTARY ENERGY PRINCIPLE BASED ON THE POLAR DECOMPOSITION

It suffices now to integrate by parts in (80) and to assume the equilibrium equations (24) and (47) satisfied to reduce the principle to a complementary energy form

$$\int_V (t_{jj} - r_{mm} - R)dV + \int_{S_1} n_m t_{mn} \bar{u}_j dS \quad \text{stationary} . \tag{81}$$

Because the rotational degrees of freedom are still free and rotational equilibrium is still cared for by the principle itself, satisfaction of the translational equilibrium equations requires only 'first order' stress functions. For instance

$$t_{mj} = e_{mrs} D_r A_{sj} \tag{82}$$

satisfies equations (24) in the absence of body loads and introduces a tensor A_{sj} of functions that generate the Piola stresses by their first derivatives. Then, provided we satisfy *a priori*

$$e_{mrs} n_m D_r A_{sj} = t_j \quad \text{on} \quad S_2 , \tag{83}$$

and consider the auxiliary variables (76) to represent

$$r_{qm} = \frac{1}{2} (e_{qrs} \alpha_{jm} + e_{mrs} \alpha_{jq}) D_r A_{sj} , \tag{84}$$

we obtain a principle governing the first order stress functions. A satisfactory treatment of the natural boundary conditions requires that (83) be incorporated into (81) by means of Lagrangian multipliers; this transforms (81) into

$$\int_V (t_{jj} - r_{mm} - R) dV + \int_S n_m u_j t_{mj} dS - \int_{S_2} u_j t_j dS \text{ stationary} \tag{85}$$

where the u_j are imposed displacements on S_1 but free Lagrangian multipliers, that are of course identified with the surface displacements, on S_2. Preparatory calculations based on (78),(79),(82) and (84), yield

$$\delta t_{jj} = e_{jrs} D_r \delta A_{sj}$$

$$\delta r_{mm} = e_{hrs} \alpha_{jp} D_r A_{sj} \delta\omega_{np} + e_{mrs} \alpha_{jm} D_r \delta A_{sj}$$

$$\delta R = e_{mrs} h_{mn} (\alpha_{jp} D_r A_{sj} \delta\omega_{np} + \alpha_{jn} D_r \delta A_{sj}) \ .$$

From them, the variational equations associated with the skew-symmetric variations $\delta\omega_{np}$ (rotational equilibrium equations) are obtained in the form

$$e_{mrs} D_r A_{sj} \{\alpha_{jp}(\delta_{mn}+h_{mn}) - \alpha_{jn}(\delta_{mp}+h_{mp})\} = 0 \ . \tag{86}$$

The variational equations associated to δA_{sj} follow from an integration by parts as

$$e_{mrs} D_r \{\alpha_{jn}(\delta_{mn}+h_{mn})\} = 0 \quad \text{all } (s,j) \ . \tag{87}$$

They are equivalent to the equations obtained from the polar decomposition (64) on eliminating the displacements by cross-differentiation. Hence they can be considered as the local integrability conditions for a displacement field. The surface terms, containing variations of the stress functions, are collected below:

$$\int_S \{e_{jrs} n_r - e_{mrs} n_r \alpha_{jn} (\delta_{mn}+h_{mn})\} \delta A_{sj} dS + \int_S e_{mrs} n_m u_j D_r \delta A_{sj} dS = 0 \tag{88}$$

and the second integral requires manipulation to remove the derivatives of the variations. To this effect we use Stokes theorem

$$\oint_C \vec{b} \cdot d\vec{s} = \int_{S'} \text{rot } \vec{b} \cdot \vec{n} \ dS \ ,$$

according to which the circulation of vector $\vec{b}$ around a closed contour c is equal to the flux of rot $\vec{b}$ through the surface S' bordered by this contour. Let S'' denote a second surface bordered by the same contour and modify the orientation of the normal on it so that it points away from the volume V comprized between both surfaces. Then

$$\oint_c \vec{b}\cdot d\vec{s} = -\int_{S''} \text{rot } \vec{b}\cdot\vec{n}\ d\ S$$

and subtracting both formulas,

$$\int_S \text{rot } \vec{b}\cdot\vec{n}\ d\ S = \int_S e_{mrs} n_m D_r b_s dS = 0$$

where S is the complete surface bounding the volume. Thus, if in the second integral of (88) we observe that

$$u_j D_r \delta A_{sj} = D_r(u_j \delta A_{sj}) - D_r u_j \cdot \delta A_{sj} \quad ,$$

only the second term will contribute. The natural boundary conditions resulting from the free variations δA_{sj} on S are consequently:

$$e_{jrs} n_r - e_{mrs}\{n_r \alpha_{jn}(\delta_{mn}+h_{mn}) + n_m D_r u_j\} = 0 \ . \tag{89}$$

Since

$$e_{mrs} n_m D_r u_j = (\vec{n} \times \text{grad } u_j)_s$$

they involve only surface derivatives of the data $\bar{u}_j$ on S_1. They are also seen to be identically satisfied by using the polar decomposition

$$\alpha_{jn}(\delta_{mn}+h_{mn}) = D_m \xi_j = \delta_{mj}+D_m u_j$$

giving

$$e_{jrs}n_r - e_{mrs}n_r\delta_{mj} - e_{mrs}n_rD_mu_j - e_{mrs}n_mD_ru_j = 0 \ .$$

The two first terms cancel and, exchanging the dummy subscripts r and m in the third and noting that $e_{rms} = - e_{mrs}$,

$$e_{mrs}\cdot n_m(D_ru_j - D_ru_j) = 0 \ .$$

8. LINEARIZED VERSION OF THE COMPLEMENTARY ENERGY PRINCIPLE

Assuming very small rotations, we can write

$$\alpha_{im} = \delta_{im} + \omega_{im} \tag{90}$$

where

$$\omega_{im} = - \omega_{mi} \quad \text{and} \quad |\omega_{mi}| \ll 1 \ . \tag{91}$$

The following approximations can then be introduced. To lowest order

$$r_{qm} = \frac{1}{2}(t_{qm} + t_{mq}) \tag{92}$$

as arguments of the complementary energy density and

$$r_{mm} - t_{jj} = t_{mj}\omega_{jm} \ . \tag{93}$$

The linearized version of principle (81) reduces then to

$$-\int_V (t_{mj}\omega_{jm}+R)dV + \int_{S_1} n_m t_{mj}\bar{u}_j dS \quad \text{stationary} \tag{94}$$

where the approximation

$$t_{qm} = t_{mq} \tag{95}$$

becomes a variational result. Hence, as for the non linear case, we need only first order stress functions in order to free the translational equilibrium equations. In approximate solutions it becomes possible to discretize the rotational equilibrium conditions.

REFERENCES

[1] FRIEDRICHS, K.O., *Nachr. Adad. Wiss. Göttingen,* A36, 1929, 13.

[2] COURANT, R. and HILBERT, D., *Methoden der Mathematischen Physik.,* Vol. 1, Chap. 4, Springer, 1937.

[3] TREFFTZ, E., *Math. Mech.,* 13, 1933, 160.

[4] LANGHAAR, H.L., *J. Franklin Inst.,* 256-15, 1953, 255.

[5] REISSNER, E., *J. Math. Phys.,* 32, 1953, 129.

[6] GREEN, A.E. and ADKINS, J.E., *Large Elastic Deformations,* Clarendon Press, 1960.

[7] PRAGER, W., *Introduction to Mechanics of Continua,* Ginn, 1961.

[8] PIPES, L.A., *J. Franklin Inst.,* 274, 1962, 198.

[9] LEVINSON, M., *J. Appl. Mech.,* 87, 1966, 826.

[10] WASHIZU, K., *Variational Methods in Elasticity and Plasticity,* Pergamon Press, 1968.

[11] ZUBOV, L.M., *Prikl. Mat. Meh.,* 34, 1970, 241.

[12] FRAEIJS de VEUBEKE, B.M., "Large Displacement Formulations for Elastic Bodies," Air Force Flight Dynamics Laboratory, *Technical Report AFFDL-TR-70-34,* 1970.

MATRIX STRUCTURAL ANALYSIS

B.M. Fraeijs de Veubeke

University of Liège, Belgium

SUMMARY—A survey is made of the basic matrix equations discretizing the strain-displacement, equilibrium and constitutive equations of linear elasticity theory. The finite element approach requires in addition a discretized form of the transitional equations between subdomains and a progressive treatment from the element level to the assembled structural level.

Special attention has been given to a rational incorporation of body loads and internal degrees of freedom.

The direct stiffness method, force method and combined methods are related to discretized formulations of variational principles. The saddle point character of the many-field principles is put in evidence.

1. INTRODUCTION

As a result of any type of discretization of a continuous structure, there are essentially 3 groups of matrix equations describing its behaviour:

- the kinematic group $S^T q = e$ (1)

  describing the relations between (generalized) displacements and strains,

---

Original publication, Lecture Notes for the International Research Seminar on 'The Theory and Application of Finite Element Methods,' held at the University of Calgary, Alberta, Canada, July 26 - August 10, 1973.

- the static group $S\ s = p$ (2)

  describing relations between (generalized) stresses and applied loads,

- the constitutive equations relating stresses and strains. For a linear elastic body they assume the form

  $$s = J\ e \qquad \text{or} \qquad e = F\ s \tag{3}$$

  where J is a positive definite matrix of generalized elastic moduli and $F = J^{-1}$ is a positive definite matrix of compliances.

Equations (1) constitute a discretized form of the linearized strain measures of a continuum

$$\begin{bmatrix} \varepsilon_{11} \\ \varepsilon_{22} \\ \varepsilon_{33} \\ \gamma_{23} \\ \gamma_{31} \\ \gamma_{12} \end{bmatrix} = \begin{bmatrix} D_1 & 0 & 0 \\ 0 & D_2 & 0 \\ 0 & 0 & D_3 \\ 0 & D_3 & D_2 \\ D_3 & 0 & D_1 \\ D_2 & D_1 & 0 \end{bmatrix} \begin{bmatrix} u_1 \\ u_2 \\ u_3 \end{bmatrix}$$

presented here in cartesian coordinates, where $D_i = \partial/\partial x_i$, and whose matrix form will be written as

$$\varepsilon = D\ u\ . \tag{4}$$

Equations (2) are the corresponding discretizations of the continuum equilibrium equations

$$D^T\ \sigma + \phi = 0 \tag{5}$$

relating the elements

$$(\tau_{11} \quad \tau_{22} \quad \tau_{33} \quad \tau_{23} \quad \tau_{31} \quad \tau_{12}) = \sigma^T$$

of the stress tensor and the vector ϕ of body loads, and

$$N^T \sigma = t \tag{6}$$

which are surface equilibrium equations, where t is the vector of surface tractions. Matrix N derives from the matrix differential operator D by replacing D_i by n_i, the direction cosines of the outward normal to the surface under consideration. Equations (3) are the discretized form of generalized relations between stresses and strains

$$\sigma = H \, \varepsilon \tag{7}$$

where the symmetrical positive definite Hooke matrix H contains the 21 elasticity moduli of the material that may vary from point to point in the case of a non homogeneous body.

Solving these equations under prescribed boundary conditions constitutes the aim of matrix structural analysis. When the discretization is performed by finite element methods, the matrix equations (1), (2) and (3) are first obtained for each individual element. The ensuing problem of connecting the elements together, in order to establish the same equations at the global structural level, becomes also one of the important steps in matrix structural analysis. Considering the continuum to be divided into a finite number of subdomains later to be discretized and become the finite elements, the connection problem involves discretization of the transition conditions between face + and face - of each interface:

$$u^+ = u^- \tag{8}$$

$$t^{+} + t^{-} = 0 \tag{9}$$

expressing respectively continuity of displacements and reciprocity of surface tractions. If necessary, those transitions conditions can be generalized to introduce dislocations and application of external surface loads at an interface. It will be observed that the presence of the same differential operator D in equations (4) and (5) of the continuum provides an indication of the existence of dual properties between kinematics and statics. This duality appears in the discretized forms (1) and (2) through the presence of the same "kinematical matrix" S.

2. KINEMATICS AT ELEMENT LEVEL

It will be assumed that the rigid body displacements are included in the degrees of freedom retained when discretizing a finite element. Even under very small strain, a region of thin-walled structure can undergo large displacements, whose major part is then of rigid body type. In the geometrically non-linear theory of elasticity the inclusion of rigid body displacements is thus imperative. The situation is different in linear elasticity, where local rotations are of the same small order of magnitude as strains. However, both intuition and experience show that the inclusion of rigid body degrees of freedom, although not absolutely necessary for convergence with respect to reduction in mesh size, is a practical requirement for good accuracy with economical sizes of elements. To the approximation of geometrically linear strain, the rigid body displacements are infinitesimal and members of a linear finite dimensional space. In other words any linear combination of rigid body displacements is itself a rigid body displacement. This leads to a matrix representation of the rigid body modes as

$$q = R\ r \tag{10}$$

where q is the vector of generalized boundary displacements of the element and the columns of R constitute a fundamental set (complete and linearly independent) of particular rigid body modes. We say then that R is a "base matrix" of the subspace of rigid body modes. The vector r is that of coefficients of an arbitrary linear combination of those columns. Let us give two examples.

Example 1. The constant strain membrane triangle (Figure 1)

The displacement field is completely determined by the local displacement vectors at the nodes. The local displacement vectors elsewhere are, by definition of the discretization, obtained by linear interpolation. The set of generalized displacements is indicated, each component is some orthogonal (covariant) projection of a nodal displacement vector on one of the sides of the triangle. A simple base matrix R is obtained by taking as columns, generalized displacements generated by rotations about the nodes 1, 2 and 3 respectively.

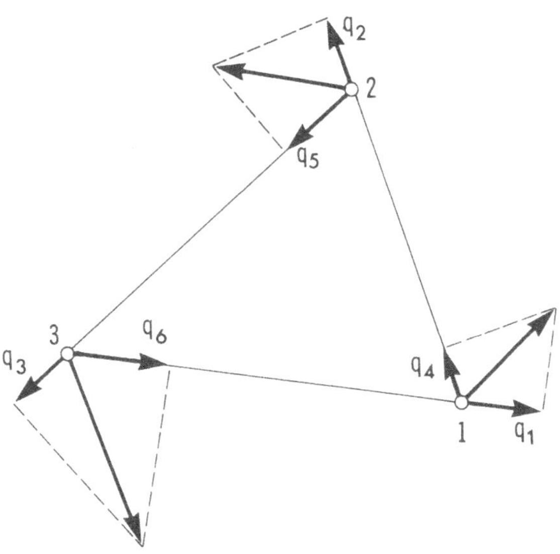

Figure 1 - Constant Strain Triangle Covariant Natural Displacements

$$R = \begin{bmatrix} 0 & 1 & 0 \\ 0 & 0 & 1 \\ 1 & 0 & 0 \\ 0 & 0 & 1 \\ 1 & 0 & 0 \\ 0 & 1 & 0 \end{bmatrix} \tag{11}$$

Example 2. The constant stress membrane triangle (Figure 2).

As the name suggests, this element is derived from a discretization of the stress field instead of the displacement field. Under constant stress the surface tractions are uniform along the edges and can be characterized by the corresponding resultants U_{12}, V_{12}, U_{23}, V_{23}, U_{31}, V_{31}, listed in that order in the generalized boundary load vector g. The conjugate boundary displacements are defined by virtual work. For instance the generalized displacement u_{12}, conjugate to the load U_{12} must be such that

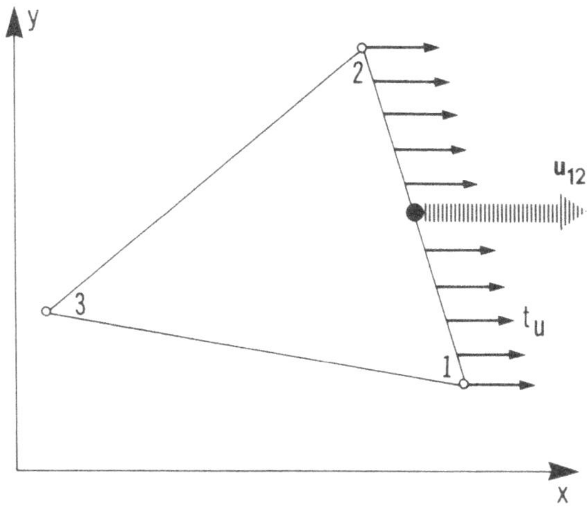

Figure 2 - Constant Stress Triangle-Generalized Boundary Loads in Global Coordinates

$$u_{12}\, U_{12} = \int_1^2 t_u\, u\, ds$$

However, since the surface traction component t_u is uniform, we have

$$t_u = u_{12}/c_{12}$$

where c_{12} is the length of side 12. Therefore,

$$u_{12} = \frac{1}{c_{12}} \int_1^2 u\, ds \tag{12}$$

and similar definitions for the other generalized displacements. From this example it can be understood that a discretization needs not necessarily be based on local values of fields but involves, in general, *linear functionals* of the field. In the present example, the displacement field is known only by its average values taken along the edges. To obtain rigid body modes, it is necessary to compute expressions like (12) for a rigid body displacement field. If, for instance, the element be rotated infinitesimally about node 1, the u component of displacement along 12 varies linearly between zero and $-(y_2 - y_1)\, d\omega$.

$$u_{12} = -\frac{1}{2}\, d\omega\, (y_2 - y_1) \quad .$$

By similar calculations we obtain, suppressing the factor $\frac{1}{2}\, d\omega$, the base matrix

$$R = \begin{bmatrix} -y_2 + y_1 & y_2 - y_1 & -y_1 - y_2 + 2\,y_3 \\ x_2 - x_1 & -x_2 + x_1 & x_1 + x_2 - 2\,x_3 \\ -y_2 - y_3 + 2\,y_1 & -y_3 + y_2 & y_3 - y_2 \\ x_2 + x_3 - 2\,x_1 & x_3 - x_2 & -x_3 + x_2 \\ -y_3 + y_1 & -y_3 - y_1 + 2\,y_2 & -y_1 + y_3 \\ x_3 - x_1 & x_3 + x_1 - 2\,x_2 & x_1 - x_3 \end{bmatrix} \tag{13}$$

The first column is the rotation mode about node 1, the second is about node 2, the third about node 3. An interesting observation is that the displacement functionals coincide for rigid body motions with the local values of displacements at the mid-side points. This is natural from virtual work considerations, since the resultant boundary loads may be considered to be applied at the "virtual nodes".

The elimination of the n(r) elements of r between the n(q) equations (10) results in a set of

$$n(q) - n(r) = n(e) \tag{14}$$

independent equations relating the elements of q

$$S^T q = 0 \tag{15}$$

S^T is a $n(e) \times n(q)$ matrix of linearly independent rows, and (15) is the necessary and sufficient condition for q to belong to the subspace of rigid body modes. By construction (10) with arbitrary r, is the general solution of (15), hence

$$S^T R = 0 . \tag{16}$$

This shows S, that has linearly independent columns, to be a base matrix for the orthogonal complementary subspace to that generated by R. Consider now the more general inhomogeneous equation

$$S^T q = e \tag{1}$$

related to (15). Since e vanishes when q belongs to the subspace of rigid body modes it is appropriate to consider e as a measure of strain of the element; its elements are called the generalized strains.

Consider now non singular transformations of the quantities r and e

$$r = (r/\hat{r})\ \hat{r} \qquad \hat{r} = (\hat{r}/r)\ r \qquad (\hat{r}/r) = (r/\hat{r})^{-1}$$

$$e = (e/\hat{e})\ \hat{e} \qquad \hat{e} = (\hat{e}/e)\ e \qquad (\hat{e}/e) = (e/\hat{e})^{-1} \quad .$$

We have

$$q = \hat{R}\ \hat{r} \qquad \hat{S}^T q = \hat{e} \quad ,$$

$$\hat{R} = R(r/\hat{r}) \qquad \hat{S}^T = (\hat{e}/e)\ S^T \quad ,$$

so that

$$\hat{S}^T \hat{R} = (\hat{e}/e)\ S^T R(r/\hat{r}) = 0 \quad .$$

This shows that R and S are only defined within equivalence classes, each member of an equivalence class being derived from another by postmultiplication by a non singular matrix.

Homogeneous constraints placed on q

$$R^{\#} q = 0 \tag{17}$$

can be derived to prevent the occurrence of rigid body modes. This implies that $R^{\#}\ R\ r = 0$ must only possess the trivial solution r = 0. In other words, the n(r) columns of $R^{\#}R$ must be linearly independent. Its number of rows, equal to that of $R^{\#}$, must be superior or equal to n(r). Hence the minimal number of constraints capable of suppressing rigid body modes is n(r). Any matrix $R^{\#}$ of size n(r) × n(r) with the property

$$R^{\#} R r = 0 \rightarrow r = 0$$

is said to define a set of external isostaticity constraints. Since premultiplication of (17) by a non singular matrix does obviously not modify the set of constraints, $R^{\#T}$ is defined within an equivalence class. It is obviously possible to select $R^{\#T}$, within its equivalence class so that $R^{\#}R$, which is non singular, becomes the identity matrix:

$$R^{\#}R = (r/r) = R^{T}R^{\#T} \quad . \tag{18}$$

Consider now the general solution of (17) when $R^{\#}$ defines a set of external isostaticity constraints. It is of the form

$$q = S^{\#T} e \tag{19}$$

$S^{\#T}$ is a base matrix of the subspace defined by (17). Each of its columns is a straining mode of the element, since rigid body modes were prevented to take place. The elements of the arbitrary vector e, that combine the columns linearly, can rightfully be called generalized strains; in fact, as the notation prefigures, the generalized strains defined by (19) can be rendered identical to those defined by (1); the condition therefore will be equation (23). As (19) provides the general solution to (17) (e arbitrary)

$$R^{\#} S^{\#T} = 0 \tag{20}$$

holds true and $R^{\#T}$ and $S^{\#T}$ are base matrices of orthogonal complementary subspaces. Once again it appears that $S^{\#T}$ is only representative of an equivalance class and that (20) is true whatever be the separate choices of $R^{\#T}$ and $S^{\#T}$ within their respective equivalence classes.

The assembled matrix $(R\ S^{\#T})$ has, according to (14), $n(r) + n(e) = n(q)$ columns, and is consequently square. Furthermore, the homogeneous equation

$$R\ r + S^{\#T}\ e = 0$$

is easily seen to possess only the trivial solution. If we premultiply by $R^{\#}$ there follows already $r = 0$ on account of (20) and the non singularity of $R^{\#}R$. Then $e = 0$ follows because of the linear independence of the columns of $S^{\#T}$. Consequently an arbitrary generalized boundary displacement vector q has a unique expansion in terms of a complete base matrix $(R\ S^{\#T})$

$$q = R\ r + S^{\#T}\ e \tag{21}$$

corresponding to a separation in rigid body modes and kinematically admissible straining modes (i.e. not violating a certain choice of isostaticity fixations). Premultiplying by S^T, there comes on account of (16)

$$S^T\ q = S^T\ S^{\#T}\ e\quad . \tag{22}$$

The matrix $S^T S^{\#T}$ is square, its number of rows, equal to that of S^T was defined to be n(e), the number of columns of $S^{\#T}$ is complementary to that of $R^{\#T}$, or $n(q) - n(r) = n(e)$ again. It is also non singular; for $S^T(S^{\#T}e) = 0$, implies that $S^{\#T}e$ belong to the complementary subspace generated by R, i.e.

$$S^{\#T}e = -\ R\ r \quad \text{for some } r\ .$$

However, such a relation was seen to imply both r = 0 and e = 0. This proves that (22) constitutes at most a non singular transformation of generalized strains between their definitions according to (1) and their definitions according to (19). Both definitions become identical by selecting $S^{\#T}$ within its equivalence class to obtain

$$S^{T}\, S^{\#T} = (e/e) = S^{\#}\, S \quad . \tag{23}$$

To determine r in the expansion (21) we premultiply by $R^{\#}$ and use (18) and (20) to obtain

$$r = R^{\#}\, q \ . \tag{24}$$

Comparison of this with (17), suggests for r the interpretation of vector of generalized displacements of the supports.

3. VIRTUAL WORK AT ELEMENT LEVEL

The q vector defines the boundary displacements of the element. It sometimes also defines unambiguously the internal displacements. There are many cases however where internal degrees of freedom are introduced with fixed boundaries (q = 0) to improve the conditioning of the stiffness matrix or better represent the effects of an eventual body loading of the element. It is extremely important to note that, for a given discretization, there is no unique way of defining the generalized displacements related to the additional internal degrees of freedom. This is because there is no unique way of defining the internal displacements which are forced by the motion of the boundaries. Between any two possible definitions b or c of vectors of generalized internal displacements belonging to the same discretization, there is a relation

$$b = c + U\,q \tag{25}$$

where U is some suitable matrix. For fixed boundaries both definitions give identical displacements but if we set c = 0, a motion of the boundaries will have different internal effects than setting b = 0. A step towards uniqueness is provided by the requirement that generalized internal displacements may be regarded as internal strains as well. In other words, b would have to vanish like e whenever the element undergoes a rigid body displacement. Suppose that, for the original choice c, we find

$$c = C\,r \quad \text{to be linked with} \quad q = R\,r$$

in the rigid body motions. We would then require U to satisfy

$$0 = (C + U\,R)\,r \quad \text{for arbitrary } r$$

and be presented the problem of solving

$$C + U\,R = 0 \quad \text{or} \quad R^T\,U^T = -\,C^T\,.$$

From (18) and the transpose of (16) the general solution of this is

$$U = -\,C\,R^{\#} + A\,S^T$$

where A is an arbitrary matrix of suitable dimensions, and we find

$$b = c - C\,R^{\#}q + A\,S^T q = c - C\,r + A\,e$$

we then say that b has a *strain-type definition.*

Introduce now the generalized loads through the virtual work equation. We distinguish between the loads applied at the boundary of the element (except along the structural boundary and in the presence of interface loading, they are internal connection loads at the structural level) and the body loads. Thus the boundary load vector g produces virtual work on q alone, while the body load vector has a virtual work associated with both internal and boundary displacements and falls in two parts: f which is conjugate to q and h which is conjugate to b. The virtual work at element level has the structure

$$V_e = q^T (g + f) + b^T h \quad . \tag{26}$$

4. STATICS AT ELEMENT LEVEL

Equilibrium of the element requires that the virtual work of loads vanishes for any displacement of rigid body type; thus, provided b has a strain-type definition, when $b = 0$ and $q = R\ r$. It follows then from (26) that

$$R^T (g + f) = 0 \tag{27}$$

represent the equilibrium conditions constraining the loads. From (16) it is found that the general solution of (27) can be written

$$g + f = S\ s \tag{28}$$

and s is a vector of generalized stresses. Premultiplying this result by q^T and considering (1) we obtain

$$q^T (g + f) = (S^T q)^T s = e^T s \tag{29}$$

showing the stresses so defined to be conjugate to the strains e. Finally the external form of virtual work (26) can be equated to an internal form

$$V_e = e^T s + b^T h = V_i \tag{30}$$

and this is a discretized version of the virtual work theorem of continuum mechanics. It will be noted that the strain-type definition of b, enables its conjugate h to be considered both as a generalized load and a generalized stress. Another consequence is that, as representing a load system, h is self-equilibrating since its virtual work vanishes under rigid body displacements. In contrast to this the body load system represented by f is generally not self-equilibrated. Being conjugate to q it is possible to balance it statically by reaction loads from the isostaticity constraints. Thus in a virtual work approach to static equilibrium

$$(R\ r)^T\ f + r^T\ j = 0$$

must hold for arbitrary r. There follows for the reaction loads

$$j = -\ R^T\ f\ . \tag{31}$$

The general solution of this, considering j to be given and f unknown, is

$$f = -\ R^{\#T}\ j + S\ s_f \tag{32}$$

the first term being a particular solution as can be verified from (18). The second term is, according to (16) in transpose, a general solution of the homogeneous equation. The expansion (32) decomposes f into a part absorbed by reactions j against the isostaticity constraints as given by (31) and a self-equilibrated part, whose generalized stresses s_f are given by (see (20) and (23))

$$s_f = S^{\#}\ f\ . \tag{33}$$

A similar expansion is valid for the generalized boundary loads, since (28) and (32) are equivalent to

$$g = R^{\#T} j + S s_g \tag{34}$$

with

$$s_f + s_g = s \quad . \tag{35}$$

In (34) we find for given g the unique values

$$j = R^T g = - R^T f \qquad s_g = S^{\#} g \quad . \tag{36}$$

The unicity of the expansion is otherwise obvious because $(R^{\#T} \; S)$ is non singular and a natural base matrix for g-space or f-space. Its transpose is the inverse of our base matrix for q-space

$$\begin{bmatrix} R^{\#} \\ \\ S^T \end{bmatrix} \begin{bmatrix} R & S^{\#T} \end{bmatrix} = \begin{bmatrix} (r/r) & 0 \\ \\ 0 & (e/e) \end{bmatrix} \tag{37}$$

as is readily verified from (18), (20), (16) and (23).

5. KINEMATICS AND STATICS AT THE STRUCTURAL LEVEL

We now affect the previously defined matrices by a bracketed subscript (i) to indicate the element to which they belong. Similar quantities without a bracket will be defined at the assembled structural level.

The operation of connecting the elements together has a kinematical aspect which consists simply in identifying each generalized boundary displacements $q_{(i)}$ of each element, with one of the elements listed in a *nodal displacement vector* q. The possibility of doing so implies of course a compatibility of definitions of generalized displacements at each node. Remember also that the nodal displacements need not be local values but simply linear functionals. The kinematical connections are then expressible as

$$q_{(i)} = L_{(i)}\, q \tag{38}$$

for each element. The Boolean matrices $L_{(i)}$ are convenient tools for presenting formulas and manipulating them. In actual software practice they are replaced by more efficient algorithms.

The statical aspect of the connections can be derived from virtual work equations:

$$V = \sum_i \left\{ q_{(i)}^T g_{(i)} + q_{(i)}^T f_{(i)} + b_{(i)}^T h_{(i)} \right\} \tag{39}$$

expresses the virtual work of all forces acting externally on all the, yet unconnected, elements. If the transitions conditions (8) and (9) in the continuum were exactly satisfied, there would be no contributions from the interfaces to the virtual work. More generally, even if some external loads were applied at the interfaces, the virtual work of internal connection loads would vanish. *This property is assumed to hold after the discretization* and, as a result, (39) as modified by (38) can be equated to the virtual work performed by the external loads (external at the structural level) only:

$$q^T \left(\sum_i L_{(i)}^T g_{(i)} + \sum_i L_{(i)}^T f_{(i)} \right) + \sum_i b_{(i)}^T h_{(i)}$$

$$= q^T g + q^T \sum_i L_{(i)}^T f_{(i)} + \sum_i b_{(i)}^T h_{(i)} \quad . \tag{40}$$

Indeed the body loads on each element are still external loads for the global structure and g condenses the external loads applied at the structural boundary and possibly at the interfaces. Simplification of (40) and consideration of the fact that it must hold for any nodal vector q, yields the relation

$$g = \sum_i L_{(i)}^T g_{(i)} \; . \tag{41}$$

At the structural boundary (41) is a discretized form of the surface equilibrium equations (6). At an interior node where there is no contribution to g from external loads, it is a discretized form of the reciprocity relations (9) between internal connecting loads.

The kinematical connection rules (38) allow to deduce the generalized boundary strains of each element from the nodal displacement vector:

$$e_{(i)} = S_{(i)}^T q_{(i)} = (S_{(i)}^T L_{(i)})\, q \; . \tag{42}$$

By definition, the generalized strain vector e of the structure is built from the sequence, in some conventional order, of the strains $e_{(i)}$

$$e^T = (e_{(1)}^T \; e_{(2)}^T \; \dots \; e_{(N)}^T) \; . \tag{43}$$

There follows that the kinematical relation (1) holds true at the structural level if a *structural kinematical matrix* S be assembled as follows

$$S = (L_{(1)}^T S_{(1)} \;\; L_{(2)}^T S_{(2)} \; \dots \; L_{(N)}^T S_{(N)}) \; . \tag{44}$$

The dual role of this structural kinematical matrix is obtained by combining equations (28) for each element

$$g_{(i)} + f_{(i)} = S_{(i)} s_{(i)}$$

into the single equation

$$\sum_i L_{(i)}^T g_{(i)} + \sum_i L_{(i)}^T f_{(i)} = \sum_i L_{(i)}^T S_{(i)} s_{(i)} \tag{45}$$

The first term is identified by (41), the second suggests the *definition*

$$f = \sum_i L_{(i)}^T f_{(i)} \tag{46}$$

of a body load vector f conjugate to q, the right hand side term involves the structural kinematical matrix (44) and the structural stress vector s

$$s^T = (s_{(1)}^T \quad s_{(2)}^T \quad \ldots \quad s_{(N)}^T) \tag{47}$$

conjugate to e. Hence (45) turns into equation (2) and at the structural level is completely analogous to (28):

$$p = g + f = S s \quad . \tag{48}$$

The virtual work theorem for the structure becomes

$$q^T p = q^T g + q^T f = q^T S s = e^T s \tag{49}$$

and is identical to (29). Both the general equilibrium equations of the structure (48) and the kinematical equations

$$e = S^T q \tag{50}$$

expressing compatibility of the structural strains are necessary for the virtual work theorem to hold.

6. TOPOLOGICAL ANALYSIS OF THE STRUCTURAL CONNECTIONS

Clearly (48) and (50) are dual systems of equations, involving the same kinematical matrix and describing the topology of interelement connection from the dual aspects of kinematics and statics. Their properties are important because of their wide range of applicability; since no constitutive equations relating stresses and strains are implied in this investigation, its results will apply to any structure where a geometrically linear measure of strain is acceptable.

The kinematical problem: solve

$$S^T q = e$$

when the strains are given and the displacements are the unknowns, is the discretized equivalent to the problem of constructing a displacement field in a continuum, knowing the strain tensor field. It is known that local and possibly global compatibility conditions must be satisfied by the strains for a solution to exist and this solution may not be unique.

The statical problem: solve

$$S s = p$$

when the loads are given and the stresses are the unknowns is known to be generally undetermined. The existence and unicity conditions of those two problems are interrelated because of their adjoint character and should consequently be treated together. We note that concerning the dimensions of the problem, we have

$$n(q) = n(p) = \nu \qquad n(e) = n(s) = \sigma \quad .$$

In equation (50) each column of S^T is, by definition, a *compatible strain vector* associated to some unit displacement vector. The columns are however not linearly independent and the homogeneous problem

$$S^T q = 0 \rightarrow q = R\, r \rightarrow S^T R = 0 \tag{51}$$

has a set of non trivial solutions, represented as in the case of a single element by linear combinations of the columns of a matrix R. These columns, which constitute a maximal set of linearly independent solutions, are $n(r) = \rho$ in number so that R is a $\nu \times \rho$ matrix. Each solution corresponds either to a rigid body displacement of the structure or a "mechanism", they are both referred to as kinematically admissible modes of the structure and the elements of r as kinematical coordinates.

Similarly in equation (48) each column of S is a self-equilibrating load vector associated to some unit stress vector. The columns are generally not linearly independent and the homogeneous problem

$$S\, s = 0 \rightarrow s = X\, x \rightarrow S\, X = 0 \tag{52}$$

has a certain number of linearly independent solutions. The general solution is a linear combination, whose coefficients are collected in x, of a fundamental set represented by the columns of X. Each solution is appropriately called a self-stressing, since it indicates a state of internal stress existing without externally applied loads. The elements of x are "redundancy coordinates". We denote $n(x) = \xi$, so that X is a $\sigma \times \xi$ matrix. Linear, non-singular transformations of the kinematical coordinates or the redundancy coordinates merely change the choice of R or X within their respective equivalence classes. Whatever be the choices, the orthogonality conditions (51) and (52) remain satisfied. The condition for problem (50) to have a solution is the orthogonality of the given strain vector e to all the solutions of the homogeneous adjoint problem S s = 0. Hence,

$$e^T X x = 0 \quad \text{for all } x \rightarrow X^T e = 0 . \qquad (53)$$

That the condition is necessary is immediately seen by premultiplying (50) by X^T and using (52) in transpose. That it is sufficient will be demonstrated by actually constructing a solution.

Similarly the existence condition for a solution of problem (48) is the orthogonality of the given load vector to all the solution of the homogeneous adjoint problem $S^T q = 0$. Hence,

$$p^T R r = 0 \quad \text{for all } r \rightarrow R^T p = 0 . \qquad (54)$$

Necessity follows from premultiplication of (48) by R^T and use of (51) in transpose. Again a solution will then be constructed. The existence condition (53) is the discretized form of the compatibility conditions to be satisfied by strains. The existence condition (54) expresses that the virtual work performed by the external loads on any kinematically admissible mode must vanish. Stated otherwise, the load vector p must be self-equilibrated. The transpose of (51) states that each column of S is a self-equilibrated load vector.

As for isolated elements, a minimal set of constraints can be selected that prevents the occurrence of kinematically admissible modes in the structure. We express the constraints in the same notation

$$R^{\#} q = 0 \qquad R^{\#} \text{ a } \rho \times \nu \quad \text{matrix with linearly independent rows.} \qquad (55)$$

Again $R^{\#}$ can be chosen within an equivalence class to satisfy

$$R^{\#} R = (\rho/\rho) = R^T R^{\#T} . \qquad (56)$$

The general solution to (55) is

$$q = W\,w \qquad W \text{ a } \nu \times (\nu-\rho) \text{ matrix with linearly independent columns.} \tag{57}$$

with the property

$$R^{\#}\,W = 0 \quad \text{or} \quad W^T\,R^{\#T} = 0 \tag{58}$$

expressing that $R^{\#T}$ and W are base matrices spanning complementary orthogonal subspaces.

Theorem 1: $(R\ W)$ is a base matrix for q-space.
It has the required number of columns, ρ for R and $\nu - \rho$ for W. The linear independence of the columns follows from the fact that the homogeneous equations

$$R\,r + W\,w = 0$$

possess only the trivial solution. Indeed, if we premultiply by $R^{\#}$ and consider (56) and (58) there follows already $r = 0$. The equations are then reduced to $W\,w = 0$, that imply $w = 0$ because of the recognized linear independence of the columns of W.

Corollary 1.1. Any nodal displacement vector has a unique expansion of the type

$$q = R\,r + W\,w\ . \tag{59}$$

The first part contains the kinematically admissible modes of the structure, the second the displacement modes compatible with the applied constraints.

Corollary 1.2. The $\sigma \times (\nu-\rho)$ matrix S^TW has linearly independent columns. Indeed by reference to (51), equation $S^T(W\,w) = 0$ has the general solution

$$W\,w = R\,r$$

but, from the theorem, this holds only for $w = 0$ and $r = 0$. Hence $S^T W w = 0$, having only the trivial solution, has linearly independent columns. Then, from $S^T q = e = S^T W w$, we conclude that the columns of $S^T W$ constitute a maximal set of linearly independent compatible strain vectors.

Working by analogy, we introduce the concept of internal isostaticity constraints or "generalized releases". They are homogeneous relations imposed on the stress vector

$$X^{\#} s = 0 \quad X^{\#} \text{ a } \xi \times \sigma \text{ matrix with linearly independent rows,} \tag{60}$$

and capable of preventing the occurrence of self-stressings. This implies that $X^{\#}X$ be non singular. Again, as $X^{\#T}$ can be selected within an equivalence class, we can obtain that

$$X^{\#}\,X = (x/x) = X^T\,X^{\#T}\ . \tag{61}$$

The general solution of equation (60) is

$$s = Y\,y \quad Y \text{ a } \sigma \times (\sigma-\xi) \text{ matrix with linearly independent columns,} \tag{62}$$

and the property

$$X^{\#}\,Y = 0 \quad \text{or} \quad Y^T\,X^{\#T} = 0\ . \tag{63}$$

Theorem 2: (X Y) is a base matrix for s-space.
It has the required number of columns, ξ for X and $(\sigma-\xi)$ for Y. Premultiplying

$$X\ x + Y\ y = 0$$

by $X^{\#}$ and using (61) and (63), there follows $x = 0$. Then $Y\ y = 0$ implies $y = 0$ because of the linear independence of the columns of Y. The columns of X and Y are thus linearly independent.

Corollary 2.1. Any generalized stress vector has a unique expansion of the type

$$s = X\ x + Y\ y \qquad (64)$$

the first part contains the self-stressings; the second those stressing modes that comply with the chosen set of releases.

Corollary 2.2. The $\nu \times (\sigma-\xi)$ matrix $S\ Y$ has linearly independent columns. Indeed, by reference to (52) the equation $S(Y\ y) = 0$ has the general solution

$$Y\ y = X\ x \quad .$$

But, by theorem 2, this can only hold for $y = 0$ and $x = 0$. Hence, $S\ Y\ y = 0$ having only the trivial solution has linearly independent columns. From $S\ s = p = S\ Y\ y$ we conclude that the columns of $S\ Y$ constitute a maximal set of linearly independent self-equilibrated loading modes.

Theorem 3: The matrix $(Y^T\ S^T W)$ is non singular.
Consider the homogeneous problem

$$Y^T\ (S^T\ W\ w) = 0\ .$$

Since Y and $X^{\#T}$ span complementary orthogonal subspaces, its general solution is

$$S^T\ W\ w = X^{\#T}\ v \qquad (65)$$

with arbitrary v. Premultiplying this by X^T we get, in view of (52) in transpose and (61) that v = 0. Then corollary 1.2. implies that also w = 0. Our homogeneous problem has only the trivial solution and the columns of $Y^T S^T W$ are linearly independent. This requires

$$\sigma - \xi \geq \nu - \rho \quad . \tag{66}$$

Conversely, consider the homogeneous problem

$$y^T (Y^T S^T W) = 0 \quad \text{equivalent to} \quad W^T (S Y y) = 0 \quad .$$

Since W and $R^{\#T}$ span complementary orthogonal subspaces, the general solution is

$$S Y y = R^{\#T} z \quad \text{where z is arbitrary} . \tag{67}$$

Premultiplication by R^T yields z = 0, in view of (51) in transpose and (56) after which Corollary 2.2. enforces y = 0. This shows that $W^T S Y$ has linearly independent columns and the matrix of the theorem linearly independent rows. Consequently

$$\nu - \rho \geq \sigma - \xi \ . \tag{68}$$

Comparison of (66) and (68) establishes that

$$\nu - \rho = \sigma - \xi \quad . \tag{69}$$

Thus $Y^T S^T W$ is square and having linearly independent rows and columns is non singular. The same is of course true of its transpose $W^T S Y$.

Corollary 3.1. $(S^T W X^{\#T})$ is a base matrix for e-space. It has the required number of columns, $\nu - \rho = \sigma - \xi$ for $S^T W$ and ξ for

$X^{\#T}$. The linear independence of all the columns is a direct consequence of the proof that problem (65) has only the trivial solution. This base matrix provides for any strain vector a unique expansion

$$e = S^T W w + X^{\#T} v \quad . \tag{70}$$

The first part collects the linearly independent compatible straining modes (for which a nodal displacement vector q exists), the second part is made of generalized dislocation modes.

Corollary 3.2. $(S\ Y \quad R^{\#T})$ is a base matrix for p-space. S Y has $\sigma - \xi = \nu - \rho$ columns and $R^{\#T}$ has ρ, making the required total of ν. Their linear independence is a consequence of the proof that problem (67) has only the trivial solution. The unique expansion

$$p = S\ Y\ y + R^{\#T} z \tag{71}$$

of an arbitrary loading vector p, divides it in a self-equilibrated part and a part reacted upon by the set of external isostaticity constraints.

Note 1: While $R^{\#T}$ and $X^{\#T}$ can be selected in their respective equivalence classes to implement (56) and (61), it is also possible to select either W or Y in their equivalence classes to reduce the matrix of theorem 3 to an identity matrix:

$$Y^T S^T W = (w/w) = (y/y) = W^T S\ Y \tag{72}$$

Note 2: Computing the virtual work through the expansions (59) and (71), we find

$$q^T p = r^T z + w^T (W^T S\ Y)\ y \tag{73}$$

Similarly the virtual work computed from (70) and (64) is

$$e^T s = v^T x + w^T (W^T S Y) y \quad . \tag{74}$$

As observed before in connection with the virtual work theorem (49), the equality between (73) and (74) is ensured when $v = 0$, which means that the strain vector is compatible, and $z = 0$, which implies that the external loading is self-equilibrated. Those expressions also exhibit the conjugate nature of the couple (r,z) and the couple (v,x). Under the conditions of Note 1, w and y would also be conjugate.

We now pass to the effective construction of solutions of problems (48) and (50) when the existence conditions are satisfied. Premultiplication of equations (50) by the non singular matrix $(X\ Y)^T$ subdivides their solution in two parts. The first

$$X^T e = X^T S^T q = (S X)^T q = 0 \quad \text{because of (52)}$$

is the known statement (53) of the existence condition for a solution. The second

$$Y^T e = Y^T S^T q$$

can be solved by calling on theorem 3. The solution is in fact unique if we add the requirement that it should satisfy the isostaticity constraints (55). The displacement vector is then of type (57) and the problem becomes

$$Y^T e = (Y^T S^T W) w \tag{75}$$

with the unique solution, based on theorem 3

$$w = (Y^T S^T W)^{-1} Y^T e \quad .$$

The final result is that, provided the compatibility conditions (52) for strains are satisfied, problem (50) has the particular solution

$$q = S^{T\#} e \qquad S^{T\#} = W (Y^T S^T W)^{-1} Y^T \tag{76}$$

satisfying the isostaticity constraints (55). If we disregard such constraints, the general solution is obviously

$$q = S^{T\#} e + R r \quad \text{arbitrary} \tag{77}$$

the sum of a particular solution and the general solution for $e = 0$, which consists of kinematically admissible displacements. This result is in fact equivalent to the expansion (59) where w has been expressed in terms of the strain vector e by means of (75). A completely analogous procedure is applicable to the solution of equation (48). After premultiplication by the non singular matrix $(R\ W)^T$ they are split into

$$R^T p = R^T S s = (S^T R)^T s = 0 \quad \text{by virtue of (51),}$$

which is a restatement of the existence condition (54), and

$$W^T p = W^T S s \quad .$$

If we here add the requirements that s should satisfy a set of releases (60), then s is of type (62) and we have a unique solution

$$y = (W^T S Y)^{-1} W^T p$$

or $$s = S^{\#} p \qquad S^{\#} = Y\ (W^T S\ Y)^{-1} W^T \ . \tag{78}$$

The general solution without releases is

$$s = S^{\#} p + X x \qquad x \text{ arbitrary} \ . \tag{79}$$

By comparing (78) to (76) we see that

$$S^{\#T} = S^{T\#} \ . \tag{80}$$

7. PSEUDO INVERSES AND PROJECTORS

Because

$$q = S^{T\#} e \tag{76}$$

is, in some sense, a solution of the singular problem

$$S^T q = e \tag{50}$$

$S^{T\#}$ is called a (oblique) pseudo-inverse of S^T. While in (76) we obtain a displacement vector for any strain vector (even **not** compatible), the strain vector obtained through substitution of this displacement vector into (50) is necessarily compatible. Hence the elimination of q between the two equations generates the product $S^T S^{T\#}$ that is not an identity matrix, as would be the case with genuine inverses. It is an operator that transforms an arbitrary strain vector into a compatible one. This operator is a projection into the subspace $X^T e = 0$; the projection rays (or vectors annihilated by the projector) are the dislocation modes $X^{\#T} v$ of expansion (70). This is to be verified by looking at the structure (76) of $S^{T\#}$ and using (63). Similarly, by eliminating e between the two equations, we obtain the product $S^{T\#} S^T$

as an operator linking two displacement vectors. Again $S^{T\#}$ is not a genuine left-inverse to S^T because the operator transforms an arbitrary displacement into one that satisfies the isostaticity constraints $R^{\#} q = 0$. It is in fact a projector of an arbitrary displacement into the subspace $R^{\#} q = 0$; the projection rays $R r$ are the kinematically admissible modes.

One way to construct an identity matrix is to complete (76)

$$q = S^{T\#} e + R r$$

in order to give back to q its arbitrariness. Replace in it

$$e = S^T q \quad \text{and} \quad R = R^{\#} q \quad \text{(justified by (56) and (58))}$$

to obtain an identity in q, whereby

$$S^{T\#} S^T + R R^{\#} = (q/q) = S S^{\#} + R^{\#T} R^T . \tag{81}$$

Proceeding in the same manner with the equations

$$s = S^{\#} p \quad \text{and} \quad p = S s$$

we construct the operators

$S^{\#} S$ projecting an arbitrary stress vector into the subspace of stresses complying with the releases $X^{\#} s = 0$ and annihilating the self-stressings;

$S S^{\#}$ projecting an arbitrary load into a subspace of self-equilibrating ones by annihilating the reactions from isostaticity constraints.

We find the identity

$$S^{\#} S + X X^{\#} = (s/s) = S^T S^{T\#} + X^{\#T} X^T . \tag{82}$$

If the base matrices for s and e are taken to be identical, which is always possible although sometimes artificial from a physical point of view, and the same is done for the base matrices of q and p; the following simplifications occur

$$X^{\#T} \equiv X \qquad Y \equiv S^T W \qquad R^{\#T} \equiv R \qquad W \equiv S\,Y \; . \tag{83}$$

Each base matrix is split into orthogonal complementary parts

$$X^T Y = 0 \qquad R^T W = 0 \quad . \tag{84}$$

The pseudo inverse $S^{\#}$ becomes the so-called *Moore-Penrose* generalized inverse of S

$$S^{t} = S^T W\,(W^T W)^{-1}\,W^T = Y\,(Y^T Y)^{-1}\,Y^T S^T \quad . \tag{85}$$

8. ALGORITHMS FOR A DIRECT DETERMINATION OF PSEUDO-INVERSES

For structures made of assembled beams the determination of redundancies was traditionally done by releases, that is by cutting members or at least some of their load transmission capabilities, until the internal forces in the released structure could be determined by statics. This procedure which largely relies for its success on physical intuition becomes slow, tedious and prone to error for the large arrays of more complicated finite elements. The automatic determination of all the self-stressing states by the computer itself is a considerable step towards salvation of one of the most time-honored method of dealing with hyperstatic systems. It was pioneered by Denke [5] and Robinson [6]. As will be shown the algorithm they introduced, which is the Jordan version of the Gauss elimination algorithm, as applied to the structural kinematical matrix, provides a constructive answer to the problem of determining all the base matrices involved in the topological analysis of section 6. The Gauss-Jordan algorithm consists in the step by step determination of a $\nu \times \nu$ non singular premultiplier P that transforms the kinematical matrix S into a matrix

$$P\,S = \begin{bmatrix} I_{\sigma-\xi} & Q \\ 0 & 0 \end{bmatrix} \tag{86}$$

having ρ last rows of zero (ρ is in fact determined by the algorithm itself). $I_{\sigma-\xi}$ is the identity matrix in $\sigma - \xi = \nu - \rho$ dimensions and Q becomes some $(\sigma-\xi) \times \xi$ matrix. Each factor of the premultiplier depends on the choice of a pivot. If in the first row of S a pivot $s_{1\alpha}$ is selected, the first factor of the premultiplier is constructed with the remaining elements of the αth column of S.

$$P_1 = \left[\begin{array}{c|c} 1/s_{1\alpha} & 0 \\ \hline -s_{2\alpha}/s_{1\alpha} & \\ \vdots & I_{\nu-1} \\ -s_{\nu\alpha}/s_{1\alpha} & \end{array}\right]$$

The αth column of P S thereby turns into a unit vector. Denoting by $s^{(1)}_{\eta\xi}$ the elements of P_1S, a pivot $S^{(1)}_{2\beta}$ is selected on the second row and with the other elements of the αth column the non singular premultiplier is constructed.

$$P_2 = \left[\begin{array}{c|c|c} 1 & -s^{(1)}_{1\beta}/s^{(1)}_{2\beta} & 0 \\ \hline 0 & 1/s^{(1)}_{2\beta} & 0 \\ \hline 0 & -s^{(1)}_{3\beta}/s^{(1)}_{2\beta} & \\ \cdot & \cdot & \\ \cdot & \cdot & I_{\nu-2} \\ \cdot & \cdot & \\ 0 & -s^{(1)}_{\nu\beta}/s^{(1)}_{2\beta} & \end{array}\right]$$

while the αth column of P_2P_1S remains the same unit vector, the βth column is now turned into the next unit vector. Whenever in pursuing this algorithm with the successive rows of the successive transforms of S a row of zeros is encountered, it is sent to the last row by a renumbering of the elements of p. The procedure terminates when the last rows to be treated are all composed of zeros. Finally the sequence (α, β, γ ...) of the columns is here transformed into the sequence (1, 2, 3 ...) by renumbering the elements of s. The final structure of the transformed S is then given by (86) and the structure of the premultiplier is

$$P_{\nu-\rho} \cdots P_2 \,.\, P_1 = P = \begin{bmatrix} N & 0 \\ T & L \end{bmatrix} \tag{87}$$

N is a non singular $(\nu-\rho) \times (\nu-\rho)$, L is non singular $\rho \times \rho$ and lower triangular, T is a $\rho \times (\nu-\rho)$. Splitting s and p to suit the partitioning of P and P S, problem (2), premultiplied by the non singular P, takes the new form

$$s_{(1)} + Q\, s_{(2)} = N\, p_{(1)} \tag{88}$$

$$0 = T\, p_{(1)} + L\, p_{(2)} \tag{89}$$

The corresponding homogeneous problem (p = 0), or self-stressing problem has clearly from (88) a solution where $s_{(2)}$ can be taken as the set of redundancy coordinates and $s_{(1)}$ is determined from its knowledge. This determines a possible matrix X in (52)

$$s_{(1)} = -\,Q\,x \qquad s_{(2)} = x \qquad X = \begin{bmatrix} -Q \\ I_\xi \end{bmatrix} \; . \tag{90}$$

Furthermore, as we can take $s_{(2)} = 0$ as set of releases, a possible $X^{\#}$ matrix is

$$X^{\#} = (0 \quad I_\xi) \tag{91}$$

and, since the general solution of (60) is simply $s_{(2)} = 0$, $s_{(1)}$ arbitrary, we can identify

$$s_{(1)} = y \qquad Y = \begin{bmatrix} I_{\sigma-\xi} \\ 0 \end{bmatrix} \; . \tag{92}$$

Equation (89) that does no more contain s, is obviously the existence condition for a solution of the equilibrium equations. Thus a possible R^T matrix in (54) is

$$R^T = (T \quad L) \; . \tag{93}$$

Finally comparing (88) to (79) we obtain a possible pseudo-inverse

$$S^{\#} = \begin{bmatrix} N & 0 \\ 0 & 0 \end{bmatrix} \; . \tag{94}$$

Confirmation of all this is obtained in the modified form taken by the dual problem (1) when we expand q as in (59) by

$$q = P^T \begin{bmatrix} w \\ r \end{bmatrix} = \begin{bmatrix} N^T & T^T \\ 0 & L^T \end{bmatrix} \begin{bmatrix} w \\ r \end{bmatrix} = \left[\begin{matrix} N^T \\ 0 \end{matrix} \middle| \, R \right] \begin{bmatrix} w \\ r \end{bmatrix} = \begin{bmatrix} N^T \\ 0 \end{bmatrix} w + R\, r$$

where R has been taken as suggested by (93). Thus we can take

$$W = \begin{bmatrix} N^T \\ 0 \end{bmatrix} \tag{95}$$

Substitution of this into problem (1) results in

$$e = S^T P^T \begin{bmatrix} w \\ r \end{bmatrix} = (P\ S)^T \begin{bmatrix} w \\ r \end{bmatrix} = \begin{bmatrix} I_{\sigma-\xi} & 0 \\ Q^T & 0 \end{bmatrix} \begin{bmatrix} w \\ r \end{bmatrix}$$

or

$$e = (I_{\sigma-\xi}\ \ Q)^T\, w \quad \text{for the compatible strains.}$$

A set of external isostaticity constraints is furnished by

$$R^{\#} = (0\ \ (L^T)^{-1}) \tag{96}$$

that verifies (56) and (58). The non singular matrix of theorem 3, turns out to be an identity matrix and formula (78) is then seen to verify the result (94).

9. COMPACT SELF-STRESSINGS

Different choices of pivots in the Gauss-Jordan algorithm lead to different sets of redundancy coordinates and matrices X describing the self-stressings of the structure. While S is sparsely populated its transformation by the Gauss-Jordan process will usually result in a fully populated Q matrix and self-stressings with a tendancy to diffuse in the whole structure. From a physical point of view this is unsatisfactory since it is now well known that compact self-stressings, involving only a few members (finite elements) of the structure are the more numerous. In fact they represent the *local* compatibility conditions for strains in the continuum and must therefore be concentrated in a few number of adjacent members. Only the *global* compatibility conditions in multiply connected domains correspond to extensive redundancy "circuits". The compact self-stressings can of course be obtained by dismantling the structure into its constitutive (finite) elements and reassembling a small number of them until a redundancy is observed. However, this procedure becomes tedious for large structures and it would be desirable to program the computer to automate the search for compact self-stressings. This would result in a sparsely populated X matrix with the attendant advantages of best-conditioning of matrix force programs. In some cases where a discretization of the stress field of the continuum is possible in terms of stress-functions, the process is trivially achieved by the nodal values of the stress functions, which are in fact local redundancy coordinates. When the stress-function approach is not viable the search of compact self-stressings via appropriate selection of pivots in the Gauss-Jordan algorithm is still a subject of research.

10. EXTERNAL HYPERSTATICITY

In many cases, especially in Civil Engineering structures, the external constraints placed on q do more than prevent the kinematically admissible displacements; they can introduce additional self-stressings. Such cases of redundancy of supports are most easily handled by considering the set of nodal displacements subjected to supporting constraints to belong to an additional virtual finite element with its own elemental kinematical matrix and strain vector. Adding this strain vector to the sequence in e, we increase n(e) but not n(q) and the number of independent self-stressings is correspondingly increased; they now include the self-stressings due to redundancy of the supports. It will be observed that only non zero strains in the additional element can induce strains in the structure, a rigid body displacement of the additional element produces a corresponding kinematical mode in the structure.

11. STIFFNESS AND FLEXIBILITY AT ELEMENT LEVEL

In approximations based on a discretization of the displacement field, the strain energy is obtained directly as a quadratic form in the boundary and internal displacement coordinates q and b:

$$W = \frac{1}{2} (q^T \; b^T) \begin{bmatrix} K_{qq} & K_{qb} \\ K_{bq} & K_{bb} \end{bmatrix} \begin{bmatrix} q \\ b \end{bmatrix} \geq 0 \; . \tag{97}$$

The symmetrical matrix of the form is known as the stiffness matrix and is non negative. In a rigid body displacement

$$q = R\,r \qquad \text{and} \qquad b = C\,r$$

the strains vanish and the strain energy also; thus

$$K_{qq} R + K_{qb} C = 0 \qquad K_{bq} R + K_{bb} C = 0 \quad . \tag{98}$$

Let da and db denote differential increases in the generalized displacements and apply the energy conservation theorem, equating the strain energy increase to the virtual work performed by the loads

$$dW = dq^T (K_{qq} q + K_{qb} b) + db^T (K_{bq} q + K_{bb} b) = dq^T (f + g) + db^T h \ .$$

Since this is valid for arbitrary dq and db, we obtain the Castigliano type formulas

$$\begin{aligned} K_{qq} q + K_{qb} b &= g + f \\ K_{bq} q + K_{bb} b &= h \quad . \end{aligned} \tag{99}$$

Let us investigate the effects of modifying the definition of internal displacement coordinates according to the formula

$$b = c + U q \quad . \tag{25}$$

The strain energy is modified into

$$\frac{1}{2} (q^T \ c^T) \begin{bmatrix} \hat{K}_{qq} & K_{qc} \\ K_{cq} & K_{bb} \end{bmatrix} \begin{bmatrix} q \\ c \end{bmatrix}$$

$$\hat{K}_{qq} = K_{qq} + U^T K_{bq} + K_{qb} U + U^T K_{bb} U$$

$$K_{qc} = K_{qb} + U^T K_{bb} = K_{cq}^T \quad .$$

The virtual work becomes

$$dq^T (g + f + U^T h) + dc^T h$$

and our stiffness formulas

$$\hat{K}_{qq} q + K_{qc} c = g + f + U^T h$$

$$K_{cq} q + K_{bb} c = h \quad .$$

The following consequences of this transformation must be stressed:

- the boundary displacement direct stiffness matrix K_{qq} is modified to $\hat{K}_{qq}$;
- the generalized body load conjugate to q is modified from f to $f + U^T h$;
- the internal stiffness matrix K_{bb} and the body load conjugate to the internal displacement remain invariant;
- the stiffness coupling $K_{qb} = K_{bq}^T$ is modified.

If the new definition of internal displacements is of strain-type, we have, by definition, c = 0 together with q = R r in rigid body displacements and the corresponding properties

$$\hat{K}_{qq} R = 0 \qquad K_{cq} R = 0 \quad .$$

Holding the boundary fixed (q = 0) and displacing the inside necessarily generates internal strains and a positive strain energy. For this reason the internal stiffness matrix K_{bb} is necessarily positive definite, which opens the possibility of solving the stiffness equations for the internal degrees of freedom. Elimination of the internal displacements is the more indicated that it can be done economically at the element level and does not impeach the interconnection problem that concerns the boundary displacements only. From (99)

$$b = K_{bb}^{-1} (h - K_{bq} q) \quad . \tag{100}$$

Substitution of this in both the strain energy and the virtual work is equivalent to an internal coordinate transformation (25) with

$$U = - K_{bb}^{-1} K_{bq}$$

and the additional advantage that

$$K_{qc} = K_{qb} - K_{qb} K_{bb}^{-1} K_{bb} = 0$$

the stiffness coupling is eliminated. Moreover, the transformation is such that c is zero under rigid body displacements, or, stated otherwise, the new internal coordinates are of strain-type definition. This follows directly from the absence of stiffness coupling; the strain energy is reduced to

$$W = \frac{1}{2} q^T K q + \frac{1}{2} c^T K_{bb} c \tag{101}$$

with

$$K = K_{qq} - K_{qb} K_{bb}^{-1} K_{bq} \tag{102}$$

and vanishing of the strain energy, which implies rigid body motion, implies $c = 0$ because K_{bb} is positive definite, while

$$K R = 0 \quad . \tag{103}$$

The same result is of course obtained from formulas (98) as modified by the transformation.

We summarize the situation after elimination of the internal degrees of freedom, as it will be applied in the sequel. Starting from the quadratic form (97) as obtained with an arbitrary definition of the internal degrees of freedom, we eliminate b by formula (100) that also enables us to retrieve the internal displacements once q has been calculated. We thus obtain the boundary-stiffness relation

$$K\ q = g + f \tag{104}$$

where K is given by (102) and, for economy of notations, f denotes the body load effectively conjugate to q

$$f = f_b - K_{qb}\ K_{bb}^{-1}\ h \tag{105}$$

where f_b stands for the body load conjugate to q under the original choice b for the internal coordinates. The stiffness matrix K is non negative. Its singular character is demonstrated by formula (103). Transforming the energy due to boundary displacements by the expansion (21) we obtain in view of (103) the positive definite form

$$\frac{1}{2}\ q^T\ K\ q = \frac{1}{2}\ e^T\ J\ e$$

where

$$J = S^{\#}\ K\ S^{\#T} \tag{106}$$

is a positive definite matrix called the core stiffness matrix. It clearly depends on the choice of e, that is on the choice of the kinematical matrix S within its equivalence class. As the following transformations show

$$s = S^{\#}\ (g + f) = S^{\#}\ K\ q = S^{\#}\ K\ (S^{\#T}\ e + R\ r) = J\ e \tag{107}$$

J is a matrix of generalized elastic moduli connecting the stresses and the strains of the element. The inverse relation

$$e = F\ s \tag{108}$$

defines the positive core flexibility matrix, or matrix of compliances. From (107)

$$f + g = S\ s = S\ J\ e = S\ J\ S^T\ q$$

we obtain the inverse of (106)

$$K = S\ J\ S^T \quad . \tag{109}$$

In connection with this result, the question arises whether it is also possible to define an extended flexibility matrix relating loads and displacements, providing in fact an inversion of the stiffness relations (104). Starting from (21) we derive

$$q = S^{\#T}\ e + R\ r = S^{\#T}\ F\ s + R\ r = S^{\#T}\ F\ S^{\#}\ (g + f) + R\ r$$

or

$$q = G\ (g + f) + R\ r \tag{110}$$

$$G = S^{\#T}\ F\ S^{\#} \quad \text{extended flexibility matrix .} \tag{111}$$

As was to be expected from the singular character of equations (104), this result holds because g + f is self-equilibrated and the existence condition (27) is satisfied, but the solution is not unique. The part of q governed by G is the one satisfying a certain choice of isostaticity constraints implicit in $S^{\#}$. From (18) and (20) we obtain

$$R^{\#}\ q = r$$

the constraint violations, or generalized support displacements, of the general solution. The converse of (111) is easily found to be

$$F = S^T G S \quad . \tag{112}$$

12. THE DIRECT STIFFNESS OR DISPLACEMENT METHOD

Having determined the stiffness matrix of each element, the stiffness relations

$$g_{(i)} + f_{(i)} = K_{(i)} q_{(i)} \qquad i = 1,2,\ldots N$$

are multiplied to the left by $L^T_{(i)}$ and added to produce in view of (38), (41) and (46)

$$p = g + f = K q \tag{113}$$

where

$$K = \sum_i L^T_{(i)} K_{(i)} L_{(i)} \tag{114}$$

K is the *master stiffness matrix* of the unsupported structure, as the right hand side demonstrates it can be obtained quite simply by a proper addressing procedure of the elemental stiffnesses. We recall that while f, the body load conjugate to q, must also be built up from the individual $f_{(i)}$ according to (46), the load g is obtained directly by computing the virtual work $q^T g$ of the loads externally applied at the structural boundaries or, as the case may be, at the interfaces. Because the structure is as yet unsupported, the master stiffness matrix is singular. If a base matrix of kinematically admissible displacements R is known, the master stiffness obeys the equation $K R = 0$. It is however tedious for a large structure to set up a base matrix R from inspection. In aerospace applications, where the structure is normally not supported, a Gauss-Jordan algorithm applied to K can again produce a base matrix R together with a pseudo-inverse of the master stiffness relations (113). With the same notations as in section 8., let the premultiplier P receive the structure

$$P = \begin{bmatrix} N & 0 \\ T & L \end{bmatrix} \quad \text{and} \quad P\,K = \begin{bmatrix} I & Q \\ 0 & 0 \end{bmatrix} .$$

However, in this case the symmetry of K induces two important properties. Noting that $P\,K\,P^T$ is also symmetrical and computing

$$(P\,K)P^T = \begin{bmatrix} I & Q \\ 0 & 0 \end{bmatrix} \begin{bmatrix} N^T & T^T \\ 0 & L^T \end{bmatrix} = \begin{bmatrix} N^T & T^T + Q\,L^T \\ 0 & 0 \end{bmatrix}$$

we obtain at once

$$N^T = N \tag{115}$$

$$T^T + Q\,L^T = 0 \qquad \text{or} \qquad T + L\,Q^T = 0 . \tag{116}$$

The premultiplication of (113) by the non singular P matrix reduces the stiffness relations to

$$q_{(1)} + Q\,q_{(2)} = N\,p_{(1)} \tag{117}$$

$$0 = T\,p_{(1)} + L\,p_{(2)} . \tag{118}$$

The first can be rewritten as

$$q = \begin{bmatrix} q_{(1)} \\ q_{(2)} \end{bmatrix} = \begin{bmatrix} N & 0 \\ 0 & 0 \end{bmatrix} \begin{bmatrix} p_{(1)} \\ p_{(2)} \end{bmatrix} + \begin{bmatrix} -Q \\ I \end{bmatrix} r$$

a pseudo-inversion of (113)

$$q = K^{\#}\,p + R\,r \tag{119}$$

where

$$K^{\#} = \begin{bmatrix} N & 0 \\ 0 & 0 \end{bmatrix} = K^{\#T} \qquad R = \begin{bmatrix} -Q \\ I \end{bmatrix} \tag{120}$$

and $q_{(2)}$ is identified with the kinematical coordinates r of the structure. The second (118) is obviously the existence condition for the solvability of (113) with respect to q, or equilibrium conditions to be satisfied by the loads to avoid excitation of the kinematically admissible displacements. As such it must be equivalent to

$$R^T p = - Q^T p_{(1)} + p_{(2)} = 0 \quad .$$

This is not contradictory with (118), which can in fact be placed in the form $L R^T p = 0$ by virtue of (116). Since L is non singular both forms are thus equivalent. In the general case the known external loads are applied on a set of coordinates distinguished by the subscript (1), in other words a set $p_{(1)}$ of external loads is given. For the complementary set, the values of displacements is known: $q_{(2)}$ is given (zero for fixed supports). We then modify the Gauss-Jordan algorithm applied to K by forbidding the choice of pivots on the columns belonging to the second set. When the algorithm stops, the premultiplier has still the same structure, but now

$$P K = \begin{bmatrix} I & Q \\ 0 & V \end{bmatrix} .$$

If V has no zero rows, the structure is at least isostatically supported. If it has zero rows it is only partially supported. The stiffness equations are immediately resolved as follows:

$$q_{(1)} = N p_{(1)} - Q q_{(2)} \tag{121}$$

gives the unknown displacements in terms of the known loads and support displacements,

$$p_{(2)} = L^{-1} (V q_{(2)} - T p_{(1)}) \tag{122}$$

yields the reaction loads in the supports.

13. THE FORCE METHOD

While the direct stiffness method completely by-passes the construction of the structural kinematical matrix, the force method is actually based on its topological analysis. This is because, as its other name indicates, the direct stiffness method selects the generalized displacements as the basic unknowns. The strains and stresses are then derived from the generalized displacements, not necessarily from matrix equations, but often by returning to the field equations in each subdomain. The force method aims directly at the stresses and begins therefore by the determination of the redundancies. We start from equation (79)

$$s = S^{\#} p + X x$$

where a pseudo-inverse $S^{\#}$ and a base matrix X have been obtained by a Gauss-Jordan algorithm, as explained in section 1.8. The strains are given through the core flexibility relations

$$e_{(i)} = F_{(i)} s_{(i)}$$

established for each element. Considering the definitions of e and s, respectively (43) and (47), for the assembled structure, the set of those relations can be presented in the form e = F s, where F is built up from diagonal $F_{(i)}$ blocks

$$F = \begin{bmatrix} F_{(1)} & 0 & 0 & 0 \\ 0 & F_{(2)} & 0 & 0 \\ 0 & 0 & \ddots & 0 \\ 0 & 0 & 0 & F_{(N)} \end{bmatrix} .$$

Thus $e = F\, S^{\#}\, p + F\, X\, x$.

The compatibility conditions for the strains were found to be expressed through equation (53) and we must have

$$X^T F\, S^{\#}\, p + (X^T F\, X)\, x = 0 \quad . \tag{123}$$

Since F is positive definite and X has linearly independent columns, this set of linear equations for the unknown redundancy coordinates is non singular and its inversion furnishes

$$x = -\,(X^T F\, X)^{-1}\, X^T F\, S^{\#}\, p$$

and the generalized stresses are given in terms of the applied loads

$$s = S^{\#}\, p - X\, (X^T F\, X)^{-1}\, X^T F\, S^{\#}\, p \quad . \tag{124}$$

Given the compatible strains which follow from $e = F\, s$, the displacements can be found from

$$q = S^{\#T}\, e + R\, r \quad . \tag{125}$$

This procedure solves the problem of the unsupported or "free" structure, as well as that of isostatic or hypostatic supports, for which r can be determined at least partially without invalidating (124). Imagine now that we incorporate an additional element whose boundary displacements $q_{(0)}$ do not introduce new generalized displacements but are taken from the existing set q. The addition in a given truss of a new bar connecting two existing nodes is the simplest example. In more complicated examples the new element can possess, in isolation, its own non empty set of self-stressings. In any case

$$q_{(0)} = L_{(0)}\, q$$

and the new element has its own set of generalized strains

$$e_{(0)} = S^T_{(0)}\, q_{(0)} = S^T_{(0)}\, L_{(0)}\, q \tag{126}$$

so that the structural kinematical matrix receives additional columns, that will be assumed to be listed in front of the others

$$S = (L^T_{(0)} \;\; S_{(0)} \;\; S_{(f)}) \quad ,$$

$S_{(f)}$ denoting the kinematical matrix of the "free" structure.

Theorem 4

The self-stressing matrix of the augmented structure can be partitioned as follows

$$\begin{bmatrix} s_{(0)} \\ s_{(f)} \end{bmatrix} = \begin{bmatrix} X_{00} & X_{0t} \\ 0 & X_{ft} \end{bmatrix} \begin{bmatrix} x_{(0)} \\ x_{(t)} \end{bmatrix} \tag{127}$$

where $x_{(0)}$ is a complete set of redundancy coordinates for the additional element in isolation. X_{00} and X_{ft} have linearly independent columns.

Proof

It is obvious that the self-stressings $s_{(0)} = X_{00}\, x_{(0)}$ of the isolated new element induce no loads on the free structure when it is being connected, so that $s_{(f)}$ does not depend on $x_{(0)}$. By definition the columns of X_{00} are linearly independent. Assume now that the columns of X_{ft} are not linearly independent. A non zero $x_{(t)}$ would then exist such that $s_{(f)} = X_{ft}\, x_{(t)} = 0$. For this $x_{(t)}$ we must have $s_{(0)} = X_{0t}\, x_{(0)} \neq 0$, otherwise the columns of X related to $x_{(t)}$ would be linearly dependent, which contradicts the definition of X. Hence this $x_{(t)}$ would generate a self-stressing in the isolated new element alone, which contradicts the assumption that $x_{(0)}$ is a complete set.

Note

If $x_{(t)}$ is properly choosen, X_{0t} itself can have a set of zero columns, the corresponding columns of X_{ft} representing self-stressings of the free structure alone. The remaining columns of X_{0t} and X_{ft} represent the new self-stressings that may appear as a result of the interconnection. The partitioning of X, considered in theorem 4, is one obtained from a Gauss-Jordan transformation in which the pivots are first selected amongst the $L^T_{(0)}\, S_{(0)}$ columns of S. It yields for $s_{(f)}$ the expansion

$$s_{(f)} = S^{\#}_{(f)}\, p + X_{ft}\, x_{(t)} \quad . \tag{128}$$

From the core flexibility relation of the free structure follows

$$e_{(f)} = F_{(f)}\, s_{(f)} = F_{(f)}\, S^{\#}_{(f)}\, p + F_{(f)}\, X_{ft}\, x_{(t)} \quad .$$

Apply now the compatibility conditions on the strains

$$X^T \begin{bmatrix} e_{(0)} \\ e_{(f)} \end{bmatrix} = \begin{bmatrix} X^T_{00} & 0 \\ X^T_{0t} & X^T_{ft} \end{bmatrix} \begin{bmatrix} e_{(0)} \\ e_{(f)} \end{bmatrix} = 0 \quad ,$$

yielding in particular

$$X_{0t}^T e_{(0)} + X_{ft}^T e_{(f)} = 0 \quad ,$$

or, in view of (127),

$$X_{0t}^T e_{(0)} + X_{ft}^T F_{(f)} S_{(f)}^{\#} p + (X_{ft}^T F_{(f)} X_{ft}) x_{(t)} = 0 \ . \tag{129}$$

Now $F_{(f)}$ being positive definite and X_{ft} having linearly independent columns, equation (129) can be solved for the unknown $x_{(t)}$. When the result is substituted into (127), there follows

$$s_{(f)} = \{S_{(f)} - X_{ft}(X_{ft}^T F_{(f)} X_{ft})^{-1} X_{ft}^T F_{(f)} S_{(f)}^{\#}\} p$$

$$- X_{ft}(X_{ft}^T F_{(f)} X_{ft})^{-1} X_{0t}^T e_{(0)} \ . \tag{130}$$

This establishes the stresses in the original structure as a result of the application of external loads and the application of given strains in the additional element. Obviously we need only consider loads p applied to the set q complementary to $q_{(0)}$. As no flexibility of the additional element is involved, this formula is applicable to the problem of the hyperstatically supported structure. The additional element is then the virtual element suggested in section 10, incorporating the set of displacements $q_{(0)}$ which are either fixed ($q_{(0)} = 0$ and thus $e_{(0)} = 0$) or whose values are prescribed. An interesting feature of the force method, which is in fact related to reciprocity properties and consequently to virtual work, is the Pasternak reduction principle. Evaluating the internal form of the virtual work theorem, it can be verified from (123) and (124) in the simplified form

$$s = s^{\#} + X\, x \qquad (s^{\#} = S^{\#}\, p)$$

$$e = F\, s = e^{\#} + F\, X\, x \qquad (e^{\#} = F\, s^{\#})$$

$$x = - (x^T\, F\, X)^{-1}\, X^T\, e^{\#}$$

that

$$e^T\, s = e^{\#T}\, s = e^T\, s^{\#}\,. \tag{131}$$

In other words, the virtual work can be evaluated by reducing either the strains, or the stresses, to their particular statically determined part.

14. THE COMBINED METHOD

In the combined method, stresses and displacements are solved simultaneously. Elimination of the strains between equations (1) and (3)

$$F\, s - S^T\, q = 0 \tag{132}$$

and equation (2), generate a combined linear system

$$\begin{bmatrix} F & S^T \\ S & 0 \end{bmatrix} \begin{bmatrix} s \\ -q \end{bmatrix} = \begin{bmatrix} 0 \\ p \end{bmatrix} \tag{133}$$

whose matrix is still symmetrical.

15. DISCRETE VARIATIONAL PRINCIPLES

The formulations of the direct stiffness, combined, or force method can be presented as the Euler-Lagrange equations of discrete variational principles. The variational principle of displacements, or principle of minimum of the total potential

$$\frac{1}{2} q^T K q - q^T p \qquad \text{minimum} \tag{134}$$

features the strain energy expressed in terms of the generalized displacements and the potential energy $- q^T p$ of the loads p assumed to be known and invariant (dead loads). If the displacements actually occuring are incremented by "variations" δq, the conditions under which the total energy remains stationary to first order in the variations is

$$\delta q^T (K q - p) = 0$$

and, since the increments are arbitrary, we obtain equations (113) of the direct stiffness method. The increment of the total potential to second order (the second variation)

$$\delta q^T K \delta q$$

is non negative and the total potential is thus a minimum. More general boundary conditions consist in specifying loads on a subset $p_{(1)}$ of p and displacements on the complementary subset $q_{(2)}$ of q. The principle takes the form

$$\frac{1}{2} q^T_{(1)} K_{11} q_{(1)} + q^T_{(1)} K_{12} q_{(2)} + \frac{1}{2} q^T_{(2)} K_{22} q_{(2)} - q^T_{(1)} p_{(1)} \quad \text{minimum.}$$

The third term is in fact a specified constant that could be dropped from the principle, were it not of interest to adopt the Lagrangian multiplier technique and restore to $q_{(2)}$ its freedom by augmenting the functional with the term $-p^T_{(2)} (q_{(2)} - u_{(2)})$. Here $p_{(2)}$ is a set of Lagrangian multipliers, $u_{(2)}$ the set of specified values of $q_{(2)}$. The Euler equations pertaining to variations on $q_{(1)}$ and $q_{(2)}$ are then respectively

$$K_{11} q_{(1)} + K_{12} q_{(2)} - p_{(1)} = 0$$

$$K_{21}\, q_{(1)} + K_{22}\, q_{(2)} - p_{(2)} = 0 \qquad (K_{21} = K_{12}^T) \quad .$$

The lagrangian multipliers are thus identified with the loads $p_{(2)}$ reacting against the imposition of displacements. If the conditions $q_{(2)} = 0$ prevent any admissible kinematical freedom of the structure, K_{11} is non singular and

$$q_{(1)} = K_{11}^{-1}\, (p_{(1)} - K_{12}\, q_{(2)}) \tag{135}$$

an equation that solves the problem of determining the remaining displacements in terms of the given data. Also

$$p_{(2)} = K_{21}\, K_{11}^{-1}\, (p_{(1)} - K_{12}\, q_{(2)}) + K_{22}\, q_{(2)} \tag{136}$$

an equation determining the reaction loads. This situation, that often prevails, is to be compared with the solution given previously by equations (121) and (122).

Consider now the Friedrichs [11] transformation of principle (134) by which the strain energy is now expressed in terms of strains through use of (109)

$$\frac{1}{2}\, q^T\, K\, q = \frac{1}{2}\, q^T\, S\, J\, S^T\, q = \frac{1}{2}\, (S^T q)^T\, J\, (S^T q) = \frac{1}{2}\, e^T\, J\, e$$

and the kinematical equations (1) are incorporated through a set of s Lagrangian multipliers

$$\frac{1}{2}\, e^T\, J\, e + s^T\, (S^T\, q - e) - q^T\, p \qquad \min_q\, [\, \max_s\, (\min_e)\,] \tag{137}$$

variations on e produce as Euler equations

$$J\, e - s = 0 \tag{138}$$

which, in view of (3), identifies the multipliers with the stresses associated with the strains in the constitutive equations. Considering the positive definiteness of J, it is clear that for fixed q and s, (137) is a minimum with respect to that choice of e. Variations on s produce as Euler equations the kinematical constraints (1); variations on q the equilibrium equations (2). Consequently this 3-field variational principle contains all three basic equations (1), (2) and (3) of our problem as Euler equations. It remains to justify its minimum-maximum character with respect to the choices on q and s. Accepting the constitutive equations (3) or (138) as satisfied a priori and substituting the minimizing choice of e into (137), we obtain

$$- \frac{1}{2} s^T F s + s^T S^T q - q^T p \qquad \underset{q}{\min} \ \underset{s}{(\max)} \tag{139}$$

Indeed, it is clear that the quadratic form in the stresses is negative definite and produces a maximum of the functional for the choice corresponding to the variational derivative (Euler equation) with respect to s:

$$- F s + S^T q = 0 . \tag{140}$$

This is of course equivalent to the compatibility equations (1) as modified by the constitutive equations. The variational derivative of (139) with respect to q yields again the equilibrium equations (2). The 2-field principle (139) is *canonical* in the sense given to it by Friedrichs. In the context of linear elasticity it is better known as the Reissner principle. Should we substitute for s its maximizing value $s = J S^T q$, taken from (140), the principle reverts to the one given by (134), where the choice of q is minimizing.

If on the contrary we write (139) in the equivalent form

$$- \frac{1}{2} s^T F s + q^T (S s - p) \quad \min_q (\max_s)$$

and assume the equilibrium equations (2) to be satisfied a priori, we obtain the single-field principle

$$- \frac{1}{2} s^T F s \qquad \max_s \tag{141}$$

By changing the sign this turns into the minimum complementary energy principle. A direct application of this principle requires that we solve the equilibrium equations. When this has been done by a suitable definition of redundancy coordinates, we can substitute equation (79) and obtain

$$\frac{1}{2} (p^T S^{\#T} + x^T X^T) F (S^{\#} p + X x) \qquad \min_x \quad , \tag{142}$$

whose Euler equations are identical to (123). Consequently as our initial single-field principle (134) is related to the direct stiffness method, the dual single-field principle (141) is related to the force method. The variational equations of the Reissner two-field principle, equations (140) and (2) are seen to correspond to the final system (133) of the combined method. A second two-field principle can be derived from the general principle (137) by observing that the a priori satisfaction of the equilibrium equations (2) reduces it to

$$\frac{1}{2} e^T J e - s^T e \qquad \min_e \quad .$$

Its direct applicability requires again that the equilibrium equations be solved, it then appears in the new form

$$\frac{1}{2} e^T J e - e^T (S^{\#} p + X x) \quad \max_x [\min_e] \ . \tag{143}$$

Indeed, its variational derivative with respect to e is

$$J\ e - S^{\#}\ p - X\ x = 0 \tag{144}$$

and, when this minimizing value of e is substituted, it turns into the functional (142) of x with a change of sign, whereby x becomes maximizing. The direct variational derivative of (143) with respect to x produces the compatibility equations

$$X^T\ e = 0\ .$$

When they are solved simultaneously with (144)

$$\begin{bmatrix} J & X \\ X^T & 0 \end{bmatrix} \begin{bmatrix} e \\ -x \end{bmatrix} = \begin{bmatrix} S^{\#}\ p \\ 0 \end{bmatrix} \tag{145}$$

we have a new type of combined method. All the discrete variational principles so reviewed have corresponding formulations for the continuum. When the displacement, strain tensor or stress-tensor fields they contain are replaced by discrete representations, some of the constraints represented by equilibrium, compatibility or constitutive equations are relaxed. The principle of minimum total potential relaxes only the equilibrium equations. Its dual, the principle of minimum complementary energy relaxes only the compatibility conditions. Reissner's principle relaxes both, but retains exact constitutive equations. The second two-field principle relaxes compatibility and constitutive equations but satisfies rigorously equilibrium.

REFERENCES

Engineering work in the field of matrix structural analysis was done by Langefors and Argyris:

[1] LANGEFORS, B., "Theory of Aircraft Structural Analysis", *Zeitschrift f. Flugwissenschaften*, 6, 1958, 282-291.

[2] ARGYRIS, J.H., "Die Matrizen Methoden der Statik" *Ingenieur Archiv*, 25, 174, 1957.

[3] ARGYRIS, J.H. and KELSEY, S., "Energy Theorems and Structural Analysis", *Butterworth Scientific Publications*, 1960.

The general concept of redundancy as intensity of a compact self-stressing state is illustrated in this publication of Argyris and some of his earlier work. It can be traced back to Ebner and Koeller:

[4] EBNER, H. and KOELLER, H., "Zur berechnung des kraftverlaufes in versteiften Zylinderschalen", *Luftfahrtforschung*, 14, (12), 1937.

The algorithmic determination of redundancy coordinates in complex structures, essential to the application of the force method, is due to Denke and to Robinson:

[5] DENKE, P.H., "A general digital computer analysis of statically indeterminate structures", *NASA Tech. Note D-1666*, 1962.

[6] ROBINSON, J., *Structural Matrix Analysis for the Engineer*, J. Wiley & Sons, 1966.

For further developments of the algorithm to improve conditioning, see:

[7] ROBINSON, J., HAGGENMACHER, G.W. and CONTINI, R., "Static force and displacement methods as eigenvalue problems", *Proc. IUTAM Symposium on High Speed Computing of Elastic Structures*, B. Fraeijs de Veubeke (Ed.), Les Congrès et Colloques de l'Université de Liège, Place du 20 Août, 16, B-4000, Liège, Belgium, 1971.

The concepts of pseudo-inverse matrices are discussed from a mathematical point of view in:

[8] MOORE, E.H., *Bull. Amer. Math. Soc.*, 26, 1920, 394-395.

[9] PENROSE, R., "A generalized inverse for matrices", *Proc. Cambridge Phil. Soc.*, 51, 1955, 406-413.

[10] MILNE, R.D., "An oblique matrix pseudo-inverse", *SIAM Jnl. Appl. Math.*, 16-5, 1968, 931-944.

The Friedrichs transformation of functionals and its related set of variational principles can be found in:

[11] FRIEDRICHS, K.O., "Ein verfahren der variationsrechnung", *Nachrichten der Ges. der Wiss. Gottingen*, 1929, 13-20.

[12] COURANT, R. and HILBERT, D., "Methods of Mathematical Physics", *Interscience*, Vol. I, 1966, 231-274.

Its application to linear elasticity theory and finite element methods in:

[13] FRAEIJS de VEUBEKE, B., "Diffusion des inconnues hyperstatiques dans les voilures à longerons couplés", *Bulletin 25, Service Technique de l'Aéronautique*, Chaussée de Waterloo, 72, Bruxelles, 1951.

[14] FRAEIJS de VEUBEKE, "Displacement and equilibrium models in the finite element method", *Stress Analysis*, Chapter 9, O.C. Zienkiewicz and G. Holister, (Eds.), Wiley, 1965.

DIFFUSIVE EQUILIBRIUM MODELS

B.M. Fraeijs de Veubeke

University of Liège, Belgium

SUMMARY—Equilibrium models are characterized by an internal stress field in equilibrium and surface tractions that should obey the diffusivity requirement of reciprocity with adjacent elements at the interfaces. The paper discusses the construction of stiffness matrices for such elements and the difficulties associated with the presence of *mechanisms*.

The use of first order or second order stress functions is presented together with the possibility of solving the final equations by the Force Method through a *direct flexibility* method.

Three methods are presented for getting rid of the mechanisms: the classical *superelement technique*, a method based on the *discretization of rotational equilibrium* and *quasi-diffusivity*.

2.1 *Discretization of Stresses and Conjugate Displacements*

There are several methods for constructing finite elements such that the internal equilibrium equations are rigorously satisfied together with the transition condition for surface tractions at the interfaces (equations 9 of reference 1). A direct discretization of the stress field (reference 2) presents some advantages; it is the most natural way of introducing body loading modes, gravitational loads in statics and inertia loads in dynamics. Coupled

---

Original publication, Lecture Notes for the International Research Seminar on 'The Theory and Application of Finite Element Methods,' held at the University of Calgary, Alberta, Canada, July 26 - August 10, 1973.

with virtual work considerations, it also allows an easy introduction of the displacement information that subsists, despite the fact that the discretized strains, obtained from the constitutive equations, are not, as a rule, compatible and consequently do not permit integration of a continuous displacement field. Finally, because precisely of the displacement information, it allows the construction of a stiffness matrix and the use of the same direct stiffness method of resolution that so naturally belongs to the conforming displacement models based on a direct discretization of the displacement field.

Hence a single resolution software becomes applicable to both types of models. It must be recognized, however, that this procedure may in certain cases be less efficient than a force method, because of the very nature of the topology of interconnexions.

The possible existence of kinematical modes in addition to the rigid body modes, which would normally impose restrictions on the loading of the element, is an inconvenience of diffusive equilibrium models that is no more severe, and is in fact completely similar, to the problems of C_1 continuity encountered with displacement models of Kirchhoff plate bending theory. They can be overcome by the same technique and should not present any obstacle to their use.

Let the stress field be described by the vector $\sigma(x_i)$ of reference 1, and consider the discretization

$$\sigma(x_i) = S(x_i)\ s + H\ (x_i)h \qquad (1)$$

where s and h are columns of discretization parameters, playing the same role at element level as the corresponding s and h in reference 1.

$S(x_i)$, not to be confused with the kinematical matrix S (that has constant elements) and $H(x_i)$ are matrices of assumed approximation functions, usually polynomials. The equilibrium equations (5) of reference 1 are satisfied if the approximation functions are so chosen that

$$D^T S(x_i) = 0 \ , \tag{2}$$

$$\phi(x_i) = -[D^T H(x_i)]h \ . \tag{3}$$

Hence each coordinate of s represents a stress field in equilibrium without body loads; each coordinate of h a certain body loading mode in equilibrium with the discretized stress-field.

The next important step is to analyze the corresponding surface traction distribution at the boundary E of the element. In general the surface traction vector will be described by

$$t(x_i) = N^T \sigma(x_i) = N^T S(x_i)s + N^T H(x_i)h \qquad x_i \in \partial E \ , \tag{4}$$

in term of the stress parameters. We need however another description in terms of generalized boundary loads, which are the elements of the column matrix g in

$$t(x_i) = T(x_i)g \qquad x_i \in \partial E \ . \tag{5}$$

Each column of $T(x_i)$, which corresponds to a unit vector for g, is a *surface traction mode*. Clearly any distribution of surface tractions generated by a set of stress parameters in (4) must be represented in (5) by some appropriate set of generalized boundary loads. In other words the generalized boundary loads are entirely determined by the stress parameters:

$$g = S\ s + H\ h \tag{6}$$

in such a way that we have the identity

$$T(x_i)S\ s + T(x_i)H\ h \equiv N^T S(x_i)s + N^T H(x_i)h \qquad x_i \in \partial E$$

or, separately,

$$T(x_i)S \equiv N^T S(x_i) , \tag{7}$$

$$T(x_i)H \equiv N^T H(x_i) . \tag{8}$$

In (6), S and H are matrices of constants, known as the *load connection matrices*. As will be seen from its properties, S is in fact the kinematical matrix of the element, as defined in reference 1, and the determination of the load connection matrices follows immediately, as examples show, from the definitions adopted for the generalized boundary loads.

The converse property is generally not true: a given set of generalized boundary loads does not determine uniquely s and h. However, before discussing the invertibility of (6), let us define the set of boundary displacements conjugate to g by the virtual work argument:

$$g^T q = \int_{\partial E} t^T(x_i)u(x_i)dS .$$

Substitution of (5) and identification of both sides furnishes

$$q = \int_{\partial E} T^T(x_i)u(x_i)dS . \tag{9}$$

The surface traction modes appear as weighting functions in the definition of generalized displacements. A similar argument is applied to the virtual work of body loads:

$$V = \int_E \phi^T(x_i)u(x_i)dV .$$

Substitution of (3) followed by an integration by parts transforms it into

$$V = -h^T \int_E [D^T H(x_i)]^T u(x_i) dV$$

$$= -h^T \int_{\partial E} [N^T H(x_i)]^T u(x_i) dS + h^T \int_E H^T(x_i) D u(x_i) dV .$$

We now use (8) and our result (9) to obtain

$$V = -(H\ h)^T q + h^T \int_E H^T(x_i) D\ u\ (x_i) dV .$$

This shows the virtual work of body loads to consist of two parts

$$V = f_b^T q + h^T b \tag{10}$$

as considered in reference 1. The first is the virtual work due to boundary displacements, with generalized body loads

$$f_b = -H\ h . \tag{11}$$

The second is due to internal displacements without boundary displacements, and the stress parameter set h may also be considered as a set of body loads. The conjugate internal displacements are then defined by

$$b = \int_E H^T(x_i) D u(x_i) dV . \tag{12}$$

It becomes quite clear from this definition that b vanishes when the element undergoes a rigid body motion, since then the strains $\varepsilon = D\ u(x_i)$ vanish everywhere. Thus, in the terminology of reference 1, b has a strain-type definition and can be looked upon as defining either a set of internal generalized displacements or a set of generalized strains.

Observe that the numerical value of q and b will be the only information available for displacements. It will not be possible to derive from this information local values of the displacement field without having recourse to an interpretation scheme based on additional assumptions. Different interpretation schemes may lead to differences in local values for the same q and b. It is convenient to express this situation by qualifying q and b as *weak* quantities. By contrast, h and s are *strong* quantities because, once they are numerically evaluated, local values of the stress field are available from equations (1). Similarly the knowledge of g determines everywhere the surface traction on the element's boundary.

The opposite situation exists in the conforming displacement models, where the knowledge of q and b determines the local values of the displacement field, hence q and b are strong, while their conjugates g and h are to be considered as weak. There is of course a natural strong interpretation scheme for the stresses, that consists in applying the constitutive equations. It is however subject to well known difficulties due to lack of single-valuedness at the interfaces.

Turning back to equation (6) and using (11), we are presented with equation (28) of reference 1

$$g + f_b = S\ s \tag{13}$$

whereby we may recognize in S a kinematical matrix of the element. In trying to solve it for the generalized stresses s we must look for possible solutions of the homogeneous adjoint problem

$$S^T\ q = 0 \tag{14}$$

that represent the admissible kinematical motions

$$q = R\ r \tag{15}$$

of the element. The identification of rigid body modes is a trivial matter since we need only input the rigid body displacement fields in the definition (9) to obtain them. More generally the generalized boundary strains e, conjugate to s, are obtained from (9) as

$$e = S^T q = \int_{\partial E} [T(x_i)S]^T u(x_i)dS . \quad (16)$$

Then, using (7), we obtain the following definitions, equivalent through integration by parts and consideration of (2),

$$e = \int_{\partial E} S^T(x_i) \; N \; u(x_i)dS = \int_E S^T(x_i)Du(x_i)dV . \quad (17)$$

Each rigid body motion we input in (16) results in $Du(x_i) \equiv 0$ and consequently $e = 0$. Hence all rigid body modes are certainly solutions of (14). If we find additional linearly independent solutions, they are *mechanisms* of the element and must be taken care of by techniques to be described later.

The existence condition

$$R^T \; (g + f_b) = 0$$

for the inversion of (13) contains the global equilibrium conditions to be satisfied by the boundary loads g and the part f of body loads, together with additional restrictions in the presence of mechanisms. The inversion of (13) is of type

$$s = S^* \; (g + f_b) + X \; x \quad (19)$$

where the first term is a particular solution involving some pseudo-inverse S^* of S; the second term is the general solution of $S\,s = 0$.

In reference 1 it was assumed that, at the element level, the kinematical matrix S had linearly independent columns and this implies that the homogeneous problem had only the trivial solution $s = 0$. This is because S was defined as a base matrix, orthogonal complement of R in q-space.

In the present definition of S we still have the orthogonality property

$$S^T \; R = 0 \quad (20)$$

stemming from (14) and (15). However, in cases of polynomial approximations of high order, we may encounter non trivial solutions of $S\ s = o$. They correspond to self-stressings within the element, where neither body-loads nor surface tractions are present. Then, considering the property

$$S\ X = 0 \tag{21}$$

we conclude from the definition of generalized boundary strains (16) that

$$X^T\ e = (S\ X)^T\ q = 0 \tag{22}$$

and the set e is subject to compatibility constraints. Looking at this situation from another viewpoint, we may distinguish in S a set of linearly independent columns and an additional set of columns which are linear combinations of the first. The corresponding additional rows of S^T generate through (16) additional strains which are linear combinations of the strains due to the independent rows of S^T.

2.2 *Construction of the Stiffness Matrix*

As a rule the body loads are specified and h can be considered as known, together with $f_b = -\ H\ h$. We now apply the complementary energy principle with the assumption that the boundary displacements q are specified:

$$\frac{1}{2}\ (s^T\ h^T) \begin{pmatrix} F_{ss} & F_{sh} \\ F_{hs} & F_{hh} \end{pmatrix} \begin{pmatrix} s \\ h \end{pmatrix} - q^T (S\ s + H\ h) \quad \min_{s}\ . \tag{23}$$

The first term is the stress-energy of the element expressed in terms of the stress parameters, the second is the complementary potential energy containing the forces reacting against the imposition of the boundary displacements. Minimization with respect to the unknown stress parameters s yields the compatibility condition:

$$F_{ss}\ s + F_{sh}\ h = S^T q \ . \tag{24}$$

Solving this for s

$$s = F_{ss}^{-1}\ (S^T\ q - F_{sh}\ h)$$

and substituting into (6) we find

$$g + f = K\ q \tag{25}$$

where

$$K = S\ F_{ss}^{-1}\ S^T \tag{26}$$

$$f = -\ H\ h + S\ F_{ss}^{-1}\ F_{sh}\ h = f_b + S\ F_{ss}^{-1}\ F_{sh}\ h\ . \tag{27}$$

K is the proper stiffness matrix of the equilibrium element, the solutions of the homogeneous problem $K\ q = 0$ are clearly the same as those of $S^T\ q = 0$ and thus given by (15). The body loads effectively conjugate to the boundary displacements are given by (27) and, being expressed in terms of h, are known. If (27) is compared with equation (135) of reference 1, it appears that the elimination of the displacements b has implicitly caused a change of definition of those internal coordinates. By reference to the general consequences of such a change, as described in Section 1.10 of reference 1, the transformation is

$$b = c + F_{hs}\ F_{ss}^{-1}\ S^T\ q\ . \tag{28}$$

The new internal coordinates c still have a strain-type definition, since the last term vanishes by virtue of (20) for rigid body displacements.

It is of course of interest to obtain the numerical information about the internal displacements. This can be done by a modification of the variational principle in which the internal displacements b are considered to be specified instead of specifying the body loads.

The last term of (23) is now replaced by

$$- q^T (g + f_b) - b^T h = - q^T S s - b^T h$$

and the minimization is applicable to h as well as to s.

The minimization with respect to s yields the same equation as before. With respect to h we obtain

$$b = F_{hs} s + F_{hh} h = F_{hs} F_{ss}^{-1} (S^T q - F_{sh} h) + F_{hh} h . \tag{29}$$

This equation allows b to be computed as soon as q is determined.

The coherence between (28) and (29) is noteworthy.

2.3 *Connections by Weak Displacements and Diffusivity*

In principle the problem of connecting the boundaries of adjacent elements is solved by the technique of identification of the generalized boundary displacement vector $q_{(i)}$ of each element with a subset $L_{(i)} q$ of the nodal displacements q. The resulting properties are however quite different from those obtained from conforming displacements models. In the latter case the identification results in a rigorous satisfaction of the interface transition conditions $u^+ = u^-$ (equation (8) of reference 1); it is the conformity property.

In the case of equilibrium models, what can be obtained is a rigorous satisfaction of the second transition conditions $t^+ + t^- = 0$ (equation (9) of reference 1).

This reciprocity of surface tractions, that we call diffusivity, requires however that the Boolean matrices $L_{(i)}$ connect the weak boundary displacements pairwise across a given interface. Hence each generalized boundary load and its conjugate weak displacement should belong to a single face of the element. Then the reciprocity of the boundary loads across an interface will cause the complete surface traction modes they represent to be reciprocal.

In very special cases only, the Kirchhoff plate bending elements are typical examples, some of the generalized boundary loads pertain to a vertex and are added up by the connexion process (indeed the corner loads at angular points of plate bending elements should add up to zero at interior nodes where a concentrated load is applied from the outside.)

This particular topology of weak nodal connections represents some loss of efficiency of the direct stiffness method when applied to diffusive equilibrium elements, since, in general, the number of nodal displacements of the structure will be larger for the same number of elements.

Conversely the number of self-stressings will in general be smaller and this can lead to a reappraisal of the force method as an efficient tool for solving the global problem.

2.4 *An Example. The Membrane with Linear Stress Variation*

In cartesian (x, y) coordinates with

$$\sigma^T(x, y) = (\sigma_x \ \sigma_y \ \tau_{xy}) ,$$

$$s^T = (\sigma_1, \sigma_2, \ldots, \sigma_7) ,$$

$$h^T = (\eta_1 \ \eta_2) ,$$

we adopt a linear stress distribution

$$\sigma_x = 2(\sigma_3 + \sigma_6 x + 3\sigma_7 y) ,$$

$$\sigma_y = 2(\sigma_1 + 3\sigma_4 x + \sigma_5 y) ,$$

$$\tau_{xy} = -2(\sigma_2 + \sigma_5 x + \sigma_6 y) + 2(\eta_1 y + \eta_2 x) ,$$

whereby the matrices of functions in equation (1) are

$$S(x, y) = \begin{pmatrix} 0 & 0 & 2 & 0 & 0 & 2x & 6y \\ 2 & 0 & 0 & 6x & 2y & 0 & 0 \\ 0 & -2 & 0 & 0 & -2x & -2y & 0 \end{pmatrix}$$

$$H(x, y) = \begin{pmatrix} 0 & 0 \\ 0 & 0 \\ 2y & 2x \end{pmatrix} .$$

The body loads are given by

$$\phi_x = -\left(\frac{\partial \sigma_x}{\partial x} + \frac{\partial \tau_{yx}}{\partial y}\right) = -2\eta_1$$

$$\phi_y = -\left(\frac{\partial \tau_{xy}}{\partial x} + \frac{\partial \sigma_y}{\partial y}\right) = -2\eta_2$$

so that the σ_i parameters are clearly devised to satisfy equilibrium without body loads; the body loading modes are uniform distributions.

In a triangular element the surface tractions are given along the side 12 by

$$t_{x12} = \frac{y_{21}}{c_{12}} \sigma_x - \frac{x_{21}}{c_{12}} \tau_{yx} ,$$

$$t_{y12} = \frac{y_{21}}{c_{12}} \tau_{xy} - \frac{x_{21}}{c_{12}} \sigma_y$$

where c_{12} is the length of the side and, for concision, the notation

$$x_{ij} = x_i - x_j \qquad y_{ij} = y_i - y_j$$

is used. Clearly the surface tractions have linear variation and are entirely determined by their local values at the end points:

$$t_{x12}(x_1, y_1) = \frac{y_{21}}{c_{12}} \sigma_x (x_1, y_1) - \frac{x_{21}}{c_{12}} \tau_{xy} (x_1, y_1)$$

and similar expressions for

$$t_{y12}(x_1, y_1), \quad t_{x12}(x_2, y_2), \qquad t_{y12}(x_2, y_2) \ .$$

As illustrated on Figure 1 the generalized boundary loads pertaining to side 12 are defined as the resultants of triangular surface traction distributions generated by those local values:

$$g_{x12} = \frac{1}{2} c_{12} \ t_{x12} \ (x_1, y_1) \ ,$$

$$g_{x21} = \frac{1}{2} c_{12} \ t_{x21} \ (x_2, y_2) \ ,$$

$$g_{y21} = \frac{1}{2} c_{12} \ t_{x12} \ (x_2, y_2) \ ,$$

$$g_{y12} = \frac{1}{2} c_{12} \ t_{y12} \ (x_1, y_1) \ .$$

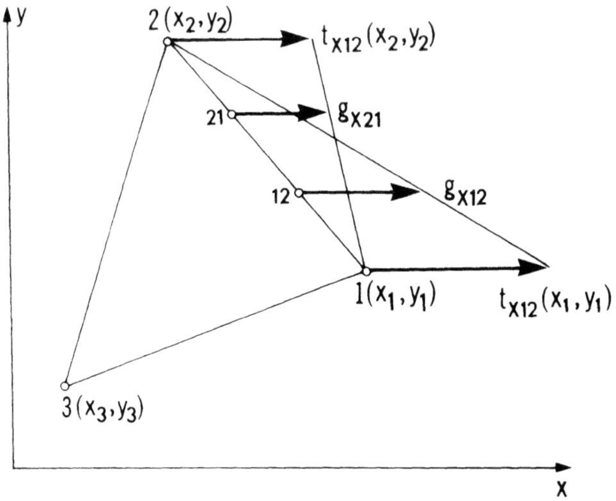

Figure 1

Substitution of the surface tractions by their values in terms of the stresses furnishes for instance

$$g_{x12} = y_{21}(\sigma_3+\sigma_6 x_1+3\sigma_7 y_1) + x_{21}(\sigma_2+\sigma_5 x_1+\sigma_6 y_1) - x_{21}(\eta_1 y_1+\eta_2 x_1) \ .$$

And, if g_{x12} is listed as the first coordinate of the generalized boundary load vector g, we obtain as first row of S in equation (6)

$$(0 \quad x_{21} \quad y_{21} \quad 0 \quad x_1 x_{21} \quad x_1 y_{21} + y_1 x_{21} \quad 3y_1 y_{21})$$

and as first row of H

$$(-y_1 x_{21} \quad -x_1 x_{21}) \ .$$

The matrices S and H are completed by similar calculations for the other coordinates of g.

As a matter of fact once the coordinates for side 12 have been determined one can simply use a cyclic permutation of the indices to find the others.

If u(x, y) denotes the x component of the displacement field, the virtual work expression

$$\int_1^2 t_{x12}(x, y)u(x, y)ds = q_{x12}g_{x12} + q_{x21}g_{x21} ,$$

can be used to interpretate the weak generalized boundary displacements conjugate to g_{x12} and g_{x21}. If the coordinate s along the side 12 is counted from 1 towards 2, we have

$$t_{x12} = t_{x12}(x_1, y_1)\,\frac{c_{12}-s}{c_{12}} + t_{x12}(x_2, y_2)\,\frac{s}{c_{12}}$$

or

$$t_{x12} = \frac{2g_{x12}}{c_{12}}\,\frac{c_{12}-s}{c_{12}} + \frac{2g_{x21}}{c_{12}}\,\frac{s}{c_{12}} .$$

This value substituted into the integral yields, by comparison with the right-hand side

$$q_{x12} = \int_1^2 \frac{2}{c_{12}}\,\frac{c_{12}-s}{c_{12}}\,u\,ds \qquad q_{x21} = \int_1^2 \frac{2}{c_{12}}\,\frac{s}{c_{12}}\,u\,ds$$

showing the type of weighted averages of u(x, y) along side 12 that are implied. Similar averages are obtained for the other weak conjugates.

Turning to the definition of the internal displacements, we consider the virtual work of body loads

$$\iint (u\,\phi_x + v\,\phi_y)\,dxdy = -\,2\eta_1 \iint u\,dxdy - 2\eta_2 \iint v\,dxdy ,$$

where the integrals are to be extended over the area of the triangle.

Clearly the coefficients of η_1 and η_2 are not strain-type definitions for the internal coordinates. Since the matrix H(x, y) is known, we can immediately apply the general result (12) to find

$$b = 2 \begin{pmatrix} \iint y \left(\frac{\partial u}{\partial y} + \frac{\partial v}{\partial x} \right) dxdy \\ \\ \iint x \left(\frac{\partial u}{\partial y} + \frac{\partial v}{\partial x} \right) dxdy \end{pmatrix}$$

as internal conjugate of strain-type definition to h. This shows that the partial integration transformations must be

$$-2 \iint u\, dxdy = -2 \oint y\ (m\ u + \ell\ v)\ ds + 2 \iint y \left(\frac{\partial u}{\partial y} + \frac{\partial v}{\partial x} \right) dxdy$$

$$-2 \iint v\, dxdy = -2 \oint x\ (m\ u + \ell\ v)\ ds + 2 \iint x \left(\frac{\partial u}{\partial y} + \frac{\partial v}{\partial x} \right) dxdy$$

and it is simple matter to verify that, as proven by the general theory, the contour integrals provide the virtual work of the generalized body loads $f_b = -H\ h$ on the boundary displacements.

Now that the S matrix is determined, the mechanisms of the element can be established from the solutions of $S^T\ q = 0$, that are not simply rigid body motions. The rigid body mode solutions are of course immediately found by injecting the rigid body displacement fields for u(x, y) and v(x, y) into the definitions of the weak displacements q. The values obtained are identical to the local displacements at the point of application of the conjugate resultants. That is why we may consider points like 12, of coordinates

$$x_{(12)} = \frac{2}{3}\, x_1 + \frac{1}{3}\, x_2 \quad , \quad y_{(12)} = \frac{2}{3}\, y_1 + \frac{1}{3}\, y_2$$

and 21, of coordinates

$$x_{(21)} = \frac{1}{3}x_1 + \frac{2}{3}x_2 \quad , \quad y_{(21)} = \frac{1}{3}y_1 + \frac{2}{3}y_2$$

as *virtual* nodes.

There are in this case 3 linearly independent rigid body modes and consequently 3 global equilibrium constraints between the 12 generalized boundary loads (in the absence of body loads).

12 - 3 = 9 of those loads should thus be independent. However the number of independent stress parameters on which they depend is only n(s) = 7. The two additional constraints between the boundary loads indicate the presence of two kinematical freedoms allowing the element to distort without stresses. They can be illustrated as follows (Figure 2).

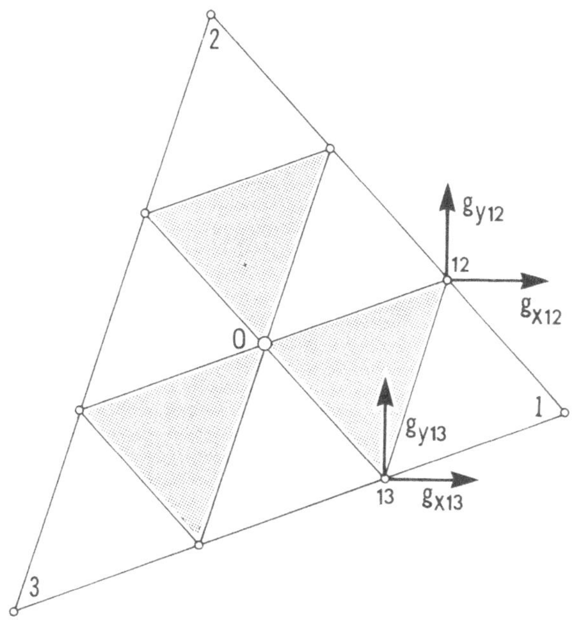

Figure 2

We isolate the rows of S corresponding, in succession, to g_{x12}, g_{x13}, g_{y12} and g_{y13}, the loads applied to the virtual nodes 12 and 13:

$$\begin{bmatrix} 0 & x_{21} & y_{21} & 0 & x_1x_{21} & x_1y_{21}+y_1x_{21} & 3y_1y_{21} \\ 0 & x_{13} & y_{13} & 0 & x_1x_{13} & x_1y_{13}+y_1x_{13} & 3y_1y_{13} \\ -x_{21} & -y_{21} & 0 & -3x_1x_{21} & -(x_1y_{21}+y_1x_{21}) & -y_1y_{21} & 0 \\ -x_{13} & -y_{13} & 0 & -3x_1x_{13} & -(x_1y_{13}+y_1x_{13}) & -y_1y_{13} & 0 \end{bmatrix}$$

and compute the resultant moment of those loads with respect to the "center" 0 of coordinates $\frac{1}{3}(x_1+x_2+x_3)$, $\frac{1}{3}(y_1+y_2+y_3)$. This resultant is

$$\frac{1}{3}\,(-y_{13}g_{x_{12}} + y_{21}g_{x_{13}} + x_{13}g_{y_{12}} - x_{21}g_{y_{13}}) \; .$$

It is immediately verified that the corresponding linear combinations of rows of S vanish identically. We thus have in

$$-y_{13}g_{x_{12}} + y_{21}g_{x_{13}} + x_{13}g_{y_{12}} - x_{21}g_{y_{13}} = 0 \; , \qquad (30)$$

and two analogous relations based on moment equilibrium of the loads applied on the pairs of virtual nodes (21, 23) and (31, 32), equilibrium constraints corresponding to *mechanisms* within the element.

By virtual work interpretation the mechanisms consists in a rigid body rotation of *one* of the skeleton triangles (hatched on Figure 2) about 0. For instance (30) is the virtual work due to a rotation of the skeleton (0, 12, 13), whereby the boundary displacements are zero except for $q_{x_{12}}$, $q_{x_{13}}$, $q_{y_{12}}$, $q_{y_{13}}$ which are respectively proportional to $(-y_{13}, y_{21}, x_{13}, -x_{12})$. The verification that such a set of displacements satisfies $S^Tq = o$ is implicitly the same as the verification of (30).

Obviously two only of the mechanisms discovered are linearly independent, a linear combination being a rigid body rotation of the whole element about 0.

2.5 *A Rib Element for Membrane Reinforcement*

This second example is instructive in that it shows a case of paramount importance of body loads and also the necessity of applying the principles underlying the general theory and not their formulas. We take the plane stress case in which the membrane parts are reinforced by ribs capable of transmitting axial loads, shear loads and bending moments.

The rib element is idealized as a beam and devised to match exactly the membrane element developed in the preceding section.

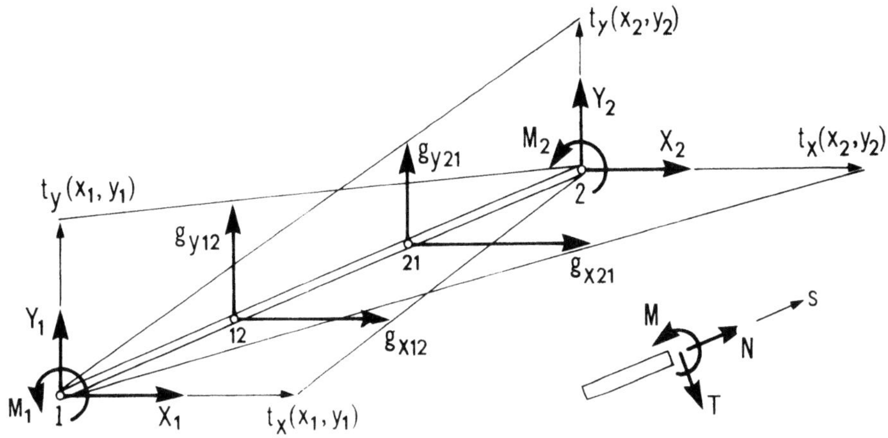

Figure 3

The boundaries of the element are punctual and the generalized boundary loads consist of end forces and moments:

$$g^T = (X_1 \; Y_1 \; M_1 \; X_2 \; Y_2 \; M_2) \; .$$

The conjugate displacements are strong because of the punctual character of the boundaries and consist of the end components of displacement and rotation

$$q^T = (u_1\ v_1\ \theta_1\ u_2\ v_2\ \theta_2)\ .$$

The loads transmitted to the element by the membranes, that can be attached on either side, are here to be considered as body loads and we can adopt as body load stress parameters

$$h^T = (g_{x12}\ g_{x21}\ g_{y12}\ g_{y21})\ .$$

To simplify formulas we use the following notations

$$\xi = \frac{x_2 - x_1}{c_{12}}\ , \qquad \eta = \frac{y_2 - y_1}{c_{12}} \quad \text{and } c = c_{12}\ ,$$

also $t_x(s)$ and $t_y(s)$ instead of t_{x12} and t_{y12}. The equilibrium equations that must be satisfied are

$$\frac{dN}{ds} = \xi\ t_x(s) + \eta\ t_y(s)\ ,$$

for the axial load N in the beam,

$$\frac{dT}{ds} = \eta\ t_x(s) - \xi\ t_y(s)$$

for the shear load T ,

$$\frac{dM}{ds} = -\ T$$

for the bending moment M.

Replacing

$$t_x(s) = t_x(x_1,y_1)\ \frac{c-s}{c} + t_x(x_2,y_2)\ \frac{s}{c} = \frac{2}{c}\ \frac{c-s}{c}\ g_{x12} + \frac{2}{c}\frac{s}{c}\ g_{x21}\ ,$$

and, similarly,

$$t_y(s) = \frac{2}{c}\frac{c-s}{c} g_{y12} + \frac{2}{c}\frac{s}{c} g_{y12} ,$$

we can integrate from $s = 0$ (in x_1, y_1) to s, finding:

$$c^2(N(s)-N_1) = (2cs-s^2)(\xi g_{x12}+\eta g_{y12}) + s^2(\xi g_{x21}+\eta g_{y21}) ,$$

$$c^2(T(s)-T_1) = (2cs-s^2)(\eta g_{x12}-\xi g_{y12}) + s^2(\eta g_{x21}-\xi g_{y21}) , \tag{31}$$

$$c^2(M_1-M(s)-T_1 s) = (cs^2 - \frac{1}{3}s^3)(\eta g_{x12}-\xi g_{y12}) + \frac{1}{3}s^3(\eta g_{x21}-\xi g_{y21}) .$$

The stress-energy of the beam follows then by taking $N(s)$ and $M(s)$ from those results and integrating

$$\frac{1}{2}\int_0^c \left[\frac{N^2(s)}{EA} + \frac{M^2(s)}{EI}\right] ds .$$

The cross-section A and moment of inertia I of the beam are usually constant, but, if variable, numerical integration instead of analytical may have to be used. It is also possible to include the shearing energy, if so desired. The result is a quadratic form in the variables represented by h and in the variables N_1, T_1 and M_1 that may thus be taken to represent the stress parameters s of the general theory:

$$s^T = (N_1 \; T_1 \; M_1) .$$

We are thereby presented with the flexibility matrices F_{ss}, $F_{sh} = F_{hs}^T$ and F_{hh} of the theory.

The end-values in (x_2, y_2) of the axial load, shear load and bending moment are obtained from equations (31) by setting $s = c$

$$N_2 = N_1 + \xi\,(g_{x12}+g_{x21}) + \eta\,(g_{y12}+g_{y21}) ,$$

$$T_2 = T_1 + \eta\,(g_{x12}+g_{x21}) - \xi\,(g_{y12}+g_{y21}) ,$$

$$M_2 = M_1 - cT_1 - \frac{c\eta}{3}(2\,g_{x12}+g_{x21}) + \frac{c\xi}{3}(2\,g_{y12}+g_{y21}) \; .$$

There remains to transform the end loads into the set of generalized boundary loads

$$X_1 = \eta\, T_1 - \xi\, N_1 \;, \qquad Y_1 = -\,\xi\, T_1 - \eta\, N_1 \;,$$

$$X_2 = -\,\eta\, T_2 + \xi\, N_2 , \qquad Y_2 = \xi\, T_2 + \eta\, N_2 \;,$$

$$M_1 = -\,M_1 \;, \qquad M_2 = M_2 \;.$$

The change in sign of M_1 is due to the absolute anticlockwise sense that must be preserved in the assembling process.

Those results furnish the load connexion matrices

$$S = \begin{pmatrix} -\xi & \eta & 0 \\ -\eta & \xi & 0 \\ 0 & 0 & -1 \\ \xi & -\eta & 0 \\ \eta & \xi & 0 \\ 0 & -c & 1 \end{pmatrix}$$

$$H = \begin{pmatrix} 0 & 0 & 0 & 0 \\ 0 & 0 & 0 & 0 \\ 0 & 0 & 0 & 0 \\ \xi^2-\eta^2 & \xi^2-\eta^2 & 2\xi\eta & 2\xi\eta \\ 2\xi\eta & 2\xi\eta & \eta^2-\xi^2 & \eta^2-\xi^2 \\ -\frac{2}{3}\,c\eta & -\frac{1}{3}\,c\eta & \frac{2}{3}\,c\xi & \frac{1}{3}\,c\xi \end{pmatrix}$$

and complete the analysis of the element, that is seen to be free of any mechanism.

2.6 *The Family of Membrane Equilibrium Elements*

The membrane element with linear stress variation is a member of a family of diffusive equilibrium triangles with stress fields approxi mated by polynomials of arbitrary degree. Its first member, the constant stress triangle appeared first in reference 4. It is the purpose of this section to analyze the mechanisms and self-stressing states of the members of this family as well as the propagation or inhibition of those mechanisms resulting from element interconnexion

In reference 4, it is shown how the constant stress element, though itself devoid of mechanisms, can generate them by the virtual pin-jointed character of its interface connexions. Inhibition of the mechanisms can be secured in this case, either by the presence of reinforcing ribs, or by the super element technique which consists in assembling four triangles in a quadrilateral with diagonal interfaces.

The interpretation of mechanisms discovered in Section 2.4 for the case of the first degree polynomials is of interest to discuss kinematically the inhibitions due to interconnexions. An attempt will now be made to discuss the general case. To this purpose a convenient simplification can be obtained by having recourse to oblique axes, as depicted on Figure 4.

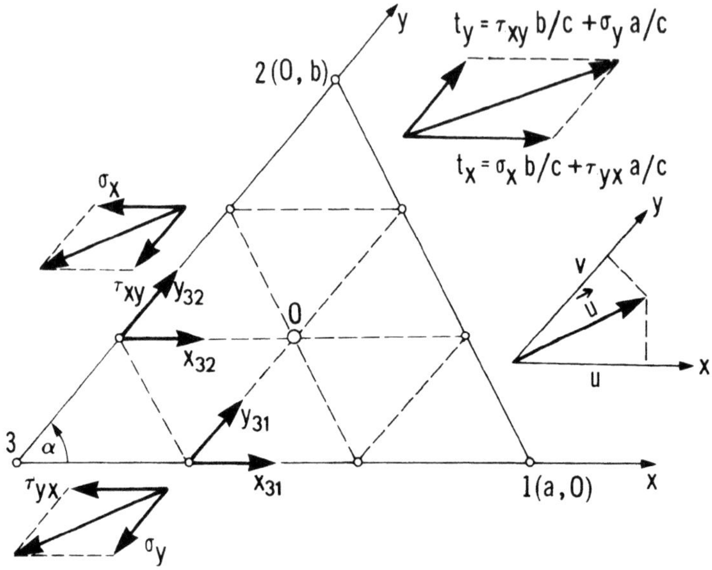

Figure 4

Provided the forces and stresses are contravariant and the displacements and strains defined in a covariant way (u and v are actually the orthogonal projections of the displacement vector respectively on the x and y axes), the basic definitions of strains are unchanged, the equilibrium equations are

$$\frac{\partial\sigma_x}{\partial x} + \frac{\partial\tau_{yx}}{\partial y} + \sin\alpha\phi_x = 0 \quad , \quad \frac{\partial\tau_{xy}}{\partial x} + \frac{\partial\sigma_y}{\partial y} + \sin\alpha\phi_y = 0 \ ,$$

$$\tau_{yx} = \tau_{xy} \ ,$$

and the stress-strain relations derive from the strain energy density as

$$\sigma_x = \sin\alpha\,\frac{\partial W}{\partial\varepsilon_x} \quad , \quad \tau_{xy} = \sin\alpha\,\frac{\partial W}{\partial\gamma_{xy}} \quad , \quad \sigma_y = \sin\alpha\,\frac{\partial W}{\partial\varepsilon_y} \ .$$

Consequently we may keep

$$\sigma_x = 2(\sigma_3 + \sigma_6 x + 3\sigma_7 y)$$

$$\sigma_y = 2(\sigma_1 + 3\sigma_4 x + \sigma_5 y)$$

$$-\tau_{xy} = 2(\sigma_2 + \sigma_5 x + \sigma_6 y)$$

as a linear stress field approximation in equilibrium without body loads. Along the side 31 (y = o) we have a tangent surface traction distribution

$$-\tau_{yx} = 2\sigma_2 + 2\sigma_5 x$$

from which we calculate the generalized load X_{31} which is proportional to the local value of the distribution x = o:

$$X_{31} = \frac{a}{2}\, 2\sigma_2 = a\sigma_2 \ .$$

Similarly along the side 32 (x = o) we have the tangent distribution

$$-\tau_{yx} = 2\sigma_2 + 2\sigma_6 y$$

and

$$Y_{32} = \frac{b}{2}\, 2\sigma_2 = b\sigma_2 \ .$$

To compute the resulting moment of loads acting on the hatched skeleton triangle we need not evaluate X_{32} nor Y_{31}, whose lines of action pass through the centre 0. The moment

$$\frac{1}{3} \sin \alpha (X_{31} b - Y_{32} a) = 0$$

vanishes whatever the value of σ_2, confirming our previous calculation in cartesian axes. However, we may now recognize that this property is a consequence of the reciprocity of tangential surface tractions at the corner 3, a property that should manifest itself for models of higher polynomial approximations.

For polynomials of second degree we should first establish a choice of generalized boundary loads that has the virtual node property. Figure 5 shows an arbitrary parabolic distribution t(x) on the straigth segment from x = -m to x = +m.

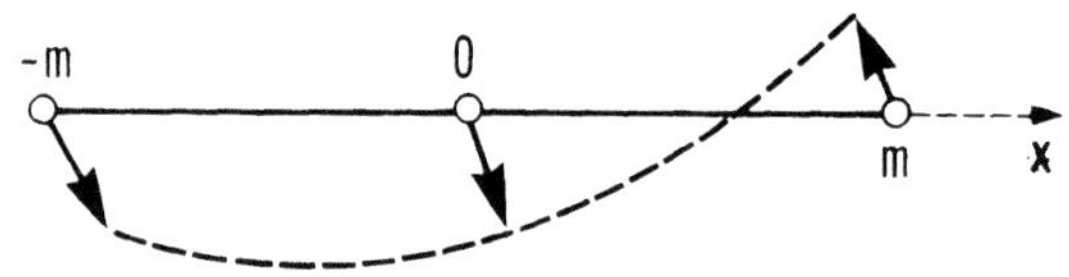

Figure 5

It is the superposition of three distributions generated by the local values at the ends and in the middle

$$t(x) = t(-m)\,\frac{x(x-m)}{m^2} + t(o)\,\frac{m^2-x^2}{m^2} + t(m)\,\frac{x(x+m)}{2m^2} .$$

The load due to the first part is

$$g_{-1} = t(-m) \int_{-m}^{m} \frac{x(x-m)}{2m^2}\,dx = \frac{m}{3}\,t(-m) .$$

Its moment with respect to 0

$$t(-m) \int_{-m}^{m} \frac{x^2(x-m)}{2m^2}\,dx = -\frac{m^2}{3}\,t(-m) = -m\,g_{-1} ,$$

so that the node at which the resultant should be placed is x = -m or, precisely, the left-end. The load due to the second part is

$$g_o = t(o) \int_{-m}^{m} \frac{m^2-x^2}{m^2} dx = \frac{4}{3} m\, t(o)$$

and its moment is clearly zero, placing the resultant at the centre. Similarly the third load is

$$g_1 = \frac{m}{3} t(m) \quad \text{and is placed at the end} \quad x = m .$$

The distribution is now completely determined by the three loads

$$t(x) = g_{-1} \frac{3x(x-m)}{2m} + g_o \frac{3(m^2-x^2)}{4m} + g_1 \frac{3x(x+m)}{2m}$$

and, by virtual work, the corresponding functionals for the displacements are

$$u_{-1} = \frac{3}{2m} \int_{-m}^{m} x(x-m)u(x)dx \quad , \quad u_o = \frac{3}{4m} \int_{-m}^{m} (m^2-x^2)u(x)dx \ ,$$

$$u_1 = \frac{3}{2m} \int_{-m}^{m} x(x+m)u(x)dx \ .$$

In an arbitrary plane rigid body displacement field, the distribution of u, orthogonal projection of the local displacement on the direction of $t(x)$, is of type $u(x) = \alpha + \beta x$. Computation of the functionals shows that

$$u_{-1} = \alpha - \beta m \ , \quad u_0 = \alpha \ , \quad u_1 = \alpha + \beta m \ .$$

This is a control on the virtual node property: "In a rigid body motion the displacement functionals take the same values as the local displacements at the points of application of the loads".

We now apply this representation to the quadratic stress equilibrium triangle, as shown on Figure 6.

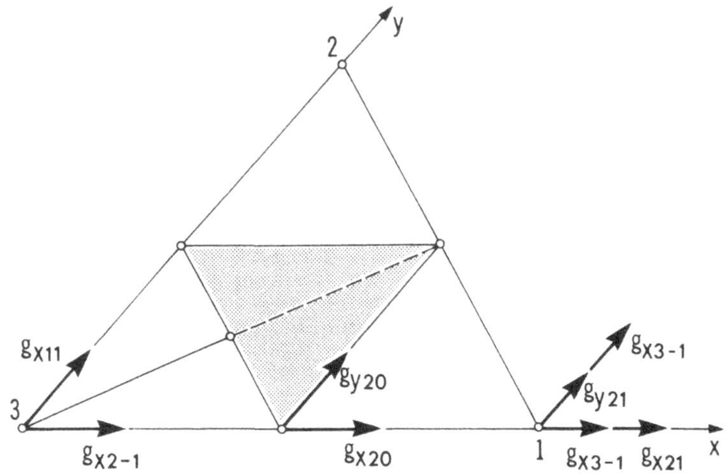

Figure 6

The notation is as follows, the first subscript orients the load, the second indicates the boundary by listing the number of the opposite vertex, the third is -1, 0 or 1 according to the node position in the positive sense of description of the boundary. Care must be taken to distinguish at a corner the pair of loads belonging to either boundary. The stress field in equilibrium without body loads now contains 12 parameters

$$\sigma_x = 2(\sigma_3+\sigma_6x+3\sigma_7y) + 2\sigma_{10}x^2+6\sigma_{11}xy+12\sigma_{12}y^2 ,$$

$$\sigma_y = 2(\sigma_1+3\sigma_4x+\sigma_5y) + 12\sigma_8x^2+6\sigma_9xy + 2\sigma_{10}y^2 ,$$

$$-\tau_{xy} = 2(\sigma_2+\sigma_5x+\sigma_6y) + 3\sigma_9x^2+4\sigma_{10}xy+3\sigma_{11}y^2 .$$

Along 31 we have

$$-\tau_{xy} = 2\sigma_2 + 2\sigma_5x + 3\sigma_9x^2 ,$$

and, proportional to this value at x = o,

$$g_{x2-1} = \frac{a}{3}\, 2\sigma_2 \ .$$

Along 32

$$-\tau_{xy} = 2\sigma_2 + 2\sigma_6 y + 3\sigma_{11} y^2 \ ,$$

and

$$g_{y11} = \frac{1}{3}\, 2\sigma_2 \ .$$

Again we see that at the corner 3, the relation

$$bg_{x2-1} = ag_{y11}$$

must hold for any stress field. Once more it expresses the reciprocity of tangential surface tractions at this corner. This equilibrium condition corresponds to a mechanism that links the corner to a rotation centre belonging to a central core of the element. Considering that only infinitesimal rotations are implied, this centre may be taken at any place along the line joining the corner to the midpoint of side 12. We so obtain the graphical representation of Figure 7, showing the 3 mechanisms of the element.

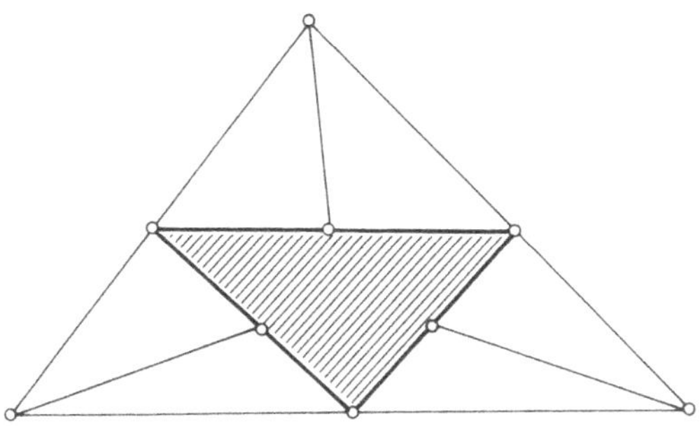

The three mechanisms of the quadratic stress field element

Figure 7

The number of mechanisms $n(m)$ for any polynomial degree n can be estimated from the difference between the number of generalized boundary loads $n(g)$ and the number of stress parameters $n(\sigma)$

$$n(g) - n(\sigma) = n(m) + 3 ,$$

since this difference contains the 3 rigid body degrees of freedom. We find easily that $n(g) = 6(n+1)$ and $n(\sigma) = \frac{1}{2}(n+1)(n+6)$, so theoretically, $n(m) = \frac{1}{2}n(5-n)$.

This gives correctly 0 for the constant stress triangle, 2 for the linear stress triangle where the 3 corner reciprocity conditions contain the statement of global rotational equilibrium, 3 for the quadratic stress, 3 for the cubic stress; but decreases to 2 for the quartic, 0 for the quintic and becomes afterwards negative. Since the number of mechanisms must remain equal to 3 as a generalization of the previous argument clearly shows, we have an indication of a gradual appearance of self-stressings, wherein some combinations of the σ_i parameters do no more generate surface tractions. In the end we get the following picture:

polynomial degree	0	1	2	3	4	5	n ,
number of mechanisms	0	2	3	3	3	3	3 ,
number of self-stressings	0	0	0	0	1	3	$\frac{1}{3}(n-2)(n-3)$.

The analogous analysis for tetraedral diffusive equilibrium elements is considerably more complicated. The results obtained so far indicate that, although they possess a larger number of mechanisms, they can be inhibited satisfactorily using similar techniques as employed in the two-dimensional case.

Figure 8 illustrates on the linear and quadratic cases, how all mechanisms are inhibited by the superelement technique consisting in assembling 3 elements to form a new triangle.

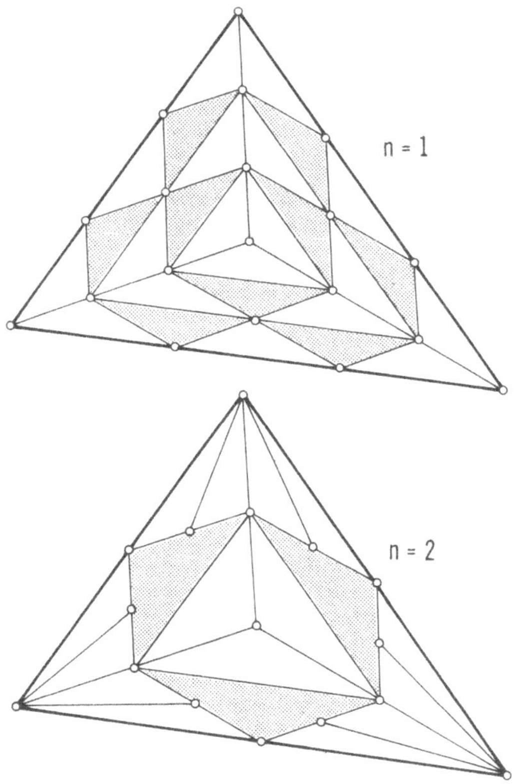

Figure 8

Figures 9 and 10 illustrate the second superelement technique that produces a quadrilateral from four triangles by diagonal subdivision. In this case there remains one mechanism of four-bar linkage type in which the pivots A, B, C and D can be fixed, showing that all other nodes, except the internal E, F, G and H remain fixed. Consequently there is no restriction on the loading applied at the boundary of the superelements. External loads applied on E, F, G and H need a constraint to maintain equilibrium.

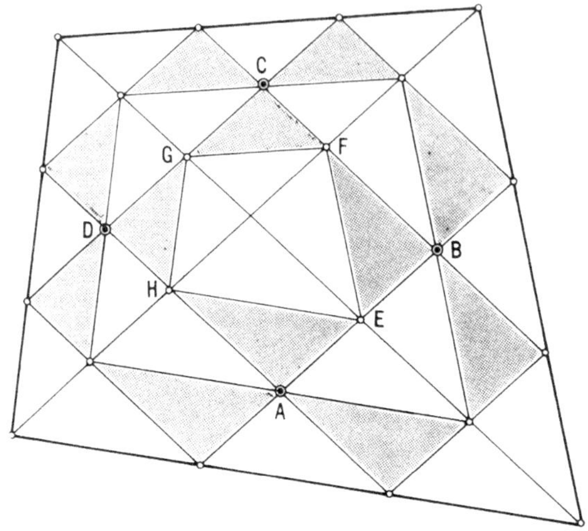

Figure 9 - Quadrilateral Superelement n = 1

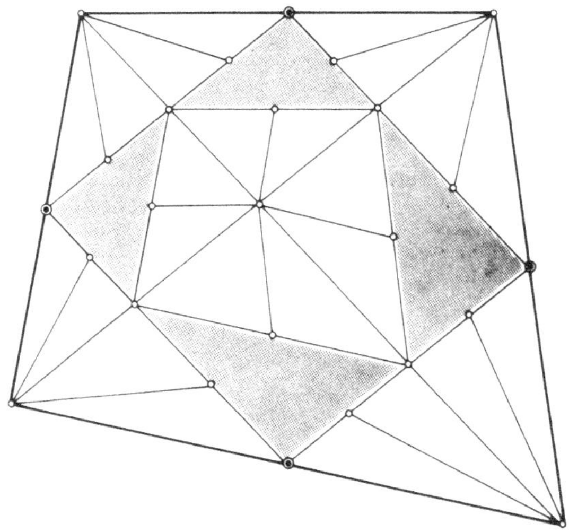

Figure 10 - Quadrilateral Superelement n = 2

2.7 *Stress Functions*

General equilibrium fields without body loads can be generated by stress functions, the parametrisation of which constitutes another approach to the construction of equilibrium models.

The general theory of stress functions is best explained by reference to their interpretation in terms of stress resultants.

If S denotes a surface bounded by the contour (c) on which a positive sense of circulation has been defined, there is on S a positive sense for the outward normal $\vec{n}$ corresponding to the application of Stokes theorem. Consider the resultant of the surface traction t_1 applied on the surface

$$\iint_S t_1 \, dS = \iint_S n_p \tau_{p1} dS \ .$$

For any other surface S', bounded by the same contour (c) and such that the volume between S and S' belongs to the continuum and is not acted upon by body loads, the resultant must be the same, otherwise the volume would not be in equilibrium. It may be argued therefore that the resultant depends only on the contour (c) and is measured by the circulation of some vector $\vec{A}_1$

$$\iint_S n_p \tau_{p1} dS = \oint_{(c)} \vec{A}_1 \cdot ds = \oint_{(c)} A_{1m} dx_m \ .$$

Applying Stokes theorem to the circulation, we get

$$\iint_S n_p \tau_{p1} dS = \iint_S e_{pqr} n_p D_q A_{1r} dS \ ,$$

where e_{pqr} denotes the alternating tensor and $D_q = \partial/\partial x_q$. By comparison

$$\tau_{p1} = e_{pqr} D_q A_{1r} \ .$$

If this is repeated for the other components of surface traction, we obtain

$$\tau_{pi} = e_{pqr} D_q A_{ir}$$

where A_{ir} is a tensor of *first order stress functions*. The skew symmetry of the alternating tensor shows immediately that

$$D_p \tau_{pi} = e_{pqr} D_p D_q A_{ir} = 0 \ . \tag{33}$$

The first order stress functions satisfy automatically the translational equilibrium equations. It remains to satisfy the rotational ones.

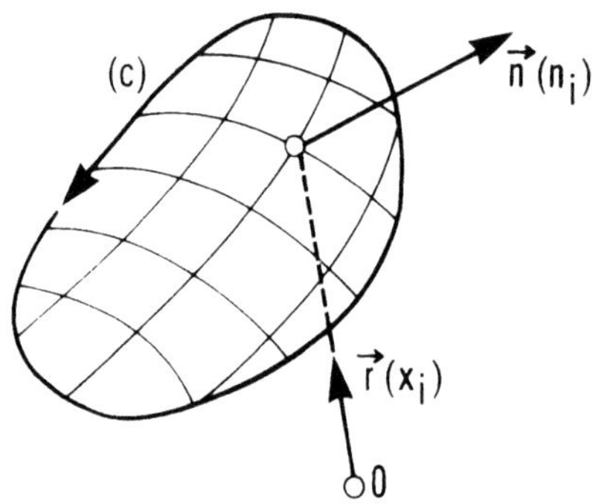

Figure 11

Rotational equilibrium can be expressed in a similar way by stating that the resulting moment of surface tractions on S with respect to the origin of coordinates depends only on the circulation of vectors $\vec{B}_p$ along (c):

$$\iint_S e_{pqr} x_q t_r dS = \oint_{(c)} B_{pm} dx_m \ .$$

On the left replact $t_r = n_j \tau_{jr}$, on the right apply Stokes theorem

$$\oint_{(c)} B_{pm} dx_m = \iint_S e_{jum} n_j D_u B_{pm} dS \ .$$

By comparison,

$$e_{pqr} x_q \tau_{jr} = e_{jum} D_u B_{pm} \ . \tag{34}$$

If the differential operator D_j is applied on both sides of this result

$$e_{pqr} D_j (x_q \tau_{jr}) = e_{jum} D_j D_u B_{pm} = 0 \ .$$

Expanding the left-hand side

$$e_{pqr}(\delta_{jq}\tau_{jr}+x_q D_j \tau_{jr}) = 0 \ .$$

We know the second term to be zero by virtue of (33), there remains

$$e_{pqr}\tau_{qr} = 0 \quad \text{or} \quad \tau_{rq} = \tau_{qr} \ , \tag{35}$$

which is the classical result of linearized rotational equilibrium. Let us now substitute (32) into (34)

$$e_{jum}(e_{pqr}x_q D_u A_{rm} - D_u B_{pm}) = 0 \ ,$$

and replace

$$x_q D_u A_{rm} = D_u(x_q A_{rm}) - \delta_{uq} A_{rm}$$

to obtain

$$e_{jum}D_u(e_{pqr}x_q A_{rm}-B_{pm}) = e_{jum}e_{pur}A_{rm} = A_{mm}\delta_{jp}-A_{jp} \ .$$

Setting $j = p$ in this formula

$$e_{pum}D_u(e_{pqr}x_q A_{rm}-B_{pm}) = 2A_{mm} \ .$$

After back-substitution of this value of A_{mm}, there comes finally

$$A_{jp} = \frac{1}{2}\, e_{jum}D_u(B_{pm}-e_{pqr}x_q A_{rm}) \ . \tag{36}$$

A tensor ϕ_{pm} of *second order stress functions* is usually defined by

$$A_{jp} = e_{jum}D_u\phi_{pm} \ , \tag{37}$$

so that, by comparison with (36),

$$B_{pm} = 2\phi_{pm} + e_{pqr} x_q A_{rm} \tag{38}$$

and by substitution into (32)

$$\tau_{pi} = e_{pqr} e_{ium} D_q D_u \phi_{rm} \ . \tag{39}$$

The symmetry obtained for τ_{pi} implies the same symmetry for the second order stress tensor:

$$\phi_{mr} = \phi_{rm} \ . \tag{40}$$

As other useful formula it may be noticed that

$$D_j A_{jp} = 0 \tag{41}$$

follows immediately from (37)

The second order stress functions are indeterminate to the extent of the addition of arbitrary functions U_m such that

$$\phi_{rm} = \frac{1}{2} (D_r U_m + D_m U_r) \ . \tag{42}$$

The arbitrariness in the first order stress functions is of type

$$A_{ir} = D_r V_i \ . \tag{43}$$

2.8 *The Complementary Energy Principle in Terms of First Order Stress Functions*

The two-dimensional case being both of direct interest to the development of membrane equilibrium elements and a guide to the interpretation of formulas in three dimensions, will be considered first.

The translational equilibrium equations without body loads

$$\frac{\partial \sigma_x}{\partial x} + \frac{\partial \tau_{yx}}{\partial y} = 0 \quad , \quad \frac{\partial \tau_{xy}}{\partial x} + \frac{\partial \sigma_y}{\partial y} = 0$$

are satisfied by the use of the first order stress functions A and B

$$\sigma_x = -\frac{\partial B}{\partial y} \;,\quad \tau_{yx} = +\frac{\partial B}{\partial x} \;,\quad \tau_{xy} = \frac{\partial A}{\partial y} \;,\quad \sigma_y = -\frac{\partial A}{\partial x} \;. \tag{44}$$

We will assume the stress-energy density

$$\Phi\left(\sigma_x \,,\, \frac{\tau_{xy}+\tau_{yx}}{2} \,,\, \sigma_y\right)$$

to be transformed by those formulas. Since the rotational equilibrium condition $\tau_{yx} = \tau_{xy}$ is not included in (44), it is considered as a constraint

$$\frac{\partial B}{\partial x} - \frac{\partial A}{\partial y} = 0 \tag{45}$$

and incorporated into the principle with a Lagrangian multiplier ω:

$$\int_S \left\{\Phi - \omega\left(\frac{\partial B}{\partial x} - \frac{\partial A}{\partial y}\right)\right\} dS - \int_{(c)} \left\{\bar{u}\left(-\ell \frac{\partial B}{\partial y} + m \frac{\partial B}{\partial x}\right) + \bar{v}\left(\ell \frac{\partial A}{\partial y} - m \frac{\partial A}{\partial x}\right)\right\} ds \quad \min A,\, B \;. \tag{46}$$

As the notation suggests, the displacements u and v are assumed to be prescribed along the contour (c) bounding the simply convected region S. The partial derivatives of the stress-energy density are

$$\varepsilon_x = \partial\Phi/\partial\sigma_x \;,\quad \varepsilon_{xy} = \partial\Phi/\partial\tau_{xy} = \partial\Phi/\partial\tau_{yx} = \varepsilon_{yx} \;,\quad \varepsilon_y = \partial\Phi/\partial\sigma_y$$

and include the symmetry of the strain tensor as a result of expressing this energy as a symmetrical function of τ_{yx} and τ_{xy}.

With

$$\delta A = \alpha \qquad \text{and} \qquad \delta B = \beta$$

as simplified notations for the variations of the stress functions, the variation of the principle yields

$$\int_S \left(- \varepsilon_x \frac{\partial \beta}{\partial y} + (\varepsilon_{yx} - \omega) \frac{\partial \beta}{\partial x} + (\varepsilon_{xy} + \omega) \frac{\partial \alpha}{\partial y} - \varepsilon_y \frac{\partial \alpha}{\partial x} \right) dS$$

$$- \int_{(c)} \left(- \bar{u} \frac{\partial \beta}{\partial s} + \bar{v} \frac{\partial \alpha}{\partial s} \right) ds = 0 ,$$

in which use was made (Figure 12) of the property

$$\ell \frac{\partial}{\partial y} - m \frac{\partial}{\partial x} = \frac{\partial}{\partial s} .$$

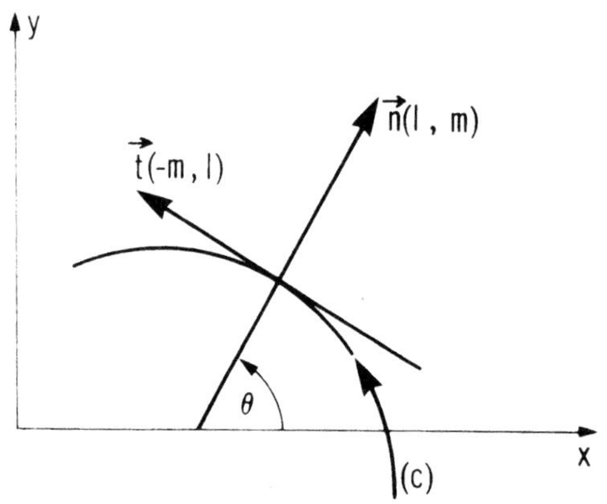

Figure 12

After integration by parts, the Euler equations for α and β, turn out to be the Beltrami equations for the integration of ω, that, as could be suspected from virtual work arguments, is identified with the material rotation

$$\beta \to \frac{\partial \omega}{\partial x} = - \frac{\partial \varepsilon_x}{\partial y} + \frac{\partial \varepsilon_{yx}}{\partial x}$$

$$\alpha \to \frac{\partial \omega}{\partial y} = - \frac{\partial \varepsilon_{xy}}{\partial y} + \frac{\partial \varepsilon_y}{\partial x} . \qquad (47)$$

The elimination of ω between those equations is the local compatibility condition on the strains. The boundary integral

$$\int_{(c)} \left(\beta(-m\varepsilon_x + \ell\varepsilon_{yx} - \ell\omega) + \alpha(m\varepsilon_{xy} - \ell\varepsilon_y + m\omega) + \bar{u}\frac{\partial\beta}{\partial s} - \bar{v}\frac{\partial\alpha}{\partial s} \right) ds = 0 , \tag{48}$$

requires a further integration by parts on the two last terms. This operation produces no contour contribution because of the single-valuedness of α and β and the single-valuedness of the data $\bar{u}$ and $\bar{v}$. Thus the arbitrariness of α and β along (c) furnishes the natural boundary conditions

$$\begin{aligned} \ell\omega &= - m\varepsilon_x + \ell\varepsilon_{yx} - \partial\bar{u}/\partial s , \\ m\omega &= - m\varepsilon_{xy} + \ell\varepsilon_y - \partial\bar{v}/\partial s . \end{aligned} \tag{49}$$

Turning to the local natural axes $(\vec{n}, \vec{t})$, they can be written

$$\omega = \varepsilon_{tn} - \frac{\partial\bar{u}_n}{\partial s} + \dot{\theta}\bar{u}_t , \tag{50}$$

$$\left(\dot{\theta} = \frac{d\theta}{ds}\right)$$

$$\varepsilon_{tt} = \frac{\partial\bar{u}_t}{\partial s} + \dot{\theta}\bar{u}_n . \tag{51}$$

The extension of the same principle to 3 dimensions is

$$\int_R \left\{ \Phi(\tau) + e_{jmn}\omega_m\tau_{nj} \right\} dV - \int_{\partial R} \bar{u}_j n_n \tau_{nj} \, dS \ \min , \tag{52}$$

where the 3 ω_m are Lagrangian multipliers for the rotational equilibrium constraints $\tau_{nj} = \tau_{jn}$, and where the stresses should be expressed in terms of the first order stress tensor A_{jr} ensuring translational equilibrium

$$\tau_{nj} = e_{nqr} D_q A_{jr} . \tag{53}$$

Again the stress energy density is expressed as a symmetrical function of the τ_{xj} and the stress-strain relations

$$\varepsilon_{ij} = \partial\Phi/\partial\tau_{ij}$$

then contain the symmetry statement $\varepsilon_{ji} = \varepsilon_{ij}$. With $\alpha_{jr} = \delta A_{jr}$, the variation of the principle (52) yields

$$\int_R \left(\varepsilon_{nj} e_{nqr} D_q \alpha_{jr} + e_{jmn} \omega_m e_{nqr} D_q \alpha_{jr} \right) dV - \int_{\partial R} \bar{u}_j n_n e_{nqr} D_q \alpha_{jr} dS = 0 .$$

Replacing

$$e_{jmn} e_{nqr} = \delta_{qj} \delta_{rm} - \delta_{qm} \delta_{rj}$$

and integrating by parts, the volume integral

$$- \int_R (\alpha_{jr} e_{nqr} D_q \varepsilon_{nj} + \alpha_{jm} D_j \omega_m - \alpha_{jj} D_m \omega_m) dV = 0$$

gives as Euler equation for α_{jm}

$$e_{nqm} D_q \varepsilon_{nj} + D_j \omega_m - \delta_{jm} D_q \omega_q = 0 . \tag{54}$$

Setting j = m and using the symmetry of the strain tensor

$$2D_q \omega_q = e_{nqm} D_q \varepsilon_{nm} = 0$$

and the divergence of ω_q is zero. This simplifies (54), that turns out to be the Beltrami equations for the integration of the material rotation vector ω_m

$$D_j \omega_m = e_{nmq} D_q \varepsilon_{nj} . \tag{55}$$

The elimination of the components ω_m between those equations furnishes the six ($T_{mp} = T_{pm}$) local compatibility conditions on the strains

$$T_{pm} = e_{jpr} e_{nmq} D_r D_q \varepsilon_{jn} = e_{jpr} D_r D_j \omega_m = 0 . \tag{56}$$

The surface integral

$$\int_{R} (\alpha_{jr} e_{nqr} n_q \varepsilon_{nj} + \alpha_{jm} n_j \omega_m - \alpha_{jj} n_m \omega_m - e_{nqr} n_n \bar{u}_j D_q \alpha_{jr}) dS = 0$$

requires, as in the two-dimensional case, a transformation on the last term containing the specified boundary displacements. With

$$-e_{nqr} n_n \bar{u}_j D_q \alpha_{ir} = -e_{nqr} n_n D_q (\bar{u}_j \alpha_{jr}) + e_{nqr} n_n \alpha_{jr} D_q \bar{u}_j$$

the contribution of the first right-hand side term to the surface integral vanishes; it is the flow through a closed surface ∂R of the rotational of a vector $(\bar{u}_j \alpha_{jr})$. There follows

$$\int_{\partial R} \left(\alpha_{jr} e_{mqr} (n_q \varepsilon_{mj} + n_m D_q \bar{u}_j) + \alpha_{jm} n_j \omega_m - \alpha_{jj} n_m \omega_m \right) dS = 0 \ .$$

As could be expected the gradients of the boundary data are taken in the tangent plane to the surface; the "surface derivatives"

$$e_{mqr} n_m D_q \bar{u}_j = \bar{\beta}_{rj} \ , \tag{57}$$

being the components of the vector product of the outward normal and the gradient of $\bar{u}_j$, so that

$$n_r \bar{\beta}_{rj} \ . \tag{58}$$

The natural boundary conditions associated with α_{jr} are thus

$$\bar{\beta}_{rj} - e_{mqr} n_m \varepsilon_{qj} + n_j \omega_r - \delta_{jr} (n_m \omega_m) = 0 \ . \tag{59}$$

Observe that by multiplication with n_r and use of (58) this is identically verified, so that the number of independent conditions is only 6. Setting $r = j$ we obtain by symmetry of ε_{qj}

$$\bar{\beta}_{rr} = 2(n_r \omega_r) \ , \tag{60}$$

yielding the rotation about the normal to the surface in terms of the boundary data. This result, substituted into (59), furnishes

$$n_j \omega_r = \frac{1}{2} \delta_{rj} \bar{\beta}_{qq} + e_{mqr} n_m \varepsilon_{qj} - \bar{\beta}_{rj} \; . \tag{61}$$

Multiplying this by n_j ($n_j n_j = 1$) we obtain a result generalizing equation (50)

$$\omega_r = \frac{1}{2} n_r \bar{\beta}_{qq} - n_q \bar{\beta}_{rq} + e_{mqr} n_m n_p \varepsilon_{qp} \tag{62}$$

and representing boundary conditions of equations (55) for the integration of the material rotation. Three other boundary conditions, generalizing (51) are obtained by elimination of ω_r between (62) and (61). They can be placed in the form

$$b_{jr} = (\delta_{pj} - n_p n_j)(e_{mpr} n_m \varepsilon_{qp} - \bar{\beta}_{rp} + \frac{1}{2} \delta_{rp} \bar{\beta}_{qq}) = 0 \; , \tag{63}$$

$$n_j b_{jr} \equiv 0 \; , \qquad n_r b_{jr} \equiv 0 \; .$$

2.9 *Membrane Equilibrium Models and First Order Stress Functions*

Suppose the first order stress functions A and B of Section 2.8 approximated by polynomials and take as example the second degree case that will generate linear stress variations:

$$A = \alpha_1 x + \alpha_2 y + \alpha_3 x^2 + 2\alpha_4 xy + \alpha_5 y^2 \; ,$$

$$B = \beta_6 x + \beta_7 y + \beta_8 x^2 + 2\beta_9 xy + \beta_{10} y^2 \; .$$

The stress energy vanishes only if $\sigma_x \equiv 0$, $\sigma_y \equiv 0$ and $\frac{1}{2}(\tau_{xy} + \tau_{yx}) \equiv 0$, because the density of energy is a positive definite quadratic form in those variables.

Thus a zero energy state is characterized by

$$\frac{\partial B}{\partial y} \equiv 0 \quad , \quad \frac{\partial A}{\partial x} \equiv 0 \qquad \text{and} \qquad \frac{\partial B}{\partial x} + \frac{\partial A}{\partial y} \equiv 0 \quad .$$

The first two conditions yield $B = B(x)$ and $A = A(y)$; the third

$$\frac{dB}{dx} = - \frac{dA}{dy} = \beta \qquad \text{a constant} \ .$$

Consequently the only zero energy state is one for which

$$\sigma_x \equiv 0 \quad , \quad \sigma_y \equiv 0 \quad , \quad \tau_{yx} = \gamma = - \tau_{xy} \quad ,$$

$$A = - \gamma_y \qquad , \qquad B = \gamma_x \ ,$$

(the undetermined constants of integration in A and B which do not influence the state of stress were set equal to zero).

Let s denote in the present context the set of all stress parameters in a discretization by polynomials of A and B. For instance in the linear stress distribution model considered,

$$s^T = (\alpha_1 \ \alpha_2 \ \alpha_3 \ \alpha_4 \ \alpha_5 \ \beta_1 \ \beta_2 \ \beta_3 \ \beta_4 \ \beta_5) \ .$$

The stress energy will be a quadratic form $\frac{1}{2} s^T F s$, in which the flexibility matrix F is nonnegative but singular because the homogeneous equation

$$F \ s = 0$$

will, as was just demonstrated, possess exactly one nontrivial solution

$$s^T_{(o)} = (0 \ -\gamma \ 0 \ 0 \ 0 \ \gamma \ 0 \ 0 \ 0 \ 0) \ . \tag{64}$$

Let S denote the load connexion matrix between the set of generalized boundary loads g, defined as in Section 2.4, and our stress parameters:

$$g = S\ s\ .$$

As we know that the stress tensor generated by first order stress functions satisfies the translational equilibrium equations, this will be true at the global level for the boundary loads. Thus if

$$q = q_{(x)} \quad \text{and} \quad q = q_{(y)}$$

denote the rigid body translation modes

$$q^T_{(x)}\ g = 0 \quad \text{and} \quad q^T_{(y)}\ g = 0$$

will be satisfied for any value of the set s and consequently

$$S^T\ q_{(x)} = 0 \quad , \quad S^T\ q_{(y)} = 0\ .$$

It will now be shown that for any polynomial order of approximation those are the only two linearly independent solutions of $S^T\ q = 0$.

The proof follows from an analysis of the homogeneous adjoint problem S s = 0 in which we seek the sets of stress parameters that generate no boundary loads and thus no surface tractions (selfstressings). The surface tractions are given by

$$\begin{aligned} t_x &= \ell\ \sigma_x + m\ \tau_{yx} = -\ \ell\ \frac{\partial B}{\partial y} + m\ \frac{\partial B}{\partial x} = -\ \frac{\partial B}{\partial s}\ , \\ t_y &= \ell\ \tau_{xy} + m\ \sigma_y = \ell\ \frac{\partial A}{\partial y} - m\ \frac{\partial A}{\partial x} = +\ \frac{\partial A}{\partial s}\ . \end{aligned} \tag{65}$$

They vanish if A and B keep constant values along the boundary (since we can add arbitrary constants to A and B we may require A and B to be zero along the boundary). Now, a polynomial that vanishes along a triangular boundary is necessarily of the form

$$P(x,y) = c_{12}(x,y) \quad c_{23}(x,y) \quad c_{31}(x,y) \quad Q(x,y) \ , \tag{66}$$

where

$$c_{12}(x,y) = (x_2-x_1)(y-y_1) - (y_2-y_1)(x-x_1) = 0 \ ,$$

$$c_{23}(x,y) = (x_3-x_2)(y-y_2) - (y_3-y_2)(x-x_2) = 0 \ ,$$

$$c_{31}(x,y) = (x_1-x_3)(y-y_3) - (y_1-y_3)(y-y_3) = 0 \ ,$$

are the cartesian equations of the sides of the triangle.

Thus, if n denotes the degree of the polynomials required for the stresses, n+1 is the degree of the polynomials approximating A and B and n-2 the degree of the Q(x,y) polynomials. The $\frac{(n-1)n}{2}$ coefficients of Q(x,y) will generate the same number of self-stressings for A and for B; the total number of self-stressings will thus be

$$n(x) = n(n-1) \ .$$

The number of stress parameters in A or B will be $\frac{(n+2)(n+3)}{2} - 1$, discounting the arbitrary constant. The total number of stress parameters is

$$n(s) = (n+2)(n+3) - 2 \ .$$

Since S is a $n(q) \times n(s)$ matrix we have the equality

$$n(s) - n(x) = n(q) - n(r)$$

where n(r) is the number of independent solutions of $S^T q = 0$. As

$$n(q) = n(g) = 6(n+1) \ ,$$

we find

$$n(r) = n(q)-n(s)+n(x) = 6(n+1)-(n+2)(n+3)+2+n(n-1) = 2$$

as announced.

We must now introduce a constraint between the stress parameters with the double objective of rendering the flexibility matrix nonsingular and ensuring global rotational equilibrium. Intuitively this goal will be achieved by taking $\omega(x,y) = \theta$ a constant in the constraining term

$$\int \omega\left(\frac{\partial B}{\partial x} - \frac{\partial A}{\partial y}\right) dS ,$$

of the variational principle (46). By the Stokes transformation this is equivalent to

$$\theta \int\left(\frac{\partial B}{\partial x} - \frac{\partial A}{\partial y}\right) dS = \theta \oint A\ dx + B\ dy .$$

A second transformation yields, taking into account the single-valuedness of A and B,

$$- \theta \oint (x\ dA + y\ dB) = - \theta \oint (x\ t_y - y\ t_x)ds .$$

The contour integral is recognized as the moment of the traction surface applied along the element boundary with respect to the origin of coordinates. The constraint stemming from the variation of θ in the principle is thus equivalent to the global rotational equilibrium of the element:

$$\oint (x\ t_y - y\ t_x)ds = 0 . \tag{67}$$

As a consequence, if the constraint is satisfied the boundary loads will produce no virtual work on a rigid body rotation and since they are already in translational equilibrium and are not subject to any other constraint, the resulting stiffness matrix of the element will be sound. To implement (67) we consider its equivalent form

$$\oint A\ dx + B\ dy = 0 \tag{68}$$

which, after substitution of the polynomials for A and B, will give the coefficients of the linear homogeneous form in the stress parameters that must vanish:

$$w^T s = \sum_1^{n+1} w_{\alpha i}\, \alpha_i + \sum_1^{n+1} w_{\beta i}\, \beta_i = 0 . \tag{69}$$

The general solution of this constraint can be written in the form

$$s = T\, t \tag{70}$$

where t is a set of "free" stress parameters, one less in number than n(s). Since (70) satisfies (69) for arbitrary s we have of course

$$w^T\, T = 0 . \tag{71}$$

The complementary energy principle can now be written as

$$\frac{1}{2}\, t^T (T^T\, F\, T) t - q^T\, S\, T\, t \qquad \min_t$$

the boundary displacements being assumed to be prescribed.

The variational equations of this principle are

$$T^T\, F\, T\, t = (ST)^T\, q \tag{72}$$

and it remains to show that our second goal is achieved, namely that the modified flexibility matrix $(T^T\, F\, T)$ has become non-singular. Assume there is a non zero set of free stress parameters $t = t_{(o)}$ such that

$$T^T\, F\, T\, t_{(o)} = 0 .$$

Then, because F is nonnegative and F s = 0 has the only solution $F\, s_{(o)} = 0$, we should have

$$T\, t_{(o)} = s_{(o)} \;.$$

Premultiplying w^T and using (71) we also should have

$$w^T\, s_{(o)} = 0 \;. \tag{73}$$

However, looking at the structure of $s_{(o)}$ in (64), its product by w^T requires only the computation of the coefficient $w_{\alpha 2}$ and $w_{\beta 1}$ which are

$$w_{\alpha 2} = \oint y\, dx = A \quad , \quad w_{\beta 1} = \oint x\, dy = -\, A \;,$$

where A denotes the area of the triangle. Finally,

$$w^T\, s_{(o)} = -\, 2\, \gamma\, A \neq 0$$

and the contradiction with (73) demonstrates the nonsingularity of $T^T\, F\, T$. Solving (72) and substituting into the load connexion equation:

$$g = S\, s = S\, T\, t = K\, q \;, \tag{74}$$

with a stiffness matrix

$$K = (ST)(T^T\, F\, T)^{-1}(ST)^T \;. \tag{75}$$

As the analysis has shown there are no more mechanisms; the only solution of $K\, q = 0$ are the rigid body modes of the element.

In conclusion, by discretizing rotational equilibrium and securing only the global rotational equilibrium, we can produce a family of diffusive membrane elements of arbitrary polynomial degree. They have "sound" stiffness matrices that allows them to be assembled by any standard direct stiffness software. The average rotation of the element is a by-product of the analysis

$$\theta = \frac{w^T}{w^T w} (F\ T(T^T\ F\ T)^{-1}T^T - I)S^T q\ . \tag{76}$$

By comparison with the same element satisfying rigorously rotational equilibrium but containing mechanisms, it may be said that the mechanisms have been locked but the loads in the locking constraints removed by allowing a divorce between τ_{yx} and τ_{xy}.

Note: Remembering the static-geometric analogies between membrane equilibrium and plate-bending displacement models, an analogous procedure can be formulated to relax C_1 continuity of conforming plate bending elements.

2.10 *Direct Flexibility Method with First Order Stress Functions*

Equations (65) show that a simple or C_o continuity of the first order stress functions is sufficient to obtain diffusivity, that is complete reciprocity of surface tractions between adjacent elements. This observation suggests a completely different method of assembling elements that, by comparison with the direct stiffness method, should be called the direct flexibility method.

In this method the continuity of the first order stress functions is organized by identification of local boundary values of these functions with a set of nodal values. This requires, as in the case of displacement models, a distinction between the set c of parameters that determines the values taken by A and B along the element boundary and the set x that subsists with zero boundary values and is responsible for self-stressings.

It must also be observed that the addition of an arbitrary constant α_o to A and β_o to B is essential for transforming the necessary diffusivity requirement: continuity of dA/ds and dB/ds, into the equivalent one of continuity of A and B. Thus α_o and β_o are additional parameters entering the set c, although they are not properly speaking stress parameters. For these reasons, the flexibility matrix in the stress energy

$$\frac{1}{2}\,(c^T\ x^T)\begin{pmatrix} F_{cc} & F_{cx} \\ F_{xc} & F_{xx} \end{pmatrix}\begin{pmatrix} c \\ x \end{pmatrix} \tag{77}$$

is not the same as our F of Section 2.10.

The boundary loads, which may have the same definitions as before and generate by virtual work the same definition of boundary displacements, depend solely on the surface tractions, that is on the set c and are therefore expressible as

$$g = C\ c\ . \tag{78}$$

Finally the global rotational equilibrium condition being expressible as in (68) by a boundary integral also constitutes a constraint

$$w^T\ c = 0\ , \tag{79}$$

for the set c alone (the coefficients vector w is of course not the same as the one of Section 2.9).

The variational principle at the element level

$$\frac{1}{2}\,c^T\ F_{cc}c + c^T\ F_{cx}\ x + \frac{1}{2}\,x^T\ F_{xx}x - \theta\ w^T\ c - q^T\ C\ c \quad \min_{c,x} \tag{80}$$

furnishes in addition to (79) the equations

$$F_{cc}c + F_{cx}x = \theta w + C^T\ q\ , \tag{81}$$

$$F_{xc}c + F_{xx}x = 0\ . \tag{82}$$

The flexibility matrix F_{xx} is nonsingular; indeed the only state of stress of zero energy was seen to consist of a uniform distribution of $\tau_{yx} = -\tau_{xy}$ and belongs to the stress distributions generated by c since it does not vanish at the boundary. The self-stressing states can thus be eliminated at element level and this produces the important flexibility matrix

$$F = F_{cc} - F_{cx} F_{xx}^{-1} F_{xc} \quad , \tag{83}$$

in the relation

$$F\ c = \theta\ \omega + C^T\ q\ . \tag{84}$$

Interelement continuity of A and B is now organized by Boolean matrices $M_{(i)}$ just as for conforming displacement models. In the relation

$$c_{(i)} = M_{(i)}\ c \tag{85}$$

a set $c_{(i)}$ belongs to one element, the set c is now that of nodal values. Then, if $F_{(i)}$ denotes the matrix (83) of element (i), the global flexibility matrix F can be assembled directly as

$$F = \sum_i M^T_{(i)}\ F_{(i)} M_{(i)} \tag{86}$$

and the global stress energy is

$$\frac{1}{2}\ c^T\ F\ c\ . \tag{87}$$

The elemental rotational equilibrium constraints

$$w^T_{(i)} c_{(i)} = w^T_{(i)} M_{(i)} c \qquad (i = 1,2,\ldots,N)$$

are translated into a final set of constraints on the nodal values

$$w^T\ c = 0\ . \tag{88}$$

Because of the continuity of A and B no interface loads are accepted and the loading of the structure is applied at the boundaries. We can select in the elemental relations

$$g_{(i)} = C_{(i)}\, M_{(i)}\, c$$

those generalized loads $g_{(2)}$ that are given on the boundary and obtain

$$g_{(2)} = C_{(2)}\, c\ . \qquad (89)$$

For the complementary set $q_{(1)}$ of prescribed displacements, the virtual work expression will be

$$q^T_{(1)}\, g_{(1)} = q^T_{(1)}\ \ {}_{(1)}\, c\ . \qquad (90)$$

The complementary energy principle at the structural level is then

$$\frac{1}{2}\, c^T\, F\, c - \omega^T W^T\, c - q^T_{(1)} C_{(1)} c + q^T_{(2)} (g_{(2)} - C_{(2)} c)\ \min_{c}$$

where the set $q_{(2)}$ plays the role of Lagrangian multipliers incorporating the equilibrium equations (89). The variational equations

$$F\, c - W\, \omega = C^T_{(1)} q_{(1)} + C^T_{(2)} q_{(2)}$$

$$W^T\, c = 0 \qquad (91)$$

$$C_{(2)} c = g_{(2)}$$

are to be solved for the unknowns c, ω and $q_{(2)}$. This will give the stress information but very little about the displacements, except the average rotations ω of the elements and the boundary displacements. If interior displacements are required, they can be obtained as follows. In the elemental equations

$$F_{(i)} c_{(i)} - \theta_{(i)} w_{(i)} = C^T_{(i)} q_{(i)}\ ,$$

substitute both Boolean connexion matrices $M_{(i)}$ and $L_{(i)}$,

$$F_{(i)} M_{(i)} c - \theta_i w_{(i)} = C^T_{(i)} L_{(i)} q \ .$$

Premultiply by $M^T_{(i)}$ and sum over all elements to receive

$$F\ c - W\ \omega = C^T\ q\ , \tag{92}$$

$$C = \sum_i L^T_{(i)}\ C_{(i)} M_{(i)}\ . \tag{93}$$

In (92) the left-hand side is known and the problem is the kinematical one to determine the displacements from already compatible strains. Note that in the case of multiply connected domains the statement that the continuity of dA/ds and dB/ds is equivalent to that of A and B is not strictly true. Barriers must be devised that restore simple connectivity. Across each barrier A and B will generally differ from a constant.

The efficiency of the direct flexibility method is linked to the topological nature of interelement connexions. In the linear stress variation models it was observed that the direct stiffness procedure involves 4 generalized displacements per interface. The same case treated by local first order stress function values involves 6 of those per interface (2 at vertices and 1 at mid-side for each stress function).

However it may well be true that, in case complete displacement information is desired, the procedure employing the stiffness matrix is more economical than the present direct flexibility one.

An interesting observation is that the local internal values of discretized stress functions represent compact self-stressings (reference 6).

2.11 *Membrane Equilibrium Models and the Airy Stress Function*

The Airy stress function ϕ is a second order stress function that permits to implement both the translational and the rotational equilibrium equations. Its relation with our first order stress functions A and B is

$$\frac{\partial \phi}{\partial x} + A = 0 \quad , \quad \frac{\partial \phi}{\partial y} + B = 0 \quad , \tag{94}$$

and is the very root of the first analogy between plate stretching (membrane elements in equilibrium) and Kirchhoff plate bending.

The Airy function is the analogue of the transverse bending deflection, and A and B the analogues of the bending rotations of transverse cross-sections of the plate, so that (94) are the analogues of the Kirchhoff-Love assumption that normals to the middle surface of the plate in the reference configuration remain normal in the deformed configuration (zero transverse shearing strains).

We now have

$$\sigma_x = \frac{\partial^2 \phi}{\partial y^2} \quad , \quad \tau_{yx} = - \frac{\partial^2 \phi}{\partial x \partial y} = \tau_{xy} \quad , \quad \sigma_y = \frac{\partial^2 \phi}{\partial x^2} \quad , \tag{95}$$

and the equilibrium field without body loads in the example of section 2.4 is actually generated by the Airy function

$$\phi = \sigma_1 x^2 + 2\sigma_2 xy + \sigma_3 y^2 + \sigma_4 x^3 + \sigma_5 x^2 y + \sigma_6 xy^2 + \sigma_7 y^3 \ .$$

As a consequence the triangular membrane elements generated by complete polynomial approximations of the Airy function will present the difficulties associated with the mechanisms analyzed in Section 2.6.

The purpose of introducing an Airy function is not only to generate automatically a discrete stress field in equilibrium without body loads but also to allow the application of a direct flexibility method. Indeed the surface tractions along the boundary of an element are given by

$$t_x = \frac{\partial^2 \phi}{\partial s \partial y} \quad , \quad t_y = \frac{\partial^2 \phi}{\partial s \partial x} \; , \tag{96}$$

or, if we resolve them in a normal component t_n and a tangential one t_s, by

$$t_n = \frac{\partial^2 \phi}{\partial s^2} + \dot{\theta} \frac{\partial \phi}{\partial n} \quad , \quad t_s = - \frac{\partial^2 \phi}{\partial s \partial n} + \dot{\theta} \frac{\partial \phi}{\partial s} \; , \tag{97}$$

so that a sufficient condition for diffusivity is the continuity of ϕ and its two slopes at interfaces; this is generally termed C_1 continuity of the Airy function. C_1 continuity is equivalent to the necessary conditions represented by (96) or (97) if the domain is simply connected, because we may add to the Airy function an arbitrary linear function (reference 8).

The problem of organizing C_1 continuity of a discretized Airy function is the exact analogue of the similar problem for the transverse deflection in Kirchhoff plate bending and can be solved by the same techniques. In fact the superelement technique illustrated by Figures 8, 9 and 10 is applicable in a version that employs the local values of ϕ and its two slopes at the vertices and the normal slope at mid-sides (Figure 13). This illustrates the case n = 1, but the procedure is general (reference 3).

For multiply-connected domains care should be taken to introduce constant jumps of unknown magnitude for the Airy function and its two slopes across the barriers.

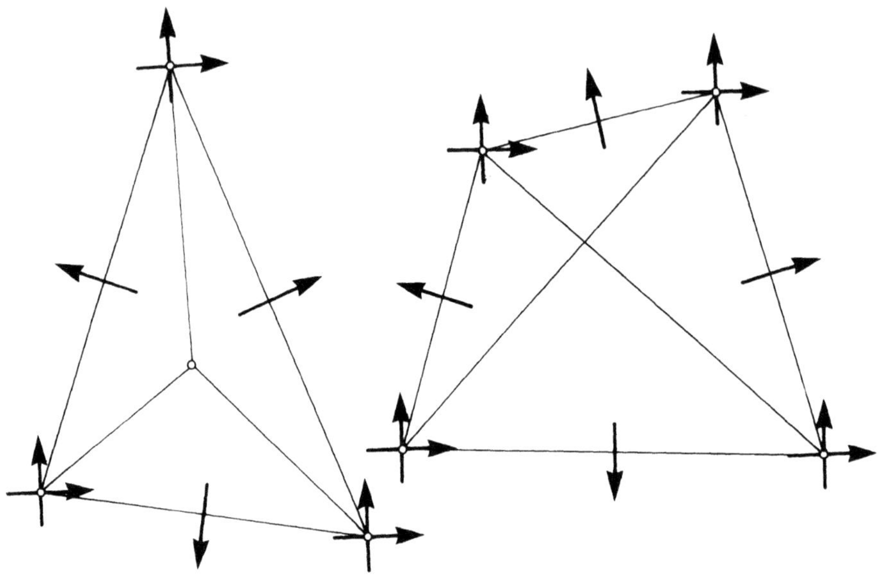

Figure 13 - Typical Superelements for C_1 Continuity of a Cubic Airy Function

2.12 *Quasi-Diffusivity*

Another method for the suppression of mechanisms in completely equilibrated models consists in abandoning exact diffusivity by avoiding the transmission of self-equilibrated surface traction distributions. The principle relies on the observation that in complete polynomial approximations of the Airy function there are some combinations of coefficients that produce a zero value of the function along the boundaries of the element. Such combinations are directly obtained as in the case of first order stress functions by

$$\phi(x,y) = C_{12}(x,y)C_{23}(x,y)C_{31}(x,y)Q(x,y) \ .$$

If n is the polynomial degree of stresses, n+r that of the Airy function, Q(x,y) is an arbitrary polynomial of degree n-1. Then, for straight boundaries ($\dot{\theta} = 0$) we shall have

$$t_n = \frac{\partial^2 \phi}{\partial s^2} = 0$$

and the surface tractions are purely tangential. It follows necessarily that, since they must form a self-equilibrated system, they are already self-equilibrated along each side of the triangle.

In the case n = 1 they are linearly distributed along each side and vanish at the mid-points. Such states of stress of the element will be termed zero-tangential states. There is only one such state for n = 1, 3 for n = 2, etc. By an appeal to the de Saint-Venant principle, such zero-tangential distributions will not exert an appreciable influence on the stress state of neighbouring elements.

Taking the case n = 1, if we minimize the stress energy of the element at its own level with respect to the zero-tangential state, we are left with 7 - 1 = 6 independent stress parameters. The boundary loads will be reduced to 3 per side, because only one tangential load is now required. The 9 boundary loads being self-equilibrated the number of independent loads is reduced to 9 - 3 = 6, which corresponds to the number of stress parameters. There will be no mechanisms.

More precisely let

$$\phi = \alpha_1+\alpha_2 x+\alpha_3 y+\alpha_4 x^2+2\alpha_5 xy+\alpha_6 y^2+\alpha_7 x^3+\alpha_8 x^2 y+\alpha_9 xy^2+\alpha_{10} y^3 \ ,$$

and consider the system of equations

$$\phi_1 = \phi(x_1, y_1) \quad , \quad \phi_2 = \phi(x_2, y_2) \quad , \quad \phi_3 = \phi(x_3, y_3)$$

$$\phi_{x1} = \frac{\partial \phi}{\partial x}(x_1, y_1) \ , \ \phi_{x2} = \frac{\partial \phi}{\partial x}(x_2, y_2) \ , \ \phi_{x3} = \frac{\partial \phi}{\partial x}(x_3, y_3)$$

$$\phi_{y1} = \frac{\partial \phi}{\partial y}(x_1, y_1) \ , \ \phi_{y2} = \frac{\partial \phi}{\partial y}(x_2, y_2) \ , \ \phi_{y3} = \frac{\partial \phi}{\partial y}(x_3, y_3)$$

determining the local values of ϕ and its two slopes at the vertices of the triangle. To this we add at some interior points, that may conveniently be taken to be $x_o = \frac{1}{3}(x_1+x_2+x_3)$, $y_o = \frac{1}{3}(y_1+y_2+y_3)$, the equation

$$\phi_o = \phi(x_o,y_o) .$$

This system of 10 equations in the 10 unknowns α_i can be inverted and expressed in terms of the local values:

$$\phi(x,y) = \phi_o F_o(x,y) + \phi_1 F_1(x,y) + \cdots + \phi_{y3} F_{y3}(x,y) .$$

There follows that $F_o(x,y)$ is an Airy function generating the zero-tangential state. As a matter of fact, changing the interior point changes the other shaping functions by addition of a term in $F_o(x,y)$. The stress energy of the element is of the form

$$\frac{1}{2} c^T F_{cc} c + \phi_o f^T c + \frac{1}{2} \phi_o^2 \alpha,$$

$$c^T = (\phi_1 \; \phi_2 \; \phi_3 \; \phi_{x1} \; \phi_{x2} \; \phi_{x3} \; \phi_{y1} \; \phi_{y2} \; \phi_{y3})$$

and α is not zero. Minimization with respect to ϕ_o gives the stress-energy

$$\frac{1}{2} c^T F c$$

with

$$F = F_{cc} - \frac{1}{\alpha} ff^T .$$

Along the boundary 12 the loads may be defined as

$$T_{12} = \int_1^2 t_s ds \qquad N_{12} = \int_1^2 t_n ds$$

and

$$M_{12} = \int_1^2 t_n s ds ,$$

where the arc length is counted from the mid-side. The generalized displacements follow by virtual work.

The load connexion matrix equation is of type

$$g = C\ c$$

and the minimization of complementary energy

$$\frac{1}{2}\ c^T\ F\ c - q^T\ C\ c$$

yields the usual equation

$$F\ c = C^T\ q$$

hence

$$g = (C\ F^{-1}\ C^T)q$$

with a sound stiffness matrix.

Alternatively we may apply the direct flexibility method with nodal values of ϕ and its slopes.

We notice that in this case ϕ is C_o continuous but not quite C_1 because of the possible incompatibility between the slopes due to the locally minimized zero-tangential states.

For n = 2, we have 5 more parameters, 2 of them are devoted to the 2 new zero tangential states, the 3 other may be used to fix the midside values of ϕ which is seen to insure again the C_o continuity of ϕ.

We may note that this procedure has an immediate analogue in Kirchhoff plate bending.

REFERENCES

[1] FRAEIJS DE VEUBEKE, B.M., *Matrix Structural Analysis*, Lecture Notes presented at the Calgary International Research Seminar, July, 1973.

[2] FRAEIJS DE VEUBEKE, B.M., "Displacement and Equilibrium Models in the Finite Element Method", Chapter 9 of *Stress Analysis*, edited by Zienkiewicz and Holister, Wiley, 1965.

[3] SANDER, G., "Application of the Dual Analysis Principle", Proceedings IUTAM Colloquium, Liège, August, 1970. Les Congrès et Colloques de l'Université de Liège, 16 place du 20 août, Liège.

[4] FRAEIJS DE VEUBEKE, B.M., "Upper and Lower Bounds in Matrix Structural Analysis", *AGARDograph 72*, Pergamon Press, 1964.

[5] FRAEIJS DE VEUBEKE, B.M., "Bending and Stretching of Plates. Special Models for Upper and Lower Bounds", *AFFDL-TR-66-80*, 1966, p. 863-886.

[6] FRAEIJS DE VEUBEKE, B.M. and ZIENKIEWICZ, O.C., "Strain Energy Bounds in Finite Element Analysis by Slab Analogies", *J. of Strain Analysis*, Vol. 2, No. 4, 1967.

[7] FRAEIJS DE VEUBEKE, B.M., "Basis of a Well-Conditioned Force Program via the Southwell Slab Analogies", USAF Report *AFFDL-TR-67-80*, 1967.

[8] FRAEIJS DE VEUBEKE, B.M., "Static-Geometric Analogies and Finite Element Models", NATO Advanced Study Institute Lectures on *Finite Element Methods in Continuum Mechanics*, Lisbon, 1971, Alabama University Press, 1973.

[9] SANDER, G. and BECKERS, P., "Improvements of Finite Element Solutions for Structural and Nonstructural Applications".

[10] FINZI, B., "Integrations delle equasione indefinite della meccanica dei sistemi continui", Rend. Linc., Vol. 19, 1934.

[11] SOUTHWELL, R.V., "On the Analogies Relating Flexure and Extension of Flat Plates", *Quart. J. Mech. Appl. Math.*, Vol. 3, 1950.

[12] SCHAEFFER, H., "Die vollständige Analogie Schiebe-Platte", Abh. Braunschweigischen Wiss. Ges., Vol. 8, 1956.

VARIATIONAL PRINCIPLES AND THE PATCH TEST

B.M. Fraeijs de Veubeke

Univeristy of Liège, Belgium

SUMMARY—The patch test is shown to be contained in the variational formulations of the finite element methods at the assembling level, all of which require the vanishing of the virtual work of interface connection loads. By a systematic introduction of stress generating functions, attention is drawn to the fact that any given finite element model can be assembled in two different ways: either by identification of a set of boundary displacements (leading to the direct stiffness method), or by identification of a set of local stress function values (leading to the dual direct flexibility method). Looking at any conjugate couple (generalized displacement—generalized surface traction) at an interface, one is strongly transmitted, the other weakly. Discretization of the zero virtual work condition at an interface of plate bending models, by means of Legendre polynomial expansions, allows a systematic construction of so-called 'non-conforming' elements that pass the patch test. They are in fact identified with weakly conforming, but strongly diffusive, hybrids, and the lowest degree element (quadratic) is in fact the Morley constant-moment element. Examples are given for higher degree displacement fields.

The case of plate stretching elements can be handled by duality, the difficulties being here associated with the requirements for diffusivity. Non-diffusive elements that pass the zero interface virtual work test can be constructed systematically and are identifiable with weakly diffusive, but strongly conforming, hybrids of the type first proposed by T.H.H. Pian.

Original publication, *Int. Jnl. for Num. Meth. in Engng.*, Vol. 8, John Wiley & Sons, 1974, pp. 783-801.

INTRODUCTION

There is a functional that generates all the equations of linear elasticity theory in the form of variational derivatives and natural boundary conditions. Its original construction [12] followed the method proposed by Friedrichs [13, 14] for one-dimensional problems. It can be considered as a very general three-field principle, from which most, if not all, finite element models can be derived through specializing assumptions [6].

With surface tractions $\bar{t}_j$ given at the boundary ∂E of the simply connected domain E of the element, the functional is

$$\int_E \left(W(\varepsilon) + \tau_{ij} \left(\frac{D_i u_j + D_j u_i}{2} - \varepsilon_{ij} \right) - X_j u_j \right) dE - \int_{\partial E} \bar{t}_j u_j dS \qquad (1)$$

$$\min_u \{ \max_\tau (\min_\varepsilon) \} .$$

The strain energy density W is assumed to be a homogeneous, positive, definite, quadratic form of the elements ε_{ij} of the strain tensor.

The variational derivatives are, with respect to the strains,

$$\tau_{ij} = \frac{\partial W}{\partial \varepsilon_{ij}} \qquad (2)$$

which are the linear elastic constitutive equations; with respect to the stresses τ_{ij}

$$\varepsilon_{ij} = \frac{1}{2} (D_i u_j + D_j u_i) \qquad (3)$$

the strain-displacement relations.

The variations on the displacement field occur both in the interior of E, generating the volume equilibrium equations

$$D_i \tau_{ij} + X_j = 0 \; , \qquad (4)$$

and on the boundary ∂E, where they produce 'natural' surface equilibrium equations

$$n_i \tau_{ij} = \bar{t}_j \quad \text{on} \quad \partial E , \tag{5}$$

involving the direction cosines n_i of the outward normal.

Should we specify the displacements $\bar{u}_j$ on the boundary, instead of the surface tractions, the variations on the displacements become limited to the interior of E but the variations on the stress field do now take place on the boundary as well as in the interior. This appears clearly in the corresponding functional

$$\int_E \left\{ W(\varepsilon) - \tau_{ij}\varepsilon_{ij} - u_j (D_i \tau_{ij} + X_j) \right\} dE + \int_{\partial E} n_i \tau_{ij} \bar{u}_j dS \tag{1a}$$

$$\min_u \{ \max_\tau (\min_\varepsilon) \}$$

that has the same variational derivatives as the preceding one, but other natural boundary conditions obtained from the variation of $t_j = n_i \tau_{ij}$ on ∂E,

$$u_j = \bar{u}_j \quad \text{on} \quad \partial E . \tag{5a}$$

In many cases the boundary data may be 'mixed' in the sense that the boundary is made of the union of partial boundaries $\partial_\alpha E$

$$\partial E = \cup \partial_\alpha E ,$$

on each of which the data, given componentwise, are not necessarily all of the displacement type or all of the surface traction type.

The mixing of boundary data is however limited by the condition that a single functional exists, having such data involved in natural boundary conditions. Rotational equilibrium is assumed to hold everywhere by virtue of the symmetry $\tau_{ji} = \tau_{ij}$ of the stress tensor. Still more general variational principles are known, wherein rotational equilibrium is contained as a natural variational result [15]. They provide finite element models circumventing the difficulties associated with C_1 continuity [16]; for the sake of simplicity they will not be considered here.

THE CANONICAL PRINCIPLE

Since equations (2) are invertible, they provide the exact minimizing choice for the strains in the three-field principle

$$\varepsilon_{ij} = \frac{\partial \Phi}{\partial \tau_{ij}}, \qquad (6)$$

where

$$\Phi(\tau) = \varepsilon_{ij}\tau_{ij} - W \qquad (7)$$

is the stress energy density, or complementary energy density. Moreover this optimal choice does not impose any constraints on the remaining fields of stresses and displacements and has no influence on the boundary conditions (the strain field is confined to the interior of E).

There is little sense therefore in keeping the strain field for discretization purposes. Following the ideas of Friedrichs, the stresses are first conceived as Lagrangian multipliers liberating the strain-displacement differential constraints, and equations (2) identify those multipliers through the fact that they satisfy the constitutive equations. Elimination of the strains through their minimizing choice (6) leads to the two-field principle called canonical (in the Hamilton-Jacobi sense) or involutory by Friedrichs and later known as the Hellinger-Reissner principle. According to the nature of the boundary data it can be given in either of the following forms:

$$\int_E (\tau_{ij} D_i u_j - \Phi(\tau) - X_j u_j) dE - \int_{\partial E} \bar{t}_j u_j dS \quad \min_u (\max_\tau), \qquad (8)$$

$$- \int_E \{\Phi(\tau) + u_j(D_i\tau_{ij}+X_j)\}dE + \int_{\partial E} n_i\tau_{ij}\bar{u}_j dS \ , \tag{8a}$$

with the same mixing possibilities of boundary data.

VARIATIONAL DERIVATION OF TRANSITION CONDITIONS

Ideally, at an interface between two adjacent finite elements the displacement field should be continuous

$$(u_j)_+ = (u_j)_- \ , \tag{9}$$

and the surface tractions in equilibrium, writing $t_j = n_i\tau_{ij}$,

$$(t_j)_+ + (t_j)_- = 0 \ . \tag{10}$$

Then, provided the coefficients of the constitutive equations (elastic moduli) are themselves continuous at the interface, the transition conditions (9) and (10) will induce continuity of the stress and strain fields. The exact transition conditions (9) and (10) are obtained variationally as follows.

Consider the elements E_+ and E_-, whose partial boundaries F_+ and F_- are the faces of an interface F. If the elasticity problem for E is solved with $\bar{t}_{j+}$ as boundary data on F_+ and similarly the elasticity problem for E_- with $\bar{t}_{j-}$ on F_-, we must assume the data to be coherent; which means, in this case, in equilibrium on F

$$(\bar{t}_j)_+ + (\bar{t}_j)_- = 0 \ .$$

In fact, opportunity can be taken of the existence of the interface to introduce a prescribed external interface loading $\hat{t}_j$, in which case the equilibrium condition is nonhomogeneous

$$(\bar{t}_j)_+ + (\bar{t}_j)_- = \hat{t}_j \ .$$

It can be solved by setting

$$(\bar{t}_j)_+ = \frac{1}{2}\hat{t}_j + t_j \quad , \quad (\bar{t}_j)_- = \frac{1}{2}\hat{t}_j - t_j \ . \tag{11}$$

While at each element level $(\bar{t}_j)_+$ and $(\bar{t}_j)_-$ are fixed values, at the assembled level t_j is an unknown internal force. Thus, if the elasticity problem is to be solved by a variational principle, the functional being now extended to the union of the domains E_+ and E_-, the variation of the terms depending on t_j with respect to t_j must disappear:

$$\delta \int_{F_+} (\bar{t}_j u_j)_+ dS + \delta \int_{F_-} (\bar{t}_j u_j)_- dS = \int_F \delta t_j (u_{j+} - u_{j-}) dS = 0 \ . \tag{12}$$

If t_j is unconstrained (not discretized) this results in the requirement of *strong conformity* (9).

Conversely, assume the data on F_+ and F_- to be respectively $\bar{u}_{j+}$ and $\bar{u}_{j-}$. If coherent they must satisfy continuity $\bar{u}_{j+} = \bar{u}_{j-}$.

More generally, if opportunity is taken to introduce a prescribed dislocation

$$\bar{u}_{j+} - \bar{u}_{j-} = \Delta u_j$$

this can be solved by setting

$$\bar{u}_{j+} = \frac{1}{2}\Delta u_j + u_j \quad , \quad \bar{u}_j = -\frac{1}{2}\Delta u_j + u_j \ , \tag{13}$$

with u_j an internal unknown at the assembled level. Then we must have

$$\delta \int_{F_+} (t_j \bar{u}_j)_+ dS + \delta \int_{F_-} (t_j \bar{u}_j)_- dS = \int_F \delta u_j (t_{j+} + t_{j-}) dS = 0 \tag{14}$$

and, for unconstrained u_j, this results in the requirement of *strong diffusitivity* (10).

Note that in formulas (12) and (14) the repeated subscript j does not necessarily mean application of the summation convention as used in the formulation of the variational principles.

In the case of mixed data the 'connectors' at an interface may well be $\bar{t}_j$ for one particular value of j and $\bar{u}_j$ for another.

As a final observation, the fact that there are two faces at an interface, allows the imposition of both dislocations and external loading. It is however necessary in this case to add potentials to the functional considered at the assembled level. In the case of $\bar{t}_j$ as connector we add to the functional the dislocation potential $+\int_F t_j \Delta u_j dS$. Then collecting the terms depending on t_j

$$-\int_{F_+} (\bar{t}_j u_j)_+ \, dS - \int_{F_-} (\bar{t}_j u_j)_- \, dS + \int_F t_j \Delta u_j dS \quad ,$$

substituting (11) and taking variations on t_j, we obtain

$$u_{j+} - u_{j-} = \Delta u_j \; .$$

In the case of $\bar{u}_j$ as connector we add the load potential

$$-\int_F \hat{t}_j \, \frac{\bar{u}_{j+} + \bar{u}_{j-}}{2} \, dS \; .$$

Then, collecting the terms depending on u_j

$$\int_{F_+} (t_j \bar{u}_j)_+ \, dS + \int_{F_-} (t_j \bar{u}_j)_- \, dS - \int_F \hat{t}_j \, \frac{\bar{u}_{j+} + \bar{u}_{j-}}{2} \, dS \quad ,$$

substituting (13) and taking variations on u_j

$$t_{j+} + t_{j-} = \hat{t}_j \; .$$

In the sequel we consider only unloaded and undislocated interfaces.

THE PURE MODELS

If we use only unconstrained $\bar{t}_j$ connectors for a finite element, the relevant variational principle is given by (8) and, as was just established, there is a requirement for strong conformity (9) at the boundaries.

Assume that this requirement can be satisfied by a discretization of the displacement field of the type

$$u_j(x) = q_m Q_{mj}(x) + b_n B_{nj}(x) \; , \tag{15}$$

where the set of independent generalized boundary displacements q_m describes completely the boundary displacement field through the shaping functions $Q_{mj}(x)$ and an additional set of bubble coordinates b_n is eventually needed to complete the description of the interior field. By definition, the bubble functions are such that

$$B_{nj}(x) = 0 \quad \text{for} \quad x \in \partial E \quad .$$

Each partial boundary $\partial_\alpha E$ that is the face of an interface has its displacements governed by a subset $\{q_m | m \in m_\alpha\}$ of the q_m, which are usually, though not necessarily, local boundary displacements. Necessary and sufficient conditions for strong conformity are then that

$$q_{m+} = q_{m-} \qquad m \in m_\alpha$$

$$Q_{mj+}(x) = Q_{mj-}(x) \quad \text{when} \quad x \in \partial_\alpha E_+ = \partial_\alpha E_- \quad .$$

The common values of q_m across the interface are known as the 'nodal displacements' and are usually chosen as the basic unknowns of the global discretized elasticity problem. For the stress field, it was already observed, and it appears clearly by looking at the functional (8), that its choice is confined to the interior domain of each element.

Unconstrained variations of the τ_{ij} produce the Euler equations

$$\frac{\partial \Phi}{\partial \tau_{ij}} = \frac{1}{2} (D_i u_j + D_j u_i) \tag{16}$$

that can be solved for the stresses and constitute in fact their optimal (maximizing) choice. Since this places no restrictions on the selection of the displacement field, there is little sense in making other choices, in particular in constructing mixed models with a separately discretized stress field. Thus models using $\bar{t}_j$ as unconstrained connectors may advantageously be set up by the simpler principle of variation of displacements:

$$\int_E \{W(Du) - X_j u_j\} \, dE - \int_{\partial E} \bar{t}_j u_j dS \quad \min_u , \tag{17}$$

to which (8) reduces on the acceptange of (16). The notation W(Du) obviously means that the arguments of the strain energy density must be calculated from the displacement field through equations (3). This result was presented earlier [6] in the form of a 'limitation principle' in the use of two-field functionals.

The elements constructed in this fashion may be referred to as 'kinematical and strongly conforming'.

Strong diffusivity would occur by virtue of the equilibrium properties (11) postulated from the connectors, if the internal stress field were related to the connectors by (5). This equation, however, holds only for unconstrained displacement fields at the boundary. Because of the discretization (15), the equation is replaced by its weak form

$$\int_{\partial E} Q_{mj}(x)\{n_i \tau_{ij} - \bar{t}_j\} dS = 0 . \tag{18}$$

As the stress field will ultimately be known through (16), (18) shows that numerical values will be received for the weak connectors by

$$\bar{g}_m = \int_{\partial E} Q_{mj}(x)\bar{t}_j dS \ . \tag{19}$$

No local (strong) information will ever by obtained for the $\bar{t}_j$.

The weak connectors are the generalized boundary loads conjugate, in the sense of virtual work, to the generalized boundary displacements q_m. From these considerations it appears that kinematical, strongly conforming models of finite elements are only weakly diffusive.

Take now the converse case in which use is made of unconstrained $\bar{u}_j$ connectors for all boundaries; the relevant variational principle being now given by (8a). It was seen that this poses a requirement of strong diffusivity (10). To implement this, the stress field must be discretized in such a way that it has a representation of the form

$$\tau_{ij} = g_m T_{mij}(x) + h_n B_{nij}(x) \ . \tag{20}$$

The h_n are bubble co-ordinates, the stress-bubble functions being by definition such that they generate no surface tractions on the boundaries:

$$n_i B_{nij}(x) = 0 \quad \text{for} \quad x \in \partial E \ .$$

The surface tractions are completely described by the generalized boundary loads g_m

$$t_j = g_m T_{mj}(x) \quad \text{with} \quad T_{mj}(x) = n_i T_{mij}(x) \ . \tag{21}$$

For any partial boundary that is the face of an interface, they depend on a subset of the g_m and diffusivity follows from the conditions

$$g_{m+} + g_{m-} = 0 \qquad m \in m_\alpha$$

$$T_{mj+}(x) = T_{mj-}(x) \quad \text{for} \quad x \in \partial_\alpha E \quad .$$

Looking now at the behaviour of such models from the viewpoint of conformity we meet with a completely different situation. It is clear from the functional (8a) that, by contrast with the former case, it is the displacement field whose choice is now confined to the interior domain. But if we accept an unconstrained choice we do this time impose restrictions on the choice of the stress field. It must satisfy the volume equilibrium equations (4). Assuming that it remains possible to organize strong diffusivity under such restrictions we obtain an element that can be referred to as 'statically admissible and strongly diffusive'.

Because u_j disappears completely the functional case in (4) is satisfied, the relevant variational principle becomes in fact the so-called complementary energy principle, and there is no proper relationship between the connectors and an internal field. There is however a weak information available on displacements.

On the one hand weak displacement connectors

$$\bar{q}_m = \int_{\partial E} T_{mj}(x)\bar{u}_j dS$$

can be defined, conjugate in the virtual work sense to the g_m, conforming ($\bar{q}_{m+} = \bar{q}_{m-}$) by virtue of the property postulated from the $\bar{u}_j$, and for which numerical values will be received. As a matter of fact they can be used as nodal displacement unknowns in the same direct stiffness method of approach as for kinematical models [6]. On the other hand numerical values can also be obtained for the weak generalized displacements conjugate to a given body. For those reasons the statically admissible and strongly diffusive models can be considered as weakly conforming. The strain energy bounds that can be obtained from the two types of models described in this section are well known [6, 10]. Used in alternate discretizations of the same problem, the dual analysis principle [17], they provide useful assessments of the discretization errors by a quantitative estimate of the energy convergence.

THE HYBRID MODELS

Cases are known for which discretization by complete polynomials leads to difficulties in organizing strong conformity (e.g., Kirchhoff plate bending) or strong diffusitivity with equilibrated stress fields (e.g., plate stretching). One way to overcome those difficulties has been in the assembling of super-elements [2, 7-9, 17]. Another has been the use of hyper-conforming elements available for high degree polynomials [2]. Not only are such elements expensive in number of degrees-of-freedom, but the fact that they make use of unnecessary continuity requirements at their interfaces restricts the application of a standard connection software whenever the elements are not coplanar, the interface is a discontinuity surface of material properties, or subject to an interface load.

If one is prepared to accept the loss of the bounding properties of the pure elements, hybrid models may provide less expensive solutions.

Consider again the case of $\bar{t}_j$ connectors on all boundaries together with the maximizing choice (16) of the internal stress field, so that the relevant variational principle is given by (17). This time however the $\bar{t}_j$ are constrained by a discrete representation

$$\bar{t}_j = \bar{g}_m T_{mj}(x) \qquad x \in \partial E \quad . \tag{22}$$

This procedure follows in fact the pattern (20) and (21) except that the surface tractions are only defined on the boundary. There is no difficulty therefore in taking $T_{mj+}(x) = T_{mj-}(x)$ on the interfaces and strong diffusivity of the surface tractions will follow from $\bar{g}_{m+} + \bar{g}_{m-} = 0$.

As the variations on the connectors are now constrained, strong conformity of the internal displacement fields will be replaced by the weak conformity

$$\int_F T_{mj}(x)(u_{j+} - u_{j-})dS = 0 \ . \tag{23}$$

Accordingly, the internal displacement fields need no more be discretized in the form (15) but their unknown coefficients are determined, partly, by the boundary displacements q_m conjugate to the $\bar{g}_m$ of (22)

$$q_m = \int_{\partial E} u_j(x) T_{mj}(x) dS \ . \tag{24}$$

Coefficients not determined by (24) are considered as internal degrees-of-freedom eliminated by direct minimization of the energy at the element level. Such elements may be termed 'kinematical and weakly conforming'.

Another category of hybrid elements, in fact the first introduced into the literature by T.H.H. Pian, is that of 'statically admissible and weakly diffusive' elements. By a procedure completely analogous to the preceding one, the complementary energy principle is used with discretized $\bar{u}_j$ connectors:

$$\bar{u}_j = \bar{q}_m Q_{mj}(x) \qquad x \in \partial E \ . \tag{25}$$

Because the shaping functions are only defined on the boundary, there is no difficulty in taking $Q_{mj+}(x) = Q_{mj-}(x)$ on the interfaces and strong conformity of the boundary displacements follows from $\bar{q}_{m+} = \bar{q}_{m-}$.

As the variations on the connectors are now constrained, strong diffusitivity is traded for the weak diffusitivity

$$\int_F Q_{mj}(x)(t_{j+} + t_{j-})dS = 0 \ . \tag{26}$$

The internal equilibrated stress field is determined partly by the weak boundary loads conjugate to $\bar{q}_m$

$$g_m = \int_{\partial E} Q_{mj}(x) n_i \tau_{ij} dS \ . \tag{27}$$

MIXED MODELS

Mixed elements arise most naturally in the presence of mixed connectors. To give an example from plate bending theory, the connectors may be split into a displacement type $(\partial/\partial n)\bar{w}$ for the normal slope of the plate deflection and $\overline{K}_n$ the Kirchhoff shear load together with the corner loads $\overline{Z}$.

The first, if undiscretized will require the existence of an internal bending moment field with strong diffusitivity. The second requires the existence of an internal transverse displacement field w with strong conformity. The difficulties associated with C_1 continuity, that is the strong conformity of both w and $(\partial/\partial n)w$ will be avoided. The relevant variational principle will then be the proper combination of (8) and (8a). Most of the mixed elements presented in the literature are of this, or a similar, nature [4, 18, 19].

Such mixed models share with the kinematical models the advantage of dealing simply with the problem of arbitrary body loadings. Those are automatically converted by virtual work into weak conjugates to the displacement degrees-of-freedom. The plate bending case is a good showcase because of the many different transverse pressure distributions that must be cared for in practice. By contrast the statically admissible elements require a search for special solutions to the non-homogeneous equilibrium equations that are not always easy to find.

DIRECT STIFFNESS AND DIRECT FLEXIBILITY

It appears that there are finally two methods for assembling any type of finite element model. The best known is that using the nodal displacement identification procedure

$$q_{m+} = q_{m-} = \text{a nodal displacement} \ . \tag{28}$$

It was used from the beginning for the kinematical and strongly conforming models, where the q_m are strong, and later for the statically admissible and strongly diffusive models [6], where the q_m are weak. It applies as well to the hybrid and mixed models. It produces a direct method for assembling the elemental stiffness matrices and is known therefore as the direct stiffness method.

Each q_m having its conjugate g_m, weak when q_m is strong and *vice versa*, another assembling method would consist in using the relation

$$g_{m+} + g_{m-} = 0 \ . \tag{29}$$

There are two drawbacks to this. One is lack of symmetry, the other is that (29) is valid only for externally unloaded interfaces. To this second drawback one may object that (28) in turn is only valid for undislocated interfaces, although the possibility of loading an interface is more useful in practice than the prescription of a dislocation. The drawback of symmetry can be overcome by the use of stress functions. It is easily shown that C_1 continuity of the Airy function and C_0 continuity of the Southwell stress functions are sufficient conditions of strong diffusivity in the respective cases of plate stretching and plate bending [10]. This allows the assembly of models with strongly diffusive connectors by the rule

$$c_{m+} = c_{m-} = \text{nodal stress function value} \tag{30}$$

and corresponds to a direct method for assembling elemental flexibility matrices, producing a particularly simple version of the Force method known as the direct flexibility method. P. Beckers [11] was the first to point out that this method applied to all models with suitable definitions of weak stress function values and established the relations between the g_m and the stress functions. There are limitations to this method: the presence of body loads, interface loading and multiple connectivity of the assembled structure require dislocations in the nodal stress function values.

INTERFACE VIRTUAL WORK

Consider two finite elements with interface F. Whatever the particular variational principle and associated type of element used, the terms containing the connectors are all of the type

$$T_j = \int_{F_+} (u_j t_j)_+ dS + \int_{F_-} (u_j t_j)_- dS \quad (j \text{ not summed})$$

where the bar indicating the connector is either on u_j or on t_j.

If t_j is a discretized connector:

$$\bar{t}_{j_+} = g_m T_{mj}(x), \quad \bar{t}_{j_-} = -g_m T_{mj}(x), \qquad x \in F, \quad m \in m_F,$$

$$T_j = g_m \int_F T_{mj}(x)\{u_{j+}(x) - u_{j-}(x)\} dS = 0$$

because each integral vanishes as a result of variations on the independent g_m. Similarly, if u_j is a discretized connector:

$$\bar{u}_{j+} = \bar{u}_{j-} = q_m U_{mj}(x), \quad x \in F, \quad m \in m_F ,$$

$$T_j = q_m \int_F U_{mj}(x)\{t_{j+}(x)+t_{j-}(x)\}dS = 0$$

because each integral vanishes as a result of variations on the independent q_m. The result $T_j = 0$ holds *a fortiori* when the connectors are undiscretized because of the strong diffusivity or conformity which is then required. The vanishing of each of its terms causes the total interface virtual work to vanish:

$$IVW = \int_{F_+} (t_j u_j)_+ dS + \int_{F_-} (t_j u_j)_- dS = 0 . \quad (j \text{ summed}) \tag{31}$$

As we now consider the two assembled elements to constitute a substructure and add a new element to it, the virtual work at the new interface will vanish by the same arguments. Thus, using a frontal assembling method, it turns out that, whatever the type of elements, the virtual work vanishes at each interface of the assembled structure.

This result, which satisfies engineering intuition, is a key ingredient in the verification of the patch test and is in face equivalent to the statement that there is no strain energy gained or lost at the interfaces, a variational viewpoint of the patch test adopted by Strang [3].

THE ZERO INTERFACE VIRTUAL WORK CONDITION FOR PLATE BENDING ELEMENTS

Conjunction of the zero interface virtual work condition and the dual connector systems (displacements or stress functions) allows a systematic construction of elements that will pass the patch test and possess the associated convergence characteristics.

We take an example where the difficulties in obtaining strong conformity are noteworthy: the Kirchhoff plate bending problem.

For one face of an interface (here an arc between two angular points 1 and 2) the virtual work is

$$\int_1^2 \left(wQ_n - \frac{\partial w}{\partial n} M_n - \frac{\partial w}{\partial s} M_{nt} \right) ds \tag{32}$$

where w is the transverse plate deflection representing the displatement field, $\partial w/\partial n$ the slope in the direction of the local outward normal ($\underline{n}$ makes an angle θ with a fixed axis Ox), $\partial w/\partial s$ the slope in the direction of the local tangent, oriented from 1 to 2. Q_n the shear load, M_n the normal bending moment, M_{nt} the twisting moment are representative of the surface tractions. We represent them [10] by means of the Southwell vector stress function (U, V).

If there is an internal self-equilibrated stress field, it is given by

$$M_{xx} = \frac{\partial V}{\partial y} , \quad M_{xy} = -\frac{1}{2}\left(\frac{\partial V}{\partial x} + \frac{\partial U}{\partial y}\right) , \quad M_{yy} = \frac{\partial U}{\partial x} , \tag{33}$$

as bending moments field,

$$Q_x = \frac{\partial \Omega}{\partial y} , \quad Q_y = -\frac{\partial \Omega}{\partial x} , \quad \Omega = \frac{1}{2}\left(\frac{\partial V}{\partial x} - \frac{\partial U}{\partial y}\right) \tag{34}$$

as shear loads field. On the boundary, where the vector is locally resolved in its normal component U_n and the tangential component U_t, there results

$$M_n = \frac{\partial U_t}{\partial s} + \dot{\theta} U_n , \quad M_{nt} = -\Omega - \frac{\partial U_n}{\partial s} + \dot{\theta} U_t , \tag{35}$$

$$\Omega = \frac{1}{2}\left(\frac{\partial U_t}{\partial n} - \frac{\partial U_n}{\partial s} + \dot{\theta} U_t\right) , \quad Q_n = \frac{\partial \Omega}{\partial s} \tag{36}$$

and the Kirchhoff shear load

$$K_n = Q_n + \frac{\partial M_{nt}}{\partial s} = 2\frac{\partial M_{nt}}{\partial s} + \frac{\partial M_n}{\partial n} + \dot{\theta}(M_n - M_t) = -\frac{\partial}{\partial s}\left(\frac{\partial U_n}{\partial s} - \dot{\theta}U_t\right). \tag{37}$$

In these formulas $\dot{\theta} = d\theta/ds$ measures the curvature of the boundary.

Equations (35), (36) and (37) continue to hold if the surface tractions are only defined on the boundary. Formula (32) now takes one of the following forms, derived from each other by integration by parts:

$$\int_1^2 \left\{ w\frac{\partial \Omega}{\partial s} - \frac{\partial w}{\partial n}\left(\frac{\partial U_t}{\partial s} + \dot{\theta}U_n\right) + \frac{\partial w}{\partial s}\left(\Omega + \frac{\partial U_n}{\partial s} - \dot{\theta}U_t\right)\right\} ds\ , \tag{38}$$

$$= w\Omega\Big|_1^2 + \int_1^2 \left\{ -\frac{\partial w}{\partial n}\left(\frac{\partial U_t}{\partial s} + \dot{\theta}U_n\right) + \frac{\partial w}{\partial s}\left(\frac{\partial U_n}{\partial s} - \dot{\theta}U_t\right)\right\} ds\ , \tag{39}$$

$$= w\left(\Omega + \frac{\partial U_n}{\partial s} - \dot{\theta}U_t\right)\Big|_1^2 + \int_1^2 \left\{ -\frac{\partial w}{\partial n}\left(\frac{\partial U_t}{\partial s} + \dot{\theta}U_n\right) - w\frac{\partial}{\partial s}\left(\frac{\partial U_n}{\partial s} - \dot{\theta}U_t\right)\right\} ds\ , \tag{40}$$

$$= w\Omega + U_n\frac{\partial w}{\partial s} - U_t\frac{\partial w}{\partial n}\Big|_1^2 + \int_1^2 \left\{ U_t\left(\frac{\partial^2 w}{\partial s \partial n} - \dot{\theta}\frac{\partial w}{\partial s}\right) - U_n\left(\frac{\partial^2 w}{\partial s^2} + \dot{\theta}\frac{\partial w}{\partial n}\right)\right\} ds\ . \tag{41}$$

Form (40) would be the one normally associated with the treatment of boundary or transition conditions from the viewpoint of variation of displacements. Under the integral sign, w and $\partial w/\partial n$ are respectively conjugate to the Kirchhoff shear load K_n and the normal bending moment M_n to be either equated to prescribed boundary values or transmitted to adjacent elements, while the first term is a reminder that the distribution of the twisting moment M_{nt} is equivalent to its contribution $\partial M_{nt}/\partial s$ to K_n plus the end values it takes in 1 and 2.

Form (41) is the one normally associated with the complementary energy principle. Under the integral sign the conjugates to U_t and U_n are boundary deformations, twist for U_t and transverse curvature for U_n; they are either equated to prescribed values or transmitted to adjacent elements. If U and V are known along all boundaries, Ω is known at the angular points and the first term in (41) will be the one ensuring single valuedness at angular points of w and its first derivatives.

The condition of zero interface virtual work will now be written by adding the contributions of F_+ and F_-. If in doing so we adopt a common orientation for the normal and the tangent, say that on F_+, we must simply make the difference between the formulas (38), (39), (40) or (41) for F_+ and F_-. This difference will be denoted by Δ and, using (39), the condition is, assuming a straight boundary $\dot{\theta} = 0$,

$$\Delta \left[w\Omega \Big|_1^2 + \int_1^2 \left\{ - \frac{\partial w}{\partial n} \frac{\partial U_t}{\partial s} + \frac{\partial w}{\partial s} \frac{\partial U_n}{\partial s} \right\} ds \right] = 0 . \tag{42}$$

Denote by 2a the length of the interface and introduce the reduced variables

$$\sigma = \frac{s}{a} \quad , \quad \nu = \frac{n}{a} \quad , \quad \Omega^* = a\Omega \; , \tag{43}$$

so that (42) becomes

$$\Delta \left[w\Omega^* \Big|_{-1}^{+1} + \int_{-1}^{+1} \left\{ - \frac{\partial w}{\partial \nu} \frac{\partial U_t}{\partial \sigma} + \frac{\partial w}{\partial \sigma} \frac{\partial U_n}{\partial \sigma} \right\} d\sigma \right] = 0 . \tag{44}$$

Assuming a discretization by polynomials, we now introduce the expansions in Legendre polynomials

$$P_m(\sigma) = \frac{1}{2^m m!} \frac{d^m}{d\sigma} (\sigma^2 - 1)^m . \tag{45}$$

The first Legendre polynomials are

$$P_0(\sigma) = 1 \quad , \quad P_2(\sigma) = \frac{3\sigma^2 - 1}{2} \; , \quad P_4(\sigma) = \frac{35\sigma^4 - 30\sigma^2 + 3}{8}$$

$$P_1(\sigma) = \sigma \; , \quad P_3(\sigma) = \frac{5\sigma^3 - 3\sigma}{2} \quad . \tag{46}$$

Take

$$\frac{\partial w}{\partial \sigma} = \sum_{0} p_m P_m(\sigma) \quad , \quad \frac{\partial w}{\partial \nu} = \sum_{0} q_m P_m(\sigma) \; ,$$
$$\frac{\partial U_n}{\partial \sigma} = \sum_{0} u_m P_m(\sigma) \quad , \quad \frac{\partial U_t}{\partial \sigma} = \sum_{0} v_m P_m(\sigma) \; . \tag{47}$$

As Legendre polynomials satisfy the differential equation

$$\frac{d}{d\sigma}\left\{(\sigma^2-1)\,\frac{dP_m}{d\sigma}\right\} = m(m+1)P_m \; , \tag{48}$$

we can adopt as first integrals

$$Q_{m+1}(\sigma) = \int P_m(\sigma)d\sigma \; ,$$

those which for $m > 0$, vanish at the ends of the interval

$$Q_{m+1}(\sigma) = \frac{\sigma^2-1}{m(m+1)}\,\frac{dP_m}{d\sigma} \qquad m > 0 \; , \tag{49}$$

while for $m = 1$ we make the particular choice

$$Q_1(\sigma) = \sigma. \tag{50}$$

It is useful to recall that the Legendre polynomials themselves take the following end values:

$$\begin{aligned} P_m(\pm 1) &= 1 \qquad \text{for m even} \; , \\ &= \pm 1 \qquad \text{for m odd} \; . \end{aligned} \tag{51}$$

Integration of the expansions consequently furnishes

$$\begin{aligned} w &= p_{-1}+p_0\sigma + \sum_{1} p_m Q_{m+1}(\sigma) \; , \\ U_n &= u_{-1}+u_0\sigma + \sum_{1} u_m Q_{m+1}(\sigma) \; , \\ U_t &= v_{-1}+v_0\sigma + \sum_{1} v_m Q_{m+1}(\sigma) \; . \end{aligned} \tag{52}$$

Taking into account the orthogonality property

$$\int_{-1}^{1} P_m(\sigma)P_n(\sigma)d\sigma = \frac{2}{2m+1}\,\delta_{mn}\,, \tag{53}$$

the zero interface virtual work conditions (44) takes the form

$$\Delta[p_{-1}(\Omega^*(1)-\Omega^*(-1))+p_0(\Omega^*(1)+\Omega^*(-1)+2u_0)-2q_0v_0+\sum_1\frac{2}{2m+1}(p_m u_m - q_m v_m)] = 0. \tag{54}$$

It becomes simpler yet if we recognize in the conjugates to p_{-1}, p_0 and q_0 reduction elements of the resultant of the total load transmitted across the interface. Computed on F_+, the total bending moment is

$$B = \int_{-a}^{a} M_n ds = \int_{-1}^{+1} \frac{\partial U_t}{\partial\sigma}\, d\sigma = U_t\Big|_{-1}^{+1} = 2v_0\;; \tag{55}$$

the total vertical load

$$V = -M_{nt}\Big|_{-a}^{a} + \int_{-a}^{a} K_n ds = -M_{nt}\Big|_{-1}^{1} - \int_{-1}^{1} \frac{\partial}{\partial\sigma}\left(\frac{\partial U_n}{\partial s}\right) d\sigma = -\left(M_{nt} + \frac{\partial U_n}{\partial s}\right)\Big|_{-1}^{1}$$

$$= \Omega\Big|_{-1}^{1} = \frac{1}{a}\,\{\Omega^*(1)-\Omega^*(-1)\}\;; \tag{56}$$

the total twisting moment about the mid-point of the interface

$$T = -sM_{nt}\Big|_{-a}^{a} + \int_{-a}^{a} sK_n ds = -sM_{nt}\Big|_{-a}^{a} - \int_{-a}^{a} s\,\frac{\partial^2 U_n}{\partial s^2}\, ds$$

$$= -s\left(M_{nt} + \frac{\partial U_n}{\partial s}\right)\Big|_{-a}^{a} + \int_{-a}^{a} \frac{\partial U_n}{\partial s}\, ds = s\Omega + U_n\Big|_{-a}^{a} \tag{57}$$

$$= \Omega^*(1)+\Omega^*(-1)+2u_0\;.$$

Computed on F_- with the orientation of normal and tangent prevailing on F_+, these formulas undergo a change of sign.

Finally (54) becomes

$$\Delta\left[p_{-1}aV+p_0T-q_0B + \sum_1 \frac{2}{2m+1}(p_m u_m - q_m v_m)\right] = 0 \qquad (58)$$

and we can discuss its application to several types of plate bending models. We note that (58) is a sum of products of generalized conjugate displacements and surface tractions. Denoting by qc the generic term, the conformity condition is

$$\Delta q = 0 \quad \text{or} \quad q_+ = q_- = q$$

and the diffusivity condition

$$\Delta c = 0 \quad \text{or} \quad c_+ = c_- = c$$

showing, by comparison with (29), the restoration of symmetry accompanying the use of the stress functions. In any type of model one of the factors in the product will be a connector. If q is the connector, $\Delta q = 0$ *a priori* and

$$\Delta(qc) = q\Delta c \ .$$

Then, *provided q is an independent discretization parameter of the element,* its variation will result in $\Delta c = 0$. The converse implication $\Delta c = 0 \rightarrow \Delta q = 0$ will follow if c is an independent discretization parameter of the element. However, since any discretization is limited as regards the degree of the polynomials involved, the infinite sum in (58) is always truncated so that either conformity or diffusivity, or even both, are weak.

STATICALLY ADMISSIBLE AND STRONGLY DIFFUSIVE PLATE BENDING MODELS

These are in fact the simplest models. In the case of triangular shapes complete polynomials of degree n are used for the vector stress function (U, V) defining the self-equilibrating internal stress field. Strong diffusivity follows from C_0 continuity of U and V at the interfaces and this requirement is known, by analogy with the kinematical models of plate stretching, to be easily enforced by a suitable selection of local values on the boundary. Bubble stress modes appear for $n \geq 3$ (their exact number is (n-1)(n-2)) which, being unconnected, are to be eliminated by energy minimization of the isolated element. The local stress function values on the boundary are independent and their variations produce the weak conformity properties

$$\Delta p_{-1} = 0 \quad \text{and} \quad \Delta p_m = 0 \;, \quad \Delta q_m = 0 \quad \text{for } 0 \leq m \leq n-1 \;, \tag{59}$$

where, by virtue of the expansions,

$$p_m = \frac{2m+1}{2} \int_{-1}^{1} \frac{\partial w}{\partial \sigma} P_m d\sigma = \frac{2m+1}{2} \left[wP_m \Big|_{-1}^{1} - \int_{-1}^{1} w \frac{dP_m}{d\sigma} d\sigma \right] , \tag{60}$$

$$q_m = \frac{2m+1}{2} \int_{-1}^{1} \frac{\partial w}{\partial \nu} P_m d\sigma \; . \tag{61}$$

A remark is needed here concerning the vanishing *a priori* of $\Delta\Omega$ at both ends of the interface which is necessary for the results $\Delta p_{-1} = 0$ and $\Delta p_0 = 0$. The C_0 continuity of U and V along the interface is not sufficient to this effect; from (35) we obtain at an end point of the interface

$$(M_{nt})_+ = -\Omega_+ - \left(\frac{\partial U_n}{\partial s} \right)_+$$

and, keeping on F_- the same orientation of normal and tangent

$$(M_{nt})_- = +\Omega_- + \left(\frac{\partial U_n}{\partial s} \right)_- \; .$$

Whence the reciprocity condition 5

$$(M_{nt})_+ + (M_{nt})_- = -\Delta\Omega - \Delta \frac{\partial U_n}{\partial s} = 0 .$$

And the result follows since the second term vanishes already by the continuity of U_n.

The singlevaluedness of Ω at a corner is moreover necessary in order that, at the assembled level, the jump in M_{nt} remains equal to the externally applied concentrated transverse load.

The lowest value of the polynomial degree in the stress functions is n = 1 and Figure 1(a) shows the choice of local values ensuring C_0 continuity. The weak conformity properties are reduced to $\Delta p_{-1} = 0$, $\Delta p_0 = 0$, $\Delta q_0 = 0$.

From (52) and the fact that $Q_{m+1}(\pm 1) = 0$ for $m \geq 1$, it follows that $\Delta p_{-1} = 0$ and $\Delta p_0 = 0$ require the transverse deflections to be transmitted at the corner points. From (61) with m = 0, it appears that $\Delta q_0 = 0$ requires transmittion of the *average normal slope*. This is represented symbolically on Figure 1(b).

It will be noted that the connection is just sufficient to transmit the resultant of the boundary loads across. In fact the element is somewhat deficient because $K_n \equiv 0$ and no additional solution can be found to account even for a uniform transverse pressure distribution without altering the existing boundary loading modes. The external load can however be replaced in a sense of best statically equivalence by concentrated loads at the nodal points of the mesh. The case n = 2 is better known and operational [20, 21]. Figure 2(a) shows the choice of local stress functions value leading to C_0 continuity and use of a direct flexibility solution at the assembled level.

We now have the additional weak conformity properties $\Delta p_1 = 0$ and $\Delta q_2 = 0$ expressing the continuity of

$$p_1 = \frac{3}{2}\left(w(1)+w(-1) - \int_{-1}^{1} w d\sigma\right)$$

$$q_1 = \frac{3}{2}\int_{-1}^{1} \sigma \frac{\partial w}{\partial \nu} d\sigma \; .$$

Considering that the corner deflections are already continuous, the first simply implies the additional continuity of the average interface deflection. The second can be used as such or be combined with the $\Delta q_0 = 0$ condition in the equivalent continuity of the pair [21]

$$\int_{-1}^{1} (1+\sigma) \frac{\partial w}{\partial} d\sigma \quad , \qquad \int_{-1}^{1} (1-\sigma) \frac{\partial w}{\partial \nu} d\sigma \; .$$

Figure 2(b) shows symbolically this weak conformity property, by which the elements of this type can be assembled by the direct stiffness method.

Here an additional solution can be found that does not disturb the boundary loading modes and equilibrates a uniform transverse pressure [21].

This special solution is worked out in local co-ordinates and involves a second degree bending moments field. As pointed out by Morley in a private communication, it would be logical to remove the dependency of the particular solution on the choice of local co-ordinates by adding the two cubic bubble stress modes in U and V (that also generate a second degree bending moment field), their contribution being settled by energy minimization.

Obviously there are no difficulties in raising arbitrarily the value of n for this class of elements and the weak connection properties required for application of the standard direct stiffness procedure follows automatically from working out (6) and (61) up to m = n-1.

Figures 3(a) and (b) show the case n = 3.

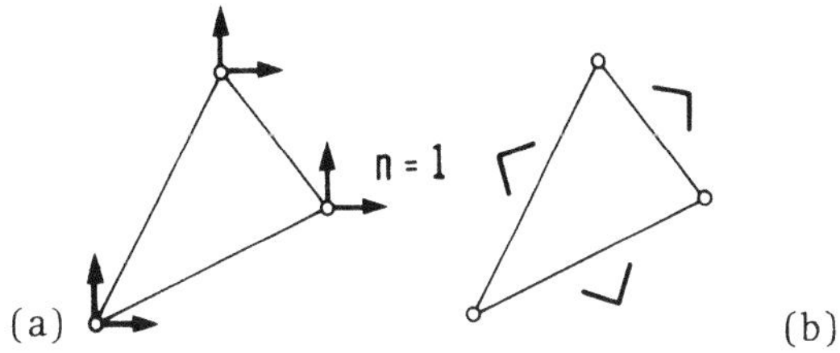

Figure 1

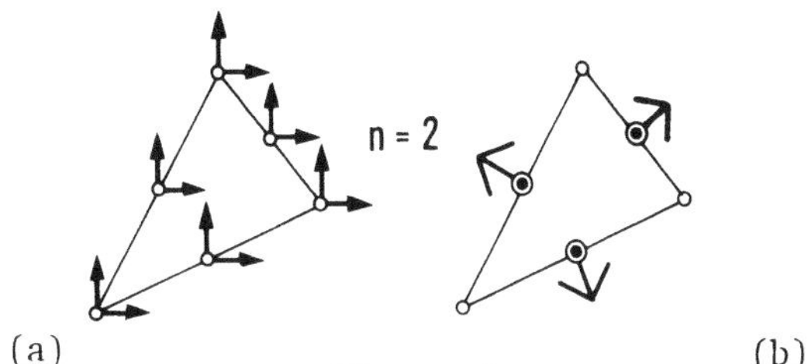

Figure 2

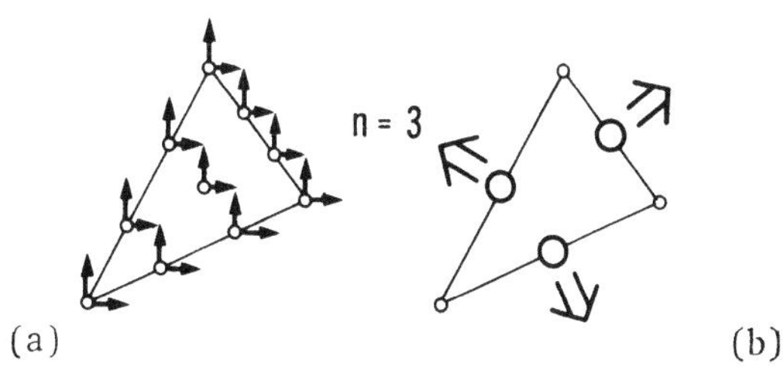

Figure 3

Legend:

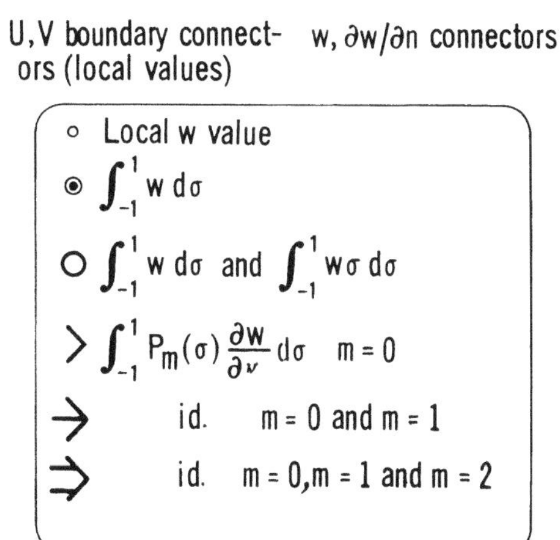

Statically admissible and strongly diffusive plate bending elements

KINEMATICAL, WEAKLY CONFORMING PLATE BENDING MODELS

With kinematical models we enter the area of difficulty in obtaining strong conformity. The converse of the procedure followed in the preceding section fails essentially because the boundary displacements required for strong conformity cannot be made independent. Their number exceeds the number of independent coefficients available in a complete polynomial of specified degree.

By discretizing the stress functions introduced as connectors at the boundary we construct hybrids of the kinematical, weakly conforming variety.

Take first for w in a triangle a complete polynomial of degree $n = 2$ and linear variations for U and V along the boundary. The summation in (58) is then truncated to $m < 2$ for both reasons, but all conjugates are present up to $m = 1$. Figure 4(a) shows the same organization of C_0 continuity of the stress functions as in Figure 1(a) but they need not (although they can in this case) be defined in the interior domain. The weak conformity properties are the same as before and may in fact be used as illustrated on Figure 1(b). However, and this is the origin of the concept of non-conforming [2,3] or delinquent [1] elements, the change in context allows a re-interpretation of the weak displacement connectors in terms of local values. Indeed, while in the former case the displacement coefficients p_m and q_m were simply unknown for $m > 1$, they are now *a priori* set equal to zero because of the existence of the internal displacement field and its limitation to a second degree polynomial. Thus the average normal slope can be identified here with its constant value along the interface, in particular the mid-point value (Figure 4(b)). The element becomes in fact one proposed by Morley [5], already recognized as passing the patch test [2], and exhibiting convergence in practice.

The case where w is a complete cubic produces new elements. The truncated expansions of w and its normal slope are:

$$w = p_{-1} + p_0\sigma + p_1 \frac{\sigma^2-1}{2} + p_2 \frac{\sigma(\sigma^2-1)}{2} , \tag{62}$$

$$\frac{\partial w}{\partial \nu} = q_0 + q_1\sigma + q_2 \frac{3\sigma^2-1}{2} . \tag{63}$$

There are two choices for the stress functions on the boundary that allow the 10 coefficients of the displacement field to be expressed in terms of generalized boundary displacements without violating the zero interface virtual work condition. In the first, the tangential component of the stress function is only allowed linear variations but the normal component has quadratic variations.

Hence in (58) $v_1 = 0$ but not u_1. The resultant weak conformity requirements are:

$\Delta p_{-1} = 0$	$\Delta p_0 = 0$	implying as we know the continuity of w at corner points
$\Delta q_0 = 0$		implying continuity of $\int_{-1}^{1} \frac{\partial w}{\partial \nu} d\sigma$,
$\Delta p_1 = 0$		implying continuity of $\int_{-1}^{1} w \, d\sigma$.

Inspection of (62) reveals that, since the p_2 term vanishes for $\sigma = 0$, the last condition can be replaced by the continuity of w at the mid-point.

On the other hand (63) does not reveal any local slope that would be related to q_0 only. Hence it is not always possible (nor is it necessary) to obtain a weak connection interpretation in terms of local displacements only. Figures 5(a) and (b) illustrate the two connection schemes. The cubic whose nine generalized boundary displacements, just defined, are zero (a pseudo-bubble function) turns out to be

$$w = \gamma\{L_1L_2(L_2-L_1)+L_2L_3(L_3-L_2)+L_3L_1(L_1-L_3)\}$$

when expressed in area co-ordinates. This 10th degree-of-freedom is elimination (γ is determined as a linear combination of the nine boundary displacements) by energy minimization of the isolated element. It should be observed that, because this 10th degree-of-freedom vanishes at the barycentre, the deflection at this point may not be taken to represent the internal degree-of-freedom, as Figure 5(b) would seem to suggest.

In the second choice (Figures 6(a) and (b)) the normal component of the vector stress function is restricted to linear variations and the tangential one to quadratic, whereby $u_1 = 0$ but not v_1.

The continuity requirement for p_1 is traded for the continuity of q_1 of $\int_{-1}^{1} \sigma(\partial w/\partial\nu)d\sigma$. Inspection of (63) reveals that, since the q_2 term vanishes at the Gauss points $\sigma = \pm 1/\sqrt{(3)}$, continuity of the normal slopes at these points is equivalent to both $\Delta q_0 = 0$ and $\Delta q_1 = 0$. Again a pseudo-bubble degree-of-freedom, connected with the barycentre, must be energy eliminated.

Suppose now that each of the connected elements is in a state of uniform strain, so that $p_2 = 0$ and $q_2 = 0$. Then for the first of our new cubic displacement models, p_{-1}, p_0, p_1 and q_0 are continuous by virtue of the connections and there will be *exact conformity* provided q_1 is also continuous. This will be the case if the two elements have the *same* state of uniform strain, because the twist $q_1 = \partial^2 w/\partial\nu\partial\sigma$ is one of the strain measures.

Similarly for the second model where p_{-1}, p_0, q_0 and q_1 are already continuous by virtue of the connections, exact conformity will follow from the condition of common uniform strain because $p_1 = \partial^2 w/\partial\sigma^2$ will be a common curvature.

This illustrates the verification of the patch test as it was originally conceived by Irons [1]. Nonconforming elements pass the patch test when they become exactly conforming under a common state of uniform strain.

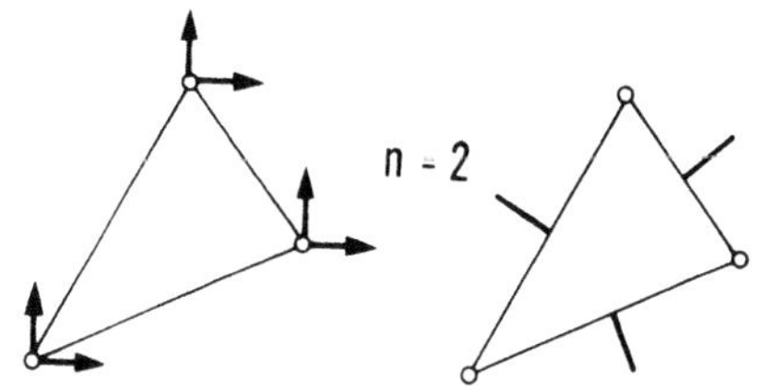

(a) (b)

Figure 4

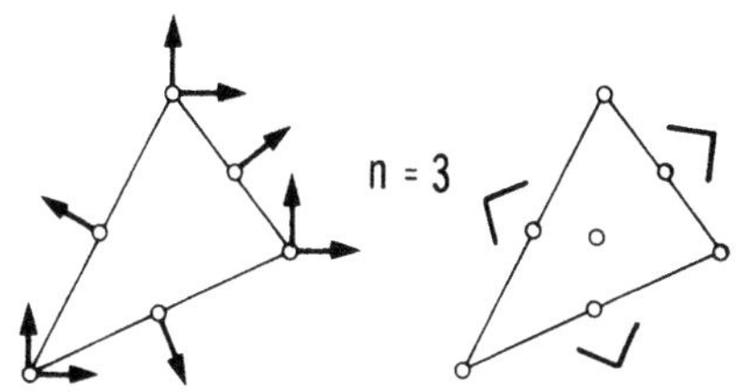

(a) (b)

Figure 5

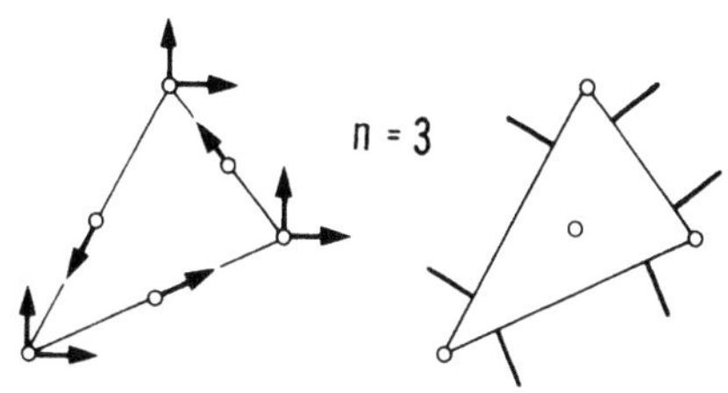

(a) (b)

Figure 6

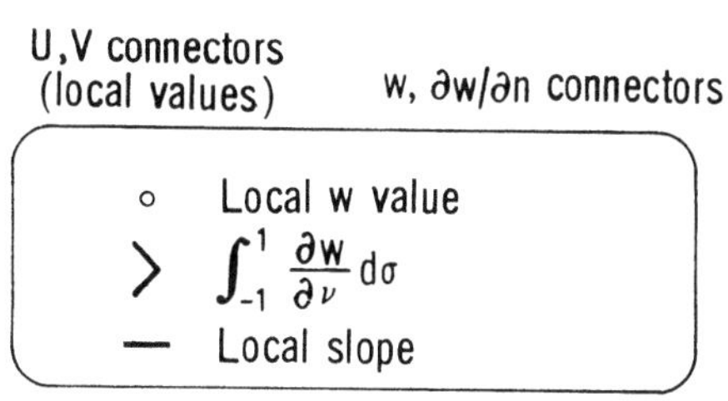

Kinematically and weakly conforming plate bending elements

CONCLUSIONS

The patch test is most easily verified by expressing the zero interface virtual work condition with the help of stress functions and is in fact identical to the construction of hybrid models. Clearly as the degree of polynomial approximation is raised, elements can be constructed that pass a higher order patch test, e.g., they become exactly conforming under a state of uniform strain variation and this property must be related to a higher rate of convergence.

Examples were only given for Kirchhoff plate bending and triangular elements because this problem has received much attention in the past. The extension to rectangular elements is straightforward. Moreover the method followed is directly applicable to the plate stretching case, using the Airy stress function. The hybrid models of Pian type, or non-diffusive elements, constructed in this way may prove to be useful substitutes for the super-elements required to enforce strong diffusivity.

The three-dimensional case is decidedly more difficult because of the complexity and lack of uniqueness of the stress function tensor.

REFERENCES

[1] IRONS, B.M. and RAZZAQUE, A., "Experience with Patch Test for Convergence of Finite Elements", *The Mathematical Foundations of the Finite Element Method with Applications to Partial Differential Equations*, edited by A. K. Aziz, Academic Press, New York, 1972, pp. 557-587.

[2] CIARLET, P.G., "Quelques méthodes d'éléments finis pour le problème d'une plaque encastrée", *Colloque Int. sur les Méthodes de Calcul Scientifique et Technique*, IRIA, 1973, pp. 66-86.

[3] STRANG, G. and FIX, G., *An Analysis of the Finite Element Method*, Prentice-Hall, Englewood Cliffs, New Jersey, 1973.

[4] PIAN, T.H.H. and TONG, P., "Basis of Finite Element Methods for Solid Continua", *Int. J. Num. Meth. Engng.*, Vol. 1, 1969, pp. 3-28.

[5] MORLEY, L.S.D., "The Constant-Moment Plate Bending Element", *J. Strain Analysis*, Vol. 6, 1971, pp. 20-24.

[6] FRAEIJS DE VEUBEKE, B.M., "Displacement and Equilibrium Models in the Finite Element Method", in *Stress Analysis*, edited by O.C. Zienkiewicz and G.S. Holister, Wiley, London, 1965, Chapter 9.

[7] SANDER, G., "Bornes supérieures et inférieures dans l'analyse matricielle des plaques en flexion-torsion", *Bull. Soc. Roy. Sci.*, Liège, Vol. 33, 1964, pp. 456-494.

[8] FRAEIJS DE VEUBEKE, B.M., "Bending and Stretching of Plates", *Conf. Matrix Meth. Struct. Mech.*, Air Force Flight Dyn. Lab., Wright-Patterson A.F.B., Ohio, 1965, pp. 863-885.

[9] CLOUGH, R.W. and TOCHER, J.L., "Finite Element Stiffness Matrices for Analysis of Plates in Bending", *Conf. Matrix Meth. Struct. Mech.*, Air Force Flight Dyn. Lab., Wright-Patterson A.F.B., Ohio, 1965, pp. 515-545.

[10] FRAEIJS DE VEUBEKE, B.M. and ZIENKIEWICZ, O.C., "Strain Energy Bounds in Finite Element Analysis by Slab Analogy", *J. Strain Analysis*, Vol. 2, 1967, pp. 265-271.

[11] BECKERS, P., "Les fonctions de tension dans la méthode des éléments finis", *Thèse de Doctorat*, Collection des Publications de la Faculte des Sciences Appliquees de l'Université de Liège, No. 41, 1973.

[12] FRAEIJS DE VEUBEKE, B.M., "Diffusion des inconnues hyperstatiques dans les voilures à longerons couplés", Bulletin 25, du Service Technique de l'Aéronautique, Bruxelles, 1952.

[13] FRIEDRICHS, K.O., "Ein verfahren der Variationsrechnung", Nachrichten der Gas-Wiss, zu Göttingen, 1929, pp. 13-20.

[14] COURANT, R. and HILBERT, D., *Methoden der Mathematischen Physik*, Springer, Berlin, 1931, Vol. 1, Chapter 4.

[15] FRAEIJS DE VEUBEKE, B.M., "A New Variational Principle for Finite Elastic Displacements", *Int. J. Engng. Sci.*, Vol. 10, 1972, pp. 745-763.

[16] FRAEIJS DE VEUBEKE, B.M., "Diffusive Equilibrium Models", *Univ. Calgary Lecture Notes*, Int. Research Seminar on the Theory and Applications of the Finite Element Methods, 1973.

[17] SANDER, G., "Application of the Dual Analysis Principle", *IUTAM Colloq. High Speed Comptng Elastic Struc.*, Vol. 1, 1971, pp. 167-207, (Publ. Colloques et Congrès de l'Université de Liège, Place du XX Août, 16, B-4000 Liège, Belgium).

[18] CONNOR, J., "Mixed Models for Plates", in *Finite Element Techniques in Structural Mechanics*, edited by H. Tottenham and C. Brebbia, Chapter 5.

[19] SAMUELSSON, A., "Mixed Finite Element Methods in Theory and Applications", *Lecture Notes of the Finite Element Course in Tirrenia,* 1973.

[20] MORLEY, L.S.D., "A Triangular Equilibrium Element with Linearly Varying Bending Moments for Plate Bending Problems", *J. Roy. Aeron. Soc.,* Vol. 71, 1967, pp. 715.

[21] FRAEIJS DE VEUBEKE, B.M. and SANDER, G., "An Equilibrium Model for Plate Bending", *Int. J. Solids Struct.,* Vol. 4, 1968, pp. 447-468.

STRESS FUNCTION APPROACH

B.M. Fraeijs de Veubeke

University of Liege, Belgium

SUMMARY—The stress function approach allows in principle to construct systematically finite elements satisfying stress equilibrium in finite regions and surface traction reciprocity at their interfaces. Some difficulties are experienced with the requirements of C_1 continuity of the stress functions, when rigorous rotational equilibrium is desired. They disappear when only first order stress functions are used, whereby translational equilibrium is maintained, but rotational equilibrium must be separately enforced by a Lagrange multiplier technique. The paper discusses in detail this new generation of finite elements with discretized rotational equilibrium. In particular the extent to which rotational equilibrium can be enforced without inducing mechanisms. Because diffusivity involves only C_1 continuity of the first order stress functions, the new elements lend themselves very flexibly to isoparametric coordinate transformations.

1. FIRST ORDER STRESS FUNCTIONS

Following the Euler-Cauchy assumption the short-range internal forces in a continuum are representable by surface integrals of surface tractions. Thus a simply connected volume of the continuum will be in translational equilibrium if

$$\iint_S \vec{t}\, dS + \iiint_V \rho\, \vec{g}\, dV = 0 \qquad (1)$$

Original publication, *Proceedings of the First World Congress on Finite Element Methods*, October 1975, Robinson and Associates, Horton Road, Woodlands, Wimborne, Dorset BH21 6NB, England.

where $\vec{t}$ is the surface traction vector acting on the boundary of the volume, ρ the specific mass of the continuum and $\vec{g}$ an accelaration of gravitational or inertial origin. The application of this equilibrium property to an infinitesimal volume leads to the well known Euler-Cauchy formula, expressed here in Cartesian coordinates,

$$t_j = n_i \, \sigma_{ij} \tag{2}$$

where n_i are the components of the unit outward normal $\vec{n}$ to a facet of the continuum and σ_{ij} the stress tensor.

Let us now consider the equilibrium property for a finite volume under the assumption $\vec{g} = 0$. Drawing a simple closed contour c on the boundary surface S that separates it into two parts S_1 and S_2 and reversing the orientation of the normal on S_2, the equilibrium property takes the form

$$\iint_{S_1} t_j \, dS = \iint_{S_2} t_j \, dS \qquad (j = 1,2,3) \; .$$

This equality will hold for any surface S that is limited by the same contour c and on which the orientation of the normal $\vec{n}$ follows the same "screw" convention with respect to a prescribed sense of circulation on c. Consequently the surface integrals will only depend on the circulation of some tensor field on c

$$\iint_S t_j \, dS = \oint_c A_{mj} \, dx_m \; . \tag{3.a}$$

In the left hand side we substitute (2), and transform the right hand side by Stokes theorem to obtain

$$\iint_S n_i \, \sigma_{ij} \, dS = \iint_S n_i \, e_{ipq} \, D_p \, A_{qj} \, dS \quad (j = 1,2,3)$$

where e_{ipq} is the alternating tensor and D_p stands for $\partial/\partial x_p$.

Because this relation will hold for an arbitrarily small and arbitrarily oriented surface element dS, it follows that

$$\sigma_{ij} = e_{ipq} D_p A_{qj} \quad . \tag{3.b}$$

It is indeed immediately verified that this expression of the stress tensor identically satisfies the local translational equilibrium equations without inertial loading

$$D_i \sigma_{ij} = 0 \qquad (j = 1,2,3) \tag{4}$$

since

$$e_{ipq} D_i D_p \equiv 0$$

on account of the skew symmetry of the alternating tensor.

We call A_{qj} a *tensor of first order stress functions*, because the stresses are combinations of their derivatives.

2. DIFFUSIVITY AND C_0 CONTINUITY OF THE FIRST ORDER STRESS FUNCTIONS

From (2) and (3.b) follows for the surface tractions

$$t_j = \partial_q A_{qj} \tag{5}$$

where

$$\partial_q = e_{ipq} n_i D_p \tag{6}$$

are surface derivative operators; they are the components of a vector operator $\vec{n} \times \text{grad}$ that projects the gradient in the surface normal to $\vec{n}$. It is clear therefore that, provided the first order stress function tensor A_{qj} is continuous across an interface separating regions where it receives different approximations, the surface tractions will also be continuous. Taking into account the opposite character of outward normals to the regions meeting at the interface, it may be concluded that C_0 continuity of the first order stress function tensor at the interface guarantees the property of diffusivity of the surface tractions, that is the action-reaction reciprocity rule of the short-range internal forces.

3. SECOND ORDER STRESS FUNCTIONS

A similar approach can be followed as regards rotational equilibrium The resultant moment with respect to the origin of the surface tractions on a finite volume has the components

$$\iint_S e_{umj}\, x_m\, t_j\, dS \qquad (u = 1,2,3)\ .$$

It must also vanish in the absence of inertial loads and, by similar reasoning we conclude that, for all surfaces S limited by the same contour c,

$$\iint_S e_{umj}\, x_m\, t_j\, dS = \oint_c B_{qu}\, dx_q \qquad (u = 1,2,3)\ . \qquad (7.a)$$

Again, substituting (2) to the left, using Stokes theorem on the right, and comparing the integrands

$$e_{umj}\, x_m\, \sigma_{ij} = e_{iqp}\, D_p\, B_{qu} \qquad \forall(u,i) \quad . \qquad (7.b)$$

If the operator D_i is applied to this equation, the right-hand side vanishes because of the skew symmetry of the alternating tensor. On the left-hand side we note that $D_i\, x_m = \delta_{im}$ and obtain

$$e_{uij}\, \sigma_{ij} + e_{umj}\, x_m\, D_i\, \sigma_{ij} = 0 \quad (u = 1,2,3)\ .$$

Thus, if the stresses are already in translational equilibrium, we obtain the well known local rotational equilibrium conditions

$$e_{uij}\ \sigma_{ij} = 0 \qquad (u = 1,2,3) \tag{8}$$

equivalent to the symmetry property of the stress tensor

$$\sigma_{ji} = \sigma_{ij} \quad . \tag{9}$$

Since precisely the translational equilibrium conditions (4) are satisfied by the use of first order stress functions, equations (8) are transformed into

$$e_{uij}\ e_{ipq}\ D_p\ A_{qj} = 0$$

or, since

$$e_{uij}\ e_{ipq} = \delta_{jp}\ \delta_{uq} - \delta_{jq}\ \delta_{up} \tag{10}$$

into

$$D_j\ A_{uj} - D_u\ A_{pp} = 0 \qquad (u = 1,2,3) \ . \tag{11}$$

The problem is now to obtain a general solution of (11).

To this effect we go back to (7.b) where (3.b) is now substituted

$$e_{ipq}\ [e_{umj}\ x_m\ D_p\ A_{qj} - D_p\ B_{qu}] = 0 \quad .$$

Transforming

$$x_m\ D_p\ A_{qj} = D_p\ (x_m\ A_{qj}) - \delta_{mp}\ A_{qj}$$

and using the formula of type (10), this can be placed in the form

$$e_{ipq} D_p [e_{umj} x_m A_{qj} - B_{qu}] - \delta_{iu} A_{qq} + A_{ui} = 0 \quad .$$

Thus, setting

$$\phi_{qu} = B_{qu} - e_{umj} x_m A_{qj} \tag{12}$$

we find

$$A_{ui} = e_{ipq} D_p \phi_{qu} + \delta_{iu} A_{qq} \quad .$$

In particular, setting $u = i$,

$$A_{uu} = e_{upq} D_p \phi_{qu} + 3 A_{qq}$$

whereby

$$A_{qq} = - \frac{1}{2} e_{upq} D_p \phi_{qu} \tag{13}$$

that, after back substitution, finally expresses the first order stress function tensor in terms of a second order stress function tensor ϕ_{qu}

$$A_{uj} = e_{jpq} D_p \phi_{qu} - \frac{1}{2} \delta_{ju} e_{mpq} D_p \phi_{qm} \quad .$$

It is easily verified that this satisfies equations (11).

After a change of notation for the subscripts

$$A_{qj} = e_{jrs} D_r \phi_{sq} - \frac{1}{2} \delta_{qj} e_{mrt} D_r \phi_{tm} \tag{14}$$

and, substituting into (3.b), the stresses are given by

$$\sigma_{ij} = e_{ipq} e_{jrs} D_p D_r \phi_{sq} - \frac{1}{2} e_{ipj} e_{mrt} D_p D_r \phi_{tm} \quad .$$

The second term is clearly skew symmetrical in the subscripts i and j and will disappear if we write

$$\sigma_{ij} = \frac{1}{2}(\sigma_{ij} + \sigma_{ji})$$

an operation that is licit since we know that rotational equilibrium is now satisfied. Then, by also exchanging the dummy subscripts p and q, and also q and s in the first term, there comes

$$\sigma_{ij} = e_{ipq}\, e_{jrs}\, D_p\, D_r \left(\frac{\phi_{qs} + \phi_{sq}}{2}\right)$$

so that the stresses depend only on the symmetrical part of the second order stress function tensor. This establishes that the result of B. Finzi [1]

$$\sigma_{ij} = e_{ipq}\, e_{jrs}\, D_p\, D_r\, \phi_{qs} \qquad \phi_{sq} = \phi_{qs} \tag{15}$$

for stresses satisfying both translational and rotational equilibrium is general. Because of the assumed symmetry of the tensor

$$e_{mrs}\, \phi_{sm} = 0$$

and (14) is seen to reduce to

$$A_{qj} = e_{jrs}\, D_r\, \phi_{sq} \quad . \tag{16}$$

In particular we have now that

$$A_{qq} = D_r\,(e_{qrs}\, \phi_{sq}) = 0 \quad . \tag{17}$$

4. THE INDETERMINACY OF THE FINZI TENSOR

For a given stress field in equilibrium without inertial loads the Finzi tensor is not unique. In other words, the homogeneous problem

$$\sigma_{ij} = e_{ipq} \, e_{jrs} \, D_p \, D_r \, \phi_{qs} \equiv 0$$

has non trivial solutions. They can be found by observing that since

$$\oint A_{mj} \, dx_m = 0 \qquad\qquad \oint B_{qi} \, dx_q = 0$$

must be satisfied for any closed curve drawn within the simply connected continuum, the integrands must be perfect differentials of single valued functions

$$A_{mj} = D_m \, \Omega_j \tag{18}$$

$$B_{qi} = D_q \, V_i \quad . \tag{19}$$

If this is used to transform (12)

$$\phi_{qi} = D_q \, V_i - e_{imj} \, x_m \, D_q \, \Omega_j = D_q \, (V_i - e_{imj} \, x_m \, \Omega_j) + e_{iqj} \, \Omega_j \quad .$$

Exchanging the subscripts q and i and using $e_{qij} = - e_{iqj}$

$$\phi_{iq} = D_i \, (V_q - e_{qmj} \, x_m \, \Omega_j) - e_{iqj} \, \Omega_j \quad .$$

As ϕ_{qi} may be taken to be symmetrical, the arithmetic mean of ϕ_{qi} and ϕ_{iq} gives

$$\phi_{qi} = \frac{1}{2} \, (D_q \, U_i + D_i \, U_q) \tag{20}$$

where

$$U_i = V_i - e_{imj}\, x_m\, \Omega_j \tag{21}$$

while, by substraction,

$$0 = D_q\, U_i - D_i\, U_q + 2\, e_{iqj}\, \Omega_j \quad .$$

This result, multiplied by e_{mqi}, gives

$$2\, \Omega_m = e_{mqi}\, D_q\, U_i \tag{22}$$

Hence, if the second order stress functions derive from a vector field U_i in the same manner as strains from a displacement field, no stresses are generated and the first order stress functions are the gradients of the associated field of rotations. As a matter of fact if the U_i are considered as displacements the conditions $\sigma_{ij} \equiv 0$ are equivalent to the de Saint-Venant compatibility conditions for strains, and the equations

$$D_m\, \Omega_j = A_{mj} = e_{jrs}\, D_r\, \phi_{sm}$$

are equivalent to the Beltrami equations for the integration of the rotation field from the knowledge of the strain field.

The indeterminacy of the Finzi tensor may be used to reduce it to its non-diagonal components. Since

$$\hat{\phi}_{qs} = \phi_{qs} - \frac{1}{2}\,(D_q\, U_s + D_s\, U_q)$$

generates the same stresses, a U_i field can always be constructed from the 3 equations

$$\phi_{11} = D_1\, U_1 \qquad \phi_{22} = D_2\, U_2 \qquad \phi_{33} = D_3\, U_3$$

so that $\hat{\phi}_{qs}$ has its diagonal terms indentically zero. The non-diagonal terms constitute the stress functions of Morera. Conversely it is easily verified that a U_i field can always be constructed from the 3 equations

$$\phi_{12} = \frac{1}{2}(D_1U_2 + D_2U_1) \quad \phi_{23} = \frac{1}{2}(D_2U_3 + D_3U_2) \quad \phi_{31} = \frac{1}{2}(D_3U_1 + D_1U_3)$$

so that the non diagonal terms of $\hat{\phi}_{qs}$ vanish identically. The diagonal terms then constitute a system of stress functions of Maxwell.

5. DIFFUSIVITY AND C_1 CONTINUITY OF THE SECOND ORDER STRESS FUNCTIONS

From (2), (6) and (15) the surface tractions generated by the second order stress functions are

$$t_j = \partial_q [e_{jrs} D_r \phi_{qs}] \quad .$$

Any cartesian derivative is expressible in terms of surface derivatives and of the derivative Δ along the normal itself

$$D_r = n_r \Delta - e_{ruv} n_u \partial_v \quad . \tag{23}$$

This formula is justified by the fact that, multiplied by $n_p e_{pqr}$ it reduces to the definition

$$e_{pqr} n_p D_r = - e_{pqr} e_{ruv} n_p n_u \partial_v = - n_q n_p \partial_p + n_p n_p \partial_q = \partial_q$$

of the surface derivative ($n_p \partial_p \equiv 0$ and $n_p n_p = 1$), while, multiplied by n_r, it reduces to the definition of the normal derivative

$$n_r D_r = \Delta \quad . \tag{24}$$

The expression of surface tractions in terms of normal and surface derivatives of the Finzi tensor is thus

$$t_j = \partial_q [e_{jrs} n_r \Delta\phi_{qs}] + \partial_q [n_j \partial_s \phi_{qs} - n_s \partial_j \phi_{qs}] \ . \quad (25)$$

It shows that diffusivity at an interface requires continuity not only of the Finzi tensor itself but also of its normal derivative. This has an important bearing on the development of isoparametric stress elements because, if an isoparametric coordinate transformation will preserve C_0 continuity at interfaces, it will in general not preserve continuity of the normal derivatives and will consequently destroy diffusivity.

6. THE TWO-DIMENSIONAL STRESS FUNCTIONS

H. Shaefer's [4] treatment of the two-dimensional case is attractive by its elegance and simplicity. The basic idea is that, although stress functions of type (20) are normally unproductive, they are able to generate stresses if the vector field U_i contains singularities. Consider then a plate of middle surface $x_3 = 0$ that is not loaded on its faces $x_3 = \pm \varepsilon$. The plate is considered to be embedded in the indefinite continuum and, while the vector field U_i is regular outside of the plate, it presents singularities inside of it. In fact, if we are content with the knowledge of internal stresses integrated throughout the thickness, we may go to the limit $\varepsilon \to 0$ and consider the middle plate as a *sheet of discontinuity* of the field U_i and its rotational Ω_m. Thus we introduce the following functions of the inplane variables (x_1, x_2)

$$\lim_{\varepsilon\to 0} [\Omega_i(x_1,x_2,-\varepsilon) - \Omega_i(x_1, x_2,+\varepsilon)] = \Omega_i(x_1,x_2) \quad (26)$$

$$\lim_{\varepsilon\to 0} [U_i(x_1,x_2,-\varepsilon) - U_i(x_1,x_2,+\varepsilon)] = U_i(x_1,x_2) \quad (27)$$

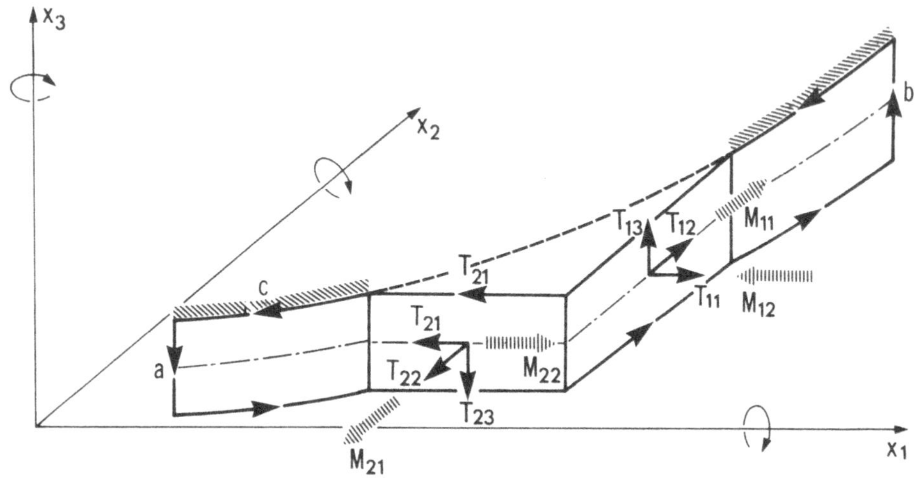

Figure 1

and consider the closed contour c of Figure 1, defined by a curve (a,b) drawn in the middle plane. For any integrand we replace

$$\lim_{\varepsilon\to 0} \oint_c f(x_1,x_2,x_3)\, ds = \int_a^b \lim_{\varepsilon\to 0} [f(x_1,x_2,-\varepsilon) - f(x_1,x_2,+\varepsilon)]ds \tag{28}$$

and apply the considerations on the computation of resultant stresses as outlined in sections 1 and 3. The total force components due to short range internal forces acting on the cross-section of the plate defined by the contour c are given by

$$\int_a^b T_{1j}dx_2 - T_{2j}dx_1 = \lim_{\varepsilon\to 0} \oint_c A_{mj}dx_m = \int_a^b D_1\Omega_j dx_1 + D_2\Omega_j dx_2 \tag{29}$$

where use has been made of (18) and (26). Since this holds for an arbitrary curve (a,b) we must have

$$T_{1j} = D_2\Omega_j \qquad T_{2j} = -D_1\Omega_j \quad . \tag{30}$$

Similarly the total moment of these forces with respect to the cartesian axes will be given by

$$\int_a^b (M_{22} - x_2T_{23})dx_1 + (-M_{12} + x_2T_{13})dx_2 = \lim_{\varepsilon\to 0} \oint_c B_{m_1}dx_m$$
$$= \int_a^b D_1(U_1 + x_2\Omega_3)dx_1 + D_2(U_1 + x_2\Omega_3)dx_2$$

$$\int_a^b (-M_{21} + x_1T_{23})dx_1 + (M_{11} - x_1T_{13})dx_2 = \lim_{\varepsilon\to 0} \oint B_{m_2}dx_m$$
$$= \int_a^b D_1(U_2 - x_1\Omega_3)dx_1 + D_2(U_2 - x_1\Omega_3)dx_2$$

$$\int_a^b (x_2T_{21} - x_1T_{22})dx_1 + (x_1T_{12} - x_2T_{11})dx_2 = \lim_{\varepsilon\to 0} \oint_c B_{m_3}dx_m$$
$$= \int_a^b D_1(U_3 - x_2\Omega_1 + x_1\Omega_2)dx_1 + D_2(U_3 - x_2\Omega_1 + x_1\Omega_2)dx_2$$

after the use of (19), (21) and (26) and (27).

The first equation furnishes the pair of results

$$M_{22} - x_2T_{23} = D_1U_1 + x_2D_1\Omega_3 \qquad -M_{12} + x_2T_{13} = D_2U_1 + \Omega_3 + x_2D_2\Omega_3$$

that can be simplified by noting that the terms still containing the coordinates x_i cancel by virtue of (30), leaving

$$M_{22} = D_1U_1 \quad \text{and} \quad M_{12} = -D_2U_1 - \Omega_3 \quad .$$

Similar simplifications occur in the two other pairs. The complete set of results may be split into two groups:

The plate stretching group

$$T_{11} = D_2\,\Omega_1 \qquad T_{21} = -\,D_1\,\Omega_1$$
$$T_{12} = D_2\,\Omega_2 \qquad T_{22} = -\,D_1\,\Omega_2 \qquad (31)$$

satisfying identically the translational membrane equilibrium equations

$$D_1\,T_{11} + D_2\,T_{21} = 0 \qquad D_1\,T_{12} + D_2\,T_{22} = 0 \qquad (32)$$

so that Ω_1 and Ω_2 are first order stress functions.

Also, from the third moment equation

$$\Omega_1 = D_2\,U_3 \qquad \Omega_2 = -\,D_1\,U_3 \qquad (33)$$

from which follows $T_{12} = T_{21}$, or equilibrium of rotation.

U_3 is a second order stress function. It is in fact the Airy stress function as easily recognized by the equations

$$T_{11} = D_2^2\,U_3 \qquad T_{21} = T_{12} = -\,D_1\,D_2\,U_3 \qquad T_{22} = D_1^2\,U_3 \qquad (34)$$

obtained after the elimination of the first order functions.

The plate bending group

$$M_{11} = D_2\,U_2 \qquad -\,M_{21} = D_1\,U_2 - \Omega_3$$
$$-\,M_{12} = D_2\,U_1 + \Omega_3 \qquad M_{22} = D_1\,U_1 \qquad (35)$$
$$T_{13} = D_2\,\Omega_3 \qquad T_{23} = -\,D_1\,\Omega_3$$

satisfying identically the plate bending equilibrium equations

$$D_1 M_{11} + D_2 M_{21} = T_{13} \qquad D_1 M_{12} + D_2 M_{22} = T_{23}$$
$$D_1 T_{13} + D_2 T_{23} = 0 \quad . \tag{36}$$

As in the limiting process the property

$$\Omega_3 = \frac{1}{2}(D_1 U_2 - D_2 U_1) \tag{37}$$

is preserved, it is observed that

$$M_{11} = D_2 U_2 \qquad M_{22} = D_1 U_1 \qquad M_{12} = M_{21} = -\frac{1}{2}(D_1 U_2 + D_2 U_1) \tag{38}$$

so that the pair (U_1, U_2) is to be recognized as the Southwell stress function vector [2],[5],[9].

It is already clear from the expression of surface tractions on facets perpendicular to the axes that reduction to two-dimensional problems has not alleviated the C_1 continuity requirements on the stress functions when complete equilibrium is postulated.

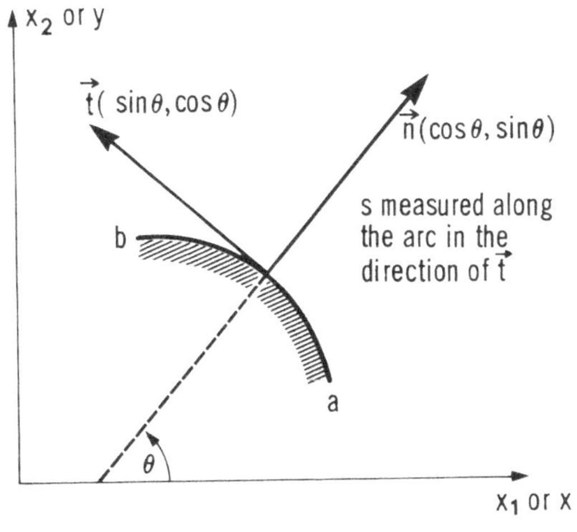

Figure 2

It becomes even clearer in local coordinates n and s (see Figure 2). In the case of the Airy function, hence forward denoted by ϕ instead of U_3, the normal and tangential components of the surface tractions (integrated throughout the thickness) are respectively

$$T_n = \frac{\partial^2 \phi}{\partial s^2} + \dot{\theta} \frac{\partial \phi}{\partial n} \qquad T_t = - \frac{\partial^2 \phi}{\partial n \partial s} \qquad \left(\dot{\theta} = \frac{d\theta}{ds} \right) . \tag{39}$$

If the curve ab is an interface, the continuity of the normal derivative is essential for the reciprocity of tangential loads and even necessary for that of normal loads if there is curvature present.

In the case of the Southwell vector, local resolution into its normal component U_n and tangential component U_t gives

$$M_n = \frac{\partial}{\partial s} U_t + \dot{\theta} U_n \tag{40}$$

for the bending moment about the $\vec{t}$ axis

$$M_{nt} = - \frac{1}{2} \left(\frac{\partial}{\partial s} U_n + \frac{\partial}{\partial n} U_t \right) + \frac{1}{2} \dot{\theta} U_t \tag{41}$$

for the twisting moment and

$$T_n = \frac{\partial \Omega}{\partial s} = \frac{1}{2} \frac{\partial}{\partial s} \left(\frac{\partial}{\partial n} U_t - \frac{\partial}{\partial s} U_n + \dot{\theta} U_t \right) \tag{42}$$

for the transverse shear load. The continuity of the normal derivative of U_t is necessary to implement that of M_{nt} and T_n. This complicates the development of equilibrium elements for plate bending based on Reissner's theory.

7. KIRCHHOFF PLATE BENDING

It is rather remarkable that the Kirchhoff assumption according to which fibers originally normal to the middle plane remain normal to the deformed middle surface, that is purely kinematical, should reduce diffusivity at an interface to C_0 continuity of the Southwell vector. The twisting moment distribution along the arc ab becomes equivalent, by virtual work considerations, to corner loads (positively oriented as the transverse axis) equal to the local value of M_{nt} in a and - M_{nt} in b, together with an equivalent additional transverse shear $\frac{\partial}{\partial s} M_{nt}$.

The total or Kirchhoff shear load becomes thus

$$K_n = T_n + \frac{\partial M_{nt}}{\partial s} = - \frac{\partial^2}{\partial s^2} U_n + \frac{\partial}{\partial s} (\dot{\theta} U_t) \qquad (43)$$

and, as was already the case for the bending moment M_n, requires only for its diffusivity that U_n and U_t be continuous. Moreover the corner loads at an internal boundary node automatically add up to zero. To show this we note that at a corner of a region (say region 1 of Figure 3) the total corner load in m, resulting

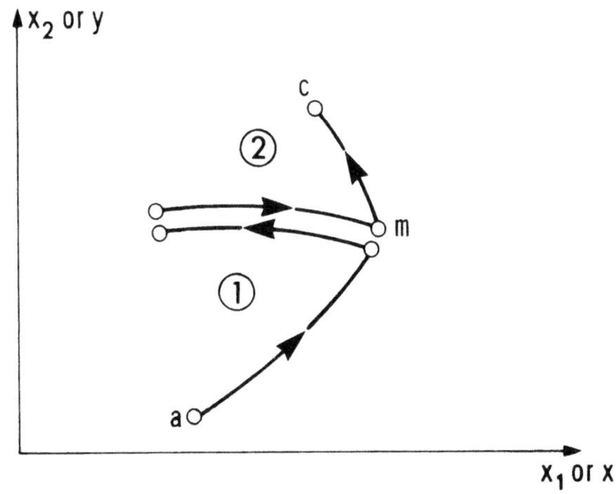

Figure 3

from the replacement of the twisting moment distributions along am and mb, is

$$\lim_{\substack{a\to m\\ b\to m}} \{(M_{nt})_b - (M_{nt})_a\} = \Delta M_{nt}$$

or the jump in twisting moment as we turn around the corner in the positive sense of description of the boundary.

Writing the twisting moment in the form

$$M_{nt} = -\Omega - \frac{\partial}{\partial s} U_n + \dot{\theta}\, U_t \tag{44}$$

and recognizing that for any given region where the stress vector is uniquely defined, Ω is also single-valued, the jump in twisting moment is equivalent to the jump

$$\Delta\left(-\frac{\partial}{\partial s} U_n + \dot{\theta}\, U_t\right) \tag{45}$$

Let region 2 become adjacent to region 1 with bm as interface, the quantity $-\frac{\partial}{\partial s} U_n + \dot{\theta}\, U_t$ is single-valued at the interface if the stress vector is continuous (but the twisting moments will not be reciprocal because in general the boundary values of Ω for region 2 will differ from those of region 1). Hence, in adding up the corner loads in m, the contributions due to the interface in (45) will cancel and the formula remains valid for the combined corner load of the two regions, the jump being measured at the corner of the combined regions. Obviously, if m becomes an internal node, there is complete cancellation of the contributions of all regions meeting in this point.

A deeper understanding of the reduction to C_0 continuity is provided by the following virtual work consideration. Closer inspection of the plate bending group of equations (35) shows that U_1, U_2 and Ω_3 form a first order stress functions group with diffusivity under C_0 continuity if we discard the rotational equilibrium condition

$$M_{12} = M_{21} \tag{46}$$

and consequently the necessity for Ω_3 to be the rotational

$$\Omega_3 = \frac{1}{2} (D_1 U_2 - D_2 U_1)$$

of the stress function vector (U_1, U_2). This equilibrium condition is the result of the virtual work equation

$$\int_{-\varepsilon}^{\varepsilon} \omega_3 (\sigma_{12} - \sigma_{21}) \, dx_3 = 0$$

where ω_3 is the material rotation of the continuum about the third axis:

$$\omega_3 = \frac{1}{2} (D_1 u_2 - D_2 u_1) \quad .$$

For small plate thickness we retain the first terms of a Taylor expansion of the displacements, which are antisymmetrical with respect to the middle plane

$$u_1 = x_3 \alpha_1 (x_1, x_2) \qquad u_2 = x_3 \alpha_2 (x_1, x_2)$$

and obtain the virtual work condition in the form

$$\frac{1}{2} (D_1 \alpha_2 - D_2 \alpha_1) (M_{12} - M_{21}) = 0 \quad .$$

If the transverse fiber twist

$$\frac{1}{2} (D_1 \alpha_2 - D_2 \alpha_1) = D_3 \omega_3$$

is unconstrained, the virtual work condition will require the satisfaction of (46) and consequently introduce C_0 continuity. However, under the Kirchhoff assumption there is no transverse shear

$$\alpha_1 + D_1\, u_3 = 0 \qquad \alpha_2 + D_2\, u_3 = 0$$

and the virtual work condition is satisfied without the need of (46) because the transverse twist becomes identically zero.

The first plate bending equilibrium model based on Kirchhoff's theory, although originally developed differently [7],[8], corresponds in fact to full quadratic polynomial approximations of the stress function vector in triangular regions (linearly varying bending moments and uniform Kirchhoff shear loads). As was first pointed out by J. Robinson, [13], it may be subjected to an isoparametric transformation, using the same interpolation functions (associated for instance to the nodal values at vertices and mid-edges) for in-plane displacements, leading to parabolically curved boundaries without loss of diffusivity. This statement remains true for complete polynomial approximations of higher degree.

8. DISCRETIZATION OF ROTATIONAL EQUILIBRIUM

The property of symmetry of the stress tensor, resulting from local rotational equilibrium, is so firmly taken for granted that little thought was spent up to now on the possibility of relaxing this constraint by discretization, in the same manner as translational equilibrium is relaxed in ordinary displacement elements. Such a relaxation, obtained by a discretization of the material rotation, allows the use of first order stress functions to enforce exact translational equilibrium. This method not only simplifies the construction of stress elements by a reduction to

C_0 continuity but opens the field to three- or two-dimensional stress elements with curved boundaries. It is furthermore applicable to the geometrical non linearity introduced by the consideration of finite elastic displacements [12]. A complete discussion of the two-dimensional case will be given, which is an improved version of an unpublished paper presented in Calgary.

Reverting to the more usual (x,y) notation, the translational equilibrium equations

$$\frac{\partial \sigma_x}{\partial x} + \frac{\partial \tau_{yx}}{\partial y} = 0 \qquad \frac{\partial \tau_{xy}}{\partial x} + \frac{\partial \tau_y}{\partial y} = 0 \tag{47}$$

are solved by using the first order stress function vector (A,B)

$$\sigma_x = \frac{\partial A}{\partial y} \qquad \tau_{yx} = - \frac{\partial A}{\partial x} \qquad \tau_{xy} = \frac{\partial B}{\partial y} \qquad \sigma_y = - \frac{\partial B}{\partial x} \; . \tag{48}$$

From this we find for the surface tractions

$$t_x ds = \sigma_x dy - \tau_{yx} dx = dA \qquad t_y ds = \tau_{xy} dy - \sigma_y dx = dB \; . \tag{49}$$

The rotational equilibrium condition

$$\tau_{xy} - \tau_{yx} = \frac{\partial A}{\partial x} + \frac{\partial B}{\partial y} = 0 \tag{50}$$

is incorporated as a constraint in the complementary energy principle by means of a Lagrangian multiplier $\omega(x,y)$ so that, for given boundary displacements $(\bar{u},\bar{v})$

$$\iint \left\{ \Phi + \omega \left(\frac{\partial A}{\partial x} + \frac{\partial B}{\partial y} \right) \right\} dxdy - \oint \bar{u}\, dA + \bar{v}\, dB \tag{51}$$

should be stationary with respect to arbitrary variations on A, B and ω. The stress energy Φ will be considered as a positive definite function of the arguments $\{\sigma_x, \frac{1}{2}(\tau_{xy} + \tau_{yx}), \sigma_y\}$ so that the constitutive equations (stress-strain relations) will be

$$\varepsilon_x = \frac{\partial \Phi}{\partial \sigma_x} \qquad \varepsilon_{xy} = \frac{\partial \Phi}{\partial \tau_{xy}} = \frac{\partial \Phi}{\partial \tau_{yx}} = \varepsilon_{yx} \qquad \varepsilon_y = \frac{\partial \Phi}{\partial \sigma_y} \quad . \tag{52}$$

It should be observed that the symmetry of the strain tensor is a consequence of the fact that the stress energy is a symmetric function of the stresses, without requiring that (50) be satisfied. Equation (50) is obviously the Euler equation resulting from variations on ω. The Euler equations due to variations on A and B (the arguments of Φ are to be replaced by (48) in the variational principle) are respectively

$$\frac{\partial \omega}{\partial x} = -\frac{\partial \varepsilon_x}{\partial y} + \frac{\partial \varepsilon_{xy}}{\partial x} \qquad \frac{\partial \omega}{\partial y} = -\frac{\partial \varepsilon_{xy}}{\partial y} + \frac{\partial \varepsilon_y}{\partial x} \tag{53}$$

and are recognized to be the Beltrami equations for the integration of ω, that must be identified with the material rotation. After the transformation

$$-\oint \bar{u}\, dA + \bar{v}\, dB = \oint A\, d\bar{u} + B\, d\bar{v}$$

justified by single valuedness of (A,B) and the data $(\bar{u},\bar{v})$ the boundary terms are found to be

$$\oint \delta A \,\{d\bar{u} + (\sin\theta\varepsilon_x - \cos\theta\varepsilon_{xy} + \omega\cos\theta)\, ds\} = 0$$

$$\oint \delta B \,\{d\bar{v} + (\sin\theta\varepsilon_{xy} - \cos\theta\varepsilon_y + \omega\sin\theta)\, ds\} = 0 \quad .$$

By linear combinations, corresponding to the resolution of the stress function vector into its local components A_n and A_t, there follows

$$\cos\theta \frac{d\bar{u}}{ds} + \sin \frac{d\bar{v}}{ds} + \omega = \sin\theta\cos\theta(\varepsilon_y - \varepsilon_x) + (\cos^2\theta - \sin^2\theta)\varepsilon_{xy} = \varepsilon_{nt}$$

$$-\sin\theta \frac{d\bar{u}}{ds} + \cos\theta \frac{d\bar{v}}{ds} = \cos^2\theta\varepsilon_y + \sin^2\theta\varepsilon_x - 2\sin\theta\cos\theta\varepsilon_{xy} = \varepsilon_t \quad .$$

Or, finally, also resolving the boundary displacement vector into its local components,

$$\frac{d\bar{u}_n}{ds} - \dot{\theta}\,\bar{u}_t = \varepsilon_{nt} - \omega \qquad \frac{d\bar{u}_t}{ds} + \dot{\theta}\,\bar{u}_n = \varepsilon_t \tag{54}$$

After substitution of (52) and (48) into (53) and (54) we obtain, together with (50), a system of three partial differential equations for the unknowns (A,B) and ω and two boundary conditions. As, by elimination of ω between equations (53), the compatibility of strains will be satisfied in a simply connected domain, a displacement field will exist with boundary values corresponding to the data.

At an interface between two regions, we must assume that the data for the connectors $(\bar{u},\bar{v})$ are coherent, that is $\bar{u}_n$ and $\bar{u}_t$ must be single-valued. The left-hand sides of (54) are then also single-valued and C_0 continuity of the stress function vector will conduce to single-valuedness of the quantities $(\varepsilon_{nt} - \omega,\ \varepsilon_t)$.

As the variation of ω does not appear in the boundary terms, the continuity of $\omega(x,y)$ is not required at interfaces. Finite elements will now be considered in which the ω field is discretized together with the stress function vector.

9. THE ZERO ENERGY STRESS STATE

As we assume the energy density Φ to be everywhere a positive definite function of its arguments, the stress energy $\iint \Phi \, dx\, dy$ will vanish if and only if

$$\sigma_x \equiv 0 \qquad \sigma_y \equiv 0 \qquad \tau_{xy} + \tau_{yx} \equiv 0 \quad .$$

In terms of first order stress functions this implies

$$\frac{\partial A}{\partial y} = 0 \qquad \text{or} \qquad A = A(x)$$

$$\frac{\partial B}{\partial x} = 0 \qquad \text{or} \qquad B = B(y)$$

and finally

$$\frac{\partial B}{\partial y} = \frac{\partial A}{\partial x} = \gamma \qquad \text{that is} \qquad \tau_{xy} = -\tau_{yx} = \gamma$$

a constant since it must be both a function of y alone and x alone. There is thus one zero energy stress field, or arbitrary uniform intensity, representing in its purest form the violation of rotational equilibrium.

10. GLOBAL ROTATIONAL EQUILIBRIUM ELEMENTS

It will be shown that if A and B are discretized as complete polynomials of degree $n+1 \geq 2$ (n is the degree of the generated stress field) and if the Lagrangian multiplier (the material rotation ω) is reduced to a constant $\omega(x,y) = \theta$ (so that the element will only satisfy global rotational equilibrium), the stiffness matrix of simple triangular element is well-behaved. For definiteness some of the characteristics of the element will be explicited in the lowest degree case $n = 1$.

For this case

$$A = \alpha_0 + \alpha_1 x + \alpha_2 y + \alpha_3 x^2 + 2\alpha_4 xy + \alpha_5 y^2$$

$$B = \beta_0 + \beta_1 x + \beta_2 y + \beta_3 x^2 + 2\beta_4 xy + \beta_5 y^2 \quad .$$

In forming a stress parameter vector s, we keep only the active parameters, discarding the "idle" parameters α_0 and β_0 (that produce no stresses). We may chose for instance

$$s^T = (\alpha_1, \alpha_2 \ldots \alpha_5 \mid \beta_1 \ldots \beta_5)$$

and note that the vector s_0 associated to the zero energy case is

then

$$s_0^T = \frac{1}{2\Delta}(1 \quad 0 \quad 0 \quad 0 \quad 0 \mid 0 \quad 1 \quad 0 \quad 0 \quad 0) \quad . \tag{55}$$

The convenience of having it normed so that

$$\alpha_1 = \beta_2 = \gamma = \frac{1}{2\Delta} \qquad \Delta = \iint_E dx\, dy$$

where Δ is the area of the triangular element, will appear later. For linear elastic constitutive equations we will have

$$\iint_E \Phi\, dx\, dy = \frac{1}{2} s^T F s \tag{56}$$

where, on account of the existence of a zero energy state, the flexibility matrix will not be positive definite but merely non negative. In fact (Appendix A) $s_0^T F s_0 = 0$ induces

$$F s_0 = 0 \tag{57}$$

and s_0 will be the only (normed) non trivial solution of the homogeneous equation $F s = 0$.

A constant multiplier enforces global rotational equilibrium of the element as can be seen from

$$\iint_E \left(\frac{\partial A}{\partial x} + \frac{\partial B}{\partial y}\right) dx \quad dy = \oint Ady - Bdx = \oint xdB - ydA = \oint (xt_y - yt_x) ds=0 \tag{58}$$

In the discretization

$$\iint_E \left(\frac{\partial A}{\partial x} + \frac{\partial B}{\partial y}\right) dxdy = w^T s \tag{59}$$

a linear norm in the active stress parameters; explicitly for $n = 1$

$$w^T = (\Delta \quad 0 \quad 2\iint_E x\,dxdy \quad 2\iint_E y\,dxdy \quad 0 \mid 0 \quad \Delta \quad 0 \quad 2\iint_E x\,dxdy \quad 2\iint_E y\,dxdy) \tag{60}$$

From (60) and (55) follows the essential property

$$w^T s_0 = 1 \tag{61}$$

that obviously holds for arbitrary polynomial degree n.

To obtain the complete discretized form of the variational principle (51) we must still express

$$\oint \bar{u}\,dA + \bar{v}\,dB = g^T q$$

in terms of a boundary loads vector g with attendant virtual work definitions of conjugate displacements, collected in the (weak) boundary displacements vector q.

The definition of boundary loads follows the pattern of equilibrium models. Reciprocity of boundary loads on each interface the element can have with a neighbour must entail complete diffusivity. As an example for the case $n = 1$ we can define the loads g_{X12} and g_{X21} of the boundary 12 (Figure 4) as the resultants of the two linear interpolation functions of the surface traction t_x. By virtual work the conjugate displacements are

$$u_{12} = \frac{2}{c_{12}} \int_1^2 \left(1 - \frac{s}{c_{12}}\right) u(s)ds \qquad u_{21} = \frac{2}{c_{12}} \int_1^2 \frac{s}{c_{12}}\, u(s)ds$$

$$c_{12} = \int_1^2 ds$$

where s is measured along the boundary from 1 towards 2.

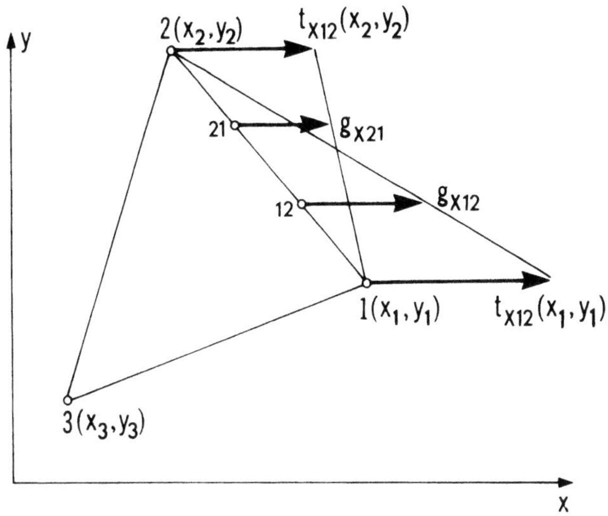

Figure 4

Similar definitions are valid for g_{y12} and g_{y21}, to define uniquely the linearly varying t_y traction, and for the conjugate vertical displacements v_{12} and v_{21}.

From the very definitions adopted for the boundary loads one obtains their value in terms of the active stress parameters

$$g = S\ s \tag{62}$$

whence finally

$$\oint \bar{u}\ dA + \bar{v}\ dB = q^t\ S\ s \tag{63}$$

where, applying the variational principle, it must be considered that the vector q is prescribed. Collecting the results (63), (59) and (56) the stationarity of

$$\frac{1}{2}\, s^T\ F\ s + \theta\ w^T\ s - q^T\ S\ s$$

with respect to the choice of s and θ, implies

$$F\,s + \theta\,w = S^T\,q \tag{64}$$

$$w^T\,s = 0\ .$$

Although F is singular, the banded symmetrical matrix of the complete linear system

$$\begin{pmatrix} F & w \\ w^T & 0 \end{pmatrix} \begin{pmatrix} s \\ \theta \end{pmatrix} = \begin{pmatrix} S^T q \\ 0 \end{pmatrix} \tag{65}$$

is non singular.

Indeed the homogeneous system

$$F\,s + \theta\,w = 0 \qquad w^T\,s = 0$$

possesses only the trivial solution as can be seen by premultiplying the first equation by s_0^T, whereby, on account of (57) and (61), we must have $\theta = 0$. Then $F\,s = 0$ has a non trivial solution $s = s_0$, but it fails to satisfy the second equation.

Thus the reciprocal matrix exists and is of the form

$$\begin{pmatrix} F & w \\ w^T & 0 \end{pmatrix}^{-1} = \begin{pmatrix} F^{\#} & v \\ v^T & \mu \end{pmatrix} \qquad F^{\#} = (F^{\#})^T\ .$$

As a right inverse its elements must satisfy

$$F\,F^{\#} + w\,v^T = I \qquad F\,v + \mu\,w = 0 \qquad w^T\,v = 1$$

and we find that $\mu = 0$ and $v = s_0$

so that

$$F\,F^{\#} + w\,s_0^{T} = I \qquad F^{\#}\,w = 0 \tag{66}$$

and the solution is

$$s = F^{\#}\,S^{T}\,q \qquad \theta = s_0^{T}\,S^{T}q\;. \tag{67}$$

In practice, of course, the pseudo-inverse $F^{\#}$ of F is obtained by numerical inversion of the banded matrix. The second of equations (67) gives the average rotation experienced by the element as a result of the given boundary displacements. When the first is substituted into (62) we obtain the stiffness matrix of the element

$$g = K\,q \qquad K = S\,F^{\#}\,S^{T} \tag{68}$$

and it remains to verify that it is well-behaved in the sense that the only solutions of

$$K\,q = 0 \tag{69}$$

are the rigid body displacement modes of the element.

As will be shown, the translational rigid body modes stem from the solutions of $S^{T}q = 0$; the rotational mode from $F^{\#}S^{T}q = 0$.

11. SELF-STRESSING STATES AND THE SOLUTIONS OF $S^{T}q = 0$

The stresses generated by A and B satisfy identically the local translational equilibrium equations. As a consequence the boundary loads g should also be in global translational equilibrium. By virtual work this can be formulated as

$$g^T q_x = 0 \qquad g^T q_y = 0 \quad \text{for any s vector}$$

where q_x and q_y are rigid body translation modes. In view of (62) this shows that

$$S^T q_x = 0 \qquad S^T q_y = 0 \tag{70}$$

are solutions of $S^T q = 0$. A direct proof also follows from (63) by inputting $\bar{u} = 1$ and $\bar{v} = 0$ and observing that the left hand side vanishes for any single valued A and similarly for $\bar{u} = 0$, $\bar{v} = 1$ and single valued B.

There are no other solutions (linearly independent) of $S^T q = 0$ than those translation modes as can be seen by counting the number of solutions n(x) of the adjoint problem

$$S\ s = 0 \quad . \tag{71}$$

As S is a n(g) x n(s) matrix we know from algebra that the number of linearly independent solutions of $S^T q = 0$ is

$$m = n(x) + n(g) - n(s)$$

and must be shown to be equal to 2.

Problem (71) is the problem of self-stressing states of the element; it enquires about the active stress parameters that would produce a state of stress whose surface tractions vanish at the boundary.

From (49) such a situation implies

$$dA = 0 \qquad dB = 0 \quad \text{along the boundary.}$$

If we add the idle parameters α_0 and β_0, so as to form complete polynomials for the stress functions, there is no restriction in setting

$$A = 0 \qquad B = 0 \quad \text{along the boundary}$$

as characteristic equations describing the self-stressing states. In polynomial approximations, using areal coordinates for simplicity of presentation, each stress function corresponding to a self-stressing state will be of the form

$$L_1\ L_2\ L_3\ P(L_1,\ L_2,\ L_3)$$

where P is a linear combination of all monomials in $(L_1,\ L_2,\ L_3)$ of degree $n+1-3 = n-2$. In other terms P will be a complete polynomial in (x,y) of degree $n-2$ containing $\frac{1}{2}\ n(n-1)$ terms. Since there are two stress functions the total number of different self-stressing states is

$$n(x) = n(n-1) \ .$$

We have none in the model with linearly varying stresses, then 2 for parabolically varying stresses, 6 for quartic variations, etc... The number of boundary loads is $2(n+1)$ per side, hence

$$n(g) = 6(n+1) \ .$$

The number of active stress parameters

$$n(s) = (n+2)(n+3) - 2$$

so that in the end we find indeed $m = 2$, irrespective of the degree n of the stress variations.

The $n(n-1)$ linearly independent solutions $s=x_{(i)}$ of (71) have the important property

$$S\ x_{(i)} = 0 \rightarrow w^T\ x_{(i)} = 0 \quad . \tag{72}$$

This follows from writing (59) in the form

$$w^T\ s = \iint_E \frac{\partial A}{\partial x} + \frac{\partial B}{\partial y} \quad dxdy = \oint Ady - Bdx$$

and remembering that A = 0 and B = 0 along the boundaries when s is a self-stressing $x_{(i)}$.

12. THE ROTATIONAL RIGID BODY MODE

Equations (67) show $F^{\#}$ to be non negative. For, premultiplying the first by $F^{\#}$ and taking the second into account, we obtain

$$F^{\#}\ F\ F^{\#} = F^{\#} \quad . \tag{73}$$

Hence, for an arbitrary vector e,

$$e^T\ F^{\#}\ e = (F^{\#}\ e)^T\ F(F^{\#}e) \geq 0 \quad \text{since F is non negative.}$$

This important property, in turn, shows that the solutions of

$$K\ q = S\ F^{\#}\ S^T\ q = 0$$

for which

$$q^T\ K\ q = (S^Tq)^T\ F^{\#}(S^Tq) = 0$$

are (theorem of Appendix A) those of the linear system

$$F^{\#}\ S^Tq = 0 \quad . \tag{74}$$

We have already seen that $S^Tq = 0$ furnishes only the two translational modes. As it again follows from equations (67), w is, up to arbitrary scale factor, the only solution of $F^{\#} e = 0$, so that the other solutions of (69) depend on those of

$$S^T q = w \quad . \tag{75}$$

The existence conditions for a particular solution of this singular problem are precisely satisfied by (72), the general solution consisting merely in adding arbitrary translations. The nature of any particular solution is clear from the solution (67) of the variational problem: there are no stresses involved (s=0) but there is a uniform rotation of the element

$$\omega = \theta = s_0^T w = 1 \quad .$$

This achieves the proof that the stiffness matrix is well behaved.

13. IMPROVEMENT OF ROTATIONAL CONDITIONING FOR n > 1

When the degree of polynomial approximation increases above n=1 a lack of balance will develop between the rigorous satisfaction of translational equilibrium and the minimum acceptable state of rotational equilibrium. On the other hand, a too generous improvement of rotational equilibrium may lead to the appearance of mechanisms. This is well known since the earliest attempts at constructing stress elements satisfying rigorously all equilibrium conditions. Taking the case n=1 and implementing exactly equation (50), which is equivalent to the use of an Airy second order stress function represented by a complete cubic, we are left with 7 active parameters only ($\alpha_1 = \beta_2, \alpha_3 = \beta_4, \alpha_4 = \beta_5$).

Thus, as n(s) = 7, n(g) = 12 and n(x) = 0, there follows

$$m = 5 \quad .$$

On the other hand F is now positive definite (no non trivial zero energy state).

Three of the solutions of $S^T q = 0$ are the rigid body modes, two other solutions are mechanisms.

In the general case n > 1, the degree of the Airy function is n+2 so that, discounting its 3 idle parameters (those of the linear part),

$$n(s) = \frac{(n+3)(n+4)}{2} - 3$$

$$n(g) = 6(n+1) \quad \text{as before} .$$

The self-stressing states are now given by Airy functions that are zero together with their normal derivative at the boundary, hence of type

$$\phi = L_1^2 \, L_2^2 \, L_3^2 \, P(L_1, L_2, L_3)$$

P is a complete polynomial of degree n+2-6 = n-4

and $$n(x) = 0 \qquad \text{for } n \leq 3$$

$$n(x) = \frac{1}{2}(n-3)(n-2) \qquad \text{for } n > 3 .$$

In all cases n > 1 it is found that

$$m = 6 \ .$$

Hence, in addition to the 3 rigid body modes, there are exactly always 3 mechanisms. They may be shown to be of the "floating corner" type.

Although one can live with such mechanisms and inhibit them, either by the composite element technique, or by imposing suitable constraints at the assembled level, it appears more satisfactory to avoid them, inasmuch as, by sticking to first order stress functions we do not thereby inhibit the extension to isoparametricity.

Let us now increase the rotational conditioning of the element by taking as multiplier a complete polynomial of degree r. It is easily seen that for n = 1 the use of a complete linear polynomial (r=1) is equivalent to rigorous rotational equilibrium and would introduce mechanisms. Hence we will impose the restriction $r \leq n-1$ and show that the stiffness matrix remains well behaved. Thus rotational conditioning can be improved in parallel with the improvement of compatibility conditions resulting of a choice of higher degrees of n.

The constraints on the active stress parameters will now be

$$\iint_E \left(\frac{\partial A}{\partial x} + \frac{\partial B}{\partial y}\right) \omega(x,y)\ dxdy = s^T W h = 0 \qquad (76)$$

for an arbitrary vector h, listing the parameters (coefficients) of the polynomial $\omega(x,y)$. The following properties may be proved

1) $W\ h = 0 \rightarrow h = 0$ (77)

that is to say, the columns of matrix W are linearly independent. If they are not, some non identically zero polynomial $\omega(x,y)$ would exist such that

$$\iint_E \left(\frac{\partial A}{\partial x} + \frac{\partial B}{\partial y}\right) \omega\ dxdy = 0$$

for arbitrary polynomials A and B of degree $n+1 \geq r+2$. The result should hold in particular for the choice

$$\frac{\partial A}{\partial x} + \frac{\partial B}{\partial y} = \omega$$

because polynomials A and B of degree r+2 can be so chosen that the terms of degree r+1 on the left-hand side

cancel, leaving an arbitrary polynomial of degree r.
For such a choice however

$$\iint_E \omega^2 \, dxdy = 0 \rightarrow \omega \equiv 0$$

and the contradiction proves the point.

2) The matrix of the new variational equations to be solved

$$\begin{pmatrix} F & W \\ W^T & 0 \end{pmatrix} \begin{pmatrix} s \\ h \end{pmatrix} = \begin{pmatrix} s^T q \\ 0 \end{pmatrix}$$

is non singular.
The proof consists again in showing that the homogeneous problem

$$F\,s + W\,h = 0 \qquad W^T\,s = 0$$

has only the trivial solution. Premultiplying the first equation by s^T and using the second equation we must have

$$s^T\,F\,s = 0$$

and, since F is non negative,

$$F\,s = 0 \quad .$$

We cannot adopt the solution $s = s_0$ (the only non trivial one because

$$W^T\,s_0 \neq 0$$

as one of the columns of W, or a linear combination of them, is our former w and $w^T s_0 = 1$. Hence we must take s=0. The first equation then reduces to W h = 0 and can only be satisfied by h = 0, as seen under 1).

The matrix has thus an inverse

$$\begin{pmatrix} F^{\#} & V \\ V^T & Q \end{pmatrix} \qquad (F^{\#})^T = F^{\#} \qquad Q^T = Q$$

such that

$$F\,F^{\#} + W\,V^T = I \qquad FV + W^T Q = 0 \qquad F^{\#} W = 0 \qquad W^T V = I \tag{78}$$

and the solution of the variational equations is

$$s = F^{\#}\,S^T\,q \qquad h = V^T\,S^T\,q\;. \tag{78}$$

The last result furnishes the information about the rotation of the element.

Again from equations (78) we obtain the characteristic pseudo-inverse property

$$F^{\#}\,F\,F^{\#} = F^{\#}$$

and $F^{\#}$ (not the same as the former one) is still non negative.

3) As a result of the last property we are again led to investigate the q vectors satisfying

$$S^T\,q = W\,h \tag{79}$$

as representing the rigid body modes or mechanisms of

problem

$$K\,q = S\,F^{\#}\,S^T\,q = 0$$

distinct from the two translational modes. Equation (79) is in fact one of the variational equations for $s = 0$.

The uniform rotation mode is certainly a solution, since $WH = w$ for a suitable choice of h and the existence conditions of a particular solution are then again implemented by (72). Other solutions are undesirable, and the stiffness matrix will be well behaved if at least one of the existence conditions

$$x^T_{(i)}\,W\,h = 0 \tag{80}$$

is violated for each other (linearly independent) choice of h. Given n, this will be shown to be true when r has its maximal value n-1 and, a fortiori, if it is taken to be smaller.

Let $n = 2$ and $r = 1$. We must prove that in each of the cases $\omega = x$ and $\omega = y$ (corresponding to choices of h linearly independent of the choice $\omega = 1$) a self-stressing state (A and B vanishing at the boundaries) can be found such that

$$\iint_E \left(\frac{\partial A}{\partial x} + \frac{\partial B}{\partial y}\right) \omega\, dxdy = - \iint_E \left(A\,\frac{\partial \omega}{\partial x} + B\,\frac{\partial \omega}{\partial y}\right) dxdy \tag{81}$$

does not vanish.

The polynomial degree of A and B is here at most equal to 3, and we dispose of only two different self-stressings. For $\omega=x$ we select

$$A = L_1\,L_2\,L_3$$

and obtain for the right-hand side of (81)

$$-\iint_E L_1\ L_2\ L_3\ dxdy < 0$$

since the bubble function $L_1\ L_2\ L_3$ is positive at every interior point of E. The same bubble function is used for B in the case ω=y with the same result.

Let now n=3 and r=2. The preceding choices of self-stressings remain valid for the same cases but we must now investigate in addition the cases $\omega = x^2$, xy and y^2. On the other hand the degree of A and B may be raised to 4.

For $\omega = x$, select $A = x\ L_1\ L_2\ L_3$ and obtain

$$-\ 2 \iint_E x^2\ L_1\ L_2\ L_3\ dxdy < 0\ .$$

For $\omega = xy$, take $A = y\ L_1\ L_2\ L_3$ and $B = x\ L_1\ L_2\ L_3$, to find

$$-\iint_E (x^2 + y^2)\ L_1\ L_2\ L_3\ dxdy < 0\ .$$

Finally for $\omega=y^2$ the choice $B = y\ L_1\ L_2\ L_3$ gives

$$-\iint_E y^2\ L_1\ L_2\ L_3\ dxdy < 0\ .$$

This systematic technique can obviously be extended indefinitely to higher values of n and furnishes the proof that the stiffness matrix remains well behaved for $r \leq n-1$.

In conclusion it may be said that the existence of a new family of two-dimensional membrane elements has been established. They share with the pure equilibrium elements the properties of rigorous translational equilibrium and diffusivity at the boundaries. They have the advantage of being formulated in terms of first order stress functions only, with simple C_0 continuity requirements and the attendant possibilities of isoparametric coordinate transformations to curved boundaries. They distinguish themselves from pure equilibrium models by the fact that they satisfy only average rotational equilibrium equations. The degree of enforcement of the rotational equilibrium may follow one step behind the degree of approximation of the stress distribution or may be lower, leaving a trade-off possibility between rotational equilibrium and strain compatibility.

14. AN ALTERNATIVE APPROACH TO THE CONSTRUCTION OF THE STIFFNESS MATRIX

Take the case n = 1. As the stresses are known to vary linearly, they are uniquely determined by the 3 corner values of the unsymmetrical stress tensor

$$\begin{pmatrix} \sigma_x & \tau_{yx} \\ \tau_{xy} & \sigma_y \end{pmatrix} .$$

Consider then the stress vector σ containing this information

$$\sigma^T = (\sigma_{x_1}\ \sigma_{x_2}\ \sigma_{x_3}\ \tau_{yx_1}\ \tau_{yx_2}\ \tau_{yx_3}\ \tau_{xy_1}\ \tau_{xy_2}\ \tau_{xy_3}\ \sigma_{y_1}\ \sigma_{y_2}\ \sigma_{y_3}) .$$

The information enables the surface tractions and consequently the generalized boundary (interface) loads to be computed. It furnishes a 12 × 12 matrix G such that

$$g = G\ \sigma .$$

Up to now equilibrium was disregarded entirely. From the rigid body modes

$$q = R\ r \qquad\qquad R \text{ a } 12 \times 3 \quad r \text{ arbitrary}$$

of the element, the global equilibrium conditions are

$$R^T g = R^T G\sigma = 0 \quad .$$

They are enforced by means of a Lagrange vector multiplier λ in the complementary energy principle

$$\frac{1}{2}\ \sigma^T\ \Phi\ \sigma + \lambda^T\ R^T\ G\ \sigma - q^T\ G\ \sigma \qquad \text{minimum.}$$

The minimizing equations are

$$\begin{pmatrix} \Phi & G^T R \\ R^T G & 0 \end{pmatrix} \cdot \begin{pmatrix} \sigma \\ \lambda \end{pmatrix} = \begin{pmatrix} G^T q \\ 0 \end{pmatrix}$$

and, from the previous theory, the matrix is invertible, yielding

$$\sigma = \Phi^{\#}\ G^T q$$

and finally

$$g = K\ q \quad \text{with} \quad K = G\ \Phi^{\#}\ G^T \quad .$$

No consideration was given here to the existence of the first order stress functions, but it is easily established that the translational equilibrium equations are rigorously satisfied. Starting from the divergence theorem

$$\int (\sigma_x \frac{\partial u}{\partial x} + \tau_{yx} \frac{\partial u}{\partial y} + \tau_{xy} \frac{\partial v}{\partial x} + \sigma_y \frac{\partial v}{\partial y})\ dx\ dy$$

$$= \oint (ut_x + vt_y)\ ds - \int \left[u\left(\frac{\partial \sigma_x}{\partial x} + \frac{\partial \tau_{yx}}{\partial y}\right) + v\left(\frac{\partial \tau_{xy}}{\partial x} + \frac{\partial \sigma_y}{\partial y}\right)\right] dx\ dy.$$

We note that the volume forces equilibrating the stresses

$$\frac{\partial \sigma_x}{\partial x} + \frac{\partial \tau_{xy}}{\partial y} = X \qquad \frac{\partial \tau_{xy}}{\partial x} + \frac{\partial \sigma_y}{\partial y} = Y$$

are at most constants. Thus if u and v are arbitrary constants

$$0 = u \oint t_x \, ds + v \oint t_y \, ds - (u X + v Y) \int dx\, dy \quad .$$

However, since global equilibrium was enforced,

$$\oint t_x \, ds = 0 \qquad \oint t_y \, ds = 0$$

and the result holds for arbitrary u and v is and only if $X = 0$ $Y = 0$. This proves that local equilibrium holds for translation. For rotation we start from the identity

$$\int [\frac{\partial}{\partial x} (x\, \tau_{xy} - y\, \sigma_x) + \frac{\partial}{\partial y} (x\, \sigma_y - y\, \tau_{yx})] \, dx\, dy = \oint (x\, t_y - y\, t_x) ds.$$

The right hand side is zero when global rotational equilibrium is enforced. Evaluating the left hand side we then find

$$\int (\tau_{xy} - \tau_{yx}) \, dx\, dy + \int (x\, Y - y\, X) \, dx\, dy = 0$$

and, since $X = 0$ and $Y = 0$, find, as before, that rotational equilibrium is only satisfied in the average.

15. RECTANGULAR ELEMENTS WITH DISCRETIZED ROTATIONAL EQUILIBRIUM

The theory established in sections 8 to 13 applies as well to elements of basic rectangular geometry. For definiteness we take the example of a rectangle with boundaries $x = \pm \ell$, $y = \pm h$ in which the unsymmetrical stress tensor is generated by bi-quadratic first order stress functions

$$A = \alpha_0 + \alpha_1 x \quad \alpha_2 y \quad + \alpha_3 xy + \alpha_4 x^2 + \alpha_5 y^2 + \alpha_6 xy^2 + \alpha_7 x^2 y + \alpha_8 x^2 y^2$$

$$B = \beta_0 + \beta_1 x + \beta_2 y + \beta_3 xy + \beta_4 x^2 + \beta_5 y^2 + \beta_6 xy^2 + \beta_7 x^2 y + \beta_8 x^2 y^2 \quad .$$

The stress tensor

$$\sigma_x \frac{\partial A}{\partial y} = \alpha_2 + \alpha_3 x + 2\alpha_5 y + 2\alpha_6 xy + \alpha_7 x^2 + 2\alpha_8 x^2 y^2$$

$$-\tau_{xy} = \frac{\partial A}{\partial x} = \alpha_1 + \alpha_3 y + 2\alpha_4 x + 2\alpha_6 y^2 + 2\alpha_7 xy + 2\alpha_8 xy^2$$

$$\tau_{xy} = \frac{\partial B}{\partial y} = \beta_2 + \beta_3 x + 2\beta_5 y + 2\beta_6 xy + \beta_7 x^2 + 2\beta_8 x^2 y$$

$$-\sigma_y = \frac{\partial B}{\partial y} = \beta_1 + \beta_3 y + 2\beta_4 x + 2\beta_6 y^2 + 2\beta_7 xy + 2\beta_8 xy^2$$

presents linear variations only along the boundaries $x = \pm \ell$, $y = \pm h$.

The active stress parameters are sequenced in the vector s

$$s^T = (\alpha_1 \ldots \alpha_8 \mid \beta_1 \ldots \beta_8) \qquad n(s) = 16 \quad .$$

The stress energy is again a quadratic form $\frac{1}{2}\, s^T F s$ with a non negative flexibility matrix possessing as unique non trivial solution to $Fs = 0$, the vector

$$s_0^T = (\gamma\ 0 \ldots 0 \mid 0\ \gamma \ldots 0) \ .$$

Along each boundary we have 4 generalized loads, the 16 generalized loads vector g being related to the active stress parameters by

$$g = S\ s \qquad S \text{ a } 16 \times 16 \text{ matrix.}$$

However, there are two self-stressing states

$$A = \lambda(x^2 - \ell^2)\ (y^2 - h^2)$$

$$B = \mu(x^2 - \ell^2)\,(y^2 - h^2)$$

and the generalized loads depend really on 16 - 2 = 14 parameters only. In fact the parameters α_8 and β_8 could be replaced by λ and μ and, g being independent of λ and μ, the S matrix would have corresponding zero columns.

Thus of the 16 generalized loads, 14 only are independent, a situation coherent with the fact that they already automatically satisfy 2 global equilibrium conditions in translation. Any additional constraint imposed on the parameters, like the one resulting from

$$\int(\tau_{xy} - \tau_{yx})\; dx\, dy = 0$$

will also be reflected as a constraint on the generalized loads, here the global rotational equilibrium constraint. Furthermore, as there are two self-stressings we may hope to improve, if we so desire, the rotational equilibrium. The situation is not fundamentally different from the previous one. We must verify whether our choice of $\omega(x,y)$ is such that, according to (81)

$$\iint_E \; A\left(\frac{\partial\omega}{\partial x} + B\,\frac{\partial\omega}{\partial y}\right) dx\, dy = 0$$

is violated for at least one of the self-stressings.

This is again the case for the choices $\omega = x$ or $\omega = y$, since

$$\iint_E \,(x^2 - \ell^2)\;(y^2 - h^2\,)\; dx\, dy < 0 \quad .$$

16. ISOPARAMETRIC COORDINATE TRANSFORMATIONS

Let (ξ,η,ζ) be global cartesian coordinates of a "primal" mesh of finite elements with plane boundaries. In polynomial approximations, the first order stress functions are piecewise analytic functions of (ξ,η,ζ), C_0 continuous at the interfaces. Consider now any isoparametric coordinate transformation from (ξ,η,ζ) to (x,y,z) using displacement type interpolation functions $P_i(\xi,\eta,\zeta)$

$$x = \xi + \Sigma\, \hat{u}_i\, P_i\, (\xi,\eta,\zeta)$$

$$y = \eta + \Sigma\, \hat{v}_i\, P_i\, (\xi,\eta,\zeta)$$

$$z = \zeta + \Sigma\, \hat{w}_i\, P_i\, (\xi,\eta,\zeta)$$

to preserve continuity of the continuum by conformity of the fixed displacements

$$(\hat{u}_i)_+ = (\hat{u}_i)_- \qquad (\hat{v}_i)_+ = (\hat{v}_i)_- \qquad (\hat{w}_i) + (\hat{w}_i)_-$$

across interfaces.

Considering (x,y,z) as new global cartesian coordinates, we obtain the "dual" mesh with curved boundaries and interfaces. The coordinate transformation is continuous at the interfaces and, considered as functions of (x,y,z), the first order stress functions remain C_0 continuous so that diffusivity is maintained. Thus there is no need whatsoever to impose to the interpolation functions P_i (ξ,η,ζ) the need to be the same as the interpolation functions used in the primal mesh to obtain C_0 continuity of the first order stress functions.

As already observed earlier, the need of C_1 continuity preservation when second order stress functions are involved, may warrant a closer investigation of whether this is at all possible by relating the interpolation functions.

REFERENCES

[1] FINZI, B., "Integratione delle equazione indefinite della meccanica dei sistemi continui", *Rend. Linc.*, 19, 1934.

[2] SOUTHWELL, R.V., "On the analogues relating flexure and extension of flat plates", *Qu. J. Mech. Appl. Math.*, 3, 1950.

[3] SCHAEFER, H., "Die Vollständige Analogie Scheibe-Platte", *Abh. Braunschwg. Wissensch. Ges.*, 8, 1956.

[4] SCHAEFER, H., "Die Drei Spannungsfunktionen des Zweidimensionalen Ebenen Kontinuums", *Oesterr. Ing. Archiv.*, X-2-3, 1956.

[5] FRAEIJS de VEUBEKE, B. and ZIENKIEWICZ, O.C., "Strain energy bounds in finite element analysis by slab analogy", *J. of Strain*, 2n°4, 1967.

[6] FRAEIJS de VEUBEKE, B., "Basis of a well conditioned force program for equilibrium models via the Southwell slab analogies", *Technical Report AFFDL-TR-67-10*, Wright-Patterson A.F.B., 1967.

[7] MORLEY, L.S.D., "A triangular equilibrium element with linearly varying bending moments for plate bending problems", *J. Roy. Aer. Soc.*, 71, 1967.

[8] FRAEIJS de VEUBEKE, B. and SANDER, G., "An equilibrium model for plate bending", *Int. J. Solids and Structures*, 4, 1968.

[9] SANDER, G., "Application of the dual analysis principle", *Proceedings IUTAM Conference on High Speed Computing of Elastic Structures*, Congrès et Colloques de l'Université de Liège, Place du XX Août, 16, 4000 Liège, 1970.

[10] FRAEIJS de VEUBEKE, B., "Static-geometric analogies", *Lectures on finite element methods in continuum mechancis*, Eds. Oden and Oliveira, University of Alabama in Huntsville, 1973.

[11] BECKERS, P., "Les fonctions de tension dans la méthode des éléments finis", *Coll. Publ. Fac. Sc. Appl.*, n° 41, Univ. Liège, 1973.

[12] FRAEIJS de VEUBEKE, B., "A new variational principle for finite elastic displacements", *Int. J. of Eng. Sc.*, 1972.

[13] ROBINSON, J., "Basis for isoparametric stress elements", *Computer Meth. in Appl. Mech. and Eng.*, 2, 1973.

[14] ROBINSON, J., *Integrated Theory of Finite Element Methods*, J. Wiley and Sons, 1973.

[15] FRAEIJS de VEUBEKE, B., "Variational principles and the patch test", *Int. J. Num. Meth. in Eng.*

APPENDIX A

A property of non negative matrices

In section 10 and later, use is made of the property that for a non negative matrix F

$$s^T F\ s = 0 \rightarrow F\ s = 0\ .$$

A simple proof is as follows. Suppose that a non zero vector s satisfies

$$s^T\ F\ s = 0 \tag{A.1}$$

and set

$$F\ s = x\ . \tag{A.2}$$

Then, since F is non negative,

$$(x + \lambda\ s)^T\ F\ (x + \lambda\ s) \geq 0 \qquad \text{for any scalar } \lambda\ .$$

Expanding and using (A.1) and (A.2)

$$x^T\ F\ x + 2\ \lambda\ x^T\ x \geq 0\ .$$

This can hold for arbitrary λ if and only if

$$x^T\ x = 0 \rightarrow x = 0$$

and the property is demonstrated.

APPENDIX B

Invariance of local rotational unbalance

The measure $\tau_{xy} - \tau_{yx}$ by which local rotational equilibrium is violated is not related to a particular orientation of cartesian axes. It is in fact invariant under a rotation of the axes.

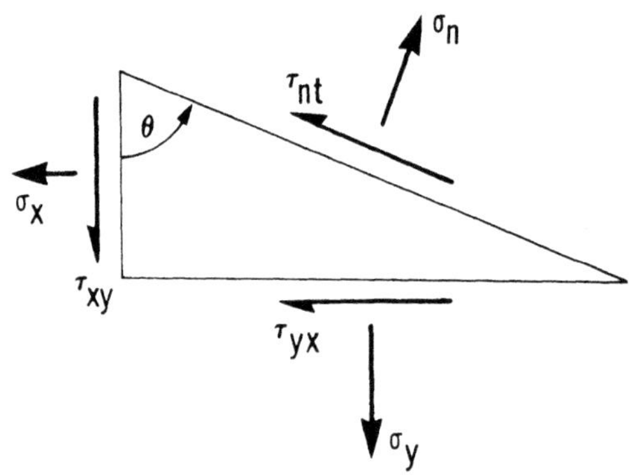

Equilibrium of the elementary triangular prism yields

$$\sigma_n = \cos^2\theta\ \sigma_x + \sin^2\theta\ \sigma_y + \sin\theta\cos\theta\ (\tau_{yx} + \tau_{xy}) \tag{B.1}$$

$$\tau_{nt} = \sin\theta\cos\theta\ (\sigma_y - \sigma_x) + \cos^2\theta\ \tau_{xy} - \sin^2\theta\ \tau_{yx}\ . \tag{B.2}$$

Increasing θ by $\pi/2$

$$-\tau_{tn} = \sin\theta\cos\theta\ (\sigma_x - \sigma_y) + \sin^2\theta\ \tau_{xy} - \cos^2\theta\ \tau_{yx}\ . \tag{B.3}$$

From these relations follow the classical tensorial transformations when the average shear stress is used

$$\sigma_n = \cos^2\theta\ \sigma_x + \sin^2\theta\ \sigma_y + 2\sin\theta\cos\theta\ \frac{\tau_{yx} + \tau_{xy}}{2}$$

$$\frac{\tau_{nt} + \tau_{tn}}{2} = \sin\theta\cos\theta\ (\sigma_y - \sigma_x) + (\cos^2\theta - \sin^2\theta)\ \frac{\tau_{yx} + \tau_{xy}}{2} .$$

Also the announced invariance property

$$\tau_{nt} - \tau_{tn} = \tau_{xy} - \tau_{yx} . \tag{B.5}$$

In the three-dimensional case we have the tensorial transformation

$$\sigma_{\alpha\beta} = T_{\alpha i}\ T_{\beta j}\ \sigma_{ij} \tag{B.6}$$

for the unsymmetrical stress tensor. The symmetrical part

$$\frac{1}{2}\ (\sigma_{\alpha\beta} + \sigma_{\beta\alpha}) = T_{\alpha i}\ T_{\alpha j}\ \frac{1}{2}\ (\sigma_{ij} + \sigma_{ji}) \tag{B.7}$$

satisfies the same transformation, while also

$$\sigma_{\alpha\beta} - \sigma_{\beta\gamma} = T_{\alpha i}\ T_{\alpha j}\ (\sigma_{ij} - \sigma_{ji}) . \tag{B.8}$$

The pseudo-vector of strict components of this antisymmetrical part is obtained through multiplication by the alternating tensor $\frac{1}{2}\ e_{\alpha\beta\gamma}$

$$\frac{1}{2}\ e_{\alpha\beta\gamma}\ (\sigma_{\alpha\beta} - \sigma_{\beta\alpha}) = a_\gamma$$

or, in detail,

$$a_1 = \sigma_{23} - \sigma_{32}\ ,\ a_2 = \sigma_{31} - \sigma_{13}, a_3 = \sigma_{12} - \sigma_{21} .$$

Noting that

$$e_{\alpha\beta\gamma}\, T_{\alpha i}\, T_{\beta j} = (\det T)\, e_{ijk}\, T_{\gamma k}$$

relation (B.8) is transformed in

$$a_{\gamma} = (\det T)\, T_{\gamma k}\, a_k \tag{B.9}$$

showing that under proper rotations (det T = 1) of the cartesian axes the rotational unbalance tensor a_k behaves as a polar (invariant) vector.

APPENDIX C

Boundary load constraints and stress parameter constraints

We can investigate somewhat differently the relationship between homogeneous boundary loads constraints, that lead to desirable global equilibrium of the element but also to undesirable mechanisms, and stress parameter constraints, that may improve rotational equilibrium of the stress field.

The relation between loads and active parameters

$$g = S\, s \tag{C.1}$$

involves the following classical algebraical equality

$$m = n(g) - n(y) = n(s) - n(x) \tag{C.2}$$

between the number n(g) of boundary loads, n(s) of active stress parameters, n(x) of independent self-stressings, solutions of the homogeneous problem

$$S\ s = 0 \qquad s = x_{(i)} \tag{C.3}$$

and n(y) of boundary displacement modes, independent solutions of the homogeneous adjoint problem

$$S^T_q = 0 \qquad q = y_{(j)}\ . \tag{C.4}$$

It was used in section 11 to establish that in two dimensional triangular membrane elements with complete polynomial approximations for the stress functions, the solutions of (C.4) were in fact restricted to the translation modes. The same was true for the example of rectangular membrane element and can be shown to be true for three-dimensional tetrahedrons, so that it seems to be a natural situation for elements generated by first order stress functions.

As

$$y^T_{(j)}\ g = (S^T\ y_{(j)})^T\ s = 0 \tag{C.5}$$

the global equilibrium conditions in translation are the necessary, but also sufficient, conditions for the existence of an inverse to (C.1)

$$s = S^{\#}\ g + \alpha_i\ x_{(i)} \tag{C.6}$$

where $S^{\#}$ is any particular pseudo-inverse to S, generating a particular set of stress parameters corresponding to given equilibrated (in translation only) boundary loads. The self-stressing intensities α_i are arbitrary and form the general solution of (C.1) for g = 0.

Any homogeneous constraint

$$q^T\ g = 0 \tag{C.7}$$

imposed on the loading induces through (C.1) a corresponding homogeneous constraint on the stress parameters

$$w^T s = 0 \tag{C.8}$$

with

$$w = S^T q \tag{C.9}$$

the constraint being effective ($w \neq 0$) if and only if q is linearly independent of the translation modes. Since

$$x^T_{(i)} w = (S x_{(i)})^T q = 0 \tag{C.10}$$

the w vectors generated by (C.9) are automatically orthogonal to all self-stressings. They span a m dimensional subspace of the n(s) dimensional space of all w vectors.

Conversely, any w vector of this subspace, used to induce a stress parameter constraint like (C.8) also induces through (C.6) a homogeneous loads constraint of type (C.7) with

$$q = S^{\#T} w \tag{C.11}$$

(C.11) is a particular inverse of (C.9) when the existence conditions for inversion, represented by (C.10), are satisfied. The general inverse is then obviously

$$q = S^{\#T} w + \beta_j y_{(j)} . \tag{C.12}$$

To produce a stress parameter constraint that improves rotational equilibrium without inducing a homogeneous loads constraint two conditions are required:

1. The vector w should be one generated by a condition of type

$$e_{mij} \int_E \omega_m \sigma_{ij} dV = \int_E [\omega_1(\sigma_{23}-\sigma_{32})+\omega_2(\sigma_{31}-\sigma_{13})+\omega_3(\sigma_{12}-\sigma_{21})]dV=0. \tag{C.13}$$

2. It must not belong to the subspace of w vectors orthogonal to all self-stressings.

In that case (C.6) shows that the constraint (C.8) is equivalent to a coupling between the loading and the self-stressing intensities

$$w^T S^{\#} g + \alpha_i w^T x_{(i)} = 0 \quad . \tag{C.14}$$

Obviously the number of such independent constraints may not exceed the number of independent self-stressings present.

THE DYNAMICS OF FLEXIBLE BODIES

B.M. Fraeijs de Veubeke

Laboratoire de Techniques, Aéronautiques et Spatiales,
Université de Liège, Belgique

ABSTRACT—The motion of a flexible body undergoing arbitrarily large rotations with respect to an inertial frame is split into a mean rigid body motion, defining a dynamical reference frame, and a relative motion taking into account the deformations.

The mean motion is usually taken to satisfy the Tisserand conditions of zero relative momentum and angular momentum, a choice that, as shown in the paper, corresponds to a minimum value of the relative kinetic energy. The condition of zero angular momentum is however non linear and introduces discretization difficulties that can be overcome by another choice.

The choice proposed in the paper minimizes the mean square of relative displacements. It preserves the zero momentum condition but linearizes the angular momentum condition in such a way that the relative displacements are representable exactly by an expansion in natural elastic vibration modes.

Hamilton's principle is used to derive all the equations, including the mean motion ones, by using the concept of quasi-coordinates. Gravitational potential and thrust vectors, as locally oriented by the body motion and deformation, are accounted for. The equations are not limited to small distortion of the body, but to small strains.

1. THE DYNAMIC REFERENCE FRAME

The formulation of the motion of a flexible body as a continuum through inertial space is unsatisfactory from several viewpoints. One is usually not interested in the details of this motion but

---

Original publication, *Int. Jnl. Engng. Sci.*, Pergamon Press, 1976, pp. 895-913.

in its main characteristics such as motion of the center of mass and, under the assumption that the deformations remain small, the history of the average orientation of the body. This last information is of course essential to pilots, real or artificial, in order to implement guidance corrections. We therefore try to define a set of cartesian mean axes accompanying the body, or dynamic reference frame, with respect to which the relative displacements, velocities or accelerations of material points due to the deformations are minimum in some global sense. If the body does not deform, any set of axes fixed into the body is of course a natural dynamic reference frame.

In principle the transfer from inertial to dynamic axes is effected in two steps: a translation $\vec{a}(t)$ of the origin G_0 to the new origin G and a rotation of the translated axes. If we think of the reference configuration C_0 as mapped by these operations, we obtain (in dotted lines on Figure 1) the dynamic reference configuration C_d. Any position vector $\vec{y}$ is carried into a corresponding new position vector $\vec{x}$ so that

$$\hat{y}_i = x_i \qquad (i = 1,2,3)$$

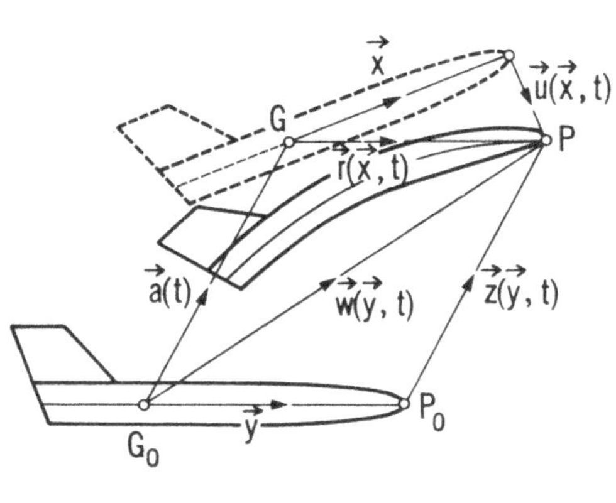

Figure 1 - Configurations of a flexible body

an equality between the inertial components of $\vec{y}$ and the dynamical components of $\vec{x}$. In matrix form we obtain the equality

$$\hat{y} = x \tag{1}$$

between the corresponding column matrices. The rotation of the axes is described by the matrix equation

$$x = U\hat{x} \quad \text{or} \quad \hat{x} = U^T x \tag{2}$$

relating the inertial and dynamic components of the same vector.

U is a rotation matrix; it is orthogonal

$$UU^T = U^T U = I \quad \text{(identity matrix)} \tag{3}$$

and its determinant is unity

$$\det U = +1 . \tag{4}$$

The obvious vector relations illustrated on Figure 1

$$\vec{w} = \vec{y} + \vec{z} = \vec{a} + \vec{r} = \vec{a} + \vec{x} + \vec{u}$$

can be stated either in terms of inertial components as

$$\hat{w} = \hat{y} + \hat{z} = \hat{a} + \hat{r} = \hat{a} + \hat{x} + \hat{u}$$

or in terms of dynamic components

$$w = y + z = a + r = a + x + u \quad .$$

As $\vec{x}$ and the "relative displacement" vector $\vec{u}$ are normally observed in the dynamic reference frame, while $\vec{w}$ and $\vec{a}$ are given in the inertial frame, the most useful form of these relations is

$$x + u = U(\hat{w}-\hat{a}) = U(\hat{y} + \hat{z} - \hat{a}) \ . \tag{5}$$

Clearly the dynamic axes will be defined by a prescription of $\hat{a}(t)$ and $U(t)$; the central considerations for selecting particular prescriptions being the resulting properties of $u(x,t)$, the field of relative displacements.

2. MEASURE OF STRAIN IN THE DYNAMIC REFERENCE FRAME

The measure of strain provided by the Green tensor

$$\hat{g}_{mn} = \frac{1}{2}\left(\frac{\partial \hat{z}_m}{\partial \hat{y}_n} + \frac{\partial \hat{z}_n}{\partial \hat{y}_m} + \frac{\partial \hat{z}_i}{\partial \hat{y}_m}\frac{\partial \hat{z}_i}{\partial \hat{y}_n}\right)$$

in the inertial frame is identical to that

$$\gamma_{mn} = \frac{1}{2}\left(\frac{\partial u_m}{\partial x_n} + \frac{\partial u_n}{\partial x_m} + \frac{\partial u_i}{\partial x_m}\frac{\partial u_i}{\partial x_n}\right) \tag{6}$$

in the dynamic frame; so that the elastic strain energy of the body with elastic moduli

$$C^{pq}_{mn} = C^{pq}_{nm} = C^{qp}_{mn} = C^{mn}_{pq}$$

is given by

$$W = \frac{1}{2}\int C^{pq}_{mn}\,\gamma_{mn}\,\gamma_{pq}\,\frac{d\mu}{\rho} \tag{7}$$

ρ being the mass per unit volume and $d\mu$ the element of mass. Denoting by

$$A = \left\{ \frac{\partial u_m}{\partial x_n} \right\} \tag{8}$$

the matrix of displacement gradients, we may write

$$\Gamma = \{\gamma_{mn}\} = \frac{1}{2} (A + A^T + A^T A) . \tag{9}$$

3. ABSOLUTE VELOCITY AND KINETIC ENERGY

If we differentiate (5) with respect to time, keeping $\hat{y} = x$ fixed,

$$\frac{\partial u}{\partial t} = \frac{dU}{dt} (\hat{y} + \hat{z} - \hat{a}) + U \left(\frac{\partial \hat{z}}{\partial t} - \frac{d\hat{a}}{dt} \right) .$$

In this expression we identify

$$\hat{v}_a = \frac{\partial \hat{z}}{\partial t} \quad \text{and} \quad \hat{v}_g = \frac{d\hat{a}}{dt}$$

as respectively the absolute velocity of a particle and the velocity of the origin of the dynamic reference frame. If we substitute

$$\hat{y} + \hat{z} - \hat{a} = \hat{x} + \hat{u} = U^T (x + u)$$

and rearrange the terms, we obtain the classical decomposition of the absolute velocity in a moving reference frame

$$v_a = v_g + [\omega] (x + u) + \frac{\partial u}{\partial t} \tag{10}$$

where

$$[\omega] = - \frac{dU}{dt} U^T = \begin{pmatrix} 0 & -\omega_3 & \omega_2 \\ \omega_3 & 0 & -\omega_1 \\ -\omega_2 & \omega_1 & 0 \end{pmatrix} = U \frac{dU^T}{dt} \tag{11}$$

is the skew symmetrix matrix of angular velocities of the dynamic axes. The skew symmetry is a direct consequence of the differentiation of (3)

$$U \frac{d}{dt} U^T + \frac{dU}{dt} U^T = 0 \ .$$

The notation conveniently suggests that this matrix is built up from the components of the pseudo-vector "angular velocity"

$$\omega = \begin{pmatrix} \omega_1 \\ \omega_2 \\ \omega_3 \end{pmatrix}$$

so that the second term of (10) is that of the dynamic components of the vector product

$$\vec{\omega} \times \vec{r} = - \vec{r} \times \vec{\omega} \ .$$

It may thus be written indifferently as

$$[\omega](x + u) = -[x + u]\omega = [x + u]^T\omega = -[\omega]^T(x + u) \ .$$

The inertial components of the velocity field induced by the rotation of the dynamical axes are given by

$$U^T[\omega]r = U^T[\omega]U\hat{r}$$

so that

$$[\hat{\omega}] = U^T[\omega]U = \frac{dU^T}{dt} U = - U^T \frac{dU}{dt} \qquad (12)$$

is the skew symmetric matrix of inertial components of the angular velocities.

If the rotation operator U is decomposed in a succession of elementary rotations defining Euler angles, such as

$$U = CBA$$

$$A = \begin{pmatrix} \cos\psi & \sin\psi & 0 \\ -\sin\psi & \cos\psi & 0 \\ 0 & 0 & 1 \end{pmatrix}$$

rotation of a (yaw) angle ψ about the third inertial axis,

$$B = \begin{pmatrix} \cos\theta & 0 & -\sin\theta \\ 0 & 1 & 0 \\ \sin\theta & 0 & \cos\theta \end{pmatrix}$$

rotation of a (pitch) angle θ about the second inertial axis as already turned by operation A ,

$$C = \begin{pmatrix} 1 & 0 & 0 \\ 0 & \cos\phi & \sin\phi \\ 0 & -\sin\phi & \cos\phi \end{pmatrix}$$

rotation of a (roll) angle ϕ about the first inertial axis as already turned by operation BA, the application of (11) yields the well known relations between angular velocities about dynamical axes and velocities of Euler angles:

$$\begin{aligned} \omega_1 &= \dot{\phi} - \dot{\psi} \sin \theta \\ \omega_2 &= \dot{\theta} \cos \phi + \dot{\psi} \sin \phi \cos \theta \\ \omega_3 &= - \dot{\theta} \sin \phi + \dot{\psi} \cos \phi \cos \theta. \end{aligned} \tag{13}$$

The Euler angles, however, are not very convenient as degrees of freedom to use in a Hamiltonian formulation.

Quasi-coordinates are more symmetrical and easier to manipulate. From (10), after some manipulation like

$$(x+u)^T[\omega]^T[\omega](x+u) = \omega^T[x+u][x+u]^T\omega$$

the kinetic energy of the body

$$T = \frac{1}{2}\int\left(\frac{\partial\hat{z}}{\partial t}\right)^T \frac{\partial\hat{z}}{\partial t}\, d\mu = \frac{1}{2}\int v_a^T v_a \, d\mu$$

can be placed in the form

$$T = \frac{1}{2} v_g^T v_g \int d\mu + v_g^T \int \frac{\partial u}{\partial t} d\mu + v_g^T[\omega] \int (x+u) d\mu$$
$$+ \frac{1}{2}\int \frac{\partial u^T}{\partial t} \frac{\partial u}{\partial t} d\mu + \omega^T \int [x+u] \frac{\partial u}{\partial t} d\mu + \frac{1}{2}\omega^T R\omega \tag{14}$$

where

$$R = \int [x+u][x+u]^T d\mu \tag{15}$$

is the matrix of inertia moments of the body about the dynamical axes in the actual configuration, that is taking into account the relative displacements u due to its flexibility.

$\int \frac{\partial u}{\partial t} d\mu$ is the vector of relative momentum of the body,

$\int [x+u] \frac{\partial u}{\partial t} d\mu$ is the vector of its relative angular momentum.

4. DECOMPOSITION OF VIRTUAL WORK. QUASI-COORDINATES

In terms of inertial components, the elementary virtual work accomplished by a force $\vec{f}$ on the virtual displacement $\vec{\delta z}$ of a material point is

$$\vec{f} \cdot \vec{\delta z} = \hat{f}^T \, \delta\hat{z} \; .$$

To obtain the expression in terms of dynamical components, begin with the differentiation of (5), noting that $\delta\hat{y} = \delta x = 0$, since the virtual displacement is that of a given material point, obtaining

$$\delta u = \delta U(\hat{y} + \hat{z} - \hat{a}) + U(\delta\hat{z} - \delta\hat{a}) \quad .$$

Hence

$$U\delta\hat{z} = U\delta\hat{a} + \delta u - \delta U U^T (x + u) \; .$$

These dynamical components of the absolute virtual displacement of a material point, can of course be given the same structure of decomposition

$$\delta h = U\delta\hat{z} = \delta p + [\delta\alpha](x + u) + \delta u \tag{16}$$

as the absolute velocity (10). (In fact the virtual displacement considerations could be replaced by virtual velocity ones.)

$$\delta p = U\delta\hat{a} \tag{17}$$

are the dynamical coordinates of the virtual displacement of the origin of dynamical axes; they will also be termed the "quasi-

$$[\delta\alpha] = - \delta U \cdot U^T = \begin{pmatrix} 0 & -\delta\alpha_3 & \delta\alpha_2 \\ \delta\alpha_3 & 0 & -\delta\alpha_1 \\ -\delta\alpha_2 & \delta\alpha_1 & 0 \end{pmatrix} = U\delta U^T \qquad (18)$$

are the "quasi-coordinates of rotation". We may note that instead of (17) we may write

$$\delta p = U\delta(U^T a) = U\delta U^T a + UU^T\delta a = \delta a + [\delta\alpha]a \qquad (19)$$

in the same manner as

$$v_g = U\hat{v}_g = U\frac{d\hat{a}}{dt} = U\frac{d}{dt}(U^T a) = \frac{da}{dt} + [\omega]a \quad . \qquad (20)$$

Finally the virtual work, as seen from the dynamic reference frame, is, using $\hat{f}^T = f^T U$ and (16)

$$f^T(\delta p + [\delta\alpha](x + u) + \delta u) = (\delta p^T + \delta\alpha^T[x + U] + \delta u^T)f. \qquad (21)$$

The interpretation of the quasi-coordinates of rotation is obtained by decomposing a perturbed rotation $U + \delta U$ into the product of the unperturbed rotation followed by the small additional rotation $I - [\delta\alpha]$. Thus we should have

$$U + \delta U = (I-[\delta\alpha])U \quad \text{or} \quad \delta U = -[\delta\alpha]U$$

and this is clearly equivalent to the definition (18). The $\delta\alpha_i$ are consequently the small angles of rotation of the dynamic frame, about its own axes, when passing from the unperturbed to a perturbed position.

The usefulness of quasi-coordinates can be seen from the possibility of expressing the variations of the arguments v_g and ω

of the kinetic energy, that are not easily related to displacement coordinates, in terms of the quasi-coordinates and their time derivatives. This permits a direct application of Hamilton's principle to derive the equations of motion of the dynamical axes themselves. The two essential formulas are

$$\delta v_g = \delta\left(U \frac{d\hat{a}}{dt}\right) = \delta U \frac{d\hat{a}}{dt} + U \frac{d}{dt} \delta\hat{a} = \delta U U^T v_g + U \frac{d}{dt} (U^T \delta p)$$

or, finally,

$$\delta v_g = [\delta\alpha] v_g + [\omega]\delta p + \frac{d}{dt} \delta p \ . \tag{22}$$

To obtain the second formula, take the time derivative of definition

$$\delta U = - [\delta\alpha] U$$

and compare it to the time derivative of definition

$$\frac{d}{dt} U = - [\omega] U$$

to obtain

$$\frac{d}{dt} [\delta\alpha] U + [\delta\alpha] \frac{dU}{dt} = [\delta\omega] U + [\omega] \delta U.$$

Postmultiplication by U^T furnishes the desired result

$$\delta [\omega] = [\omega] [\delta\alpha] - [\delta\alpha] [\omega] + \frac{d}{dt} [\delta\alpha]$$

which is equivalent to

$$\delta\omega = \frac{d}{dt} \delta\alpha + [\omega] \delta\alpha \ . \tag{23}$$

5. DECOMPOSITION OF ABSOLUTE ACCELERATION

In inertial axes the absolute acceleration of a particle is

$$\hat{j}_a = \frac{\partial}{\partial t}\hat{v}_a = \frac{\partial}{\partial t}(U^T v_a) = \frac{d}{dt}U^T v_a + U^T \frac{\partial v_a}{\partial t} \quad .$$

Hence, using definition (11), its dynamic components are given by

$$j_a = U\hat{j}_a = \frac{\partial v_a}{\partial t} + [\omega]v_a \quad . \tag{24}$$

The substitution of (10) will produce the decomposition theorem of acceleration in moving axes

$$j_a = \frac{d}{dt}v_g + [\omega]v_g + [\omega][\omega](x+u) + \left[\frac{d\omega}{dt}\right](x+u) + 2[\omega]\frac{\partial u}{\partial t} + \frac{\partial^2 u}{\partial t^2} \tag{25}$$

the first four terms being the acceleration experienced when the particle is frozen in the moving axes, the fifth being the complementary of Coriolis term, the sixth and last the relative acceleration.

The same result is obtained in the application of Hamilton's principle to the kinetic energy alone. We must have

$$\int_{t_1}^{t_2} T\, dt = -\int_{t_1}^{t_2}\int j_a^T \delta h\, d\mu \tag{26}$$

where δh, the dynamic components of the absolute virtual displacement of a particle, were already seen to satisfy a decomposition (16) completely analogous to that of the absolute velocity.

Since

$$\delta T = \int v_a^t \delta v_a\, d\mu$$

we are led to compute the variation of the absolute velocity. This can be done by taking the variation of (10) and applying the formulas (22) and (23). The following derivation is shorter. Differentiate with respect to time the relation defining δh

$$\delta\hat{w} = \delta\hat{z} = U^T \delta h \quad ,$$

obtaining

$$\delta\hat{v}_a = \frac{d}{dt} U^T \delta h + U^T \frac{\partial}{\partial t} \delta h = \delta(U^T v_a) = \delta U^T v_a + U^T \delta v_a \ .$$

Premultiplication by U and use of definitions (11) and (18) produces

$$\delta v_a = \frac{\partial}{\partial t} \delta h + [\omega]\delta h - [\delta\alpha] v_a \ . \qquad (27)$$

As $v_a^T[\delta\alpha]v_a = 0$ because of the skew symmetry of the central matrix, there comes

$$\delta \int_{t_1}^{t_2} T \, dt = \int v_a^T \delta h \, d\mu \Bigg|_{t_1}^{t_2} - \int_{t_1}^{t_2} \int \frac{\partial v_a^T}{\partial t} + v_a^T [\omega]^T \, \delta h \, d\mu$$

and, since δh must vanish at the time limits, the result indeed coincides with (26) on account of (24). Should we now make use of the decomposition (16) in (26),

$$\delta \int_{t_1}^{t_2} T \, dt = - \int_{t_1}^{t_2} \delta h^T j_a d\mu = - \int_{t_1}^{t_2} \left(\delta p^T \int j_a d\mu + \delta\alpha^T \int [x+u] j_a d\mu + \right.$$

$$\left. + \int \delta u^T j_a d\mu \right) dt \ .$$

This formula reveals the inertia forces attached to the different components of the motion of particles:

$-j_a d\mu$ to the relative motion of each particle,

$-\int j_a d\mu$ to the motion induced by a translation of the dynamical axes,

$-\int [x+u] j_a d\mu$ the inertial moment induced by the rotation of the dynamical axes.

The last two are dynamically equilibrated respectively by the vectorial sum of all external forces and their moment with respect to the origin. However the equations of momentum and angular momentum, or equations of the mean motion, are redundant. They will follow as consequence of the equations of motion of individual particles.

The interest of a decomposition into a mean motion and a relative motion lies in the simplifications that may be obtained for the relative motion by appropriate choices of a(t) and U(t). Clearly, if the body is rigid, the relative motion disappears entirely if the axes are fixed into the body. The interest is then entirely concentrated into the equations of mean motion, from which the motion of individual particles follows in a trivial manner. If the body is only slightly deformable a suitable choice of dynamical axes may reduce the local relative rotations to small quantities of the same order of magnitude as the strains, thus allowing simplifications to take place by linearization procedures.

But, even if the body is highly deformable, which can be true even for small strains, there are still choices of mean motion that are physically significant and conducive to simplifications of the equations of relative motion. Of the two main choices that will be investigated, the first minimizes the kinetic energy of the relative motion, the second minimizes the mean square of relative displacements.

6. MINIMUM KINETIC ENERGY IN RELATIVE MOTION

Assuming the absolute velocity of particles to be given, we analyze such motions of the dynamical axes that minimize the relative kinetic energy

$$T_r = \frac{1}{2}\int\left(\frac{\partial u}{\partial t}\right)^T \frac{\partial u}{\partial t}\, d\mu \ .$$

The minimum property requires that

$$\int\left(\frac{\partial u}{\partial t}\right)^T \Delta\, \frac{\partial u}{\partial t}\, d\mu \ = 0 \tag{28}$$

for all changes $\Delta(\partial u/\partial t)$ in relative velocity that are simply due to changes in the motion of the dynamical axes. The relative position of the dynamical axes at the epoch of comparison may be kept the same so that

$$\Delta u = 0 \qquad \Delta U = 0$$

and consequently

$$\Delta v_a = \Delta(U\hat{v}_a) = 0 \ .$$

The non zero changes are contributed by

$$\Delta v_g = \Delta(U\hat{v}_g) = U\,\Delta\frac{d\hat{a}}{dt} \quad \text{and} \quad \Delta\omega \ .$$

From (10) follows then

$$\Delta\,\frac{\partial u}{\partial t} = -\,\Delta v_g - [\Delta\omega](x+u) = -\Delta v_g + [x+u]\Delta\omega \ .$$

This result, substituted into (28), gives as minimum conditions

$$\int \frac{\partial u}{\partial t}\, d\mu = 0 \tag{29}$$

$$\int [x+u]\, \frac{\partial u}{\partial t}\, d\mu = 0 \ . \tag{30}$$

The first requires the relative momentum of the body to be zero, the second requires the same for the relative angular momentum. To simplify further deductions we assume the more stringent condition

$$\int (x+u)\, d\mu = 0 \tag{31}$$

at all times, which implies the satisfaction of (29) by differentiation with respect to time. It is equivalent to assume that the origin of dynamical axes is located at all times at the center of mass of the body. It is then also true that

$$\int [x+u]\, d\mu = 0 \ . \tag{32}$$

If we then take the relative velocity from (10) and substitute for it in the conditions of zero momentum and angular momentum, there comes

$$v_g \int d\mu = \int v_a\, d\mu \tag{33}$$

and, provided we replace

$$[x+u]\,[\omega]\,(x+u) = [x+u]\,[x+u]^T \omega$$

and consider the definition (15)

$$R\omega = \int [x+u]\, v_a\, d\mu \ . \tag{34}$$

Equation (33) fixes the dynamic components of the velocity of the origin of moving axes in terms of the absolute momentum of the body; for the inertial observer

$$\hat{v}_g \int d\mu = \int \hat{v}_a \, d\mu \, .$$

Equation (34) fixes the angular velocity of the moving axes capable of complying with the zero relative angular momentum of the body. The angular velocities are uniquely defined as long as the matrix of inertia moments of the actual configuration is non singular. It is in fact positive definite unless the rather absurd situation is reached where all the particles are aligned on a straight line. For the inertial observer

$$\hat{R}\hat{\omega} = \int [\hat{w}-\hat{a}]\hat{v}_a \, d\mu$$

$$\hat{R} = \int [\hat{w}-\hat{a}]\,[\hat{w}-\hat{a}]^T \, d\mu = U^T R U \, .$$

It should be observed that, while their motion is uniquely determined, the dynamic axes complying with the conditions of zero relative momentum and angular momentum, or Tisserand axes, are not unique. Any system of Tisserand axes may be deduced from another by a fixed (independent of time) rotation. In Tisserand axes the kinetic energy of the body is reduced from (14) to

$$T = \frac{1}{2} v_g^T v_g \int d\mu + \frac{1}{2} \int \frac{\partial u^T}{\partial t} \frac{\partial u}{\partial t} \, d\mu + \frac{1}{2} \omega^T R \omega \, . \qquad (35)$$

7. MINIMUM SQUARE AVERAGE OR RELATIVE DISPLACEMENTS

In the case of a rigid body it is possible to let the relative displacement field vanish entirely by fixing the moving axes in the body. This suggests a choice of moving axes for a flexible body that minimizes at all times the functional

$$\frac{1}{2}\int u^T u \; d\mu \quad \text{minimum} \; . \tag{36}$$

The corresponding requirement $\int u^T \Delta u \; d\mu = 0$ hold for all

$$\Delta u = \Delta\{-x+U(\hat{w}-\hat{a})\} = \Delta U(\hat{w}-\hat{a}) - U\Delta\hat{a}$$

due to a modification of the position of the moving axes. In view of (17) and (18) this can be written

$$\Delta u = -[\Delta\alpha](x+u) - \Delta p = [x+u]\Delta\alpha - \Delta p \; .$$

As the variations Δp and $\Delta\alpha$ are free and independent we obtain the minimizing conditions

$$\int u \; d\mu = 0$$

$$\int [x+u]u \; d\mu = \int [x]u \; d\mu = 0 \qquad ([u]u \equiv 0) \quad . \tag{38}$$

The time derivative of the first condition coincides with the Tisserand condition (29) of zero relative momentum. It coincides also with the more stringent condition (31) provided the origin of inertial axes be taken at the center of mass of the reference configuration

$$\int x \; d\mu = 0 \quad . \tag{39}$$

The time derivative of the second condition

$$\int [x] \frac{\partial u}{\partial t} \; d\mu = 0 \tag{40}$$

is a linearized version of the Tisserand condition of zero relative angular moment, in which the angular momentum of particles is computed with the reference position vector, instead of the displaced one.

For the inertial observer, condition (37) is

$$U \int (\hat{w}-\hat{a})d\mu - \int x \, d\mu = 0$$

and, in view of (39), yields the explicit value to be taken for the position vector of the origin of dynamical axes, given the actual configuration of the body

$$\hat{a} \int d\mu = \int \hat{w} \, d\mu \ . \tag{41}$$

Similarly condition (38) becomes an implicit equation for the computation of the angular position of the dynamical axes

$$\int [x] U \hat{w} \, d\mu = 0 \ . \tag{42}$$

(as the other terms $-a \int [x] d\mu$ and $- \int [x] x \, d\mu$ vanish).

The value to be found for U clearly depends only on the actual configuration of the body as measurable by the inertial observer.

The solution of this problem is surprisingly complicated and, in order to reduce it to a classical eigenvalue problem, it seems preferable to come back to a statement of the functional to be minimized in terms of quantities measured by the inertial observer. Thus we replace in functional (36)

$$u = U(\hat{w}-\hat{a})-x$$

and expand the scalar product

$$\begin{aligned} u^T u &= (\hat{w}-\hat{a})^T(\hat{w}-\hat{a}) = 2x^T U(\hat{w}-\hat{a}) + x^T x \\ &= \hat{w}^T\hat{w} + x^T x + 2x^T U\hat{a} - 2\hat{a}^T\hat{w} + \hat{a}^T\hat{a} - 2x^T U\hat{w} \ . \end{aligned}$$

The contribution of the two first terms to the functional is constant and may be dropped. That of the third term vanishes because of (39). There remains

$$-2\hat{a}^T \int \hat{w}\, d\mu + \hat{a}^T\hat{a} \int d\mu - 2 \int x^T U\hat{w}\, d\mu \qquad \text{minimum.}$$

Minimization with respect to $\hat{a}$ restitutes (41) and we are left with the problem

$$\phi = - \int x^T U\hat{w}\, d\mu \qquad \text{minimum.} \tag{43}$$

To solve it we now introduce a Rodrigues-Hamilton representation of the rotation operator

$$U = I + 2\beta[b] + 2[b][b] \ . \tag{44}$$

b is a vector invariant under the operator, since $[b]b \equiv 0$.

This shows b to be oriented as the axis of the rotation. For the same reasons $b^T U = b^T$ and, if u is any vector perpendicular to the axis of rotation,

$$b^T Uu = b^T u = 0$$

showing that the rotated vector Uu is also perpendicular to the axis. Now, if θ denotes the angle through which the body has rotated, we should have

$$u^T Uu = u^T u \cos\theta \ .$$

Computing the left-hand side $(u^T[b]u \equiv 0)$

$$u^T Uu = u^T u + 2u^T[b][b]u = u^T u + 2u^T(-b^T bI + bb^T)u$$
$$= u^T u(1-2b^T b)$$

and comparing, there follows

$$1 - 2b^Tb = \cos\theta \quad \text{or} \quad b^Tb = \sin^2\frac{\theta}{2} . \tag{45}$$

Finally the condition that $U^TU = I$, imposes the constraint

$$\beta^2 + b^Tb - 1 = 0 \quad \text{or} \quad \beta^2 = \cos^2\frac{\theta}{2} . \tag{46}$$

When substituting (44) into the functional (43) we note that

$$x^T[b]\hat{w} = - b^T[x]\hat{w}$$

and

$$x^T[b][b]\hat{w} = - b^T[x][\hat{w}]^T b$$

so that the functional may be written as a quadratic form in the constrained variables b and β

$$\phi = - \int x^T\hat{w}\, d\mu + 2\beta b^T\hat{m} + b^T\hat{M}b \tag{47}$$

with

$$\hat{m} = \int [x]\hat{w}\, d\mu \tag{48}$$

and, having taken the symmetrical part,

$$\hat{M} = \int \{[x][\hat{w}]^T + [\hat{w}][x]^T\}\, d\mu . \tag{49}$$

The minimizing conditions obtained by equating to zero the partial derivatives of the augmented functional

$$\Phi = \phi - \lambda(\beta^2 + b^Tb - 1)$$

take the form of a self-adjoint eigenvalue problem

$$\begin{pmatrix} \hat{M} & \hat{m} \\ \hat{m}^T & 0 \end{pmatrix} \begin{pmatrix} b \\ \beta \end{pmatrix} = \lambda \begin{pmatrix} b \\ \beta \end{pmatrix} . \tag{50}$$

For any eigenvalue λ and associated eigenvector, normed by satisfying the constraint (46), we find from those equations,

$$\phi = - \int x^T \hat{w} \, d\mu + \lambda$$

and the minimum of the functional is associated with the *smallest eigenvalue* of problem (50). In principle there are four eigenvalues: apart from the minimizing one, there is a maximizing one and two merely associated with a stationary value of the functional.

It is of course possible for this problem to degenerate and the case that warrants further investigation is the one corresponding to a multiple root for the smallest eigenvalue. Because then, the choice of the dynamical axes implementing (36) is not unique.

The discussion is simplified if we consider that the smallest eigenvalue and one associated eigenvector have been determined from (50). Denote by U the corresponding rotation and

$$w = U\hat{w} = a + x + u$$

the displacement as seen in this particular set of dynamic axes. We investigate the other choices of axes by considering a rotation operator V, applied *subsequently* to U, or if we prefer to start again from the reference configuration, the rotation operator VU. Following (43) the problem will then present itself as

$$- \int x^T V w \, d\mu = - \int x^T V(x+u) \, d\mu \qquad \text{minimum}$$

if due account is taken of (39).

The corresponding eigenvalue problem will be

$$\begin{pmatrix} M & m \\ m^T & 0 \end{pmatrix}\begin{pmatrix} b \\ \beta \end{pmatrix} = \lambda\begin{pmatrix} b \\ \beta \end{pmatrix} \tag{51}$$

$$m = \int [x](x+u)d\mu = \int [x]u\ d\mu$$

$$M = \int \{[x][x+u]^T + [x + u][x]^T\}\ d\mu\ .$$

From construction it is known that the values

$$b = 0 \quad \beta = 1$$

corresponding to the identity operator V = I represent an eigen-solution associated with the smallest eigenvalue. This however implies

$$m = 0 \quad 0 = \lambda$$

equations to which the system (51) is then reduced. In m = 0 we have rediscovered the minimizing property (38). The eigenvalue problem is then split into

$$Mb = \lambda b \tag{52}$$

and

$$\lambda\beta = 0\quad . \tag{53}$$

If $\lambda = 0$, that was seen to be the smallest eigenvalue, is not a mulitple root, problem (52) has positive eigenvalues and matrix M is positive definite. Conversely, if M is positive definite, the smallest eigenvalue of (50) will be simple and possess only one normed eigenvector. The solution to problem (36) will be unique.

This is the situation that prevails when the relative displacements are not unduly large, for M can be presented as the difference of two matrices of moments of inertia, one corresponding to a "half-way" configuration of the body, that is certainly positive definite, and one that takes only into account the displacements themselves and is only of second order of magnitude

$$M = 2\int\left[x + \frac{u}{2}\right]\left[x + \frac{u}{2}\right]^T d\mu - \frac{1}{2}\int [u][u]^T d\mu . \tag{54}$$

A consequence of equation (53) is that the other eigenvectors, for which $\lambda > 0$, have their β component equal to zero. In view of (46) this amounts to say that they correspond to rotations V of 180° amplitude. The three possible rotations occur about three mutually perpendicular axes, since the eigenvectors b of problem (52) are orthogonal. It is easy to see that the identity operator $V = I$ and the three rotations V_1, V_2, V_3 of 180° amplitude generated by problem (52) form an Abelian group. Assume now that $\lambda = 0$ is a double root. It implies that a unit vector n will exist such that

$$Mn = 0 \qquad n^T n = 1$$

n is parallel to the axis of the rotation V and since $\lambda = 0, \beta$ may be taken arbitrarily. We may thus consider that

$$b = n \sin\frac{\theta}{2} \qquad \beta = \cos\frac{\theta}{2}$$

$$V = I + (nn^T - I)(1 - \cos\theta) + \sin\theta\,[n] .$$

It is easily verified that ϕ remains minimum under rotations of this type *with an arbitrary value of* θ.

The example, illustrated on Figure 2, shows the possibility of non uniqueness of the minimizing choice when large elastic displacements are involved. A satellite with a rigid central axisymmetric body has two flexible massless appendages

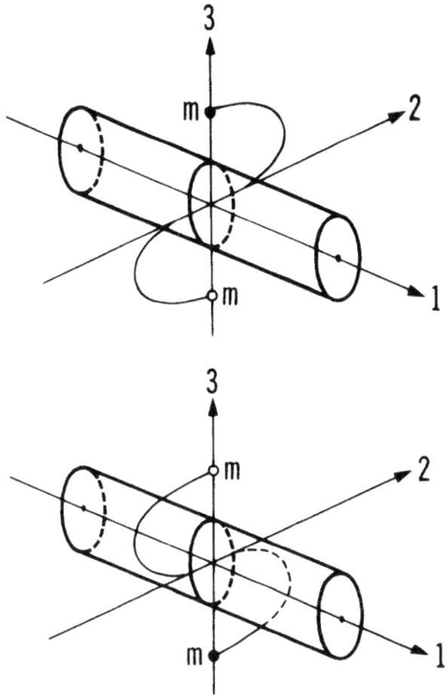

Figure 2 - Axisymmetric satellite with flexible appendages

terminated by concentrated masses. In its reference configuration the matrix of moments of inertia is

$$\begin{pmatrix} A + 2mR^2 & 0 & 0 \\ 0 & B + 2mR^2 & 0 \\ 0 & 0 & B \end{pmatrix}.$$

Suppose that in the deformed configuration the central body has not moved but that the masses of the appendages have been interchanged. We find from (48) to (49)

$$\hat{m} = 0 \quad \hat{M} = 2 \begin{pmatrix} A - 2mR^2 & 0 & 0 \\ 0 & B - 2mR^2 & 0 \\ 0 & 0 & B \end{pmatrix}.$$

The eigenvalues of problem (50) are:

$\lambda_0 = 0$ (associated to $b=0$, $\beta = 1$ and the identity operator)

$$\lambda_1 = 2(A - 2mR^2) \qquad \lambda_2 = 2(B - 2mR^2) \qquad \lambda_3 = 2B \quad .$$

Take the case $B > A$, so that λ_1 is the smallest of the last three.

If $2mR^2 < A$ the smallest eigenvalue is λ_0 and is unique. The original axes are also the ones that minimize the functional. The rotations V_1, V_2 and V_3 are of 180° respectively about the first, second and third cartesian axis; V_3 maximizes the functional.

If $2mR^2 > A$, the smallest eigenvalue is λ_1 and is again unique. The rotation V_1 of 180° about the axis of symmetry of the central body brings the displacements of the concentrated masses of the appendages back to zero, but displaces all the masses of the central body. This operation achieves the relative configuration minimizing the functional.

In the limiting case $2mR^2 = A$, $\lambda_0 = \lambda_1 = 0$ and the smallest eigenvalue is a doublet. The functional remains invariant and minimum under rotations of arbitrary amplitude about the axis of symmetry.

8. DISCRETIZATION OF BODY FLEXIBILITY

A modal analysis of the small amplitude vibrations of the free body produces the following expansion for small relative displacements, in which the summation convention on repeated indices is used and e_{ipn} denotes the alternating tensor

$$u_i(x,t) = v_i(t) + e_{ipn}\alpha_p(t)\, x_n + q_\beta(t) f_i^\beta(x) \quad .$$

The unknowns are the rigid body translation amplitudes $v_i(t)$, the small rigid body rotation amplitudes $\alpha_p(t)$ and a denumerable set of vibration amplitudes $q_\beta(t)$.

As the set of functions describing the displacement field is complete, the expansion can be used even for large relative displacements, in which case large $\alpha_p(t)$ terms induce strains because of the non linear terms in the exact strain measures (6). However, the minimizing conditions (37) and (38) are precisely satisfied by keeping the q_β terms alone.

Indeed the modal functions $f_i^\beta(x)$ have the properties

$$\int f_i^\beta \, d\mu = 0 \qquad (i = 1,2,3) \text{ all } \beta \tag{55}$$

$$e_{jqi} \int x_q f_i^\beta \, d\mu = 0 \qquad (j = 1,2,3) \text{ all } \beta \tag{56}$$

expressing their inertial orthogonality with respect to the small rigid body modes (which are natural modes of zero frequency).

Thus, considering that the origin of the reference configuration is at the center of mass,

$$\int x_n \, d\mu = 0 \qquad (n = 1,2,3)$$

the minimizing conditions (37)

$$\int u_i(x,t) d\mu = 0 \qquad (i = 1,2,3)$$

reduce to

$$v_i(t) \int d\mu = 0 \qquad (i = 1,2,3)$$

and are satisfied by setting $v_i(t) = 0$. The minimizing conditions (38)

$$e_{jqi} \int x_q u_i \, d\mu = 0 \qquad (j = 1,2,3)$$

reduce to

$$\alpha_p(t) e_{ipn} e_{jqi} \int x_n x_q \, d\mu = \alpha_p(t) \int (x_n x_n \delta_{jp} - x_j x_p) d\mu = 0.$$

The matrix of integrals is that of the inertia moments of the body in its reference configuration. It is positive definite, and the conditions can only be satisfied by taking $\alpha_p(t) = 0$.

Thus, an expansion of relative displacements limited to the natural modes of non zero frequency

$$u_i(x,t) = q_\beta(t) f_i^\beta(x) \tag{57}$$

satisfies automatically the principle of minimum square average of relative displacements and the equations of mean motion will be those associated to the corresponding choice of mean axes.

Another advantage of the discretization in natural modes is of course the existence of the orthogonality properties

$$\int f_i^\beta f_i^\gamma \, d\mu = \delta^{\beta\gamma} \int d\mu \tag{58}$$

the squared norm of a mode, or generalized mass, being here conventionally equated to the total mass of the body, and then, with $D_m = \partial/\partial x_m$

$$\int C_{mn}^{pq} D_m f_n^\beta D_p f_q^\gamma \frac{d\mu}{\rho} = 0 \qquad \text{if } \gamma \neq \beta$$

$$= \lambda_{(\beta)}^2 \int d\mu \qquad \text{if } \gamma = \beta \ .$$

The natural circular frequencies $\gamma_{(\beta)}$ are assumed to be ordered by increasing values. In practice, as interest is primarily centered on the low frequency response of the body, the expansion is truncated, leaving only a finite number of degrees of freedom.

9. INERTIA TERMS OF THE EQUATIONS OF MOTION

The expansion (57) is used to compute the absolute velocity (10)

$$v_{ai} = v_{gi} + e_{imn}\omega_m(x_n + q_\beta f_n^\beta) + \dot{q}_\beta f_i^\beta \tag{60}$$

and produces for the kinetic energy the expression

$$\begin{aligned} 2T &= \int v_{ai} v_{ai}\, d\mu \\ &= (v_{gi} v_{gi} + \dot{q}_\beta \dot{q}_\beta) \int d\mu + \omega_m \omega_m \left(\int x_n x_n\, d\mu + 2q_\beta F_{nn}^\beta + q_\beta q_\beta \int d\mu \right) \\ &- \omega_m \omega_n \left(\int x_m x_n\, d\mu + 2q_\beta F_{mn}^\beta + q_\beta q_\gamma S_{mn}^{\beta\gamma} \right) + 2A_m^{\gamma\beta} \omega_m \dot{q}_\beta q_\gamma \,. \end{aligned} \tag{61}$$

This formula contains three types of coupling coefficients resulting from the modal analysis

$$F_{mn}^\beta = \int x_m f_n^\beta\, d\mu = F_{nm}^\beta \tag{62}$$

the symmetry with respect to the lower indices being a consequence of (56).

$$S_{mn}^{\beta\gamma} = \frac{1}{2} \int (f_m^\beta f_n^\gamma + f_n^\beta f_m^\gamma)\, d\mu \tag{63}$$

symmetrical in both pairs of indices and finally a set of skew symmetric matrices, governing the gyroscopic terms

$$A_m^{\gamma\beta} = e_{imn} \int f_n^\gamma f_i^\beta\, d\mu = -A_m^{\beta\gamma} \,. \tag{64}$$

The inertia terms of the equations of motion follow from the computation of the variation

$$\delta\int_{t_1}^{t_2} T\,dt = \int_{t_1}^{t_2}\left(\frac{\partial T}{\partial v_{gi}}\,\delta v_{gi} + \frac{\partial T}{\partial \omega_i}\,\delta\omega_i + \frac{\partial T}{\partial q_\beta}\,\delta q_\beta + \frac{\partial T}{\partial \dot q_\beta}\,\delta\dot q_\beta\right)dt\ .$$

Substitution of (22) and (23) and integration by parts yield the following inertia terms:

Mean translation (coefficient of δp_i under the integral sign)

$$-\frac{d}{dt}\frac{\partial T}{\partial v_{gi}} + e_{mni}\omega_n\frac{\partial T}{\partial v_{g_m}} \qquad (i = 1,2,3)\ . \tag{65}$$

Mean rotation (coefficient of $\delta\alpha_i$)

$$-\frac{d}{dt}\frac{\partial T}{\partial \omega_i} + e_{mni}\omega_n\frac{\partial T}{\partial \omega_m} \qquad (i = 1,2,3)\ . \tag{66}$$

Deformation mode of index β (coefficient of δq_β)

$$-\frac{d}{dt}\frac{\partial T}{\partial \dot q_\beta} + \frac{\partial T}{\partial q_\beta}\ . \tag{67}$$

10. ELASTIC RESTORING TERMS OF THE EQUATIONS OF MOTION

Under the expansion in modes the deformation tensor (6) becomes

$$2\gamma_{mn} = q_\beta(D_m f_n^\beta + D_n f_m^\beta) + q_\beta q_\gamma D_m f_i^\beta D_n f_i^\gamma\ . \tag{68}$$

and the strain energy (7), due account being taken of (59)

$$W = \frac{1}{2}\,q_\beta q_\beta \lambda^2_{(\beta)}\int d\mu + \frac{1}{2}\,q_\beta q_\eta q_\zeta \Gamma^\beta_{\eta\zeta} + \frac{1}{8}\,q_\beta q_\gamma q_\eta q_\zeta \Gamma^{\beta\gamma}_{\eta\zeta}\ . \tag{69}$$

The following coupling coefficients were introduced

$$\Gamma^{\beta}_{\eta\zeta} = \int C^{pq}_{mn} D_m f^{\beta}_{n} D_p f^{\eta}_{i} D_q f^{\zeta}_{i} \frac{d\mu}{\rho} = \Gamma^{\beta}_{\zeta\eta} \tag{70}$$

$$\Gamma^{\beta\gamma}_{\eta\zeta} = \int C^{pq}_{mn} D_m f^{\beta}_{i} D_n f^{\gamma}_{i} D_p f^{\eta}_{j} D_q f^{\zeta}_{j} \frac{d\mu}{\rho} \tag{71}$$

the last one presenting the same type of symmetry as the elastic moduli

$$\Gamma^{\beta\gamma}_{\eta\zeta} = \Gamma^{\beta\gamma}_{\zeta\eta} = \Gamma^{\gamma\beta}_{\eta\zeta} = \Gamma^{\eta\zeta}_{\beta\gamma}. \tag{72}$$

The generalized elastic restoring forces appear in the deformation mode equations only as

$$-\frac{\partial W}{\partial q_{\beta}} = -\lambda^{2}_{(\beta)} q_{\beta} \int d\mu - \frac{1}{2} q_{\eta} q_{\zeta} (\Gamma^{\beta}_{\eta\zeta} + 2\Gamma^{\zeta}_{\beta\eta}) - \frac{1}{4} q_{\gamma} q_{\eta} q_{\zeta} (\Gamma^{\eta\zeta}_{\beta\gamma} + \Gamma^{\gamma\eta}_{\zeta\beta}) \tag{73}$$

For small deformations only the first term needs to be retained.

11. GRAVITATIONAL TERMS OF THE EQUATIONS OF MOTION

In the case of the gravitational potential use is made of the fact that the body dimensions are usually small compared to a characteristic length of the gravitational gradient. A truncated Taylor expansion of the specific gravitational potential is then considered, centered at the origin of dynamic axes

$$G = G(\hat{a}) - \hat{g}_m \hat{r}_m - \frac{1}{2} H_{mn} \hat{r}_m \hat{r}_n \tag{74}$$

where

$$\hat{g}_m(\hat{a}) = \frac{\partial G(\hat{a})}{\partial \hat{a}_m} \tag{75}$$

is the gravitational acceleration at the origin, and

$$\hat{H}_{mn}(\hat{\mathbf{a}}) = - \frac{\partial^2 G(\hat{a})}{\partial \hat{a}_m \partial \hat{a}_n} \tag{76}$$

the local gravity gradient tensor. In dynamical axes this becomes

$$G = G(\hat{a}) - g_m(x_m+u_m) - \frac{1}{2} H_{mn}(x_m+u_m)(x_n+u_n)$$

and, integrated over the mass of the body, produces a potential

$$P = G(\hat{a}) \int d\mu - H_{mn} \left\{ \frac{1}{2} \int x_m x_n \, d\mu + q_\beta F^\beta_{mn} + \frac{1}{2} q_\beta q_\gamma S^{\beta\gamma}_{mn} \right\} . \tag{77}$$

Unlike the inertial and strain-energy forces, the gravitational forces have fixed orientations in the inertial axes; in other terms the dynamic components H_{mn} depend on the orientation U of the dynamic axes

$$H = U\hat{H}U^T$$

and, using (18),

$$\delta H = \delta U \hat{H} U^T + U \hat{H} \delta U^T = -[\delta\alpha]H - H[\delta\alpha]^T$$

or

$$\delta H_{mn} = - e_{mij}\delta\alpha_i H_{jn} - e_{nij}\delta\alpha_i H_{jm} .$$

On the other hand from (17)

$$\delta G(\hat{a}) = - \hat{g}_i \delta \hat{a}_i = - g_i \delta p_i .$$

The contributions of the gravitational forces to the equations of motion are then established as the respective coefficients of δp_i, $\delta\alpha_i$ and δq_β appearing in $-\delta P$.

Mean translation

$$g_i \int d\mu \quad (i = 1,2,3) \tag{78}$$

Mean rotation

$$e_{mij} H_{jn} \left\{ \int x_m x_n \, d\mu + 2 q_\beta F^\beta_{mn} + q_\beta q_\gamma S^{\beta\gamma}_{mn} \right\} \quad (i = 1,2,3) \tag{79}$$

Deformation mode of index β

$$H_{mn} \{ F^\beta_{mn} + q_\gamma S^{\beta\gamma}_{mn} \} \ . \tag{80}$$

In aircraft applications, as long as the velocity of flight V is small compared to the orbital velocity $\sqrt{(R_0 g_0)}$ (R_0 mean earth radius, g_0 modulus of gravitational acceleration at this distance from the center of the earth) it is common practice to accept a "flat earth" approximation. The gravitational field is considered to be uniform and oriented as the third inertial axis. In this case, with the Euler angles previously introduced, the dynamical components g_i are

$$-g_0 \sin\theta \ , \quad g_0 \sin\phi \cos\theta \ , \quad g_0 \cos\phi \cos\theta$$

along the roll, pitch and yaw axes respectively.

In satellite applications the inertial axes are usually centered on the attracting body and oriented towards "fixed stars". Let R be the distance at which the gravitational acceleration has a known modulus g_R; then, neglecting harmonics, the potential of a unit mass is

$$G(\hat{a}) = - g_R \frac{R^2}{\rho} \quad \rho = \sqrt{(\hat{a}^T \hat{a})} = \sqrt{(a^T a)} \quad .$$

From which follows easily

$$g_i = - \frac{a_i}{\rho^3} g_R R^2 \quad H_{mn} = \left\{ \frac{3}{\rho^5} a_m a_n - \frac{1}{\rho^3} \delta_{mn} \right\} g_R R^2$$

12. THRUST FOLLOWER FORCES

Gravitational forces are so-called "dead" loads; they have components determined in inertial space and oriented independently of the deformation of the body. Propulsion forces generated by air-breathing engines or rocket thrusters are, generally speaking, "followers".

Attached rigidly to a rigid body their components remain fixed with respect to the dynamic axes. Mounted flexibly on a flexible body they are moreover influenced by the deformations. To take this into account, we assume that the thrust axis of a given propulsion unit passes through a given material point x of the body and is oriented by the local material rotation prevailing at this point.

Let dx denote a differential step taken from the point x of the thrust axis in the direction of the thrust in the reference configuration, so that

$$n = \frac{dx}{\sqrt{(dx^T dx)}}$$

are the direction cosines of the thrust vector f in the reference configuration. In the deformed state the convected unit vector will be

$$n' = \frac{dx+du}{\sqrt{[(dx+du)^T (dx+du)]}} \; .$$

Introduce the matrix A of displacement gradients, as defined in (8)

$$du = A\ dx$$

and

$$n' = \frac{(I+A)dx}{\sqrt{(dx^T dx + 2dx^T \Gamma\, dx)}} .$$

Under the assumption that the strains remain very small $2dx^T \Gamma\, dx$ is negligible before $dx^T\ dx$ and

$$n' = (I + A)n \quad .$$

In other words the local Jacobian matrix, governing the local neighborhood transformation, represents with a good approximation the local rotation operator. The dynamic components of the thrust in the deformed configuration are thus given by

$$f' = (I + A)f \ .$$

The virtual work, computed from the general formula developed in Section 4, is now expressible in terms of the known thrust vector of the reference state

$$(\delta p^T + \delta\alpha^T [x+u] + \delta u^T)(I + A)f \ .$$

From this the contributions of a thrust follower to the different equations of motion is obtained; it requires the computation of the deformation modes and their derivatives at the local attachment point.

Mean translation

$$(\delta_{mi} + q_\beta D_m f_i^\beta) f_m \qquad (i = 1,2,3) \tag{81}$$

Mean rotation

$$e_{pin}(x_n+q_\beta f_n^\beta)(\delta_{pm}+q_\gamma D_m f_p^\gamma)f_m \quad (i = 1,2,3). \tag{82}$$

If the deformations are small enough, only the linear terms in the amplitudes q_β may be retained.

Deformation mode of index β

$$f_i^\beta(\delta_{im}+q_\gamma D_m f_i^\gamma)f_m \ . \tag{83}$$

REFERENCES

[1] BUCKENS, F., *Proc. Vth Intern. Symp. Space Technol.*, Tokyo, 1963, pp. 193-203.

[2] MILNE, R.D., "Dynamics of the deformable aeroplane", *R&M 3345*, British Aeronautical Research Council, 1964.

[3] BUCKENS, F., *Astrodynamics*, 1965, pp. 327-342.

[4] FRAEIJS DE VEUBEKE, B., *AGARD Rep. 39*, 1956; See also *AGARD Manual of Aeroelasticity*, Chap. 3 of Part 1.

[5] HAWKINS, F.J., *RAE Tech. Rep. 65142*, 1965.

[6] FRAEIJS DE VEUBEKE, B., GERADIN, M., HUCK, A. and HOGGE, M.A., *Intern. Center for Mechanical Sciences, Courses and Lectures*, 126, Udine, Springer-Verlag, Wien, New York, 1962.

[7] WILLEMS, P.Y., "Attitude stability of deformable satellites", in *Attitude changes and stabilization of satellites, CNES Int. Colloq.*, Paris, 8-11 Oct. 1968.

GPSR Compliance

The European Union's (EU) General Product Safety Regulation (GPSR) is a set of rules that requires consumer products to be safe and our obligations to ensure this.

If you have any concerns about our products, you can contact us on ProductSafety@springernature.com

In case Publisher is established outside the EU, the EU authorized representative is:

Springer Nature Customer Service Center GmbH
Europaplatz 3
69115 Heidelberg, Germany

Batch number: 10607320

Printed by Printforce, the Netherlands